Handbook of
Quantile Regression

Chapman & Hall/CRC
Handbooks of Modern Statistical Methods

Series Editor

Garrett Fitzmaurice

Department of Biostatistics
Harvard School of Public Health
Boston, MA, U.S.A.

Aims and Scope

The objective of the series is to provide high-quality volumes covering the state-of-the-art in the theory and applications of statistical methodology. The books in the series are thoroughly edited and present comprehensive, coherent, and unified summaries of specific methodological topics from statistics. The chapters are written by the leading researchers in the field, and present a good balance of theory and application through a synthesis of the key methodological developments and examples and case studies using real data.

The scope of the series is wide, covering topics of statistical methodology that are well developed and find application in a range of scientific disciplines. The volumes are primarily of interest to researchers and graduate students from statistics and biostatistics, but also appeal to scientists from fields where the methodology is applied to real problems, including medical research, epidemiology and public health, engineering, biological science, environmental science, and the social sciences.

Published Titles

Handbook of Mixed Membership Models and Their Applications
Edited by Edoardo M. Airoldi, David M. Blei,
Elena A. Erosheva, and Stephen E. Fienberg

Handbook of Statistical Methods and Analyses in Sports
Edited by Jim Albert, Mark E. Glickman, Tim B. Swartz, Ruud H. Koning

Handbook of Markov Chain Monte Carlo
Edited by Steve Brooks, Andrew Gelman,
Galin L. Jones, and Xiao-Li Meng

Handbook of Big Data
Edited by Peter Bühlmann, Petros Drineas,
Michael Kane, and Mark van der Laan

Handbook of Discrete-Valued Time Series
Edited by Richard A. Davis, Scott H. Holan,
Robert Lund, and Nalini Ravishanker

Published Titles Continued

Handbook of Design and Analysis of Experiments
*Edited by Angela Dean, Max Morris,
John Stufken, and Derek Bingham*

Longitudinal Data Analysis
*Edited by Garrett Fitzmaurice, Marie Davidian,
Geert Verbeke, and Geert Molenberghs*

Handbook of Spatial Statistics
*Edited by Alan E. Gelfand, Peter J. Diggle,
Montserrat Fuentes, and Peter Guttorp*

Handbook of Cluster Analysis
*Edited by Christian Hennig, Marina Meila,
Fionn Murtagh, and Roberto Rocci*

Handbook of Survival Analysis
*Edited by John P. Klein, Hans C. van Houwelingen,
Joseph G. Ibrahim, and Thomas H. Scheike*

Handbook of Quantile Regression
*Edited by Roger Koenker, Victor Chernozhukov,
Xuming He, and Limin Peng*

Handbook of Spatial Epidemiology
*Edited by Andrew B. Lawson, Sudipto Banerjee,
Robert P. Haining, and María Dolores Ugarte*

Handbook of Missing Data Methodology
*Edited by Geert Molenberghs, Garrett Fitzmaurice,
Michael G. Kenward, Anastasios Tsiatis, and Geert Verbeke*

Handbook of Neuroimaging Data Analysis
*Edited by Hernando Ombao, Martin Lindquist,
Wesley Thompson, and John Aston*

**Handbook of Methods for Designing, Monitoring, and
Analyzing Dose-Finding Trials**
Edited by John O'Quigley, Alexia Iasonos, Björn Bornkamp

Chapman & Hall/CRC
Handbooks of Modern Statistical Methods

Handbook of Quantile Regression

Edited by

Roger Koenker
Victor Chernozhukov
Xuming He
Limin Peng

CRC Press
Taylor & Francis Group
Boca Raton   London   New York

CRC Press is an imprint of the
Taylor & Francis Group, an **informa** business
A CHAPMAN & HALL BOOK

CRC Press
Taylor & Francis Group
6000 Broken Sound Parkway NW, Suite 300
Boca Raton, FL 33487-2742

First issued in paperback 2020

© 2018 by Taylor & Francis Group, LLC
CRC Press is an imprint of Taylor & Francis Group, an Informa business

No claim to original U.S. Government works

ISBN-13: 978-1-4987-2528-6 (hbk)
ISBN-13: 978-0-367-65757-4 (pbk)

Visit the Taylor & Francis Web site at
http://www.taylorandfrancis.com

and the CRC Press Web site at
http://www.crcpress.com

To the memory of F.Y. Edgeworth

Contents

Preface

Quantile regression constitutes a family of statistical techniques intended to estimate and draw inferences about conditional quantile functions. Median regression as introduced in the eighteenth century by Boscovich and Laplace is a (central) special case. In contrast to conventional mean regression that minimizes sums of squared residuals, median regression minimizes sums of absolute residuals; quantile regression simply replaces symmetric absolute loss by asymmetric linear loss.

Since its introduction in Koenker and Bassett (1978), quantile regression has gradually been extended to a wide variety of data analytic settings, including time series, survival analysis, and longitudinal data. By focusing attention on local slices of the conditional distribution of response variables, it is capable of providing a more complete, more nuanced view of heterogeneous covariate effects. Applications of quantile regression can now be found throughout the sciences, including astrophysics, chemistry, ecology, economics, finance, genomics, medicine, and meteorology. Software for quantile regression is now widely available in all the major statistical computing environments.

Our objective for this handbook has been to provide a thorough review of recent developments of quantile regression methodology, illustrating its applicability in a wide variety of scientific settings. The intended audience for the volume is students and more senior researchers across a diverse set of disciplines.

The text consists of 22 chapters by leading scholars in the field, including reviews of recent work on resampling methods for inference, nonparametric penalty methods for function estimation, computational methods for large data applications, and Bayesian methods. Survival analysis has been a fertile growth area for quantile regression methods, and there are three chapters devoted to these developments. Causal inference has been another very active research area, and there are two chapters devoted to this topic. High-dimensional models and model selection have become obsessions in the recent statistical literature; two chapters confront these issues. Time series, longitudinal data, errors in variables, missing data, and sample selection all pose unique challenges for quantile regression modeling and are treated in four distinct chapters. Extension of quantile regression methods to multivariate and functional response are surveyed in two chapters. And finally, three chapters are devoted to surveys of applications in ecology, genomics, and finance.

We would like to offer our profound thanks to all the contributors to the handbook for their dedicated work on this project and for their prior contributions to the literature on quantile regression. We would also like to express our appreciation to Rob Calver, our editor, for his encouragement throughout the course of the project, and to Richard Leigh, our copy-editor, for his careful treatment of the manuscript. Finally, to the many others who have contributed to the development of quantile regression methods through theory and applications, we say thanks and express the hope that this volume will help to foster further developments.

Roger Koenker, Victor Chernozhukov, Xuming He, and Limin Peng
April 2017

Contributors

Victor Chernozhukov
MIT
Cambridge, Massachusetts

Xuming He
University of Michigan
Ann Arbor, Michigan

Roger Koenker
University of Illinois
Urbana, Illinois

Limin Peng
Emory University
Atlanta, Georgia

Part I

Introduction

1

A Quantile Regression Memoir

Gilbert W. Bassett Jr.

University of Illinois at Chicago, USA

Roger Koenker

University of Illinois at Urbana-Champaign, USA

CONTENTS

1.1   Long ago . . .

In the summers of 1972 and 1973 the two of us spent a lot of time playing tennis, in a successful effort to avoid working on our dissertations at the University of Michigan. Gib was working with Lester Taylor on theoretical aspects of l_1 regression, and Roger on hierarchical models for longitudinal data. Inevitably our anxiety about work intruded into the tennis conversation and there were frequent discussions of linear programming aspects of the l_1 regression problem. Gib had derived conditions under which the l_1 estimator was linear in the response vector, which explained some pathological simulation results of Taylor's. This might be called "breakdown" of the estimator due to influential design points now. More significantly, we frequently mentioned that the l_1 estimator seemed to be a regression analogue of the median since it was easily shown that essentially half the regression responses must lie above the fitted l_1 regression hyperplane and half must lie below as long as there was an intercept in the model. We also began to ask ourselves the question: if the l_1 estimator is a median regression estimator, must there not be other quantile regression estimators?

In the fall of 1973 we had both accepted positions at the University of Illinois, Gib in Chicago, and Roger in Urbana-Champaign (UIUC), but we continued to discuss our research via the dedicated "WATS" line that connected the two campuses by telephone. At some point we asked ourselves: Suppose instead of weights 1 and -1 on positive and negative residuals as in median regression, or weights 0 and -1 as in a proposal of Aigner and Chu (1968) for an extremal estimator, we used weights, τ and $\tau - 1$ for τ in $(0,1)$. Could we show that roughly τn of the observations would lie below the fitted plane and $(1-\tau)n$ above? The affirmative answer seemed to resolve our old question about how to define the rest of the regression quantiles, and we began an intensive effort to understand better how they behaved. Only later did we recall that we had both done an exercise on a one-sample version of this idea from Ferguson's decision theory text in a course given by Bruce Hill. And much, much later we discovered the univariate germ of this idea in an influential paper of Edgeworth (1888).

Neither of us was asymptotically adept, but our background in economics and Gib's thesis work did provide a useful foundation on the linear programming aspects of the problem. Rather naively, we began to attack the asymptotic theory via a combinatorial approach to the finite-sample density. Our expression for the finite-sample density did not seem to be terribly practical since it required, when there were p parameters, summation of $\binom{n}{p}$ terms involving exact fits to all "elementary subsets" of p observations. However, eventually we were able to show limiting normality of the joint density of several regression quantiles for certain replicated designs. We presented a early version of this at the Winter Meetings of the Econometric Society in San Francisco in 1974. By a fortuitous circumstance the discussant on that occasion was Joe Gastwirth, who was very encouraging and suggested that we explore connections to the then rapidly expanding robustness literature. In January 1975 we submitted our paper to *Econometrica* and in due course we received a report stating, in essence, that we had failed to make a convincing case that $\tau \neq \frac{1}{2}$ was "interesting," but perhaps a revision could be considered. We were consequently pessimistic about our prospects at *Econometrica*, so in June after some further revision we submitted the paper to the *Annals of Statistics*. The sole referee report, quoted in its entirety, was far briefer, but the message was the same:

> I regret that I cannot see any point in this paper, and therefore cannot recommend its publication. It may be of interest to compute regression analyses to minimize the sum of absolute deviations between the observed and fitted responses, and there is a fair amount of literature on this topic. But why should one consider $\tau \neq \frac{1}{2}$?

This report has continued to serve as a valuable reminder that however obvious the quantile regression idea may now appear to be, it was not always so apparent. Meanwhile, we had received some more positive feedback on the paper so we decided to prepare a revision for *Econometrica*. Steve Portnoy had joined the faculty at UIUC in the fall of 1975, and the next spring we decided to try to get his reaction to what we were doing. He was immediately enthusiastic and this encouraged us further. The new manuscript contained an extended introduction in which we tried to motivate the idea of L-statistics for regression along the lines of the work in the late 1940s and early 1950s by Mosteller and others. Bickel (1973) constituted a persuasive case for this idea, and we believed our approach had some advantages from an equivariance standpoint. In 1976 Roger moved to Bell Laboratories and was exposed over the next several years to a broad spectrum of current research in robust statistics; when our paper finally appeared in 1978, the introduction undoubtedly reflected some of this exposure. In retrospect the emphasis in our revised introduction on robust estimation of the conditional central tendency of the response was probably somewhat unfortunate since it tended to obscure the more important message concerning heterogeneity of the conditional quantile functions.

About this time Dave Ruppert and Ray Carroll began to look into the question of trimmed least-squares estimation using our approach. In Ruppert and Carroll (1980) they showed, rather surprisingly, that trimming based on residuals from preliminary estimators such as least squares had much less satisfactory asymptotic behavior than the regression quantile methods we had proposed. Later Welsh (1987) showed that a modified version of trimming using Winsorized residuals *could* succeed in giving estimators with asymptotics like that of the trimmed mean location estimator. Ruppert and Carroll, using earlier work by Jana Jurečková, also provided a much more straightforward proof of the asymptotic normality of the regression quantiles than our density-based approach. In the one-parameter regression-through-the-origin model Laplace had already derived the asymptotic behavior of the l_1 (weighted median) estimator in the early part of the nineteenth century, as we learned eventually from Steve Stigler.

We continued to work on these ideas over the next several years, and gradually others became interested as well. Jana Jurečková was enthusiastic early on, and wrote several papers extending the results of Ruppert and Carroll on trimmed least squares, emphasizing their advantage in overcoming the lack of scale invariance of the Huber M-estimator, and even proposing higher breakdown versions to avoid difficulties with influential design points. Gutenbrunner and Jurečková (1992) provided a crucial link between quantile regression ideas and rank tests as exposited by Hájek and Šidák through the formal duality of the linear programming approach. Steve Portnoy also maintained a strong interest in these ideas, and when Roger returned to UIUC from Bell Labs in 1983, they began a close collaboration. Portnoy (1984) established the tightness of the quantile regression process on $[\epsilon, 1 - \epsilon]$, and this led to further work on more general L-statistics and adaptive estimation.

At Jana's suggestion, we were invited in 1983 to an Oberwolfach meeting on quantile processes and extreme value theory. We were, to put it mildly, not notably successful in conveying our enthusiasm about the potential value of regression quantiles to the distinguished participants of this meeting. Two indelible memories of this meeting remain: the Schumann romance for concertina and piano played for the evening musicale by Henry Daniels and Richard Smith, and the comment by the conference organizer Willem van Zwet to us in the back of the lecture hall near the end of the sessions: "Erich Lehmann once told me that any good idea takes at least ten years to percolate to the surface of the field." Now, more than 40 years after the first glimmer of the idea, it is nice to see that it is still percolating.

A turning point in the econometric reception of these methods was the 1990 lecture of Gary Chamberlain at the World Congress of the Econometric Society in Barcelona and the related thesis work of Moshe Buchinsky that flipped the original "regression quantile" terminology of our 1978 paper to the current "quantile regression" usage and emphasized the heterogeneity of covariate effects interpretation of the methods.

Bibliography

D. Aigner and S.F. Chu. On estimating the industry production function. *American Economic Review*, 58:826–839, 1968.

P. J. Bickel. On some analogues to linear combinations of order statistics in the linear model. *Annals of Statistics*, 1:597–616, 1973.

M. Buchinsky, Changes in US Wage Structure 1963–87. *Econometrica*, 62:405–458, 1994.

G. Chamberlain, Quantile Regression, Censoring and the Structure of Wages. In C. Sims, editor, *Advances in Econometrics*, Elsevier, New York, 1994.

F. Y. Edgeworth. A mathematical theory of banking. *Journal of the Royal Statistical Society*, 51:113–127, 1888.

C. Gutenbrunner and J. Jurečková. Regression quantile and regression rank score process in the linear model and derived statistics. *Annals of Statistics*, 20:305–330, 1992.

S. Portnoy. Tightness of the sequence of empiric cdf processes defined from regression fractiles. In J. Franke, W. Hardle, and D. Martin, editors, *Robust and Nonlinear Time Series Analysis*. Springer-Verlag: New York, 1984.

D. Ruppert and R. J. Carroll. Trimmed least squares estimation in the linear model. *Journal of the American Statistical Association*, 75:828–838, 1980.

A. H. Welsh. The trimmed mean in the linear model. *Annals of Statistics*, 15:20–36, 1987.

2

Resampling Methods

Xuming He

University of Michigan, Ann Arbor, Michigan, USA

CONTENTS

2.1 Introduction

Regression quantile estimators solve a linear program and can be computed efficiently. The finite-sample distributions of regression quantiles can be characterized (Koenker, 2005), but they are difficult to use for statistical inference. Suppose that we have data $\{(X_i, Y_i), i = 1, \ldots, n\}$, where the conditional quantile of Y given X is of interest and assumed to be linear. The asymptotic distributions of regression quantiles are normal under mild conditions, but the asymptotic variance depends on the conditional densities of Y given $X = X_i$, which are generally unknown. Statistical inference based on the asymptotic variance is arguably difficult for a simple reason. That is, one needs to use a nonparametric estimate of the asymptotic variance that requires the choice of a smoothing parameter, and such estimates can be quite unstable. Even if the asymptotic variance is well estimated, the accuracy of its approximation to the finite-sample variance depends on the design matrix as well as the quantile level. Resampling methods provide a reliable approach to inference for quantile regression analysis under a wide variety of settings.

In this chapter, we view resampling methods as a broad class of statistical methods that use Monte Carlo simulations on a given sample. The classical bootstrap is a common resampling method in regression, but quantile regression, which can be characterized by estimating equations based on signs of the residuals, affords other forms of bootstrap. We consider two types of problems, depending on whether the design points X_i are treated as fixed or not. When the pairs (X_i, Y_i) can be considered as random draws from a population, as in the example of height (X) and weight (Y) measurements from a random sample of school children, a valid bootstrap method is to sample the pairs with replacement. Freedman

(1981) used the term "correlation model" to distinguish this from the "regression model" where the X_i are fixed. The paired bootstrap has some drawbacks. First, if the purpose of the analysis is hypothesis testing, the paired bootstrap does not generate samples under the null hypothesis, making the bootstrap method less suited for hypothesis testing. Second, probably even more importantly, the paired bootstrap tends to inflate the variance in the cases where the X_i are fixed, especially when some X_i values are outliers (Kocheringsky et al., 2003). In the recent literature, a number of resampling methods have been developed for the "regression model," and they will be discussed in the chapter as well. First we fix our notation.

Given the sample of size n, $\{(X_i, Y_i), \ i = 1, \ldots, n\}$, from the linear quantile model $Q_Y(\tau|X = x) = x^T\beta_\tau$, the τth regression quantile coefficient estimate of Koenker and Bassett (1978) is

$$\hat{\beta}_\tau = \mathrm{argmin}_{\beta \in R^p} \left\{ \sum_{i=1}^n \rho_\tau(Y_i - X_i^T\beta) \right\},$$

where $\rho_\tau(r) = r(\tau - I(r < 0))$ is the quantile loss function. A resample will refer to any sample $\{(X_i^*, Y_i^*), \ i = 1, \ldots, n\}$ that is obtained under a resampling scheme, and the resulting quantile estimate is denoted by $\hat{\beta}_\tau^*$. The distribution of $\hat{\beta}_\tau^*$ conditional on the sample $\{(X_i, Y_i), \ i = 1, \ldots, n\}$ is called the resampling distribution. In the usual case where X includes the constant 1 for the intercept term in the model, most resampling methods maintain the intercept in the resamples.

2.2   Paired bootstrap

If $\{(X_i, Y_i), \ i = 1, \ldots, n\}$ can be considered as a random sample, it is natural to use the nonparametric bootstrap by sampling from the original sample with replacement. Each bootstrap sample, denoted by $\{(X_i^*, Y_i^*), \ i = 1, \ldots, n\}$, yields a quantile estimate β_τ^* given by

$$\beta_\tau^* = \mathrm{argmin}_{\beta \in R^p} \left\{ \sum_{i=1}^n \rho_\tau(Y_i^* - X_i^{*T}\beta) \right\}.$$

As shown in Freedman (1995), the resampling distribution of $\sqrt{n}(\beta_\tau^* - \hat{\beta}_\tau)$ is asymptotically normal with mean zero and variance V, which is the same as the asymptotic variance of the sampling distribution of $\sqrt{n}(\hat{\beta}_\tau - \beta_\tau)$. If we repeat the resampling process B times independently, resulting in a sequence $\beta_{\tau_1}^*, \ldots, \beta_{\tau_B}^*$, we can then use its sample covariance to approximate the covariance matrix of $\hat{\beta}_\tau$ for large n. The approximation is valid under the condition that the conditional τ quantile of $e_i = Y_i - X_i^T\beta_\tau$ is zero, but the conditional distribution of e_i given X_i may depend on X_i. An approximate confidence interval for any component of β_τ can then be constructed from the resampling distribution.

In parametric statistics, the bootstrap is also known for its approximation power in statistical inference. For example, the coverage probabilities of the bootstrap-based confidence intervals can be made second-order accurate. The Wald-type intervals obtained from the asymptotic normality of the parameter estimates are usually first-order accurate, that is, the coverage probabilities of the resulting confidence intervals differ from the nominal level by an amount in the order of $1/\sqrt{n}$. The error can be driven down to the order of $1/n$ for some of the bootstrap methods, including the Studentized bootstrap. The higher-order accuracy of bootstrap confidence intervals, however, requires the estimators to be derived from a smooth objective function, because the theory relies on a Taylor series approximation. Such an approximation does not work for the quantile regression objective function. One question of interest is whether the bootstrap methods can be second-order accurate for quantile regression. Horowitz (1998) addressed this question by smoothing the quantile objective function. To keep the exposition simple, we shall use the median ($\tau = 1/2$) in this exposition, as in Horowitz (1998), but the same idea applies to other quantiles.

The median regression is given by minimizing

$$n^{-1} \sum_{i=1}^{n} |Y_i - X_i^T \beta| = n^{-1} \sum_{i=1}^{n} (Y_i - X_i^T \beta)[2I(Y_i - X_i^T \beta > 0) - 1].$$

By taking a bounded and differentiable kernel K satisfying $K(v) = 0$ if $v \leqslant -1$ and $K(v) = 1$ if $v \geqslant 1$, we can define the smoothed median regression estimator $\tilde{\beta}_\tau$ by minimizing

$$H_n(\beta) = n^{-1} \sum_{i=1}^{n} (Y_i - X_i^T \beta)[2K((Y_i - X_i^T \beta)/h_n) - 1],$$

for a sequence of positive numbers $h_n = o(1)$. Under suitable assumptions on h_n, the smoothed estimator is first-order equivalent to the median regression estimator, but the paired bootstrap applied to the t-statistic from the smoothed median regression estimator can be accurate up to the order of $1/(nh_n)$. The results imply that the error in the coverage probability can be of the order $n^{-1+\eta}$ for any $\eta > 0$.

In the almost sure sense, the normal approximation to $\sqrt{n}(\tilde{\beta}_\tau - \beta_\tau)$ has an error rate of $n^{-1/4}$, not the usual rate of $n^{-1/2}$ (He and Shao, 1996). In the distributional sense, however, the error rate is nearly $n^{-1/2}$, as shown by Portnoy (2012). It is possible that nearly second-order bootstrap confidence intervals can be constructed for quantile regression even without smoothing, but it is unclear how to do so.

2.3 Residual-based bootstrap

Let $e_i = Y_i - X_i^T \beta_\tau$ be the τ-quantile residuals. If the e_i (given X_i) are independent and identically distributed, or more generally exchangeable, then a simple residual bootstrap with

$$Y_i^* = X_i^T \hat{\beta}_\tau + e_i^*, \quad i = 1, \ldots, n,$$

where e_i^* is a bootstrap sample of the estimated residuals $\hat{e}_i = Y_i - X_i^T \hat{\beta}_\tau$, works in the following sense. Under the usual conditions to ensure that the sampling distribution of $\hat{\beta}_\tau$ is asymptotically normal with mean β_τ and variance-covariance V, the resampling distribution of $\hat{\beta}_\tau^*$ is asymptotically normal with mean $\hat{\beta}_\tau$ and the same variance-covariance V. Based on such a result, the bootstrap variance estimated from the variance of the resamples can be used to approximate the sampling variance (Bickel and Freedman, 1981).

In the more general case, where the residuals are not exchangeable, simple versions of residual bootstrap are unjustified. We will consider one option based on the idea of the wild bootstrap of Wu (1986) and Liu (1988). The basic idea of the wild bootstrap is to use

$$Y_i^* = X_i^T \hat{\beta}_\tau + w_i \hat{e}_i, \quad i = 1, \ldots, n,$$

where the w_i are random draws from a prespecified distribution. An important point of the wild bootstrap is that in each case, the pairing of the X_i and the estimated residual $\hat{e}_i$ is

TABLE 2.1
Comparison of nominal 90% confidence intervals at $n = 50$. Coverage is the estimated coverage probability of confidence intervals, and length (SE) gives the average lengths and their standard errors.

	β_0		β_1		β_2	
	Coverage	Length (SE)	Coverage	Length (SE)	Coverage	Length (SE)
Wild Bootstrap	87.1	5.5 (0.02)	91.1	1.5 (0.01)	89.2	5.7 (0.02)
Paried Bootstrap	93.6	7.0 (0.03)	94.6	1.7 (0.01)	95.1	7.4 (0.02)
Generalized Bootstrap	93.6	7.0 (0.03)	94.2	1.6 (0.01)	95.3	7.6 (0.03)
Wald CI	90.7	6.4 (0.02)	87.7	1.4 (0.01)	92.1	6.8 (0.02)

always maintained except that the residual is multiplied by a random weight. For linear estimators, a broad class of weight distributions such as standard normal or exponential work for the wild bootstrap, but the choice of the weight distribution needs to be τ-specific for quantile regression. Feng et al. (2011) proposed a slight modification of the wild bootstrap as follows:

$$Y_i^* = X_i^T \hat{\beta}_\tau + w_i |\hat{e}_i|, \quad i = 1, \ldots, n,$$

where the w_i are randomly drawn from a distribution G satisfying the following conditions.

1. $G(0) = \tau$, that is, the τth quantile of w_i is zero.

2. The support of G is contained in the interval $[-c_1, c_2]$, where c_1 and c_2 are positive numbers.

3. $\int_0^\infty w^{-1} dG(w) = -\int_{-\infty}^0 w^{-1} dG(w) = 1/2$.

A simple weight distribution that satisfies these conditions is the two-point mass distribution with probabilities $1 - \tau$ and τ at $w = 2(1 - \tau)$ and -2τ, respectively. Results from Feng et al. (2011) show that the wild bootstrap is more reliable than the paired bootstrap when the configuration of X_i is skewed or heavy-tailed. With the wild bootstrap it is also straightforward to obtain resamples under that null hypothesis that some of the β_τ coefficients are zero; all one needs to do is to replace $\hat{\beta}_\tau$ by an estimate under the null hypothesis.

A simple simulation example used in Feng et al. (2011) is to generate samples of size $n = 50$ from

$$y_i = \beta_0 + \beta_1 x_{1i} + \beta_2 x_{2i} + 3^{-1/2} \big[2 + \{ 1 + (x_{1i} - 8)^2 + x_{2i} \} / 10 \big] \epsilon_i,$$

for $i = 1, \ldots, n$, with $(\beta_0, \beta_1, \beta_2) = (1, 1, 1)$, where the x_{1i} are generated from the standard log-normal distribution with a fixed seed, the x_{2i} are set to 1 for the first 80% of the observations and 0 for the rest, and the ϵ_i are drawn from the t distribution with 3 degrees of freedom. Table 2.3 provides a comparison of three bootstrap methods and the Wald-type confidence intervals from the asymptotic approximation in the quantreg package in R (with the NID option). In this example, the paired bootstrap and the generalized bootstrap (described in the next section) give overly conservative confidence intervals, because they do not treat the predictors as fixed. The wild bootstrap has better performance for the slope parameters.

2.4   Generalized bootstrap

There is probably no universally adopted name for it, but Chatterjee and Bose (2005) introduced a generalized bootstrap technique that is similar to the weighted bootstrap of Barbe and Bertail (1995). In the quantile regression setting, we consider obtaining β_τ^* by minimizing

$$\sum_{i=1}^{n} w_i \rho_\tau (Y_i - X_i^T \beta)$$

over β, where $\{w_i\}$ is a random sample from some distribution such as the Multinomial$(n; 1/n, \ldots, 1/n)$ distribution or an exponential distribution. Equivalently, β_τ^* is taken as any root of

$$\sum_{i=1}^{n} w_i (I(Y_i - X_i^T \beta < 0) - \tau) X_i.$$

As shown in Chatterjee and Bose (2005), a wide range of choices for the distribution of w_i can be made for the asymptomatic validity of the generalized bootstrap, which contrasts with the choices for the wild bootstrap in the preceding section. The characteristics of the generalized bootstrap are similar to those of the paired bootstrap, but with one advantage. That is, if the original design matrix is of full rank, the pseudo-design matrix under the generalized bootstrap is also of full rank, avoiding the complication that some of the resamples from the paired bootstrap may produce singular design matrices.

Sampling weights w_i to perform inference through the quantile objective functions has been considered by Chen et al. (2008). Suppose that we wish to test the hypothesis $H_0 : \beta_\tau \in \Omega_0$, a subspace of R^p. A natural test statistic is

$$M_n = \min_{\beta \in \Omega_0} \sum_i \rho_\tau (Y_i - X_i^T \beta) - \min_{\beta \in R^p} \sum_i \rho_\tau (Y_i - X_i^T \beta).$$

The asymptotic behavior of this test statistic is expected to be a mixture of chi-squares and the distribution depends on the conditional densities of Y given X. If we draw $w_1, \ldots, w_n$ as a random sample from a positively valued distribution with mean 1 and variance 1, then the resampling distribution of

$$M_n^* = \min_{\beta \in \Omega_0} \sum_i w_i \rho_\tau (Y_i - X_i^T \beta) - \min_{\beta \in R^p} \sum_i w_i \rho_\tau (Y_i - X_i^T \beta)$$
$$- \sum_i w_i \rho_\tau (Y_i - X_i^T \hat{\beta}_{H_0}) - \sum_i w_i \rho_\tau (Y_i - X_i^T \hat{\beta}_\tau)$$

can approximate the distribution of M_n, where $\hat{\beta}_{H_0}$ is the quantile estimate under H_0 with the original sample. One point to make is that the quantity M_n^* has to be centered at $\sum_i w_i \rho_\tau (Y_i - X_i^T \hat{\beta}_{H_0}) - \sum_i w_i \rho_\tau (Y_i - X_i^T \hat{\beta}_\tau)$. Otherwise, the resampling distribution does not approximate the sampling distribution of M_n. The validity of this approach is established in Chen et al. (2008) in the special case of $\tau = 1/2$.

2.5   Estimating function bootstrap

Resampling the components of an estimating equation rather than the data points or the residuals has been proposed by Parzen et al. (1994) and Hu and Zidek (1995). Note that the quantile estimating function can be written as

$$S_n(\beta) = n^{-1/2} \sum_{i=1}^{n} X_i (I(Y_i - X_i^T \beta < 0) - \tau).$$

Conditional on X_i, the distribution of $S_n(\beta_\tau)$ for the true quantile coefficient β_τ is pivotal, that is, its distribution does not depend on β_τ. Suppose that U is a random variable with the same distribution as the pivotal distribution. Then the random variable β_U which solves $S_n(\beta_U) = U$ provides a resampling distribution for the quantile estimate.

More specifically, the distribution of U given $\{X_i\}$ is a mixture of independent Bernoulli variables B_i in the form of

$$U = n^{-1/2} \sum_{i=1}^{n} X_i (B_i - \tau). \tag{2.1}$$

The estimating equation bootstrap then proceeds as follows. For each random draw $\{B_i\}$ from the Bernoulli distribution with probability τ, we have U^* obtained from (2.1). By solving $S_n(\beta) = U^*$, we obtain β_τ^*. It is also important to note that solving the equation

$$n^{-1/2} \sum_{i=1}^{n} X_i (I(Y_i - X_i^T \beta < 0) - \tau) = U^* \tag{2.2}$$

is equivalent to finding the solution to $\sum_{i=1}^{n+1} X_i (I(Y_i - X_i^T \beta < 0) - \tau) = 0$, where the additional point $(X_{n+1}, Y_{n+1}) = (U^*/\tau, -\infty)$. One should note that this equation may not have an exact root due to the fact that the left-hand side is a jump function, but any solution that makes the left-hand side of the order of $o_p(1)$ leads to asymptotically valid results. Finding such a solution can be achieved by computing the augmented regression quantile by minimizing $\sum_{i=1}^{n+1} \rho_\tau (Y_i - X_i^T \beta)$. As a consequence, the estimating function bootstrap is computationally comparable to the paired bootstrap, but with the advantage of not resampling the design points.

2.6   Markov chain marginal bootstrap

From the computational perspective, any resampling method described in the earlier sections requires repeated calculations of the regression quantile. For problems of moderate sizes, quantile regression computations are fast, but the computational cost increases with p. To reduce the computational cost, He and Hu (2002) proposed a dimension-reducing variant of the estimating function bootstrap and called it the method marginal chain marginal bootstrap (MCMB). Instead of solving (2.2), the MCMB method takes one component of the equation at a time and iterates through p components in the following manner.

Let $\beta^{(0)}$ be the initial value taken to be the regression quantile of the original sample. For the iteration from $\beta^{(k)}$ to $\beta^{(k+1)}$, let $z_{i,j}^* = X_{i,j}(B_{i,j} - \tau)$ for $i = 1, \ldots, n$ and $j = 1, \ldots, p$, where the $B_{i,j}$ are mutually independent random draws from the Bernoulli distribution with probability τ. The first component of $\beta^{(k+1)}$, denoted by $\beta_1^{(k+1)}$, solves

$$\sum_{i=1}^{n} X_{i,1} \left\{ I(Y_i - X_{i,1}\beta_1 - \sum_{j=2}^{p} X_{i,j}\beta_j^{(k)} < 0) - \tau \right\} = \sum_{i=1}^{n} X_{i,1}(B_{i,1} - \tau).$$

As we remarked concerning the solution to (2.2), the solution to the above scalar equation is a weighted sample quantile for univariate data. The lth component of $\beta^{(k+1)}$, denoted by $\beta_l^{(k+1)}$, solves

$$\sum_{i=1}^{n} X_{i,l} \left\{ I(Y_i - X_{i,l}\beta_l - \sum_{j=1}^{l-1} X_{i,j}\beta_j^{(k+1)} - \sum_{l+1}^{p} X_{i,j}\beta_j^{(k)} < 0) - \tau \right\} = \sum_{i=1}^{n} X_{i,l}(B_{i,l} - \tau)$$

for $l = 2, \ldots, p$, sequentially. By iterating through $l = 1, \ldots, p$ within each k, and then $k = 1, \ldots, K$ for a total chain length of K for $\beta^{(k)}$, we obtain a Markov chain instead of independent realizations from the usual resampling method. When the chain length K increases with the sample size n to infinity, the Markov chain has a stationary distribution with covariance matrix matching the asymptotic covariance of the regression quantile estimator. The asymptotic theory becomes more complicated with the MCMB, but as shown in Kocheringsky et al. (2003), the MCMB method provides a computationally efficient approximation to the construction of confidence intervals (or regions). The MCMB method is available in both the quantreg package in R and the QUANTREG procedure in SAS. In the latter it is the default method for confidence interval construction when $n > 5000$ or $p > 20$.

The MCMB method does not show improved performance compared to the estimating equation bootstrap, but is a more scalable algorithm for large problems. The MCMB algorithm relies on the repeated calculations of weighted quantiles for univariate data regardless of how large p is. On the other hand, the validity of the approximation requires a reasonably large sample size, so it should be used only when n is large.

While the MCMB method reduces computation for problems with larger p, other approaches are needed to handle problems with really large n at each sample or resample. One useful method based on subsampling is given in Portnoy and Koenker (1997), which utilizes the fact that all the points above (or below) the quantile can be replaced by a single point with an appropriate weight. Kleiner et al. (2014) proposed a randomized algorithm that can be implemented in big data platforms (e.g. MapReduce), and Yi and Huang (2015) considered semismooth Newton coordinate descent algorithms for penalized quantile regression computation with big data in high dimensions. The idea of the bag of little bootstrap (Yang et al., 2014) provides a scalable bootstrap for massive data.

2.7 Resampling methods for clustered data

Longitudinal data and clustered data arise often in public health studies and in social science research. To account for the within-cluster correlation, bootstrap sampling methods may be applied to clusters. Davison and Hinkley (1997, pp. 100–102) discussed strategies for resampling, and Field and Welsh (2007) provided asymptotic theory for such methods. In the context of quantile regression, Chen et al. (2004) considered the estimating equation approach for clustered data, and Wang and He (2007) used the quantile rank score test assuming a constant within-cluster correlation structure. Hagemann (2016) introduced a procedure called the wild gradient bootstrap that is more cluster-robust than earlier methods especially when the number of clusters is small but the variabilities across clusters are high. We shall describe how the wild gradient bootstrap works.

Suppose that there are n clusters and c_i observations (X_{ik}, Y_{ik}), $k = 1, \ldots, c_i$, in the ith cluster. The resampled estimate of β_τ^* is determined by the first-order equation

$$n^{-1/2} \sum_{i=1}^{n} \sum_{k=1}^{c_i} X_{ik}(I(Y_{ik} - X_{ik}^T \beta < 0) - \tau) = W_n(\tau), \qquad (2.3)$$

where

$$W_n(\tau) = n^{-1/2} \sum_{i=1}^{n} W_i \sum_{k=1}^{c_i} X_{ik}(I(Y_{ik} - X_{ik}^T \hat{\beta}_\tau < 0) - \tau),$$

with W_i drawn from any distribution with mean 0, variance 1, and a finite qth moment for some $q > 2$. Similar to the estimating equation bootstrap discussed earlier, β_τ^* can be obtained from a weighted quantile regression. Hagemann (2016) shows that the resampling process $n^{-1/2}(\beta_\tau^* - \hat{\beta}_\tau)$ as τ varies in a closed set $T \in (0,1)$ is approximately the same as the sampling process of $n^{-1/2}(\hat{\beta}_\tau - \beta_\tau)$. This procedure is implemented in the quantreg package.

2.8   Resampling methods for censored quantile regression

Censored data arise in a wide range of applications including biomedical, social, and environmental studies. Some measurements are censored by design or by natural constraints. In other cases, a natural bound for a positive variable, say, income, can be viewed as censoring to fit linear models to the data. When the response variable Y is censored, the conditional mean of Y given the covariates X is generally nonidentifiable without a parametric likelihood specification. Quantile regression is especially attractive for modeling censored response, because many of the conditional quantile functions are identifiable under weaker conditions. There are challenges, however, in fitting censored quantile regression and performing statistical inference. In this section we consider the case of left-censored responses, that is, we observe $\{(X_i, Z_i, \delta_i)\}$ instead of $\{(X_i, Y_i)\}$, where $Z_i = \max(Y_i, C_i)$ is the left-censored value of Y_i, with C_i being a censoring variable, and $\delta_i = I(Y_i > C_i)$ is the noncensoring indicator. Under the linear quantile model of $Q_Y(\tau|X = x) = x^T \beta_\tau$, we consider the problem of inference on β_τ .

If the C_i are known, the censored quantile regression problem can be reduced to a nonlinear quantile regression problem based on $Q_Z(\tau|X = X_i) = \max(X_i^T \beta_\tau, C_i)$. In other words, β_τ can be estimated by minimizing the quantile objective function

$$\sum_{i=1}^{n} \rho_\tau(Z_i - \max(X_i^T \beta, C_i))$$

over β. This is a well-known approach taken by Powell (1986). A difficulty with this formulation of censored quantile regression is that the objective function is highly nonconvex and multimodal, and therefore there is no sure way of finding a good solution. In this framework, resampling methods have to be used with care, because it can easily occur in a nonnegligible fraction of the resamples that incorrect solutions are given from the nonconvex optimization. Some alternatives have been proposed in the literature to avoid nonconvex optimization, and we will focus on the method of Bilias et al. (2000) for Powell's estimator.

This method is based on the simple idea that if $C_i = C$, a known constant, then Powell's estimator for minimizing $\sum_{i=1}^{n} \rho_\tau(Z_i - \max(X_i^T \beta, C))$ is asymptotically equivalent to the solution that minimizes

$$\sum_{i=1}^{n} \rho_\tau(Z_i - X_i^T \beta) I(X_i^T \beta_\tau > C),$$

with the true quantile coefficient β_τ. Suppose that $\hat{\beta}_\tau$ is Powell's estimate from the original sample. Then, given any bootstrap sample $\{(X_i^*, Y^*)\}$, we can use

$$\beta_\tau^* = \operatorname{argmin}_\beta \sum_{i=1}^n \rho_\tau(Y_i^* - X_i^{*T}\beta)I(X_i^{*T}\hat{\beta}_\tau > C).$$

The same idea can be used for the estimating function bootstrap. This approach uses linear programming for each resample, but it relies on the estimate $\hat{\beta}_\tau$ being well estimated from the original sample.

A different approach to censored quantile regression is proposed by Portnoy (2003) and Peng and Huang (2008). These estimators can be viewed as generalizations to regression of the Kaplan–Meier and Nelson–Aalen estimators for univariate censored data. These two approaches have similar modeling assumptions and similar performance, so we will focus on the method of Portnoy here. With left-censored data, the idea is to start with a high quantile (with τ close to 1) that is little affected by censoring, and then sequentially update the quantile regression estimates as τ moves down from 1 to 0. At each step, one estimates the quantile level associated with each censored point, and uses that information to redistribute the probability mass of each censored point to the left and right of the point for the calculation of the next quantile regression estimator. This approach works for random censoring as well as fixed censoring, but has to two important assumptions. First, the variables C_i and Y_i are conditionally independent given X_i. Second, the conditional quantile function $Q_Y(\tau|X)$ is linear at all quantile levels τ. The second assumption can be relaxed if one uses a nonparametric estimator for the conditional distribution of Y_i given X_i in determining the redistribution of each censored point, as done in Wang and Wang (2009) and Wey et al. (2014).

From the inferential point of view, the asymptotic distribution of the censored quantile regression estimator is difficult to approximate, and therefore resampling methods are handy. An annoying aspect of the censored quantile regression analysis under resampling is that some quantiles are not estimable due to censoring but the bound on τ for estimable quantiles is not known. Suppose that we are interested in the $\tau = 0.1$ quantile, but the censored quantile estimator might not be available for 15% of the resamples. How would you then use the resampling distribution as we normally do? This issue was visited in Portnoy (2014), who argued that it is desirable for the resamples to resemble the original sample as much as possible so that few of the resamples will lead to a nonestimable τth quantile. One suggestion is to use the delete-d_n jackknife, where $d_n = c\sqrt{n}$ for a constant around 2.5 (and increases slowly with n in theory).

In practice, one should always keep track of how often the τth quantile is nonestimable under a given resampling scheme. If there is a nonnegligible fraction, it might be wise to question whether this quantile level is identifiable after all.

2.9   Bootstrap for post-model selection inference

Statistical inference is usually carried out under a specific model, and it is naturally questionable if inference is made after model selection based on the same data. Efron (2014) suggested using the bootstrap for computing standard errors and confidence intervals that take model selection into account. The idea of bagging, or bootstrap smoothing, can be used for inference for quantile-based quantities. For simplicity, assume that we have a random sample $\{(X_i, Y_i),\ i = 1, \ldots, n\}$ with $X_i \in R^p$, and we use a statistical model selection technique to find a subset of the covariates $X(S) \in R^q$ ($q \leqslant p$), and proceed to estimate the selected quantile model assuming $Q_Y(\tau|X) = X(S)^T \beta(S)$ for some $\beta(S) \in R^q$. If we use $\hat{Q}(X_0) = X_0(S)^T \hat{\beta}(S)$ as an estimate of the τ-quantile of Y given $X = X_0$, what is a good interval estimate of the quantile?

Following Efron (2014), we can use paired bootstrap samples (X_i^*, Y_i^*), $i = 1, \ldots, n$, to perform model selection, and estimate the model assuming $Q_Y(\tau|X) = X(S^*)^T \beta(S^*)$, where $X(S^*)$ is the subset of the covariates selected under each bootstrap sample, and then obtain the bootstrap estimate $Q(X_0)^* = X_0(S^*)^T \hat{\beta}(S^*)$ for the quantile of interest. A confidence interval can then be constructed from a large number of bootstrap estimates. The average over the bootstrap estimates can be used as the smoothed bootstrap estimate of the quantile of interest at X_0.

To avoid the (undesirable) impact of model selection for inference on specific parameters, a split-sample approach may be used. The procedure involves splitting the data into a training sample (of size n_1) and a testing sample (of size n_2). Model selection is undertaken with the training sample, and the estimation and inference are undertaken with the test sample. By using random splits, a confidence interval on the quantile of interest can be constructed based on the estimates from the repeated random splits. The split-sample approach appears simple, but it is not well understood how the ratio n_1/n_2 should be chosen to maximize statistical performance. If n_1/n_2 is too small, we tend to select models that are too small, missing some of the important covariates, and leading to a possible bias in estimation. If n_1/n_2 is too large, the estimates from the testing samples can suffer from high variability.

To illustrate the point, we use a Monte Carlo experiment with random samples generated from $Y = X_1 + X_2 + e$ of size $n = 100$, with $X_1 = Z_1$, $X_2 = Z_1 + 0.5Z_2$ and $e = Z_3$, with (Z_1, Z_2, Z_3) distributed as standard trivariate normal. The model selected from the training sample contains either X_1 only or X_1 and X_2, depending on whether the X_2 term is significant at the 5% level based on the *rq* function in R with default options at the median

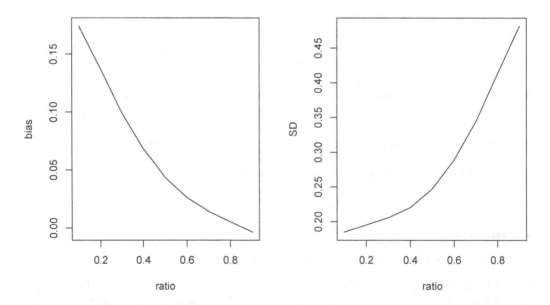

FIGURE 2.1
Bias (left) and standard deviation (right) as functions of the ratio n_1/n_2 in the split-sample experiment.

$\tau = 0.5$. The coefficient for X_1 is then estimated from the selected model using the test data. Over 500 replications with various levels of the ratio n_1/n_2, we show the bias and the standard deviation of the parameter estimate in Figure 2.1. Even in this simple case, the ratio plays an important role in the bias–variance trade-off, and a ratio around 0.4 leads to the best mean squared error.

Post-model selection inference becomes more challenging if there are a large number of potential covariates to start with. Resampling methods usually do not perform well in the high-dimensional setting, especially when the sample size n is not much larger than the dimension of the variables p; see El Karoui and Purdom (2015) and Zheng et al. (2014) for illustrations. We refer to Belloni et al. (2014) for one possible approach to valid inference for sparse quantile regression in high dimensions.

Bibliography

P. Barbe and P. Bertail. *The Weighted Bootstrap*. Springer, 1995.

A. Belloni, V. Chernozhukov, and K. Kato. Valid post-selection inference in high-dimensional approximately sparse quantile regression models. arXiv:1312.7186, 2014.

P. J. Bickel and D. A. Freedman. Some asymptotic theory for the bootstrap. *Annals of Statistics*, 9:1196–1217, 1981.

Y. Bilias, S. Chen, and Z. Ying. Simple resampling methods for censored regression quantiles. *Journal of Econometrics*, 99:373–386, 2000.

C. Chatterjee and A. Bose. Generalized bootstrap for estimating equations. *Annals of Statistics*, 33:414–436, 2005.

C. Chen, Z. L. Ying, H. Zhang, and Zhao L. Analysis of least absolute deviation. *Biometrika*, 95:107–122, 2008.

L. Chen, L. J. Wei, and M. I. Parzen. Quantile regression for correlated observations. *Proceedings of the Second Seattle Symposium in Biostatistics*, pages 51–69, 2004.

A. C. Davison and D. V. Hinkley. *Bootstrap Methods and Their Application*. Cambridge University Press, Cambridge, 1997.

B. Efron. Estimation and accuracy after model selection (with discussion). *Journal of the American Statistical Association*, 109:991–1007, 2014.

N. El Karoui and E. Purdom. Can we trust the bootstrap in high-dimension? arXiv:1608.00696, 2015.

X. Feng, X. He, and J. Hu. Wild bootstrap for quantile regression. *Biometrika*, 98:995–999, 2011.

C. A. Field and A. H. Welsh. Bootstrapping clustered data. *Journal of the Royal Statistical Society, Series B*, 69:369–390, 2007.

D. A. Freedman. Bootstrapping regression models. *Annals of Statistics*, 9:1218–1228, 1981.

D. A. Freedman. Bootstrapping quantile regression estimators. *Econometric Theory*, 11: 105–121, 1995.

A. Hagemann. Cluster-robust bootstrap inference in quantile regression models. *Journal of the American Statistical Association*, 111, 2016.

X. He and F. Hu. Markov chain marginal bootstrap. *Journal of the American Statistical Association*, 97:783–795, 2002.

X. He and Q. M. Shao. A general Bahadur representation of M-estimators and its application to linear regression with nonstochastic designs. *Annals of Statistics*, 24:2608–2630, 1996.

J. L. Horowitz. Bootstrap methods for median regression models. *Eonometrica*, 66:1327–1351, 1998.

F. Hu and J. V. Zidek. A bootstrap based on the estimating equations of the linear model. *Biometrika*, 82:263–275, 1995.

A. Kleiner, A. Talwalkar, P. Sarkar, and M. I. Jordan. A scalable bootstrap for massive data. *Journal of the Royal Statistical Society, Series B*, 76:795–816, 2014.

M. Kocheringsky, X. He, and Y. Mu. Practical confidence intervals for regression quantiles. *Journal of the Computational and Graphical Statistics*, 14:41–55, 2003.

R. Koenker. *Quantile Regression*. Cambridge University Press, Cambridge, 2005.

R. Koenker and G. W. Bassett. Regression quantiles. *Econometrica*, 46:33–49, 1978.

R. Y. Liu. Bootstrap procedures under some non-i.i.d. models. *Annals of Statistics*, 16:1696–1708, 1988.

M. I. Parzen, L. J. Wei, and Z. Ying. A resampling method based on pivotal estimating functions. *Biometrika*, 81:341–350, 1994.

L. Peng and Y. Huang. Survival analysis with quantile regression models. *Journal of the American Statistical Association*, 103:637–649, 2008.

S. L. Portnoy. Censored regression quantiles. *Journal of the American Statistical Association*, 98:1001–1012, 2003.

S. L. Portnoy. Nearly root-n approximation for regression quantile processes. *Annals of Statistics*, 40:1714–1736, 2012.

S. L. Portnoy. The jackknife's edge: Inference for censored regression quantiles. *Computational Statistics & Data Analysis*, 72:273–281, 2014.

S. L. Portnoy and R. Koenker. The Gaussian hare and the Laplacian tortoise: Computability of squared-error versus absolute-error estimators. *Statistical Science*, 12:279–300, 1997.

J. L. Powell. Censored regression quantiles. *Journal of Econometrics*, 32:143–155, 1986.

H. J. Wang and X. He. Detecting differential expressions in genechip microarray studies: A quantile approach. *Journal of the American Statistical Association*, 102:104–112, 2007.

H. J. Wang and L. Wang. Locally weighted censored quantile regression. *Journal of the American Statistical Association*, 104:1117–1128, 2009.

A. Wey, L Wang, and K. Rudser. Censored quantile regression with recursive partitioning-based weights. *Biostatistics*, 15:170–181, 2014.

C. F. J. Wu. Jackknife, bootstrap and other resampling methods in regression analysis. *Annals of Statistics*, 14:1261–1295, 1986.

J. Yang, X. Meng, and M.W. Mahoney. Quantile regression for large-scale applications. arXiv:1305.0087, 2014.

C. Yi and J. Huang. Semismooth Newton coordinate descent algorithm for elastic-net penalized Huber loss regression and quantile regression. arXiv:1509.022957, 2015.

S. Zheng, D. Jiang, Z. Bai, and X. He. Inference on multiple correlation coefficients with moderately high dimensional data. *Journal of the American Statistical Association*, 101: 748–754, 2014.

3

Quantile Regression: Penalized

Ivan Mizera

University of Alberta, Edmonton, Alberta, Canada

CONTENTS

3.1   Penalized: how?

3.1.1   A probability path

As often happens with important concepts, the idea of penalization applied to functional fitting can be traced to more origins than one. The inspiration from the probabilistic side arose in the context of "graduation," a "borrowing strength" enterprise in the actuarial context. The best exposition of the Whittaker (or, as actuarial science prefers to have it, Whittaker–Henderson) graduation is, as a matter of fact, Chapter XI of Whittaker and Robinson (1924), entitled "Graduation, or the Smoothing of Data." It also provides the best account of the earlier references of Whittaker (1922–1923, 1924); those of Henderson (1924, 1925, 1928, 1938) are to be found elsewhere.

     A process of "smoothing" is to be performed over the original values, typically those of a mortality table, burdened by observational or other errors. The objective, as Whittaker and Robinson point out, is not so much to get a smooth curve, as to "get the most probable deaths." After surveying several existing methodologies (among them one that can be regarded as a prototype for local polynomial fitting) they adopt the conventional normal model for the distribution of errors, together with the normal *a priori* distribution for what they call a *measure of roughness*, S. The subsequent use of what they consider a "fundamental theorem in the theory of Inductive Probability" (with no personal attribution) yields as the "most probable hypothesis" that "which makes $\lambda^2 S + F$ a maximum," with F signifying *fidelity* ("of the graduated to the ungraduated values, respectively").

The subsequent century of further development changed not that much. Nowadays, $\lambda > 0$ is used rather than λ^2; the word "fidelity" is still around – although, as the mathematical expression quantifies "infidelity," more specific terms like "lack-of-fit," or even better and becoming widespread, *loss*, $L(f)$, associated with fit f, are taking its place. Given that various other objectives may be pursued rather than smoothness, it is preferred to speak in a neutral fashion about *penalty*, $P(f)$. Finally, a custom developed to write the terms in the reverse order: in the modern formulation, we seek the solution of

$$L(f) + \lambda P(f) \rightsquigarrow \min_f ! \tag{3.1}$$

The result of (3.1) is sometimes called the maximum a posteriori (MAP) estimate. The prior/posterior, random-effect interpretation of (3.1) has received regular attention in the literature, in the work of Good (1971), Good and Gaskins (1971), Kimeldorf and Wahba (1970), Watson (1984), Cressie (1989, 1990), Wahba (1990a), Kent and Mardia (1994), and others.

3.1.2   Regularization of ill-posed problems

A separate line of development – probability-free, as not aligning itself with what earlier could be called, according to Stigler (1986), "combination of observations" or "reconciling inconsistent observational equations," and is now often referred to as "statistics" – recognizes as its origin the observation of Hadamard (1902) that not all problems are really well posed. As a consequence, solving them may be not easy; to this end, Tikhonov (1943, 1963) proposed a technique, or rather a family of techniques, for which he coined the catchy word *regularization*. His most successful proposal, the one referred to as the *Tikhonov regularization*, coincides in its form with (3.1).

Some used to prefer to speak about the Tikhonov–Phillips regularization instead; but then it may be of interest that Phillips (1962) originally started from the formulation

$$P(f) \rightsquigarrow \min_f ! \quad \text{subject to } L(f) \leqslant \Sigma \tag{3.2}$$

(although actually in a slightly different form, with the equality $L(f) = \Sigma$ instead of the inequality $L(f) \leqslant \Sigma$). This phrasing is particularly apt if we (somehow) know the noise level Σ; otherwise, of course, the Lagrange multiplier theory makes it quite apparent that the Tikhonov formulation (3.1) readily derives from (3.2). The choice between equality and inequality is not that substantial, as in typical cases the optimum makes the inequality constraint active, hence satisfying the equality.

Hence, it may be appropriate to reserve (3.1) for Tikhonov alone and give the specific name *Phillips regularization* to the case where the problem is formulated in the loss-constrained form (3.2) – as was done in the Russian literature, which also spoke about *Ivanov regularization* when the tasks were specified in a penalty-constrained form

$$L(f) \rightsquigarrow \min_f ! \quad \text{subject to } P(f) \leqslant \Lambda. \tag{3.3}$$

Ideologically, this alternative corresponds to the fixed-effect viewpoint: if the estimated regression function does not vary with every realization, and allows for repeated sampling (possibly with varying number of observations and different sizes of error), then it is crucial to determine the right amount of "roughness" (or whatever P aims to control) of the "true"

f: once we know it, we keep it fixed and optimize merely the goodness of fit. Gu (1998) advocates (3.3) and its tuning parameter Λ as better suited for indexing models across replications. It may be of interest that Tibshirani (1996) also starts his exposition with the penalty-constrained formulation, perhaps motivated by the connection to the nonnegative garrote of Breiman (1995).

In the view of the Lagrange multiplier theory, it is clear that Tikhonov, Phillips, and Ivanov are closely related, the essential assumption being the convexity of both L and P (see Vasin, 1970). Tibshirani (2015) studies a numerical method for the general form of (3.3) related to the "forward stagewise" algorithm of Efron et al. (2004). One reason for the popularity of the Tikhonov formulation (3.2) over the other two – as witnessed in the further development of the Whittaker graduation by Schoenberg (1964) and Reinsch (1967, 1971) – is the computational simplicity it offers in the quadratic case. This may also explain why Rudin et al. (1992) could easily embrace the Phillips formulation (3.2) (with equality) instead: in their case, there is no noteworthy computational advantage to be gained from (3.1). The case of quantile regression is similar – the reason why we recall here the alternative formulations.

3.2   Penalized: what?

3.2.1   The finite differences of Whittaker and others

Unlike the development of unpenalized regression fitting (the technology minimizing the loss function $L(f)$ alone) the initial work in the penalized domain did not see that much variety and rather adhered to the ideal of "wonderful simplicity" in the sense of Gauss: all formulations were quadratic. Both Whittaker and Tikhonov, as well as Phillips and Schoenberg, took $L(f)$ to be the sum of squared errors. For instance, the formulation of Whittaker determines the graduated values f_t from the original ones y_t via solving

$$\sum_{t=1}^{T} (y_t - f_t)^2 + \lambda \sum_{t=2}^{T-2} (f_{t-1} - 3f_t + 3f_{t+1} - f_{t+2})^2 \leftrightsquigarrow \min_f ! \qquad (3.4)$$

It may be of some little conceptual advantage to take the "average error" instead, the sum of squares divided by the number of observations: the prototypic objective function of Tikhonov regularization (3.1) is then

$$\frac{1}{n} \sum_{t=1}^{T} (y_t - e_t(f))^2 + \lambda J(Df) \leftrightsquigarrow \min_f ! \qquad (3.5)$$

Here $e_t(f)$ is the result of an *evaluation functional*, corresponding to y_t, and applied to f: a putative fitted value of f. In the graduation setting, the fits f_t form a vector, f, of the same length as that of the observations y_t; thus, $e_t(f) = f_t$. When the y_t are fitted by a function, f, of covariates x_t, then $e_t(f) = f(x_t)$.

The typical form of the penalty used in (3.5) developed into $P(f) = J(Df)$, with J a convex function, typically a norm or its square, and D a linear operator. The motivation for the particular form of the penalty is usually rather intuitive; it is fairly clear that pushing the total size of the differences/derivatives down makes fits smoother, but which particular order of derivatives has to be used may be not that apparent. As a rule, the only specific exact motivation setting apart various penalties is the limiting behavior with respect to the tuning parameter. When in (3.5) $\lambda \to 0$ (or $\Lambda \to \infty$ in the Ivanov formulation), the sum of $(y_t - f_t)^2$ approaches 0 and observations are in the limit fitted by themselves. This is even more apparent in the Phillips formulation, where for $\Sigma = 0$ we obtain an *interpolation problem*

$$P(f) \leftrightarrow \min_{f} !\quad \text{subject to } L(f) = 0 \tag{3.6}$$

whose solution, as will be recalled below, is the key to the shape of the solutions of regularization formulations considered above.

The important limit behavior we speak of is, however, that on the opposite side: if $\lambda \to \infty$, then the fit in the limit satisfies $J(Df) = 0$. Consequently, if J is a norm (or just a strictly increasing function with $J(0) = 0$), it follows that the fits lie on a graph of a function satisfying $Df \equiv 0$. If D is a difference/derivative operator of order k, then this function is constant, linear (affine), or quadratic, respectively, for $k = 1, 2, 3$. Conversely, the penalty does not influence the additive part of the fitted function for which it vanishes.

The prevailing choice in the literature, forcing the limit fits to be linear, happens to be $k = 2$, as in $(Df)_t = f_{t-1} - 2f_t + f_{t+1}$. A notable exception is Silverman (1982), who in the context of density estimation proposed to penalize the *logarithm* of the probability density and then suggested the use of the third derivative: the limiting fit is then the exponential of a quadratic function, the density of a normal distribution. Interestingly, the third-order differences were Whittaker's original choice (3.4) as well.

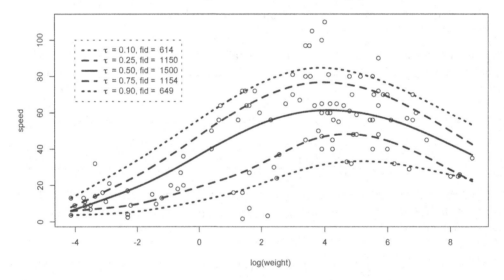

Quantile regression with cubic smoothing splines

FIGURE 3.1

Maximum recorded running speed as a function of the logarithm of body weight, fitted for various quantile indices τ as cubic smoothing splines with the appropriate sum-of-check-functions losses. Tuning parameters were selected to match fidelities with those in Figure 3.3.

3.2.2 Functions and their derivatives

The setting of graduation enjoyed relative simplicity due to the sequential, equidistant character of observations. On the other hand, in general nonparametric regression we may face complications related to possibly different distances between covariates, and their possible repetition. While differences are still possible (divided by the magnitudes of those distances), a revolutionary leap forward was accomplished by Schoenberg (1964), who recognized that the Whittaker graduation can be formulated in a very general functional setting, and pointed out its connection to splines, the effective interpolation vehicle in numerical analysis.

The new setting means that differences are replaced by derivatives: Df is now a differential operator applied to a function f, and the norm J typically involves some integral over the domain of the definition of f. This brings some technical complications, which we will not detail here: the integrals need to converge, and in general may depend on the domain of integration. The classical theory requires the penalty to have *the L^2 character*: it is a strictly increasing function of a norm in a Hilbert space, a norm that can be expressed in terms of the inner product in a suitable functional space. A canonical instance of such a penalty, related to the inner product in the Sobolev space of twice-differentiable functions with square-integrable second derivative, is

$$J(D(f)) = \int_\Omega (f''(x))^2 dx. \tag{3.7}$$

It this particular fortuitous case, when x is a real variable, the choice of the domain of integration is not essential: any $\Omega = [a, b]$ that covers all covariate points, and in particular $\Omega = [-\infty, \infty]$, gives the same result.

Once the penalty has the L^2 character, the theory retains its full strength when $e_t(f) = f(x_t)$ and the minimization involves a more general loss function

$$L((y_1, x_1, f(x_1)), \ldots, (y_T, x_T, f(x_T))) + \lambda J(Df) \leadsto \min_f ! \tag{3.8}$$

The search for the minimizer of (3.8), an infinite-dimensional problem, can be then reduced to a finite optimization in a finite-dimensional space: the so-called *representer theorem* of Kimeldorf and Wahba (1970) – we refer to the most general version given by Schölkopf and Smola (2002) – asserts that the solution is a linear combination of a finite number of basis functions. Those functions generally depend on the problem (in particular, on the covariates) and are related to an important notion of positive definite functions of two variables, the so-called *reproducing (Mercer) kernel*. As it is somewhat more straightforward to obtain the penalty from the reproducing kernel than conversely, a considerable part of the literature concerning regularization in the L^2 context concentrates on kernels rather than on penalties; the minimization of the loss is no longer "penalized" but "kernelized." See also Wahba (1990b), Green and Silverman (1994), and Ramsay and Silverman (2005).

We stress here that the thrust of the representer theorem lies in its explicit characterization, not merely in finite dimensionality. In fact, pretty much every solution to a problem of the form (3.8), with an arbitrary penalty, is characterized by the fitted values $\hat{f}(x_t)$, the values of the optimal function f at the points x_i for $t = 1, \ldots, T$; for if $\tilde{f}$ is any function such that $\tilde{f}(x_t) = \hat{f}(x_t)$, then the loss is the same for $\tilde{f}$ and $\hat{f}$, and thus by the optimality of $\hat{f}$, the values of the penalty for both functions must be the same too. Therefore, the solution of (3.8) is characterized by the T fitted values, and consequently the space of such solution is some set parametrized by those (finite number of) values, although possibly in a nonlinear way. The only problem here may arise if the function interpolating fitted values $\hat{f}(x_t)$ is not unique; but even then sometimes a convenient version can be picked up in a unique way, as in the case of the quantile smoothing splines considered below. This small exercise also shows that the shape, the contour structure of the solution (whether it piecewise linear, cubic spline, or something similar) depends, for the pointwise defined loss functions, essentially only on the penalty.

All this was certainly not an issue in the original finite-dimensional formulations of

Whittaker and others; thus, many feel that it should have remained at that. The advantages of the functional setting may not be that compelling when the only task is to process the data set at hand, but we believe that they become visible when more elaborate sampling strategies (e.g. adding new observations) are involved. A more thorough investigation of such situations may still remain to be done.

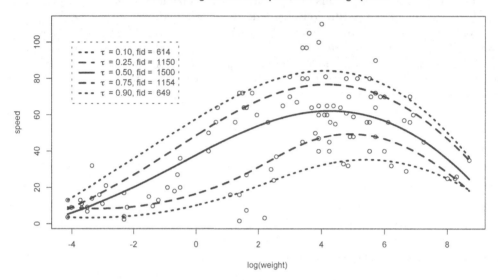

FIGURE 3.2
The same data as in Figure 3.1, now fitted by quintic smoothing splines with the sum-of-check-functions losses. Tuning parameters are again selected to yield the same fidelities in all examples.

3.2.3   Quantile regression with smoothing splines

Quantile regression loss functions formed as sums of pointwise evaluations of check functions satisfy the conditions required by the representer theorem, which then applies to the formulations where they replace the sum-of-squares loss accompanied by a penalty of the L^2 type,

$$\sum_{t=1}^{T} \rho_\tau \left(y_t - f(x_t)\right) + \lambda \int_\Omega (f^{(k)}(x))^2 \, dx \rightsquigarrow \min_f ! \tag{3.9}$$

Here $\Omega = [a, b]$ is an interval covering all covariate points and $f^{(k)}$ denotes the kth derivative (possibly in the sense of absolute continuity). The solution of the functional, infinite-dimensional optimization task (3.9) can be reduced to a finite-dimensional one of finding the coefficients c_j in an expansion $f = c_1 g_1 + \cdots + c_K g_K$, with known g_j, possibly depending on x_t. In such a case, we can also write $J(Dg)$ as $c^\top G c$, where the matrix G depends only on the basis functions g_i. The computation of a "quantile regression with smoothing splines" – the terminology is a bit confusing here, especially in view of "quantile smoothing splines" to come below; we thus adopt the term of Bosch et al. (1995) – via (3.9) is then reduced to a quadratic programming problem.

Quantile regression proposals in this direction, for the most part with $k = 2$, were put forward by Bloomfield and Steiger (1983) and others, including Nychka et al. (1995), Bosch et al. (1995), and Oh et al. (2012); the latter paid some attention to the development of dedicated computational algorithms in the iterated reweighted least-squares fashion, to eschew the full deployment of quadratic programming, a task which was at the time not that routine as it is nowadays, due to the rapid evolution of convex optimization software as documented by Koenker and Mizera (2014). More complex and general "kernelized" versions, with potentially a multidimensional scope, were proposed by Takeuchi et al. (2006) and Li et al. (2007).

Figure 3.1 shows the fit of various quantiles (the indices τ indicated in the legend) of the maximum recorded running speed of mammals as a function of the logarithm of their body weight, as cubic ($k = 2$) smoothing splines; by way of homage to Whittaker, we show also the analogous fits for $k = 3$ in Figure 3.2. The fact that we take the logarithm of the body weight in these examples, but not that of the running speed, is not meant as a dispute with Example 7.2 of Koenker (2005), featured also in Koenker et al. (1994), where the logarithm of speed is taken as well; our intention was rather to achieve some variety. The tuning parameters for the penalized fits were initially selected by eye, roughly the same λ for all τ; later they were slightly modified to match the fidelities in Figure 3.3, where the properties of quantile smoothing splines do not allow for that much flexibility in setting arbitrary fidelity values as in the other cases.

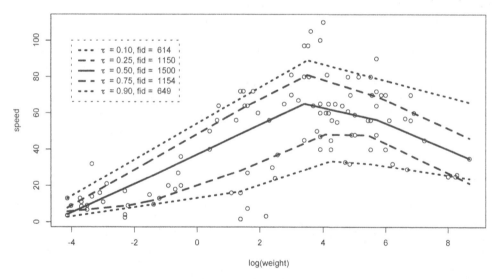

FIGURE 3.3
The maximum speed of mammals as a function of the logarithm of weight, now fitted as quantile smoothing splines, linear total-variation splines with the sum-of-check-functions loss. Tuning parameters selected to yield fidelities that can be matched in the other examples.

3.2.4   Quantile smoothing splines

The difficulties anticipated in their time by Koenker et al. (1994) in the practical use of quadratic programming for solving (3.9) with the penalty (3.7) motivated a step sideways: holding no prejudices against "kinky" statistics of piecewise linear loss functions, and at the same time aware of the potential of linear programming, Koenker et al. proposed to replace the square in the definition of the penalty (3.7) by the absolute value, thus instantiating formulation (3.9) as

$$\sum_{t=1}^{T} \rho_\tau(y_t - f(x_t)) + \lambda \int_\Omega |f''(x)| \, dx \rightsquigarrow \min_f ! \qquad (3.10)$$

Here $\Omega = [a, b]$ is again an interval covering all covariate points. By the Vitali theorem, the penalty is equal to the *total variation* of f' (not of f: note *two* derivatives) on Ω. The second derivative may be thus merely in the sense of absolute continuity and the problem can be equivalently written as

$$\sum_{t=1}^{T} \rho_\tau(y_t - f(x_t)) + \lambda \bigvee_\Omega f' \rightsquigarrow \min_f ! \qquad (3.11)$$

The original infinite-dimensional formulation can be again reduced to a finite-dimensional one, admitting a representer theorem in the spirit discussed above. Koenker et al. (1994) showed that optimal solutions of (3.9) can be found among piecewise linear f, with the breaks occurring only at covariate points x_t. The fact that the objective function of (3.11) is convex, but not strictly convex, permits in some cases also optimal solutions that are not piecewise linear; nevertheless, an equivalent piecewise linear solution, a solution with the same value of the objective function, can always be found. (In fact, this aspect is also present in the proof of a general representer theorem for L^2 penalties.) The resulting finite-dimensional problem is a linear programming task, for which there are, especially nowadays, very efficient algorithms. An example of the result can be seen in Figure 3.3.

As often happens with important ideas, the use of the L^1 norm in penalization has a complex history. In the context of multiple regression (a response vector y fitted by a linear combination $X\beta$) penalized by a simple norm penalty (a penalty without any linear operator inside it), the use of the ℓ^1 norm penalty,

$$J(\beta) = \sum_i |\beta_i|, \qquad (3.12)$$

as a replacement for the ℓ^2 penalty of ridge regression of Hoerl and Kennard (1970),

$$J(\beta) = \sum_i \beta_i^2, \qquad (3.13)$$

can be traced back to the thesis of Logan (1965) – see also Donoho and Logan (1992) – and the work of geophysicists Claerbout and Muir (1973), Taylor et al. (1979), and Santosa and Symes (1986). Chen and Donoho (1995) and Chen et al. (1998) pointed out the connection to the total-variation denoising of Rudin et al. (1992), which penalizes the total variation of f rather than that of f'; the setting of image processing is akin rather to that of graduation – the data appear in *every* pixel, there is no need for interpolation. In graduation itself, the use of absolute-value loss (median regression as the special case of (3.11) with $\tau = 1/2$) and the ℓ^1 norm/total variation in graduation was proposed by Schuette (1978). In statistics, the L^1 regularization became extremely popular under the acronym lasso, "*l*east *a*bsolute *s*election and *s*hrinkage *o*perator," introduced by Tibshirani (1996).

Quantile smoothing splines enjoy many favorable properties. They are conceptually transparent and enjoy a straightforward linear programming implementation (in any of Tikhonov, Phillips, or Ivanov formulations). They offer relatively easy mitigation of the potential complication of crossing, the anomaly that sometimes happens (especially on the boundary of or outside the domain of covariates, for instance in our Figure 3.3 for $\tau = 0.1$ and $\tau = 0.25$) when the quantile fit for τ_1 lies for certain covariate values below the fit for $\tau_2 < \tau_1$. The straightforward strategy of rearrangement (Chernozhukov et al., 2010), amounting basically to the exchange of fits in parts where they violate the desired monotonicity, does not leave any aesthetically unpleasing kinks as for smooth curves, owing to the fact that both maximum and minimum of any piecewise linear functions is again piecewise linear.

Quantile smoothing splines can be also easily adapted to potential *shape constraints* imposed on fits, like monotonicity or convexity. These shape constraints can actually be viewed themselves as a special case of penalized fits, with penalties assuming values 0 if the fits uphold them, or $+\infty$ if they do not. Such penalties can be designed to conform to the scheme $J(Df)$: for instance, J can an made the indicator (in the sense of convex analysis) of, say, a set $E = [0, +\infty)$, with $J(u) = 0$ if u belongs to E, and $J(u) = +\infty$ if it does not; combined with the operator D of the first or second derivative, such penalties respectively enforce (nondecreasing) monotonicity or convexity.

The only somewhat unfortunate property may be their name: the adjective "smoothing" may be considered misleading by those expecting in such a context smooth corners rather than kinks. However, this kind of thing may be a marketing problem for emerging brands, but not for well-established ones: everybody now knows that the three musketeers were actually four. Finally, those preferring round corners and wavy curves may opt not only for the L^2 penalties mentioned above, but also for L^1 penalties with higher-order derivatives, exposed in the next section – or just smooth the corners, if their principal objective is the same as the desideratum of Whittaker and Robinson (1924): to achieve "not that much smoothness but the most probable deaths."

3.2.5   Total-variation splines

Total variation in spline smoothing was taken further in the important paper of Mammen and van de Geer (1997). They, having no special interest in quantile regression, transferred the setting to the realm of squared-error loss, but carried the idea further by introducing the whole class of estimates obtained through solving

$$\sum_{t=1}^{T} (y_t - f(x_t))^2 + \lambda \bigvee_{\Omega} f^{(k-1)} \looparrowright \min_{f} ! \tag{3.14}$$

Recall that by the Vitali theorem, the penalty can be equivalently written as the integral, over Ω, of $|f^{(k)}|$, as the required absolute continuity of $f^{(k-1)}$ is in fact necessary for the feasibility of (3.14). Combining this with the return to quantile loss functions, we obtain

$$\sum_{t=1}^{T} \rho_\tau(y_t - f(x_t)) + \lambda \int_{\Omega} |f^{(k)}(x)| \, dx \looparrowright \min_{f} ! \tag{3.15}$$

The latter is a generalization of (3.11) involving possibly other k than 2.

Given that due to the connection to the interpolation problem (3.6) the representer theorem depends more essentially on the penalty once the loss amounts to a finite number of pointwise evaluations, it was not surprising that Mammen and van de Geer (1997) reaffirmed the representer theorem of Koenker et al. for $k = 2$, and added that it holds true also for

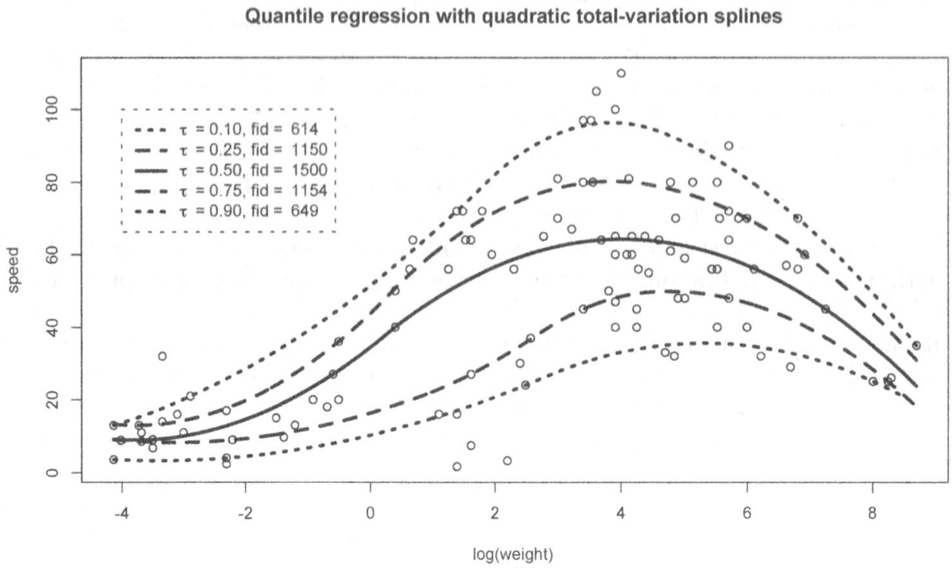

Quantile regression with quadratic total-variation splines

Legend:
- $\tau = 0.10$, fid = 614
- $\tau = 0.25$, fid = 1150
- $\tau = 0.50$, fid = 1500
- $\tau = 0.75$, fid = 1154
- $\tau = 0.90$, fid = 649

(y-axis: speed; x-axis: log(weight))

FIGURE 3.4

The same data as in all examples, now fitted by quadratic total-variation smoothing splines combined with the sum-of-check-functions loss. Tuning parameters selected again to match fidelities with those in all examples.

$k = 1$, again in the sense that the knots of the minimizing piecewise constant "spline" *can* be chosen at the covariate points. They pointed out, however, that in general this is not true for $k \geq 3$: their "weak" representer theorem established that in such a case it is known only that the solution is the spline of order $k - 1$, but the locations of knots are unknown and may not coincide with any of the covariate points x_t.

This may constitute, for $k \geq 3$, a possible difficulty for computation, particularly for squared-error loss. However, for quantile regression, the use of a relatively fine grid of "potential knots" (the original proposal of Mammen and van de Geer) renders the problem computationally quite feasible with modern methods for linear programming. In the least-squares case, Tibshirani (2014) proposed to act as if the knots were in the covariate points, using the opportunity to rename the whole technology – which differs from that of Mammen and van de Geer (1997) only for $k = 3$ and higher, with $k \leq 3$ being probably the most likely to be ever used – after the "graduation proposal" (this time in the context of time-series sequences) of Kim et al. (2009), blending the least-squares loss of Whittaker, as given by Whittaker and Robinson (1924), with the absolute-value difference penalty of Schuette (1978).

The most important achievement of Mammen and van de Geer (1997) came on the theoretic front: they showed that in the smoothness class of the functions whose total variation of the $(k-1)$th derivative is (uniformly) bounded by a constant, the total-variation splines, with appropriately selected tuning parameter, achieve the minimax $n^{-k/(2k+1)}$ rate of convergence – the rate that Donoho and Johnstone (1998) showed cannot be achieved by the estimators that are linear in the observations. These are, in particular, the estimators obtained via the L^2 penalties, but, ironically, only when coupled with the least-squares loss functions; those coupled with the quantile regression sum-of-check-functions functions are technically not linear.

Translated into the real world, the smoothness class condition essentially means the number of derivatives we believe the functions underlying the data possess. If we cannot rule out jumps, we have to opt for $k = 1$; if we believe that the functions we seek are continuous, but kinks are allowed, we choose $k = 2$; finally, if we believe that the fitted functions should be continuous and differentiable, then $k = 3$ should do. The latter, with the penalty that may be viewed as the absolute-value transmutation of that of Whittaker, may be, in fact, the best choice for practical smoothing (in the original sense of the word), as suggested by the example in Figure 3.4. The larger values of k may be prone to numerical difficulties as well as yielding fits suffering too much from the inflexibility of the higher-order polynomials involved in their construction – as possibly indicated by Figure 3.5 involving $k = 4$.

3.3 Penalized: what else?

3.3.1 Tuning

Whittaker and Robinson (1924) identify several advantages of their method of graduation, among them its superiority to the precursor of local polynomial fitting (also discussed in their book) in handling boundary data. Interestingly, they also include among its positive virtues the dependence on the tuning parameter λ, because "a satisfactory method of graduation ought to possess such elasticity, because the degree to which we are justified in sacrificing fidelity in order to obtain smoothness varies greatly from one problem to another;" to determine λ, they advise to "try two or three different values and see which gives the most satisfactory result" (an approach which was pretty much used in the examples we present here).

In the meantime, the outlook may have changed: the onus is now on the availability of "automatic" methods for the selection of tuning parameters. Theoretically oriented researchers may be inclined to feel that more work remains to be done in this respect, while practically oriented ones usually report no problems when already well-known tuning procedures are used.

The latter include various forms of cross-validation, as suggested by Hastie et al. (2008), who favor splitting the data set into relatively few groups. This *k-fold cross-validation* may suffer from some volatility caused by (typically) random allocation of observations to k groups, but is usually computationally feasible (especially for $k = 2$, but often also for the commonly used $k = 10$), and thus more often than not returns operational results.

The classical, leave-one-out, n-fold (n being the size of the data set) cross-validation used for L^2 penalties is a bit out of favor here, as its crude computational form can result in a prohibitive amount of computation time, and its refined form is not available, due to the nonlinear character of the problems with nonquadratic objectives; this applies also to its step-sibling, generalized cross-validation. It is possible, however, to employ these strategies on various linearizations of the problem, as proposed by Oh et al. (2012), Yuan (2006), and others.

The more preferred approach for the L^1 penalties seems to be via the notion of the *degrees of freedom*, the projected number of free parameters in the fitted model. As pointed out already by Koenker et al. (1994), and verified by several other sources later, for the L^1 penalty the degrees of freedom are equal to the number of exact fits, the number of fitted

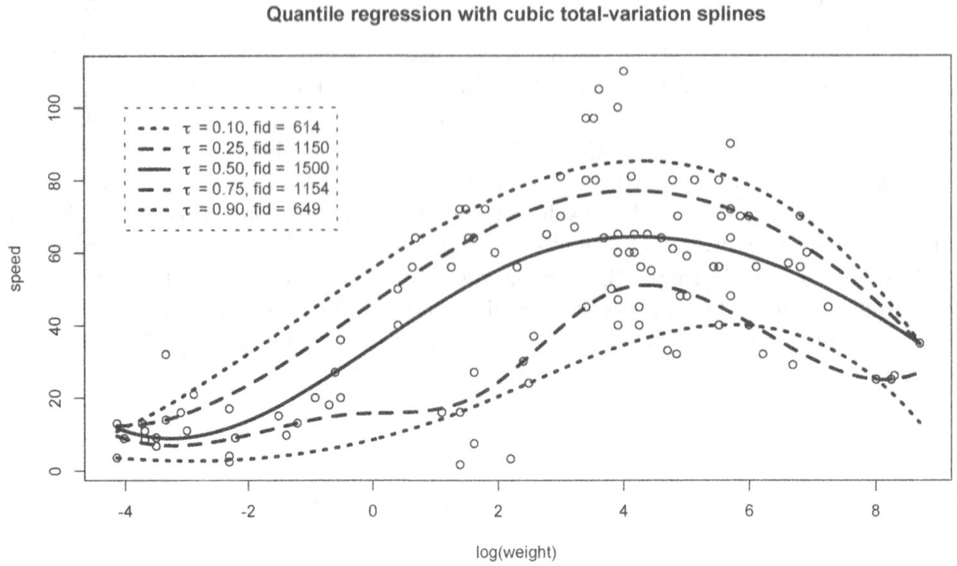

Quantile regression with cubic total-variation splines

FIGURE 3.5
Once again the same data, this time fitted by cubic total-variation smoothing splines combined with the sum-of-check-functions loss, with the same strategy for the selection of tuning parameters as in all other examples.

points with residual equal to 0. The degrees of freedom, p_λ, calculated in this way can then be plugged into the objective function of the Akaike information criterion,

$$\log\left(\frac{1}{n}\sum_{t=1}^{T}\rho_\tau\left(y_t - f(x_t)\right)\right) + \frac{p_\lambda}{n},$$

or (apparently the more favored choice) of the Schwarz information criterion,

$$\log\left(\frac{1}{n}\sum_{t=1}^{T}\rho_\tau\left(y_t - f(x_t)\right)\right) + \frac{0.5p_\lambda\log n}{n}$$

(see Koenker et al., 1994; Koenker and Mizera, 2004; Zou and Yuan, 2008). The desired tuning parameter λ is then found as the one yielding the minimum.

3.3.2   Multiple covariates

The ℓ^1 background inherent in prescriptions involving minimizing total-variation penalties yields in the one-dimensional case results consistent with the usual heuristics of sparse regression: most of the fitted parameters come out as zeros. In one-dimensional spline fitting involving total variation, "zero" means "no knot;" penalizing splines with total variation is therefore more akin to the knot deletion strategies of Smith (1982), championed by Kooperberg and Stone (1991), among others, in the context of log-spline density estimation, than to the classical smoothing splines with L^2 penalties. With smoothing splines starting with interpolatory splines and ending up with the linear fit, it is tempting to think that the fits in between pass through all possible intermediate complexity patterns; the truth is rather that the complexity of the fit remains the same, but the parameters are more shrunk toward zero.

The success of L^1 penalties in one-dimensional nonparametric regression led to the study of possible extensions in situations with multiple covariates. From the practical point of view, the most accessible path is to act one covariate at a time: either via the additive models briefly discussed in the next section, or coordinatewise, via tensor-product fits based on the univariate technologies. Thus, He et al. (1998) proposed "tensor-product quantile smoothing splines," piecewise bilinear functions defined on a rectangular grid generated by the covariate points. The bivariate proposal for $k = 1$ (when the penalty is the total variation of the function itself) also put forward in Mammen and van de Geer (1997) in the context of least-squares estimation was in a similar spirit.

The driving force of these pursuits was the belief that the promise of a behavior similar to the one-dimensional case lies in the appropriate generalization of the mathematical notion of total variation to multidimensional functions and their gradients. An overview of some of this fascinating mathematics was given by Koenker and Mizera (2004), whose aim was also to go beyond the coordinatewise approach, to achieve coordinate-independent, rotationally invariant fits.

All multivariate (and, as we have seen above, some univariate) proposals pretty much had to forgo the hopes for a representer theorem. In such situations, the alternative was to act as if such characterizations existed: to restrict the fits to finite-dimensional domains that were deemed reasonable, and study penalized fitting on those domains. In fact, similar strategies, motivated mostly by real or imaginary algorithmic problems, were also proposed in the world of L^2 penalties – but there they rather lacked the essential appeal due to the existence of principled reproducing-kernel methodology in that context.

Koenker and Mizera (2004), too, sought the fits among *triograms*, piecewise linear functions on a triangulation (Lagrange–Courant triangles in the finite-element jargon); the Delaunay triangulation was preferred due to its optimal properties and algorithmic accessibility. For this particular case, they were able to isolate a fortuitous property which amounts to the following. A general notion of the total variation of the gradient is defined in *à la* Vitali through the formula

$$\bigvee_\Omega \nabla f = \int_\Omega \|\nabla^2 f\|, \tag{3.16}$$

where the gradient and Hessian are interpreted in the sense of Schwartz distributions. To ensure that the penalty defined via (3.16) is invariant with respect to rotations, (i.e. orthogonally invariant), it is natural to require the matrix norm involved in (3.16) to be orthogonally invariant, invariant with respect to orthogonal similarity. In such a case, every total variation (3.16) with these properties reduces (up to a multiplicative constant dependent on the norm, but independent of the function and triangulation) to the same *triogram penalty* of Koenker and Mizera (2004): the sum, over all edges, of the length of the edge multiplied by the norm of the difference of gradients in the two pieces of linearity adjacent to the edge. As the latter difference is orthogonal to the edge, the finite-dimensional problem of finding *penalized triograms*,

$$\sum_{t=1}^{T} \rho_\tau(y_t - f(u_t, v_t)) + \lambda \bigvee_\Omega \nabla f \rightsquigarrow \min_{f}! \tag{3.17}$$

restricted to triograms becomes ones again a linear programming task, with the penalty in (3.17) replaced by the triogram penalty.

An example of the matrix norm used in definition (3.16) could be simply the ℓ^2 Eu-

clidean sum-of-squares, Frobenius–Hilbert–Schmidt norm; together with the quantile objective function, it would create a formulation

$$\sum_{t=1}^{T} \rho_\tau(y_t - f(u_t, v_t)) + \lambda \int_\Omega \sqrt{f_{uu}^2(u, v) + 2f_{uv}^2(u, v) + f_{vv}^2(u, v)} \, du \, dv \rightsquigarrow \min_f ! \qquad (3.18)$$

where subscripts denote partial derivatives. This can be seen as the L^1 pendant for *thin-plate splines*, the smoothing spline methodology in the realm of L^2 penalization, formulated
through

$$\sum_{t=1}^{T} \rho_\tau(y_t - f(u_t, v_t)) + \lambda \int_\Omega \left(f_{uu}^2(u, v) + 2f_{uv}^2(u, v) + f_{vv}^2(u, v) \right) du \, dv \rightsquigarrow \min_f ! \qquad (3.19)$$

However, despite the fact that the structure of the penalty in (3.18) resembles that of
the group lasso of Yuan and Lin (2006), which in its typical applications achieves "group
sparsity" (either all of the group is zero, or then no sparsity is pushed for within), this
outcome seems not to be the case for the solutions of (3.18) or (3.17). A phase-transition
phenomenon of Donoho et al. (2013) comes to mind here; also, the situation parallels that
for the finite version of (3.19) yielded by the representer theorem, in which the matrix
involved does not enjoy sparse banded structure as it did in the one-dimensional case. A
possible explanation from physics (plate mechanics) of what is happening in L^1 and L^2
penalizations defined by formulations like (3.18) and (3.19) was attempted by Balek and
Mizera (2013); see also Koenker and Mizera (2002).

Summing up, apart from coordinatewise approaches, if there is desire for rotationally
independent penalized quantile regression in two (or more) dimensions, the user can either
resort to L^2 penalization or "kernelization" in the vein of the already mentioned proposals
of Takeuchi et al. (2006) and Li et al. (2007), or can use penalized triograms relying on
linear programming. The question is always whether full nonparametric regression in many
covariates is really what is needed; some dimension reduction may rather be desirable. One
of the options may be then additive fits, briefly reviewed in the final section of this chapter.

3.3.3   Additive fits, confidence bandaids, and other phantasmagorias

We do not discuss penalizing by means of ridge (3.13) or lasso (3.12) penalties in linear
quantile regression, with possibly many covariates, as this is covered elsewhere in this book.
For the same reason, we do not give a comprehensive account of nonparametric regression
methods in quantile regression, only of certain ones involving penalization.

As for what remains and was reviewed above, in the future we are bound, in the spirit
of the apocryphal quote of Kolmogorov by Vapnik about mathematics operating inside the
thin layer between the trivial and the intractable (see Vapnik, 2006, for the exact wording),
to continue seeing incremental ramifications and modifications for the methods operating
with one covariate, and, on the other hand, it is not at all clear how much progress will be
encountered in the difficult realm of multiple covariates.

One area, however, that can still use a substantial advance even in the one-dimensional
case, is confidence bands and, more generally, inference for the penalized regression formulations considered here. This is still a pressing issue even for methods using L^2 penalties
and least-squares loss; in the context of quantile regression, and L^1 penalties in particular, the recent explosion in the post-model selection inference literature, which we do not
attempt to survey here, may perhaps eventually create a body of research work which the
area discussed here can benefit from, rather than be taken advantage of.

Some inspiring work in this direction by Wahba (1983) and Nychka (1988) was carried further in the framework of additive models by Wood (2006); the latter in turn became the departure point for Koenker (2011). Recall that in classical additive models the functions of covariates are decomposed into additive contributions of separate variables

$$f(x_1, x_2, \ldots, x_n) = f_1(x_1) + f_2(x_2) + \cdots + f_n(x_n),$$

and the latter are fitted by nonparametric methods aimed at the functions of one covariate. Not all the f_i need be present in the resulting fit; also, some of the f_i can be fitted parametrically.

An extension of this is the ANOVA fits of Gu (2002), which also add functions of two variables (and then possibly three, and so on). This are the approach adopted by Koenker (2011): the fitted functions are expanded into a sum of univariate and bivariate functions,

$$f(x_1, \ldots, x_n) = f_1(x_1) + \cdots + f_n(x_n) + f_{12}(x_1, x_2) + f_{13}(x_1, x_3) + \ldots,$$

and out of those present in the model, the univariate ones are fitted by quantile smoothing splines, and the bivariate ones by penalized triograms. Koenker (2011) then gives the methodology for calculating uniform confidence "bandaids" for the additive components of the fits.

Another possible direction of expansion could be new kinds of additive models arising in connection with the strategy adopted by Maciak and Mizera (2016) for fitting "splines with changepoints." As indicated at the end of the section discussing total-variation splines, there is no "universal Swiss army knife," no universal penalty for all levels of smoothness. As a consequence, if we, say, admit jumps in our fits, we may have to face the piecewise constant character of the rest of them, even if we believe that between the jumps our fits can be general smooth functions. In such a situation, a representer theorem is hardly feasible, as more thorough mathematical analysis typically shows that the functions that one penalty allows (making it finite) make another penalty infinite.

Therefore, the fit has to be sought as a linear combination of basis functions that include, say, both classical cubic spline basis functions together with the Heaviside step functions, with steps located at covariate points, and their first and second antiderivatives, to allow possible changes in level or derivative(s). There is a certain analogy to the proposal of Mumford and Shah (1989) in image processing here: the image is segmented into subregions with the help of a total-variation penalty, and in the subregions smoothing is performed via the classical smoothing L^2 penalty.

The extension of this approach suggested by Maciak and Mizera (2016) might take this further and allow for additive fits *in the same variable*: $f_1(x_1)$, for instance, may be further decomposed into a sum of several functions of x_1, each with different smoothness properties, which can subsequently be fitted via L^1 and L^2 penalties of different orders. These penalties may then act either separately, or they may be coupled in an L^1/L^2 penalty in the spirit of the group lasso of Yuan and Lin (2006), expressing the belief that changes in level and in derivative are bound to occur simultaneously rather than separately.

The proliferation of modal verbs in the last paragraph indicates that we have come to the last word in the title of this last section, and thus to the very end of this chapter as well.

Bibliography

V. Balek and I. Mizera. Mechanical models in nonparametric regression. In M. Banerjee, F. Bunea, J. Huang, V. Koltchinskii, and M. H. Maathuis, editors, *From Probability to Statistics and Back: High-Dimensional Models and Processes, A Festschrift in Honor of Jon A. Wellner*, volume 9 of *IMS Collections*, pages 5–19. Institute of Mathematical Statistics, Beachwood, OH, 2013.

P. Bloomfield and W. S. Steiger. *Least Absolute Deviations. Theory, Applications and Algorithms.* Birkhäuser, Boston, 1983.

R. J. Bosch, Y. Ye, and G. G. Woodworth. A convergent algorithm for quantile regression with smoothing splines. *Computational Statistics and Data Analysis*, 19:613–630, 1995.

L. Breiman. Better subset selection using nonnegative garrote. *Technometrics*, 37:373–384, 1995.

S. S. Chen and D. L. Donoho. Examples of basis pursuit. In *SPIE's 1995 International Symposium on Optical Science, Engineering, and Instrumentation*, pages 564–574. International Society for Optics and Photonics, 1995.

S. S. Chen, D. L. Donoho, and M. A. Saunders. Atomic decomposition by basis pursuit. *SIAM Journal on Computing*, 20:33–61, 1998.

V. Chernozhukov, I. Fernández-Val, and A. Galichon. Quantile and probability curves without crossing. *Econometrica*, 78(3):1093–1125, 2010.

J. F. Claerbout and F. Muir. Robust modeling with erratic data. *Geophysics*, 38:826–844, 1973.

N. Cressie. Geostatistics. *American Statistician*, 43:197–202, 1989.

N. Cressie. Reply. *American Statistician*, 44:256–258, 1990.

D. L. Donoho and I. Johnstone. Minimax estimation via wavelet shrinkage. *Annals of Statistics*, 26(8):879–921, 1998.

D. L. Donoho and B. F. Logan. Signal recovery and the large sieve. *SIAM Journal on Applied Mathematics*, 52(2):577–591, 1992.

D. L. Donoho, I. Johnstone, and A. Montanari. Accurate prediction of phase transitions in compressed sensing via a connection to minimax denoising. *IEEE Tractions on Information Theory*, 59(6):3396–3433, 2013.

B. Efron, T. Hastie, I. Johnstone, and R. Tibshirani. Least angle regression. *Annals of Statistics*, 32(2):407–499, 2004.

I. J. Good. A nonparametric roughness penalty for probability densities. *Nature*, 229:29–30, 1971.

I. J. Good and R. A. Gaskins. Nonparametric roughness penalties for probability densities. *Biometrika*, 58:255–277, 1971.

P. J. Green and B. W. Silverman. *Nonparametric Regression and Generalized Linear Models: A Roughness Penalty Approach.* Chapman & Hall, London, 1994.

C. Gu. Model indexing and smoothing parameter selection in nonparametric function estimation (with discussion). *Statistica Sinica*, 8:607–646, 1998.

C. Gu. *Smoothing spline ANOVA models.* Springer-Verlag, New York, 2002.

J. Hadamard. Sur les problèmes aux dérivées partielles et leur signification physique. *Princeton University Bulletin*, pages 49–52, 1902.

T. Hastie, R. Tibshirani, and J. Friedman. *The Elements of Statistical Learning: Data Mining, Inference, and Prediction.* Springer, New York, 2nd edition, 2008.

X. He, P. Ng, and S. Portnoy. Bivariate quantile smoothing splines. *Journal of the Royal Statistical Society B*, 60:537–550, 1998.

R. Henderson. A new method of graduation. *Tractions of the Actuarial Society of America*, 25:29–40, 1924.

R. Henderson. Further remarks on graduation. *Tractions of the Actuarial Society of America*, 26:52–57, 1925.

R. Henderson. Some points in the general theory of graduation. In *Proceedings of the International Mathematical Congress held in Toronto, August 11-16, 1924, Vol II.*, pages 815–820. University of Toronto Press, Toronto, 1928.

R. Henderson. *Mathematical Theory of Graduation*. Actuarial Society of America, New York, 1938.

A. E. Hoerl and R. W. Kennard. Ridge regression: Biased estimation for nonorthogonal problems. *Technometrics*, 12:55–67, 1970.

J. T. Kent and K. V. Mardia. The link between kriging and thin-plate splines. In *Probability, Statistics and Optimization*, pages 326–339. Wiley, New York, 1994.

S.-J. Kim, K. Koh, S. Boyd, and D. Gorinevsky. ℓ^1 trend filtering. *SIAM Review*, 51(2): 339–360, 2009.

G. S. Kimeldorf and G. Wahba. A correspondence between Bayesian estimation on stochastic processes and smoothing by splines. *Annals of Mathematical Statistics*, 41:495–502, 1970.

R. Koenker. *Quantile Regression*. Cambridge University Press, Cambridge, 2005.

R. Koenker. Additive models for quantile regression: Model selection and confidence bandaids. *Brazilian Journal of Probability and Statistics*, 25(3):239–262, 2011.

R. Koenker and I. Mizera. Elastic and plastic splines: Some experimental comparisons. In Y. Dodge, editor, *Statistical Data Analysis Based on the L1-norm and Related Methods*, pages 405–414. Birkhäuser, Basel, 2002.

R. Koenker and I. Mizera. Penalized triograms: total variation regularization for bivariate smoothing. *Journal of the Royal Statistical Society B*, 66:1681–1736, 2004.

R. Koenker and I. Mizera. Convex optimization in R. *Journal of Statistical Software*, 60 (5):1–23, 2014.

R. Koenker, P. Ng, and S. Portnoy. Quantile smoothing splines. *Biometrika*, 81:673–680, 1994.

C. Kooperberg and C. J. Stone. A study of logspline density estimation. *Computational Statistics and Data Analysis*, 12:327–347, 1991.

Y. Li, Y. Liu, and J. Zhu. Quantile regression in reproducing kernel Hilbert spaces. *Journal of the American Statistical Association*, 102:255–268, 2007.

B. F. Logan. *Properties of high-pass signals*. PhD thesis, Columbia University, New York, 1965.

M. Maciak and I. Mizera. Splines with changepoints: Additive models for functional data, 2016. Preprint.

E. Mammen and S. van de Geer. Locally adaptive regression splines. *Annals of Statistics*, 25:387–413, 1997.

D. Mumford and J. Shah. Optimal approximation of piecewise smooth functions and associated variational problems. *Communications on Pure and Applied Mathematics*, 42(5): 577–685, 1989.

D. Nychka. Bayesian confidence intervals for smoothing splines. *Journal of the American Statistical Association*, 83(404):1134–1143, 1988.

D. Nychka, G. Gray, P. Haaland, D. Martin, and M. O'Connell. A nonparametric regression approach to syringe grading for quality improvement. *Journal of the American Statistical Association*, 90(432):1171–1178, 1995.

H.-S. Oh, T. C. M. Lee, and D. W. Nychka. Fast nonparametric quantile regression with arbitrary smoothing methods. *Journal of Computational and Graphical Statistics*, 20: 510–526, 2012.

D. L. Phillips. A technique for the numerical solution of certain integral equations of the first kind. *Journal of the ACM*, 9:84–97, 1962.

J. O. Ramsay and B. W. Silverman. *Functional data analysis, second edition.* Springer, New York, 2005.

C. H. Reinsch. Smoothing by spline functions. *Numerische Mathematik*, 10:177–183, 1967.

C. H. Reinsch. Smoothing by spline functions II. *Numerische Mathematik*, 16:451–454, 1971.

L. I. Rudin, S. Osher, and E. Fatemi. Nonlinear total variation based noise removal algorithms. *Physica D*, 60:259–268, 1992.

F. Santosa and W. Symes. Linear inversion of band-limited reflection seismograms. *SIAM Journal on Computing*, 7(4):1307–1330, 1986.

I. J. Schoenberg. Spline functions and the problem of graduation. *Proceedings of the National Academy of Sciences of the USA*, 52:947–950, 1964.

B. Schölkopf and A. J. Smola. *Learning with Kernels.* MIT Press, Cambridge, MA, 2002.

D. R. Schuette. A linear programming approach to graduation. *Transactions of the Society of Actuaries*, 30:407–445, 1978.

B. W. Silverman. On the estimation of a probability density function by the maximum penalized likelihood method. *Annals of Statistics*, 10:795–810, 1982.

P. L. Smith. *Curve Fitting and Modeling with Splines Using Statistical Variable Selection Techniques.* NASA, Langley Research Center, Hampla, VA, 1982. NASA Report 166034.

S. M. Stigler. *The History of Statistics: The Measurement of Uncertainty Before 1900.* Harvard University Press, Cambridge, MA, 1986.

I. Takeuchi, Q. V. Le, T. D. Sears, and A. J. Smola. Nonparametric quantile estimation. *Journal of Machine Learning Research*, 7:1231–1264, 2006.

H. Taylor, S. Banks, and J. McCoy. Deconvolution with the ℓ_1-norm. *Geophysics*, 44(1): 39–52, 1979.

R. Tibshirani. Regression shrinkage and selection via the lasso. *Journal of the Royal Statistical Society B*, 58:267–288, 1996.

R. J. Tibshirani. Adaptive piecewise polynomial estimation via trend filtering. *Annals of Statistics*, 42(1):285–323, 2014.

R. J. Tibshirani. A general framework for fast stagewise algorithms. *Journal of Machine Learning Research*, 16:2543–2588, 2015.

A. N. Tikhonov. On the stability of inverse problems. *Doklady Akademii Nauk SSSR*, 39: 195–198, 1943.

A. N. Tikhonov. On the solution of incorrectly posed problems and the regularization method. *Doklady Akademii Nauk SSSR*, 151:501–504, 1963.

V. Vapnik. Empirical inference science: Afterword of 2006. In *Estimating of Dependences Based on Empirical Data*. Springer, New York, 2nd edition, 2006.

V. V. Vasin. Relationship of several variational methods for the approximate solution of ill-posed problems (in Russian). *Mathematical Notes*, 7:161–165, 1970.

G. Wahba. Bayesian "confidence intervals" for the cross-validated smoothing spline. *Journal of the Royal Statistical Society B*, 45(1):133–150, 1983.

G. Wahba. Comment on Cressie. *American Statistician*, 44:255–256, 1990a.

G. Wahba. *Spline Models for Observational Data*. SIAM, Philadelphia, 1990b.

G. S. Watson. Smoothing and interpolation by kriging and with splines. *Journal of the International Association for Mathematical Geology*, 16:601–615, 1984.

E. T. Whittaker. On a new method of graduation. *Proceedings of the Edinburgh Mathematical Society*, 41:63–75, 1922–1923.

E. T. Whittaker. On the theory of graduation. *Proceedings of the Royal Society, Edinburgh*, 44:77–83, 1924.

E. T. Whittaker and G. Robinson. *The Calculus of Observations*. Blackie, London, 1924.

S. N. Wood. *Generalized Additive Models: An Introduction with R*. Chapman and Hall/CRC, Boca Raton, FL, 2006.

M. Yuan. GACV for quantile smoothing splines. *Computational Statistics and Data Analysis*, 50(3):813–829, 2006.

M. Yuan and Y. Lin. Model selection and estimation in regression with grouped variables. *Journal of the Royal Statistical Society B*, 68:49–67, 2006.

H. Zou and M. Yuan. Regularized simultaneous model selection in multiple quantiles regression. *Computational Statistics and Data Analysis*, 52:5296–5304, 2008.

4

Bayesian Quantile Regression

Huixia Judy Wang

George Washington University, Washington, DC, USA

Yunwen Yang

Google Inc., Seattle, Washington, USA

CONTENTS

Bayesian approaches provide convenient alternative inference tools for quantile regression. Even though conventional quantile regression does not require any parametric distributional assumptions, a working likelihood is needed to carry out Bayesian analysis. In this chapter, we provide a review of Bayesian quantile regression methods based on different types of working likelihoods.

4.1 Introduction

Bayesian quantile regression has several advantages compared to frequentist approaches. First, with Bayesian methods, both point and uncertainty estimates can be obtained easily from the posterior sequences from Markov chain Monte Carlo (MCMC) draws. In contrast, inference based on the asymptotics for the frequentist approaches often requires estimating some unknown quantities such as the conditional density function of the response, which is a challenging task in many applications. Second, the MCMC computation in some situations can alleviate the difficulty in solving some complex optimization, for instance, minimizing the nonconvex objective function used in Powell (1986) for censored quantile regression.

Since a typical quantile regression model does not specify a parametric likelihood, a working likelihood is needed for Bayesian methods to work. In this chapter, we review and discuss some recent developments in Bayesian quantile regression approaches based on various working likelihoods, including parametric likelihood based on the asymmetric Laplace distribution, empirical likelihood, and some semiparametric and nonparametric likelihoods.

As commented in Meng (2016), a working model is a model adopted for particular purposes with the knowledge that it may be flawed in some other respects. We will review both the usefulness of those working likelihoods, and provide some empirical verification and theoretical understanding of the harm of some working models.

4.2   Asymmetric Laplace likelihood

At a given quantile level $\tau \in (0,1)$, consider the linear quantile regression model

$$Q_Y(\tau | \mathbf{X} = \mathbf{x}) = \mathbf{x}^T \boldsymbol{\beta}_\tau, \tag{4.1}$$

where Y is the response, $\mathbf{X}$ is the p-dimensional covariate with first element equal to 1, and $\boldsymbol{\beta}_\tau$ is the unknown p-dimensional quantile coefficient. Based on a random sample $\mathcal{D} = \{(y_i, \mathbf{x}_i), \ i = 1, \dots, n\}$ of $(Y, \mathbf{X})$, the coefficients $\boldsymbol{\beta}_\tau$ can be consistently estimated by

$$\hat{\boldsymbol{\beta}}_\tau = \arg \min_{\boldsymbol{\beta} \in \mathbb{R}^p} \sum_{i=1}^n \rho_\tau(y_i - \mathbf{x}_i^T \boldsymbol{\beta}), \tag{4.2}$$

where $\rho_\tau(u) = u\{\tau - I(u < 0)\}$ is the quantile loss function.

The quantile loss function $\rho_\tau(\cdot)$ in (4.2) is an asymmetric L_1 loss function, and is proportional to the negative log density of the asymmetric Laplace (AL) distribution. This connection motivated researchers to consider the AL working likelihood for Bayesian quantile regression,

$$L(\boldsymbol{\beta}_\tau; \mathcal{D}) = \frac{\tau^n (1-\tau)^n}{\sigma^n} \exp \left\{ -\sum_{i=1}^n \rho_\tau(y_i - \mathbf{x}_i^T \boldsymbol{\beta}_\tau)/\sigma \right\}, \tag{4.3}$$

where $\sigma > 0$ is a scale parameter; see, for instance, Yu and Moyeed (2001) and Geraci and Bottai (2007). Let $\pi(\boldsymbol{\beta}_\tau)$ be a prior on $\boldsymbol{\beta}_\tau$; the posterior of $\boldsymbol{\beta}_\tau$ can be written as

$$p_n(\boldsymbol{\beta}_\tau | \mathcal{D}) \propto \pi(\boldsymbol{\beta}_\tau) \exp \left\{ -\frac{\sum_{i=1}^n \rho_\tau(y_i - \mathbf{x}_i^T \boldsymbol{\beta}_\tau)}{\sigma} \right\}. \tag{4.4}$$

If an improper flat prior on $\boldsymbol{\beta}_\tau$ is used (with σ treated as known), the posterior mode is exactly the same as the conventional estimator $\hat{\boldsymbol{\beta}}_\tau$, and the resulting posterior was shown to be proper in Yu and Moyeed (2001). Choi and Hobert (2013) established specific conditions for the posterior to be proper when an improper prior is posed on $(\boldsymbol{\beta}_\tau, \sigma)$. Even though a standard conjugate prior is not available for the Bayesian quantile regression setup with AL likelihood, MCMC methods can be used for posterior sampling. The posterior of $\boldsymbol{\beta}_\tau$ is in general intractable, but the sampling can be simplified by using mixture representations of the AL distribution. Suppose that ϵ follows the AL distribution with density

$$f_{\mathrm{AL}}(\epsilon) = \frac{\tau(1-\tau)}{\sigma} \exp\{-\rho_\tau(\epsilon)/\sigma\}. \tag{4.5}$$

By Kotz et al. (2001), ϵ can be represented by a scale mixture of normals:

$$\epsilon = \sigma(\theta \nu + \gamma \sqrt{\nu} z), \quad \text{with } \theta = \frac{1 - 2\tau}{1 - \tau}, \ \gamma^2 = \frac{2}{\tau(1-\tau)},$$

where z and ν are independent standard normal and standard exponential random variables. Based on this mixture representation, Kozumi and Kobayashi (2011) proposed a three-variable Gibbs sampling algorithm assuming a proper Gaussian–inverse Gamma prior on $(\boldsymbol{\beta}_\tau, \sigma)$, and Choi and Hobert (2013) proposed a data augmentation (DA) algorithm assuming an improper prior on $(\boldsymbol{\beta}_\tau, \sigma)$. The Markov chains underlying both algorithms were shown to be geometrically ergodic at the median (Khare and Hobert, 2012; Choi and Hobert, 2013). An alternative sandwich algorithm was proposed in Khare and Hobert (2011) for $\tau = 0.5$ and extended to any $\tau \in (0,1)$ in Hobert and Khare (2016), where the algorithm was shown to converge at least as fast as the DA algorithm. Using a different normal mixture representation of the AL distribution, Tsionas (2003) suggested a Gibbs sampling algorithm assuming a flat prior on $(\boldsymbol{\beta}_\tau, \sigma)$ but the algorithm is more complicated than the aforementioned two.

The posterior mean and variance of $\boldsymbol{\beta}_\tau$ can be directly computed from the MCMC chains. Based on simulation findings, Yu and Moyeed (2001) argued that the use of the AL likelihood provides good parameter estimation even when the AL likelihood is misspecified. Sriram et al. (2013) provided a formal justification for this claim by establishing sufficient conditions for the posterior consistency of model parameter $\boldsymbol{\beta}_\tau$ under the AL working likelihood. To construct interval estimates, a common practice in the literature is to use quantiles of the posterior or normal approximation using the variance–covariance matrix of the posterior sequence; see Yu and Moyeed (2001), Li et al. (2010), Alhamzawi et al. (2012), Yue and Hong (2014), and Lum and Gelfand (2012).

Unfortunately, the posterior consistency results cannot imply the validity of the interval estimates constructed from the posterior. Theoretical studies of the properties of the pseudo-posterior based on working likelihoods can be traced back to Kim (2002) and Chernozhukov and Hong (2003). Kim (2002) established the asymptotic normality of the posterior based on a limited-information likelihood constructed from moment-based information available in the generalized method of moments framework. Chernozhukov and Hong (2003) established the consistency and validity of point estimators and confidence intervals based on the pseudo-posterior, including the nonlinear/censored quantile regression and the instrumental variable quantile regression model as examples. To our knowledge, Chernozhukov and Hong (2003) was the first work to stress the need for adjusting the variance of the pseudo-posterior to obtain valid frequentist interval estimates, and it provided a general theory for asymptotically valid interval adjustment in its Theorem 4. However, the results in this important work have been overlooked by many practitioners when performing Bayesian quantile regression analysis based on working likelihoods. Recently the need for posterior adjustment was reemphasized in Yang et al. (2016a), where the authors proposed a specific adjustment to the posterior variance–covariance obtained from (4.4), and established the asymptotic validity of the adjusted interval estimates.

Let $\boldsymbol{\beta}_\tau^0$ denote the true value of $\boldsymbol{\beta}_\tau$. Under the same conditions that guarantee the asymptotic normality of $\hat{\boldsymbol{\beta}}_\tau$, it can be shown that for $\|\boldsymbol{\beta}_\tau - \boldsymbol{\beta}_\tau^0\| = O(n^{-1/2})$, the posterior density (assuming σ is known) is

$$p_n(\boldsymbol{\beta}_\tau | \mathcal{D}) \propto \pi(\boldsymbol{\beta}) \exp \left\{ - \frac{n(\boldsymbol{\beta}_\tau - \hat{\boldsymbol{\beta}}_\tau)^T D_1 (\boldsymbol{\beta}_\tau - \hat{\boldsymbol{\beta}}_\tau) + o_p(1)}{2\sigma} \right\}, \qquad (4.6)$$

which, when the flat prior is used, is approximately normal with mean $\hat{\boldsymbol{\beta}}_\tau$ and covariance $\hat{\Gamma} \doteq \sigma D_1^{-1}/n$, where $D_1 = \lim_{n\to\infty} n^{-1} \sum_{i=1}^n f_{Y_i}(\mathbf{x}_i^T \boldsymbol{\beta}_\tau^0 | \mathbf{x}_i) \mathbf{x}_i \mathbf{x}_i^T$ with $f_{Y_i}(\cdot|\mathbf{x})$ the conditional density of Y_i given $\mathbf{X} = \mathbf{x}$. On the other hand, the asymptotic covariance of $n^{1/2}\hat{\boldsymbol{\beta}}_\tau$ is known to be $\tau(1-\tau)D_1^{-1}D_0 D_1^{-1}$ with $D_0 = \lim_{n\to\infty} n^{-1} \sum_{i=1}^n \mathbf{x}_i \mathbf{x}_i^T$ (see Koenker, 2005, Ch. 3). Clearly the posterior variance is not the right approximation to the sampling variance of $\hat{\boldsymbol{\beta}}_\tau$. Yang et al. (2016a) suggested an adjusted posterior variance

$$\hat{\Gamma}_{\text{adj}} = \tau(1-\tau)\hat{\Gamma} \left(\sum_{i=1}^n \mathbf{x}_i \mathbf{x}_i^T \right) \hat{\Gamma}/\sigma^2,$$

TABLE 4.1
Empirical coverage probabilities and empirical mean lengths for confidence intervals with nominal level 90% for model (4.7) at $\tau = 0.5$. The standard errors for EML are in the range 0.004–0.008 in this table.

Method	100×ECP $a(\tau)$	$b(\tau)$	EML $a(\tau)$	$b(\tau)$	100×ECP $a(\tau)$	$b(\tau)$	EML $a(\tau)$	$b(\tau)$
		$n = 200$				$n = 500$		
BAL$_{adj}$	91	90	0.71	1.16	90	91	0.44	0.72
BAL	86	78	0.56	0.78	84	76	0.35	0.49
Boot	**85**	87	0.59	1.04	**84**	85	0.36	0.63

ECP, empirical coverage probability; EML, empirical mean length; BAL$_{adj}$ and BAL, the Bayesian quantile regression method based on AL-type likelihood for censored data with and without posterior variance adjustment, respectively; Boot, the bootstrap-based interval estimates of Powell (1986).

which can be used to construct asymptotically valid posterior intervals using normal approximation. Unlike the unadjusted posterior variance, $\hat{\Gamma}_{adj}$ is asymptotically invariant in the value of σ and the resulting posterior intervals are less sensitive to the choice of σ. Using a similar idea, Yang et al. (2016a) also proposed a posterior variance adjustment for censored quantile regression. They showed through numerical studies that the adjusted posterior variance leads to improvement in coverage probability for quantile regression models with both complete and censored data.

We will use one example (Case 4 in Yang et al., 2016a) to demonstrate the value of posterior variance correction. Suppose that data are generated from the model

$$T_i = 2.5 + 5x_i + \{1 + (x_i - 0.5)^2\}e_i, \quad i = 1, \ldots, n, \tag{4.7}$$

where T_i is the latent response, and $e_i \sim t_3$ is independent of $x_i \sim N(0, 1)$. Due to censoring, we observe $y_i = \max(T_i, 0)$ instead of T_i, with a censoring proportion around 30%. In this scenario, the conditional quantile of T given x is $a(\tau) + b(\tau)x$ only at $\tau = 0.5$, with $a(0.5) = 2.5$ and $b(0.5) = 5$, and is nonlinear in x at the other quantiles. Table 4.1 summarizes the empirical coverage probabilities and mean lengths of confidence intervals constructed based on the AL-type working likelihood for censored data with (BAL$_{adj}$) and without (BAL) posterior variance correction, and the frequentist bootstrap-based confidence intervals (Boot) based on 100 bootstrap replications for Powell's estimator (Powell, 1986). Clearly the interval estimates from BAL have poor coverage probabilities, while the posterior variance adjustment leads to respectable performance. The frequentist bootstrap intervals in general suffer from undercoverage. Part of the issue with Boot is that we might not find the right solution through optimization for every bootstrapped data set.

The asymptotic theory in Yang et al. (2016a) does not address the model performance in the presence of high-dimensional covariates or under a broader class of models. Yang et al. (2016b) carried out a simulation study for a partially linear model where the nonparametric function is approximated by a B-spline function, and showed that the adjustment remains useful for correcting the problem in the BAL intervals for the parametric coefficients. Interval estimates on the nonparametric component of the model would require a more careful study of the bias–variance trade-off, and we hope there will be future investigation in this direction.

Finally, we would like to point out that to carry out analysis at multiple quantile levels, applying the AL likelihood would lead to several mathematically incomparable working models for the same data set.

4.3   Empirical likelihood

Parametric likelihood such as the AL is convenient for building the Bayesian quantile regression framework and for posterior sampling. One problem with parametric likelihood is that model misspecification may cause biased estimation, and inference based on an unadjusted posterior is generally not valid. Instead of specifying a parametric family for the data, Lancaster and Jun (2010) proposed a pseudo-Bayesian framework for quantile regression at a single quantile level using the exponentially tilted empirical likelihood of Schennach (2005), and Yang and He (2012) considered the empirical likelihood of Owen (1988) for multiple-quantile regression, which was extended to clustered data in Kim and Yang (2012) and spatially correlated data in Yang and He (2015). In this section we focus on the method in Yang and He (2012), which provides a convenient way to incorporate informative priors on β_τ across quantile levels.

Assume that the quantile regression model (4.1) holds for any $\tau \in (0,1)$. At a given $\tau \in (0,1)$, by the subgradient condition, the quantile coefficient estimator $\hat{\beta}_\tau$ satisfies an estimating equation

$$n^{-1} \sum_{i=1}^{n} \{\tau - I(y_i - \mathbf{x}_i^T \hat{\beta}_\tau)\}\mathbf{x}_i \approx 0. \qquad (4.8)$$

Suppose our interest is in k quantiles: $0 < \tau_1 < \cdots < \tau_k$. Denote by $\zeta^0 = (\beta_{\tau_1}^0, \ldots, \beta_{\tau_k}^0)$ the true parameters in $\mathbb{R}^{kp}$. Let $(\mathbf{x}, y, \zeta)$ be the kp-dimensional estimating functions, where $\zeta = (\beta_{\tau_1}, \ldots, \beta_{\tau_k})$, and the components of $m(\mathbf{x}, y, \zeta)$ are

$$m_{dk+j}(\mathbf{x}, y, \zeta) = x_j\{\tau_{d+1} - I(y < \mathbf{x}^T \beta_{\tau_{d+1}})\} \quad \text{for } d = 0, \ldots, k-1, j = 0, \ldots, p,$$

where x_j is the jth component of $\mathbf{x}$.

For any given ζ, the profile empirical likelihood ratio is

$$\mathcal{R}(\zeta) = \max\left\{ \prod_{i=1}^{n}(n\omega_i) \;\Big|\; \sum_{i=1}^{n} \omega_i(\mathbf{x}_i, y_i, \zeta) = 0, \omega_i \geq 0, \sum_{i=1}^{n} \omega_i = 1 \right\},$$

and the empirical likelihood function is $\mathcal{R}(\zeta)/n^n = \prod_{i=1}^{n} \omega_i(\zeta)$, where $\omega_i(\zeta) = [n\{1 + \lambda_n(\zeta)^T(\mathbf{x}, y, \zeta)\}]^{-1}$, with $\lambda_n(\zeta)$ determined by

$$\sum_{i=1}^{n} \frac{(\mathbf{x}_i, y_i, \zeta)}{1 + \lambda_n^T(\zeta)(\mathbf{x}_i, y_i, \zeta)} = 0.$$

Let $\pi(\zeta)$ be the prior specified on ζ. Yang and He (2012) proposed considering the posterior density

$$p(\zeta|\mathcal{D}) \propto \pi(\zeta) \times \mathcal{R}(\zeta), \quad \text{where } \mathcal{D} = \{(\mathbf{x}_i, y_i) : i = 1, \ldots, n\}. \qquad (4.9)$$

The $p(\zeta|\mathcal{D})$ was referred to as the posterior distribution from the Bayesian empirical likelihood (BEL) approach, though it is not really a posterior in the strict sense. In the BEL approach, an improper prior cannot guarantee a proper posterior distribution. In fact, the posterior will be improper for flat priors on ζ, so this should be avoided in practice.

Compared to the frequentist empirical likelihood inference methods, the BEL approach avoids the computationally demanding step of directly maximizing the empirical likelihood function over ζ. Instead, the BEL approach only requires evaluating the empirical likelihood function at any given ζ, which can be obtained by calculating $\lambda(\zeta)$ using the Newton–Raphson algorithm, or a modified version such as that in Wu (2004). This makes it feasible to sample from the posterior using the Metropolis–Hastings algorithm.

The framework in Yang and He (2012) makes it convenient to aggregate information across quantiles. For instance, some slopes in β_τ may appear the same across τ within a quantile region. In this case, we can impose strict equalities of β_τ to reduce the number of unknown parameters in ζ, making it less than the number of moment restrictions, which can help improve the estimation efficiency; see Qin and Lawless (1994). However, in practice it is often difficult to justify such strict equalities. Alternatively, under the Bayesian framework, we can incorporate an informative prior on the differences of neighboring β_τ in the quantile region to regularize quantile estimates.

By establishing the first-order asymptotics of the posterior distribution as $n \to \infty$, Yang and He (2012) justified the frequentist validity of the BEL approach. Let $\hat{\zeta} = \operatorname{argmax}_\zeta \{\mathcal{R}(\zeta)\}$ be the maximum empirical likelihood estimator (MELE). Under some regularity assumptions, $\hat{\zeta}$ can be shown to be consistent for ζ^0. Furthermore, for any $\|\zeta - \zeta^0\| = O(n^{-1/2})$, the posterior density of ζ has the expansion

$$p(\zeta|\mathcal{D}) \propto \exp\left\{ -\frac{1}{2}(\zeta - \hat{\zeta})^T J_n (\zeta - \hat{\zeta}) + R_n \right\}, \tag{4.10}$$

where $J_n = n V_{12}^T V_{11}^{-1} V_{12}$, $V_{11} = \Psi \otimes E(\mathbf{x}\mathbf{x}^T)$, $V_{12} = -\partial E\{(\mathbf{x}, y, \zeta)\}/\partial\zeta|_{\zeta=\zeta^0}$, Ψ is a $k \times k$ matrix with (i,j)th element $\Psi_{ij} = \tau_i \wedge \tau_j - \tau_i \tau_j$, and $R_n = o_p(1)$. The asymptotic result (4.10) suggests the following.

(1) The posterior variance from the posterior chain matches the asymptotic variance of the MELE, as given in Qin and Lawless (1994). This justifies the asymptotic validity of the BEL approach for inference, that is, the resultant posterior chain can be used to estimate the variance of the BEL estimates. This property is not shared by the Bayesian quantile regression method based on the AL working likelihood discussed in Section 4.2.

(2) When ζ has the same dimension as the estimating function $(\mathbf{x}, y, \zeta)$, the asymptotic posterior variance, J_n^{-1}, is equivalent to the asymptotic variance of the conventional quantile regression estimator obtained at each quantile level separately. On the other hand, when ζ has reduced dimensionality, the derivative in the definition of V_{12} is with respect to the reduced parameter vector, so an efficiency gain over the conventional estimator becomes possible.

The result in (4.10) is for the BEL approach with fixed priors on ζ. With any fixed prior, the posterior is basically determined by the likelihood once the sample size n is sufficiently large. However, in a practical problem with a finite sample size, an informative prior can complement the likelihood, and thus influence the posterior. Yang and He (2012) proposed using shrinking priors, which shrink with n, to study how informative priors would function together with the empirical likelihood to influence the posterior. In reality, shrinking priors can be constructed from data of a secondary source, such as records from neighboring stations in climate prediction, or when a statistical test fails to reject the null hypothesis of common slope parameters. Suppose that the logarithm of the prior density $\pi_n(\zeta)$ is twice continuously differentiable with a bounded mode $\zeta_{0,n}$, and that the matrix $J_{0,n} = -\partial^2 \log\{\pi_n(\zeta)\}/\partial\zeta^2|_{\zeta=\zeta_{0,n}} = O(n)$. Under some regularity conditions, Yang and He (2012) showed that for any $\{\zeta : \|\zeta - \zeta^0\| = O(n^{-1/2})\}$,

$$p(\zeta|\mathcal{D}) \propto \exp\left\{ -\frac{1}{2}(\zeta - \theta_{\text{post}})^T J_n (\zeta - \theta_{\text{post}}) + R_n \right\}, \tag{4.11}$$

where $J_n = J_{0,n} + nV_{12}^T V_{11}^{-1} V_{12}$, $\boldsymbol{\theta}_{\text{post}} = J_n^{-1}\big(J_{0,n}\boldsymbol{\zeta}_{0,n} + nV_{12}^T V_{11}^{-1} V_{12}\hat{\boldsymbol{\zeta}}\big)$, and $R_n = o_p(1)$.

In (4.11), the additional term $J_{0,n}$ in J_n provides helpful insights into how an informative prior can complement the empirical likelihood part and thus influence the resultant posterior.

(1) When $J_{0,n} = o_p(n)$, the empirical likelihood will dominate the prior information, making the posterior expansion the same as that with fixed priors.

(2) When $J_{0,n}$ grows faster than n, the prior will dominate the empirical likelihood.

(3) When $J_{0,n}$ grows at the same rate with n, the BEL approach will lead to consistent estimation of $\boldsymbol{\zeta}_0$ if $\|\boldsymbol{\zeta}_{0,n} - \boldsymbol{\zeta}^0\| = o_p(1)$; otherwise a bias will appear in $\boldsymbol{\theta}_{\text{post}}$, and the resultant posterior will not directly provide asymptotically valid inference. However, in the latter case, since $J_{0,n}$ is known, we can estimate $nV_{12}^T V_{11} V_{12}$ by $\hat{J}_n - J_{0,n}$ with $\hat{J}_n$ as the posterior variance from the MCMC chain, which is needed for constructing asymptotically valid confidence intervals. This correction is in the same spirit as the posterior variance adjustment discussed in Section 4.2.

Note that the posterior mean in (4.11) depends on the mode $\boldsymbol{\zeta}_{0,n}$. In some cases appropriate shrinkage priors can be chosen to eliminate the bias due to a misspecified prior mode. For instance, under the assumption of common slopes across τ, if the prior variance of the slope differences at neighboring quantiles is in the order of n^{-1} but the prior variances of the other parameters increase with n, then the posterior mean is asymptotically unbiased for $\boldsymbol{\zeta}^0$ regardless of what the prior mode $\boldsymbol{\zeta}_{0,n}$ is.

The work of Yang and He (2012) provided the basic theory for justifying the validity of the posterior inference based on the empirical likelihood working likelihood, and for incorporating informative priors across quantiles. In situations with common features across quantiles, the BEL approach with informative priors can lead to more efficient estimation, and it provides an automatic way to balance the quantile loss functions at different quantile levels.

4.4 Nonparametric and semiparametric likelihoods

In this section, we discuss a class of methods based on nonparametric or semiparametric modeling of the likelihood.

4.4.1 Mixture-type likelihood

At a given quantile level $\tau \in (0, 1)$, assume the linear regression model

$$y_i = \mathbf{x}_i^T \boldsymbol{\beta}_\tau + \epsilon_i, \quad i = 1, \dots, n, \tag{4.12}$$

where the ϵ_i are independent errors with the τth quantile zero, that is, $\int_{-\infty}^0 f_\epsilon(x)dx$ with f_ϵ denoting the density of ϵ_i. Kottas and Gelfand (2001) and Kottas and Krnjajić (2009) proposed nonparametric Bayesian methods based on Dirichlet process mixture models for (4.12) at the median and at general τ, respectively. The main idea is to approximate the error distribution f_ϵ by a nonparametric mixture of distributions with Dirichlet process priors for the mixing distributions that preserve the zero τth quantile of ϵ_i.

Specifically, let $p(\cdot; \theta)$ be a density function with τth quantile zero. Kottas and Gelfand

(2001) and Kottas and Krnjajić (2009) considered the mixture model

$$f_1(\epsilon; G) = \int p(\epsilon; \theta) dG(\theta), \quad G \sim DP(\alpha, G_0), \tag{4.13}$$

where $DP(\alpha, G_0)$ denotes the Dirichlet process with precision parameter α and base distribution G_0. The density $f_1(\cdot)$ preserves the zero τth quantile with extra variability captured through mixing on θ. For median regression, Kottas and Gelfand (2001) chose a split normal for the kernel of the mixture:

$$p(\epsilon; \theta, \phi) = f_N(\epsilon|0, \phi\theta)I(\epsilon < 0) + f_N(\epsilon|\theta/\phi)I(\epsilon > 0),$$

where $f_N(\cdot|\mu, \sigma^2)$ is the density of $N(\mu, \sigma^2)$, and $\phi > 0$ is a parameter that determines the skewness of the distribution. For general $0 < \tau < 1$, Kottas and Krnjajić (2009) chose $p(\cdot; \theta)$ to be the asymmetric Laplace density in (4.5) with scale parameter θ. By introducing a latent mixing parameter θ_i for each y_i, model (4.13) can be written as the hierarchical form

$$Y_i|\boldsymbol{\beta}_\tau, \theta_i \overset{\text{ind.}}{\sim} p(y_i - \mathbf{x}_i^T\boldsymbol{\beta}_\tau; \theta_i), \quad i = 1, \ldots, n,$$
$$\theta_i|G \overset{\text{i.i.d}}{\sim} G, \quad i = 1, \ldots, n,$$
$$G|\alpha \sim DP(\alpha, G_0)$$

with chosen priors on $\boldsymbol{\beta}_\tau$, α and G_0.

The mixture model (4.13) allows for skewness and increased variability, but the mixing does not affect the skewness of the kernel of the mixture. To allow more flexibility, Kottas and Gelfand (2001) and Kottas and Krnjajić (2009) proposed a nonparametric scale mixture of uniform densities. By Feller (1971, p. 158), any nonincreasing density g on the positive real line can be written as $g(t; G) = \int \theta^{-1}I(0 \leqslant t < \theta)dG(\theta)$, where G is a distribution on $[0, \infty)$. This result motivated Kottas and Gelfand (2001) to consider modeling the error density by the mixture model

$$f_2(\epsilon; G_1, G_2) = \int \int \kappa_\tau(\epsilon; \sigma_1, \sigma_2)dG_1(\sigma_1)dG_2(\sigma_2), \tag{4.14}$$

where $G_r \sim DP(\alpha_r, G_{r0})$, $r = 1, 2$, $\kappa_\tau(\epsilon; \sigma_1, \sigma_2) = \sigma_1^{-2}\tau I(-\sigma_1 < \epsilon < 0) + \sigma_2^{-1}(1 - \tau)I(0 \leqslant \epsilon < \sigma_2)$. In model (4.14), the positive and negative parts of the errors are modeled separately to ensure the τth quantile is zero. With latent mixing parameters σ_{1i} and σ_{2i} for each y_i, model (4.14) can be written as the hierarchical form

$$Y_i|\boldsymbol{\beta}_\tau, \theta_i \overset{\text{ind.}}{\sim} \kappa_\tau(y_i - \mathbf{x}_i^T\boldsymbol{\beta}_\tau; \sigma_{1i}, \sigma_{2i}), \quad i = 1, \ldots, n,$$
$$\sigma_{ri}|G_r \overset{\text{i.i.d}}{\sim} G_r, \quad i = 1, \ldots, n,$$
$$G_r|\alpha_r \sim DP(\alpha_r, G_{r0}), \quad r = 1, 2,$$

with chosen priors on $\boldsymbol{\beta}_\tau$, α_r, and G_{r0} for $r = 1, 2$. Posterior inference under models (4.13) and (4.14) can then be carried out by using posterior simulation methods for Dirichlet process mixture models.

Under models (4.13) and (4.14), the resulting densities have their mode exactly at the τth quantile, and they may be discontinuous at the mode. To overcome these limitations, Reich et al. (2010) proposed an alternative nonparametric method by modeling the error density as an infinite mixture of Gaussian densities that each satisfy the desired quantile constraint. To account for possible heteroskedasticity, they considered the location–scale shift model

$$y_i = \mathbf{x}_i^T\boldsymbol{\beta}_\tau + \mathbf{x}_i^T\boldsymbol{\gamma}_\tau\epsilon_i, \quad i = 1, \ldots, n, \tag{4.15}$$

where $\mathbf{x}_i^T \boldsymbol{\gamma}_\tau$ is assumed to be positive for all $\mathbf{x}_i$, and the ϵ_i are independent and identically distributed random errors with τth quantile equal to zero. Reich et al. (2010) proposed modeling the error density f_ϵ by the infinite mixture

$$f\epsilon(\epsilon|\boldsymbol{\mu}, \boldsymbol{\sigma}^2) = \sum_{k=1}^{\infty} p_k f(\epsilon|\boldsymbol{\mu}_k, \sigma_k^2, q_k), \qquad (4.16)$$

where the p_k are the mixture proportions with $\sum_{k=1}^{\infty} p_k = 1$, the base density is the quantile-restricted two-component mixture $f(\epsilon, \sigma_k^2, q_k) = q_k \phi\{(\epsilon-\mu_{1k})/\sigma_{1k}\} + (1-q_k)\phi\{(\epsilon-\mu_{2k})/\sigma_{2k}\}$ with $\phi(\cdot)$ as the standard normal density, and $q_k \in (0,1)$ is a constant ensuring that $\int_{-\infty}^{0} f(\epsilon, \sigma_k^2, q_k) d\epsilon = \tau$. This construction leads to a simple but flexible error distribution that satisfies the desired quantile constraint. To facilitate the posterior sampling, Reich et al. (2010) proposed posing priors on the mixing proportions p_k by using their stick-breaking representations, and used Metropolis–Hastings algorithms for MCMC sampling. For methods based on mixture-type likelihood, posterior inference is valid when the model assumptions such as (4.12) and (4.15) are satisfied.

4.4.2 Approximate likelihood via quantile process

For a continuous distribution, it is known that the likelihood can be fully determined if the quantile function is known or can be estimated in the entire quantile region. This motivates another class of methods for Bayesian quantile regression, which approximates the likelihood by modeling the conditional quantile regression process.

Assume that the linear quantile regression model (4.1) holds for all $\tau \in (0,1)$. Under this model, we can link the conditional density function with the linear conditional quantile function by

$$f_Y(y|\mathbf{x}) = \lim_{\delta \to 0} \frac{\delta}{\mathbf{x}^T(\boldsymbol{\beta}_{u_y+\delta_{u_y}} - \boldsymbol{\beta}_{u_y})},$$

where $u_y = \{\tau \in (0,1) : \mathbf{x}^T\boldsymbol{\beta}_\tau = y\}$. Let $0 < \tau_1 < \cdots < \tau_K < 1$ be a grid of quantile levels and define $\boldsymbol{\beta}_K = (\boldsymbol{\beta}_{\tau_1}^T, \ldots, \boldsymbol{\beta}_{\tau_K}^T)^T$. Feng et al. (2015) proposed a Bayesian quantile regression procedure based on the so-called linearly interpolated density (LID)

$$
\begin{aligned}
\hat{f}_Y(y_i|\mathbf{x}_i, \boldsymbol{\beta}_K) &= I\{y_i \in (-\infty, \mathbf{x}_i^T\boldsymbol{\beta}_{\tau_1})\}\tau_1 f_l(y_i) \\
&+ \sum_{k=1}^{K-1} I\{y_i \in [\mathbf{x}_i^T\boldsymbol{\beta}_{\tau_k}, \mathbf{x}_i^T\boldsymbol{\beta}_{\tau_{k+1}})\}\frac{\tau_{k+1} - \tau_k}{\mathbf{x}_i^T(\boldsymbol{\beta}_{\tau_{k+1}} - \boldsymbol{\beta}_{\tau_k})} \qquad (4.17) \\
&+ I\{y_i \in [\mathbf{x}_i^T\boldsymbol{\beta}_{\tau_K}, \infty)\}(1 - \tau_K)f_r(y_i), \quad i = 1, \ldots, n,
\end{aligned}
$$

where $f_l(\cdot)$ and $f_r(\cdot)$ are some prespecified left and right tail density functions. In Feng et al. (2015), f_l was taken to be the left half of $N(\mathbf{x}_i^T\boldsymbol{\beta}_{\tau_1}, \sigma^2)$ and f_r was chosen as the right half of $N(\mathbf{x}_i^T\boldsymbol{\beta}_{\tau_1}, \sigma^2)$ for a given σ^2. Jang and Wang (2015) relaxed the parametric specifications of tail densities by approximating them using extreme value theory and extended the LID method for analyzing clustered data. Based on (4.18), the approximate posterior distribution can be written as

$$\hat{p}_n(\boldsymbol{\beta}_K|\mathcal{D}) = \pi(\boldsymbol{\beta}_K) \prod_{i=1}^{n} \hat{f}_Y(y_i|\mathbf{x}_i, \boldsymbol{\beta}_K). \qquad (4.18)$$

Feng et al. (2015) suggested a truncated normal prior on $\boldsymbol{\beta}_K$ that satisfies the quantile monotonicity constraint: $\mathbf{x}_i^T\boldsymbol{\beta}_{\tau_1} < \cdots < \mathbf{x}_i^T\boldsymbol{\beta}_{\tau_K}$, $i = 1, \ldots, n$. Unlike most other methods that focus on one quantile at a time, the LID method can estimate the joint posterior distribution of multiple quantiles, leading to more efficient estimation than the conventional quantile regression method when multiple quantiles are of interest.

Instead of modeling a set of quantiles in the grid as in the LID method, Reich et al. (2011) proposed modeling the entire quantile process using Bernstein polynomials. Let β_τ^j be the coefficient associated with x_{ij}, the jth element of $\mathbf{x}_i$. Reich et al. (2011) assumed that

$$\beta_\tau^j = \sum_{m=1}^{M} B_m(\tau)\alpha_{j,m},$$

where $B_m(\tau) = \binom{M}{m}\tau^m(1-\tau)^{M-m}$ are the Bernstein basis polynomials, and $\alpha_{j,m}$ are unknown coefficients. To ensure the monotonicity of $\mathbf{x}_i^T\boldsymbol{\beta}_\tau$ in τ, one sufficient condition is $\sum_{j=1}^{p} x_{ij}\alpha_{j,m} \geqslant \sum_{j=1}^{p} x_{ij}\alpha_{j,m-1}$ for all $m > 1$. For cases with $x_{ij} \in [0,1]$ after some scaling, Reich et al. (2011) suggested a prior on $\delta_{j,m} = \alpha_{j,m} - \alpha_{j,m-1}$ to ensure quantile monotonicity. In addition, they also studied spatial quantile regression by assuming the quantile process to be a smooth function of the spatial location s, $\beta_\tau^j(s) = \sum_{m=1}^{M} B_m(\tau)\alpha_{j,m}(s)$, where the $\alpha_{j,m}(s)$ are spatially varying basis coefficients. The spatial quantile regression model, however, is difficult to implement for large spatial data. To reduce the computational demand, Reich et al. (2011) proposed a two-stage approximation approach.

Let $0 < \tau_1 < \cdots < \tau_K < 1$ be a grid of quantile levels. At each given location s_i, let $\boldsymbol{\beta}(s_i) = (\boldsymbol{\beta}_{\tau_1}(s_i), \ldots, \boldsymbol{\beta}_{\tau_K}(s_i))$ denote the set of coefficients. In the first stage, the frequentist quantile regression estimator $\hat{\boldsymbol{\beta}}_{\tau_k}(s_i)$ is obtained at the location s_i and the kth quantile level separately, for $k = 1, \ldots, K$. Based on the asymptotic normality result of $\hat{\boldsymbol{\beta}}(s_i) = (\hat{\boldsymbol{\beta}}_{\tau_1}(s_i), \ldots, \hat{\boldsymbol{\beta}}_{\tau_K}(s_i))$, Reich et al. (2011) proposed fitting the approximate model

$$\hat{\boldsymbol{\beta}}(s_i) \sim N(\boldsymbol{\beta}(s_i), \Sigma_i),$$

where the elements of $\boldsymbol{\beta}(s_i)$ are functions of the Bernstein basis coefficients $\alpha_{j,m}(s_i)$. In the second stage, MCMC methods can be used to obtain the approximate posterior distribution of the basis coefficients and consequently the spatial quantile function. Note that the approximate posterior is conditional on the Kp-dimensional initial estimator $\hat{\boldsymbol{\beta}}(s_i)$ instead of the original responses y_i. Therefore, this approximation provides a dramatic reduction in computing time when there are a large number of observations at each site s_i.

The two-stage model in Reich et al. (2011) used large-sample approximations to the likelihood, while the method in Feng et al. (2015) is for a fixed sample size. Both methods allow simultaneous analysis at multiple quantiles. However, due to the approximation of likelihoods, the theoretical validity of direct inference from the posterior draws remains unclear, and the estimation at tail quantiles may be unstable.

A few other researchers have also considered approximate likelihood through the modeling of the quantile process or multiple quantiles simultaneously. Dunson and Taylor (2005) used a substitution likelihood proposed by Lavine (1995) that is constructed through a grid of quantiles. Tokdar and Kadane (2012) proposed a semiparametric Bayesian approach for simultaneous analysis of linear quantile regression models. The method is based on the observation that when there is a univariate covariate falling into a bounded interval, the quantile monotonicity constraints can be satisfied by interpolating two monotonic functions of τ. Bayesian inference was carried out by specifying a prior on the two monotone functions via logistic transformations of a smooth Gaussian process. Reich and Smith (2013) proposed a semiparametric approach for censored quantile regression, where the quantile process is represented as a linear combination of basis functions, and priors on the basis functions are chosen to make the quantile process centered on a location–scale shift model.

4.5 Discussion

Bayesian quantile regression is an important topic that requires better understanding from both the computational and theoretical perspectives. The main challenge for Bayesian quantile regression is that the likelihood has no parametric forms, so a working likelihood is needed for the Bayesian approaches to work. We have provided a rather incomplete review of methods based on different types of working likelihood, and we would certainly hope to see more developments in this emerging area in the future.

In the current literature the most popular choice of the working likelihood is the asymmetric Laplace distribution. The AL likelihood leads to simple computation and is easy to incorporate into the Bayesian framework for different types of data; see the developments of AL-based Bayesian quantile regression in various setups in Yuan and Yin (2009), Yue and Håvard (2011), Luo et al. (2014), Li et al. (2010), Alhamzawi et al. (2012), Lum and Gelfand (2012), Lee and Neocleous (2010), Thompson et al. (2010), Hu et al. (2013), and Benoit et al. (2013), to name but a few. Even though AL-based algorithms are consistent in terms of estimation even under misspecification of the likelihood, the posterior variance has to be adjusted to provide valid inference. Yang et al. (2016a) provided variance adjustments for linear quantile regression models with complete data and data subject to fixed censoring. Whether convenient adjustment is available for more complicated models is an interesting topic to study.

The mixture-type likelihoods are more flexible than the AL likelihood for capturing general forms of skewness and tail behaviors. Similar to the AL likelihood, the mixture-type likelihoods are specified for a single quantile level of interest; consequently, no coherent likelihoods can be obtained for simultaneous analysis at multiple quantiles. In addition, the mixture-type likelihoods are constructed by modeling the densities of the regression error, so a parametric specification of the variance function is needed to account for heteroskedasticity.

In contrast to the mixture-type likelihoods, the approximate likelihoods discussed in Section 4.4.2 can account for heteroscedasticity automatically by modeling the entire quantile process and thus can be used to obtain the joint posterior distribution of multiple quantiles. Since methods based on the approximate likelihood require modeling of the entire quantile process, this complicates the theoretical justification of the posterior inference, and the results could be sensitive to poor tail estimation; see some numerical evidence in Yang and He (2015).

Unlike other methods, Bayesian empirical likelihood methods do not require the error density to be modeled; instead they are based on the profile empirical likelihood obtained to satisfy the quantile constraints. Even though the Bayesian empirical likelihood posterior is not a posterior in the strict sense, Yang and He (2012) provided an asymptotic justification for the resultant posterior inference. The Bayesian empirical likelihood is a promising approach especially for simultaneous analysis of multiple quantiles, but additional work is needed to adapt the idea to more complex situations such as for censored and longitudinal data.

Acknowledgments

The research of Wang is partially supported by the National Science Foundation CAREER award DMS-1149355 and the KAUST OSR-2015-CRG4-2582 project.

Bibliography

R. Alhamzawi, K. Yu, and D. F. Benoit. Bayesian adaptive lasso quantile regression. *Statistical Modeling*, 12(3):279–297, 2012.

D. F. Benoit, R. Alhamzawi, and K. Yu. Bayesian lasso binary quantile regression. *Computational Statistics*, 28(6):2861–2873, 2013.

V. Chernozhukov and H. Hong. An MCMC approach to classical estimation. *Journal of Econometrics*, 114(2):293–346, 2003.

H. M. Choi and J. P. Hobert. Analysis of MCMC algorithms for Bayesian linear regression with Laplace errors. *Journal of Multivariate Analysis*, 117:32–40, 2013.

D. Dunson and J. Taylor. Approximate Bayesian inference for quantiles. *Journal of Nonparametric Statistics*, 17:385–400, 2005.

W. Feller. *An Introduction to Probability Theory and Its Applications*, volume II. Wiley, New York, 2nd edition, 1971.

Y. Feng, Y. Chen, and X. He. Bayesian quantile regression with approximate likelihood. *Bernoulli*, 21:832–850, 2015.

M. Geraci and M. Bottai. Quantile regression for longitudinal data using the asymmetric Laplace distribution. *Biostatistics*, 8(1):140–154, 2007.

J. P. Hobert and K. Khare. Discussion. *International Statistical Review*, 84:349–356, 2016.

Y. Hu, R. Gramacy, and H. Lian. Bayesian quantile regression for single-index models. *Statistics and Computing*, 23:437–454, 2013.

W. Jang and H. Wang. A semiparametric Bayesian approach for joint-quantile regression with clustered data. *Computational Statistics & Data Analysis*, 84:99–115, 2015.

K. Khare and J. P. Hobert. A spectral analytic comparison of trace-class data augmentation algorithms and their sandwich variants. *Annals of Statistics*, 39:2585–2606, 2011.

K. Khare and J. P. Hobert. Geometric ergodicity of the Gibbs sampler for Bayesian quantile regression. *Journal of Multivariate Analysis*, 112:108–116, 2012.

J. Y. Kim. Limited information likelihood and Bayesian analysis. *Journal of Econometrics*, 107:175–193, 2002.

M. Kim and Y. Yang. Semiparametric approach to a random effects quantile regression model. *Journal of the American Statistical Association*, 106:1405–1417, 2012.

R. Koenker. *Quantile Regression*. Cambridge University Press, Cambridge, 2005.

A. Kottas and A. E. Gelfand. Bayesian semiparametric median regression modeling. *Journal of the American Statistical Association*, 96:1458–1468, 2001.

A. Kottas and M. Krnjajić. Bayesian semiparametric modelling in quantile regression. *Scandinavian Journal of Statistics*, 36(2):297–319, 2009.

S. Kotz, T. Kozubowski, and K. Podgorski. *The Laplace Distribution and Generalizations: A Revisit with Applications to Communications, Economics, Engineering, and Finance.* Birkhäuser, Boston, 2001.

H. Kozumi and G. Kobayashi. Gibbs sampling methods for Bayesian quantile regression. *Journal of Statistical Computation and Simulation*, 81:1565–1578, 2011.

T. Lancaster and S. J. Jun. Bayesian quantile regression methods. *Journal of Applied Economics*, 25(2):287–307, 2010.

M. Lavine. On an approximate likelihood for quantiles. *Biometrika*, 82:220–222, 1995.

D. Lee and T. Neocleous. Bayesian quantile regression for count data with application to environmental epidemiology. *Applied Statistics*, 59(5):905–920, 2010.

Q. Li, R. Xi, and N. Lin. Bayesian regularized quantile regression. *Bayesian Analysis*, 5(3): 533–556, 2010.

K. Lum and A. E. Gelfand. Spatial quantile multiple regression using the asymmetric Laplace process. *Bayesian Analysis*, 7(2):235–258, 2012.

Y. Luo, H. Lian, and M. Tian. Bayesian quantile regression for longitudinal data models. *Journal of Statistical Computation and Simulation*, 82(11):1635–1649, 2014.

X. L. Meng. Discussion: Should a working model actually work? *International Statistical Review*, 84:362–374, 2016.

A. Owen. Empirical likelihood ratio confidence intervals for a single functional. *Biometrika*, 75:237–249, 1988.

J. L. Powell. Censored regression quantiles. *Journal of Econometrics*, 32(1):143–155, 1986.

J. Qin and J. Lawless. Empirical likelihood and general estimating equations. *Annals of Statistics*, 22:300–325, 1994.

B. J. Reich and L. B. Smith. Bayesian quantile regression for censored data. *Biometrics*, 69(3):651–660, 2013.

B. J. Reich, H. D. Bondell, and H. Wang. Flexible Bayesian quantile regression for independent and clustered data. *Biostatistics*, 11(2):337–352, 2010.

B. J. Reich, M. Fuentes, and D. B. Dunson. Bayesian spatial quantile regression. *Journal of the American Statistical Association*, 106:6–20, 2011.

S. M. Schennach. Bayesian exponentially tilted empirical likelihood. *Biometrika*, 92:31–46, 2005.

K. Sriram, R. V. Ramamoorthi, and P. Ghosh. Posterior consistency of Bayesian quantile regression based on the misspecified asymmetric laplace density. *Bayesian Analysis*, 8(2): 489–504, 2013.

P. Thompson, Y. Cai, R. Moyeed, D. Reeve, and J. Stander. Bayesian nonparametric quantile regression using splines. *Computational Statistics & Data Analysis*, 54:1138–1150, 2010.

S. T. Tokdar and J. B. Kadane. Simultaneous linear quantile regression: A semiparametric Bayesian approach. *Bayesian Analysis*, 7:51–72, 2012.

E. G. Tsionas. Bayesian quantile inference. *Journal of Statistical Computation and Simulation*, 73:659–674, 2003.

C. Wu. Some algorithmic aspects of the empirical likelihood method in sampling. *Statistica Sinica*, 14:1057–1067, 2004.

Y. Yang and X. He. Bayesian empirical likelihood for quantile regression. *Annals of Statistics*, 40(2):1102–1131, 2012.

Y. Yang and X. He. Quantile regression for spatially correlated data: An empirical likelihood approach. *Statistica Sinica*, 25:261–274, 2015.

Y. Yang, H. Wang, and X. He. Posterior inference in Bayesian quantile regression with asymmetric Laplace likelihood. *International Statistical Review*, 84:327–344, 2016a.

Y. Yang, H. Wang, and X. He. Rejoinder. *International Statistical Review*, 84:367–370, 2016b.

K. Yu and R. A. Moyeed. Bayesian quantile regression. *Statistics and Probability Letters*, 54(4):437–447, 2001.

Y. Yuan and G. Yin. Bayesian quantile regression for longitudinal studies with nonignorable missing data. *Biometrics*, 66:105–114, 2009.

Y. R. Yue and R. Håvard. Bayesian inference for additive mixed quantile regression models. *Computational Statistics & Data Analysis*, 55(1):84–96, 2011.

Y. R. Yue and H. G. Hong. Bayesian tobit quantile regression model for medical expenditure panel survey data. *Statistical Modeling*, 12(4):323–346, 2014.

5

Computational Methods for Quantile Regression

Roger Koenker

University of Illinois at Urbana-Champaign, USA

CONTENTS

5.1   Introduction

The earliest computation of a median regression estimator is usually attributed to the Croatian Jesuit, Rudjer Boscovich. In 1760 Boscovich visited London and, as recounted by Stigler (1984) and Farebrother (1990), posed the computation problem to Thomas Simpson. In Boscovich's version of the problem the mean residual was constrained to be zero, a requirement that conveniently reduces the problem to finding a (scalar) weighted median. Thus, the bivariate median regression problem of minimizing the sum of absolute residuals,

$$\hat{\beta} = \mathrm{argmin}_{(b_0, b_1) \in \mathbb{R}^2} \left\{ \sum_{i=1}^{n} |y_i - b_0 - b_1 x_i| \right\},$$

is reduced to solving

$$\hat{\beta}_1 = \mathrm{argmin}_{b_1 \in \mathbb{R}} \left\{ \sum_{i=1}^{n} w_i |z_i - b_1| \right\},$$

where $z_i = (y_i - \bar{y})/(x_i - \bar{x})$ and $w_i = |x_i - \bar{x}|$, for $i = 1, \dots, n$ and $\hat{\beta}_0 = \bar{y} - \bar{x}\hat{\beta}_1$. Simpson apparently recognized that the solution to the constrained problem could be found to be

$$\hat{\beta}_1 = z_{(j*)},$$

where $z_{(1)}, \dots, z_{(n)}$ denote the order statistics of the z_i, $w_{(1)}, \dots, w_{(n)}$ denote the correspondingly ordered weights and

$$j^* = \min \left\{ j | \sum_{i=1}^{j} w_{(i)} > \tfrac{1}{2} \sum_{i=1}^{n} w_{(i)} \right\}.$$

This formulation was maintained by Laplace and became known as the *méthode de situation*.

Gauss (1809, §186), noting that the Boscovich–Laplace proposal could be generalized by removing the zero mean residual constraint and including more than a single covariate, makes several remarkably astute observations about the resulting procedure:

> It can be easily shown, that a system of values of unknown quantities, derived from this principle alone, must necessarily exactly satisfy as many equations out of the number proposed, as there are unknown quantities, so that the remaining equations come under consideration only so far as they help to *determine the choice*: if, therefore, the equation $V = M$, for example, is of the number of those which are not satisfied, the system of values found according to this principle would in no respect be changed even if any other value N had been observed instead of M, provided that, denoting the computed value by n, the differences $M - n$, $N - n$, were affected by the same signs.

In this brief passage Gauss recognizes not only that minimizing the sum of absolute residuals yields solutions determined by an exact fit of p observations when there are p parameters to be estimated, but also that these solutions are insensitive to perturbations that do not alter the signs of the residuals. Whether Gauss had further algorithmic ideas is unclear, but he seems well on the way to a full understanding of the linear programming structure of the problem. In his later memoir on least-squares fitting, Gauss (1823, §7) seems more on the defensive:

> Laplace has also considered the problem in a similar manner, but he adopted the absolute value of the error as his measure of loss. Now if I am not mistaken this convention is no less arbitrary than mine. Should an error of double size be considered as tolerable as a single error twice repeated or worse? Is it better to assign only twice as much influence to a double error or more? The answers are not self evident, and the problem cannot be resolved by mathematical proofs, but only by an arbitrary decision. Moreover, it cannot be denied that Laplace's convention violates continuity and hence resists analytic treatment, while the results that my convention leads to are distinguished by their wonderful simplicity and generality.

Perhaps by 1823 he had forgotten the wonderful simplicity and generality of the Laplace method that he had grasped so easily earlier?

In a series of papers in the 1880s Edgeworth also proposed removing the constraint on the mean residual, and defined a "plural median" generalizing the original Boscovich formulation to multiple covariates. Edgeworth (1888) suggested an ingenious geometric strategy for the case of bivariate regression that anticipated later development of the simplex algorithm. Noting that points in sample space $(x_i, y_i) \mapsto \{(\alpha, \beta) : \alpha = y_i - x_i\beta\}$ map to lines in parameter space, and thus lines through pairs of points in sample space map to points in parameter space, Edgeworth proposed starting at one of these intersections in parameter space, choosing a direction of steepest descent, and proceeding to the next intersection. Continuing in this fashion eventually leads to a solution characterized by a pair of points that determine the optimal solution. This approach can be generalized to additional covariates, indeed Edgeworth notes this himself, but rather apologetically observes that it may require "the attention of a mathematician ... with some power of hypergeometrical conception."

Edgeworth's formulation of what has become known as the "dual plot" (e.g. Rousseeuw and Hubert, 1999) incorporates the essential features of the Barrodale and Roberts (1974) algorithm for median regression. Starting from an initial basic solution consisting of an exact fit to p observations, we consider the local consequences of dropping each of the p observations and moving in either a positive or negative direction. Choosing the steepest of the possible directions of descent these choices present, we then decide how far to go by solving a one-dimensional weighted median problem of the same type as that originally formulated by Boscovich. This identifies a new observation to replace the one removed by our determination of the descent direction, and the procedure continues until we can no longer find a direction of descent. For problems of modest size, up to a few thousand observations and a few dozen parameters, this form of the algorithm is extremely efficient. However, in very large problems we now have available a new arsenal of techniques that can be adapted to various forms of larger problems. We will briefly survey some of these techniques in the sequel.

5.2   Exterior point methods

Linear programming and the associated simplex solution method emerged out of the fog of World War II, as did many other important statistical ideas. Dantzig's (2002) memoir recounts that his simplex method ideas arose in 1947 as an attempt to solve a class of military planning problems using methods similar to those he had employed in earlier work with Wald and Neyman on the Neyman–Pearson lemma. Kantorovich's 1939 contributions were not appreciated in the west until they appeared in translated form (Kantorovich, 1960). Following these developments it was quickly recognized that the median regression problem fitted nicely into the linear programming framework; Charnes et al. (1955) appears to be the first explicit use of simplex to solve the median regression problem.

The algorithm of Barrodale and Roberts (1974) was the first to exploit the bounded variables dual form of the median regression problem. The primal median regression problem can be formulated as

$$\min\{1_n^\top u + 1_n^\top v \mid y - Xb = u - v, \ (u, v) \geqslant 0\}$$

and seems a bit unwieldy since the minimization is over a $(2n + p)$-dimensional vector. In contrast, the dual problem has the simpler form,

$$\max_a \{y^\top a \mid X^\top a = \tfrac{1}{2} X^\top 1_n, \ a \in [0, 1]^n\}.$$

In effect, Barrodale and Roberts implemented a general form of the Edgeworth dual-plot strategy. Given a basic solution, which we can write as

$$b(h) = X(h)^{-1} y(h),$$

where h indexes p-element subsets of the integers $\mathbb{N} = \{1, 2, \ldots, n\}$, $X(h)$ denotes the submatrix of X consisting of the rows h, and $y(h)$ is the corresponding subvector of the response y, we need to find the direction of steepest descent. In the original Edgeworth bivariate setting this amounts to looking in one of four possible choices: starting from an intersection in the dual plot, we consider dropping one of the two basic observations in h, and moving away from the intersection along the line representing the chosen observation. Thus, we need only look at four possible directions, and among those with negative slope choose the steepest. Rather than stopping at the next adjacent vertex, the Barrodale and Roberts innovation was to continue in this direction as long as such motion reduced the objective function. This is just the weighted median problem that we have already described. When we are estimating $p > 2$ parameters the situation is essentially the same, except that we have $2p$ directions to examine in order to select the descent direction. See Bloomfield and Steiger (1983) for a more detailed investigation of simplex-based algorithms for median regression.

Modification of this approach to compute quantile regression models other than the median is straightforward. In the primal we only replace the 1_ns by appropriate asymmetric weights, in the dual we simply change the $\frac{1}{2}$ to $1 - \tau$ to obtain the τth regression quantile estimate. Some further details are provided in Koenker and d'Orey (1987). It may seem alarming that there are a continuum of problems of this form: do we really need to solve such problems for every $\tau \in [0, 1]$? Fortunately, the answer to this question is "no;" there are only a finite set of distinct solutions, and they are easily found by classical parametric linear programming methods. Given a vertex solution at a particular basic solution, $b(h)$, small changes in τ leave the vertex solution unperturbed. Eventually, larger changes in τ tilt the plane representing sufficiently so that it is no longer "tangent" to the constraint set at a unique vertex, but now coincides with the constraint set along an entire edge of that set. It is easy to compute these "critical" τs at which the solution jumps and thereby produce the entire solution path for $\tau \in [0, 1]$. Portnoy (1989) has shown that the expected number of distinct solutions along this path is $O(n \log n)$; of course in the one-sample setting there are always precisely n distinct solutions provided that observations are themselves distinct.

Similar parametric programming techniques may be employed to study the solution path for penalized smoothing problems or lasso-type penalized estimators. They have also proven very useful in inferential applications such as the inversion of the rank tests proposed by Gutenbrunner and Jurečková (1992) which can be carried out by parametric programming to produce confidence intervals for quantile regression coefficients. While it has become common to encounter paeans to the computation of the "entire regularization path," it should also be recognized that such computations quickly become burdensome in large data applications. It is a great virtue of exterior point methods like the simplex that it is easy to trace out the trajectory of solutions for parametric families of problems, but the number of distinct solutions can easily become overwhelming and in such cases we need to find better ways to approximate the path. Unfortunately, the great advances made in the development of interior point methods for linear programming and discussed in the next section do not easily lend themselves to this task.

5.3   Interior point methods

In contrast to the "exterior point" algorithms exemplified by the Edgeworth procedure and its simplex progeny that move from vertex to vertex on the exterior of the constraint set, interior point methods move from the center of the constraint set toward a vertex solution. Although prior work in the Soviet literature offered theoretical support for the idea that polynomial algorithms for linear programming could be structured in this way, Karmarker (1984) constituted a pivotal moment in the development of optimization tools for linear programs and convex problems more generally. It was quickly recognized that Karmarker's ideas were closely connected to earlier work on barrier methods for nonlinear programming as developed by Fiacco and McCormick (1968) and even earlier for linear programs by Frisch (1956).

The logarithmic barrier method of Frisch for the canonical linear program

$$\min\{c^\top x \mid Ax = b,\ x \geqslant 0\}$$

simply replaces the inequality constraints with a penalty term that forces x to stay in the positive orthant,

$$\min\left\{c^\top x - \mu \sum_{j=1}^{p} \log x_j \mid Ax = b\right\}.$$

By gradually relaxing the penalty parameter, μ, we can approach a vertex solution as $\mu \to 0$. The modified problem has the obvious advantage that it has a smooth objective that, for any fixed μ, generates Newton steps. Following the exposition in Portnoy and Koenker (1997), denoting diagonal matrices by upper-case letters corresponding to lower-case vectors (e.g. $X = \text{diag}(x)$), and letting e denote a p-vector of 1s, we can write the quadratic (Newton) problem for a direction of descent, p, starting from x as

$$\min\{c^\top p - \mu p^\top X^{-1}e + \tfrac{1}{2}\mu p^\top X^{-2}p \mid Ap = b\}.$$

Denoting a vector of Lagrange multipliers for the equality constraint by y, this problem yields first-order conditions

$$\{c - \mu X^{-1}e + \mu X^{-2}p = A^\top y, \ Ap = 0\}$$

which, multiplying through by AX^2, can be reformulated as

$$AX^2 A^\top y = AX^2 c - \mu AX^1 e.$$

Solving for y and substituting back into the first-order conditions yields a Newton direction, δ. The inherent difficulty of each step of this primal log barrier method thus lies in solving the $p \times p$ linear system in this equation. As long as p is modest, or the matrix $AX^2A^\top$ is sparse, this can be done quite efficiently.

Some improvement in performance can be achieved by exploiting both the primal and dual formulations of the problem. The dual of our canonical problem may be expressed as,

$$\max_y\{b^\top y \mid A^\top y + z = c, \ z \geqslant 0\}.$$

Optimality in the primal implies that $c - \mu X^{-1}e = A^\top y$, so we can set $z = \mu X^{-1}e$ to satisfy the dual constraint and obtain the system

$$Ax = b, \quad x \geqslant 0,$$
$$A^\top y + z = c, \quad z \geqslant 0,$$
$$Xz = \mu e.$$

The parametric trajectory $(x(\mu), y(\mu), z(\mu))$ describes the "central path" from the center of the constraint set to a solution on the boundary of the constraint set satisfying the classical complementary slackness condition, $Xz = 0$, when $\mu = 0$. As described in more detail in Portnoy and Koenker (1997), this primal–dual formulation yields a slightly perturbed version of the primal Newton step described above, but again results in a $p \times p$ linear system that requires the same computational effort at each iteration.

To complete the description of the primal–dual method we would need to specify how far to go in the direction p, how to adjust μ as we proceed along the central path, and how to stop. Each of these aspects is addressed in Portnoy and Koenker (1997, Section 4) where the bounded variables approach of Lustig et al. (1994) and Mehrotra (1992) is adapted to the quantile regression dual problem. This approach has been implemented in Fortran in several variants in the quantreg package in R (Koenker, 2015).

Comparison of performance of the modified Barrodale–Roberts algorithm by Koenker and d'Orey (1987) for quantile regression with the interior point implementation indicates that the exterior point (simplex) approach has a clear advantage for relatively small problems with sample size, n, less than a few thousand and parametric dimension, p, also modest (say, less than 20). However, for larger problems the interior point method is substantially quicker and also more accurate than Barrodale–Roberts. Accuracy of the Barrodale–Roberts solutions for large problems could be improved by periodically reinverting the current basic solution, since extensive pivoting can produce substantial accumulated error. The interior point algorithm typically requires at most only a few dozen iterations and accuracy is easily monitored by the duality gap in the primal–dual formulation.

A natural extension of the basic quantile regression problem that maintains its linear programming structure involves the imposition of additional linear inequality constraints on the model parameters. Such constraints arise in a variety of contexts, including portfolio optimization and the introduction of shape constraints in nonparametric regression. Koenker and Ng (2005) describe a modified version of the interior point method that is implemented in the quantreg package. The only potential difficulty with adding such constraints is lack of an initial feasible solution, in contrast to the original dual problem where the center of the unit cube is always feasible.

When the parametric dimension of the model is large, the original implementations of both the Barrodale–Roberts and interior point methods can be quite slow, so it is worthwhile to consider other options. The first question in such circumstances should always be: how sparse is the design matrix X? In most nonparametric applications like those encompassed by the total-variation penalized additive models described in Koenker (2011) and implemented in the quantreg function *rqss*, the design matrix is extremely sparse, typically with only 1 or 2 percent nonzero entries. In such cases sparse linear algebra comes to the rescue, and in particular sparse Cholesky factorization as described in Koenker and Ng (2003) makes the interior point approach entirely feasible even for problems with several thousand parameters.

5.4   Preprocessing

In many linear programming applications we can profitably remove dominated constraints and thereby reduce the effective dimensionality and consequently the effort required to solve the problem. In large quantile regression problems it is worthwhile to consider strategies that might be able to reduce both the column and row dimension of the design matrix. Especially in dense design settings with large column dimension, p, it is natural to consider lasso methods to reduce the column dimension in a preliminary phase. This tactic is described in some detail by Belloni, Chernozhukov, and Kato in Chapter 15, so I will not dwell on it here, instead I will briefly describe a strategy for reducing the row dimension.

In Portnoy and Koenker (1997) we considered a relatively simple strategy for reducing the row dimension of large, dense problems. Variants of this technique are likely to prove helpful in many applications. An important feature of the linear quantile regression problem – already apparent to Gauss, as we have seen – is that the subgradient condition for optimality of a solution depends only on the signs of the residuals. More explicitly, the directional derivative of the objective,

$$R(b) = \sum_{i=1}^{n} \rho_\tau(y_i - x_i^\top b),$$

in the direction δ, is

$$\partial R(b, \delta) = -\sum_{i=1}^{n} x_i^\top \delta(\tau - \text{sgn}^*(y_i - x_i^\top b, -x_i \delta)),$$

where $\text{sgn}^*(u, v) = \text{sgn}(u)I(u \neq 0) + \text{sgn}(v)I(u = 0)$. Optimality at b requires that $\partial R(b, \delta) \geqslant 0$ for all δ on the unit sphere in $\mathbb{R}^p$. Thus, if we had a way to predict that a group of observations (say, $J_L \subset \mathbb{N}$) would be below the optimal hyperplane $\hat{h}(x) = x^\top \hat{\beta}$, and another group (say, $J_H \subset \mathbb{N}$) would be above, we would also know precisely how these observations would contribute to the subgradient. This implies that we could aggregate the observations in J_L and J_H and treat them as two globbed observations with the new objective function,

$$\tilde{R}(b) = \sum_{\mathcal{J}} \rho_\tau(y_i - x_i^\top b),$$

where the index set $\mathcal{J} = \{\mathbb{N} \backslash J_L \backslash J_H, L, H\}$ and $y_j = \sum_{i \in J_j} y_i$ and $x_j = \sum_{i \in J_j} x_i$ for $j = L, H$. If the number of elements of J_L and J_H is large relative to n we have significantly reduced the row dimension of the problem.

Of course, no one is likely to do our predictions for us, but we can easily do them ourselves using a subset of m of the n observations. As shown in Portnoy and Koenker (1997), standard confidence band procedures yield bands with expected width $\mathcal{O}(p/\sqrt{m})$ and it is optimal to choose $\mathcal{O}((np)^{2/3})$ to balance coverage and the complexity of the band construction. Given a band, it is easy to determine how many of the original points lie within the band; this number M is of order $\mathcal{O}(np/\sqrt{m})$. Reestimating using the globbed sample of $M + 2 = \mathcal{O}((np)^{2/3})$ observations, we have a trial solution. It must now be verified that the globbed observations do indeed lie above or below the fitted hyperplane as predicted. If they do, we are done; if not, we can expand m and try again. The probability of failing this check, π, can be controlled, and the number of required repetitions of this cycle is a geometric random variable with expectation π^{-1}, so we can ensure that only a small number of cycles is needed. Each cycle operates on a significantly reduced sample, reducing a sample of 1 million observations, for example, to only 10,000–20,000. The entire strategy is implemented in the **pfn** option for the **rq** fitting function of the quantreg package in R.

5.5 First-order, proximal methods

However pleased we might be with the performance of interior point methods and preprocessing for large problems, there may come a time when the parametric dimension of new problems stretches the effort required for Cholesky factorization at each iteration to breaking point. When this happens it is time to reconsider first-order, gradient descent methods. Fortunately, here too we find that great progress has been made in recent years and a unified, efficient approach has emerged well suited to modern parallelized computation.

5.5.1 Proximal operators and the Moreau envelope

Proximal algorithms for convex optimization rely on additive separability of the objective function and efficient computation of optima for separable components of the problem. This structure is well adapted to a wide variety of statistical applications, including quantile regression. A brief introduction to these methods will be sketched here; for further details the reader is encouraged to consult Parikh and Boyd (2013) and the extensive references provided there.

Let $f : \mathbb{R}^n \to \mathbb{R} \cup \{\infty\}$ be a closed, proper convex function with effective domain $\mathrm{dom} f = \{x \in \mathbb{R}^n \mid f(x) < \infty\}$. The proximal operator $P_f : \mathbb{R}^n \to \mathbb{R}$ of f is

$$P_f = \mathrm{argmin}_x \{f(x) + \tfrac{1}{2}\|x - v\|_2^2\},$$

where $\| \cdot \|_2$ denotes the usual Euclidean norm. $P_f(v)$ can be interpreted as seeking to minimize f without allowing the solution to move too far away from v. By rescaling the function f, so

$$P_{\lambda f}(v) = \mathrm{argmin}_x \left\{ f(x) + \frac{1}{2\lambda}\|x - v\|_2^2 \right\},$$

we can control the relative strength of this trade-off. $P_f(v)$ can be viewed as a generalized projection: if f is simply the indicator of a convex set $\mathcal{C}$, so $f(x) = 0$ if $x \in \mathcal{C}$ and $f(x) = \infty$ otherwise, we have

$$P_f = \mathrm{argmin}_{x \in \mathcal{C}}\|x - v\|_2^2,$$

the Euclidean projection of v onto $\mathcal{C}$.

To illustrate the behavior of the proximal operator, P_f a bit further, Figure 5.1, reversed engineered from an example in Parikh and Boyd (2013), depicts the action of P_f for the function,

$$f(x) = \begin{cases} \|x\|_2 & \text{if } x_1 x_2 \geqslant 1 \\ \infty & \text{otherwise.} \end{cases}$$

Points v are mapped by the proximal operator toward the constrained optimum at $(1,1)$: when v lies outside the constraint set it is sent to the nearest boundary point, when v lies inside the constraint set it is directed toward the optimum by an amount controlled by λ.

To pursue the connection to projection a bit further, recall that the infimal convolution of two closed proper convex functions f and g on $\mathbb{R}^n$ is,

$$(f \,\square\, g)(v) = \inf_x\{f(x) + g(v - x)\}.$$

If we take $g(x) = \tfrac{1}{2}\| \cdot \|_2^2$, then,

$$M_{\lambda f}(v) = \inf_x\{f(x) + \tfrac{1}{2\lambda}\|x - v\|_2^2\},$$

is called the Moreau envelope of the function λf. We may view $M_{\lambda f}(v)$ as a smoothed, or regularized version of f and as such it has several advantages. It has domain $\mathbb{R}^n$ even when f does not, it is continuously differentiable even though f may not be, and perhaps most importantly, f and $M_{\lambda f}$ have the same minimizers. Parikh and Boyd (2013) interpret M_f in the following way: letting f^* denote the convex conjugate of f, that is $f^*(y) = \sup_x\{y^\top x - f(x)\}$, we may write

$$M_f = (f^* + \tfrac{1}{2}\| \cdot \|_2^2)^*,$$

so M_f results from adding a smooth regularization to f^*, and then transforming back to obtain a smooth approximation of f. The connections between P_f and M_f are obviously very intimate: $P_f(x)$ is the unique point that achieves the infimum of M_f, that is,

$$M_f(x) = f(P_f(x)) + \tfrac{1}{2}\|x - P_f(x)\|_2^2,$$

and

$$\nabla M_{\lambda f}(x) = \tfrac{1}{\lambda}(x - P_{\lambda f}(x)).$$

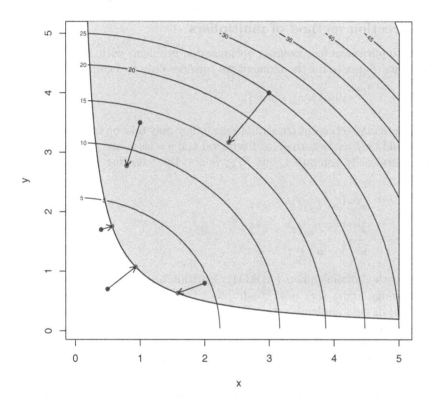

FIGURE 5.1
The proximal operator $P_f(v)$ projects points, v outside the shaded constraint set to the constraint boundary, while points inside the constraint set are mapped toward the boundary by an amount controlled by the choice of λ.

The latter expression, when rewritten as,

$$P_{\lambda f}(x) = x - \lambda \nabla M_{\lambda f}(x),$$

reveals that $P_{\lambda f}(x)$ can be viewed as a gradient step of length λ for the regularized function $M_{\lambda f}$. This interpretation also suggests the fixed point iteration,

$$x^{k+1} = P_{\lambda f}(x^k),$$

which can be shown to converge under quite general conditions. As noted by Parikh and Boyd (2013) such methods are closely related to gradient flow methods for solving differential equations, and in special cases to the well-known EM and MM algorithms that have been extensively employed in the statistics literature.

5.5.2 Alternating direction method of multipliers

It is common in statistical applications to encounter optimization problems with additively separable convex components. Suppose for the moment we consider only two components,

$$\min_{x}\{f(x) + g(x)\},$$

one, or even both, components may represent constraints since they may take on infinite values. A familiar example would be f as (negative) log likelihood and g a lasso-like parametric penalty. When P_f and P_g are easily computed, but P_{f+g} is not, the following iteration is attractive:

$$x^{k+1} = P_{\lambda f}(z^k - u^k)$$
$$z^{k+1} = P_{\lambda g}(x^k + u^k)$$
$$u^{k+1} = u^k + x^{k+1} - z^{k+1}.$$

This alternating direction method of multipliers (ADMM) algorithm has broad applicability and has been shown to converge under very mild conditions.

Fougner and Boyd (2015) discuss implementation details for an extension of the ADMM approach introduced in Parikh and Boyd (2014) to problems in the following "graph form,"

$$\min_{(x,y)}\{f(y) + g(x) \mid y = Ax\}.$$

Now, (x, y) is constrained to the graph $\mathcal{G} = \{(x, y) \in \mathbb{R}^{n+m} \mid y = Ax\}$. The modified ADMM algorithm becomes:

$$(x^{k+1/2}, y^{k+1/2}) = (P_{\lambda g}(x^k - \tilde{x}^k), P_{\lambda f}(y^k - \tilde{y}^k))$$
$$(x^{k+1}, y^{k+1}) = \Pi_A(x^{k+1/2} - \tilde{x}^k, y^{k+1/2} - \tilde{y}^k)$$
$$(\tilde{x}^{k+1}, \tilde{y}^{k+1}) = (\tilde{x}^k + x^{k+1/2} - x^{k+1}, \tilde{y}^{k+1/2} + y^{k+1/2} - y^{k+1})$$

where Π_A denotes the (Euclidean) projection operator into the set $\mathcal{G}$. This projection has a relatively simple structure and has the advantage that the linear system representing the solution need only be solved once. See Appendix A of Parikh and Boyd (2014) for full details. This contrasts sharply with interior point methods where we repeatedly need to solve linear systems involving a diagonally weight moment matrix.

ADMM algorithms, like other first-order gradient type methods, have the advantage that they are efficiently parallelizable; all we need to be able to do is compute the proximal

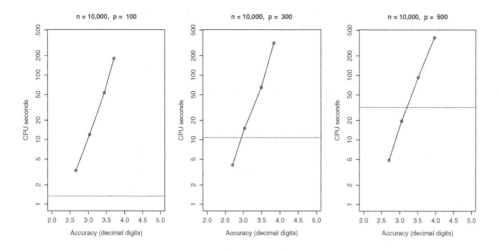

FIGURE 5.2
Accuracy versus computational effort: CPU effort (in seconds) is plotted against accuracy in the number of correct decimal digits, averaged over the p coefficients for the POGS GPU solutions to the primal quantile regression problem. Baseline accuracy is determined by the interior point solution depicted by the horizontal line, which is accurate to at least six decimal digits. Although the POGS procedure is quite quick to produce a solution with two to three digit accuracy, the effort required to produce better accuracy increases rapidly. In contrast, there is little advantage observed in the interior point timings when the convergence tolerance is relaxed.

operators for f and g and we have a gradient descent strategy based on the regularized problem, a strategy that avoids the burden of computing Cholesky factorization at each iteration. Fougner and Boyd (2015) discuss some implementation issues and Fougner (2014) describes a *C++* library with both a Matlab and an R interface. In the next subsection some computational experience with this approach for quantile regression is described.

5.5.3 Proximal performance

To evaluate performance of the ADMM approach for large quantile regression problems I have carried out some very limited tests based on the GPU implementation of Fougner (2014). These tests were conducted on an unloaded IBM x3400M3 machine running linux with an NVIDIA Tesla C2050 graphics card.

In Figure 5.5.3 we compare the timing and accuracy trade-off for the interior point solver described above and the GPU implementation of the POGS solver in three large quantile regression problems. Each setting has sample size $n = 10,000$, but p varies from 100 to 300. In each case entries of X are independent and identically standard Gaussian distributed, except for the appended intercept. By adjusting the convergence tolerance we can control the accuracy of both methods. Accuracy is measured in decimal digits relative to the interior point solution solution with the (default) tolerance of $\epsilon = 10^{-6}$, as (minus) the logarithm (base 10) of root mean squared error. The interior point solution yields essentially single precision accuracy averaged over the p estimated coefficients. This can be relatively easily evaluated by tightening the convergence tolerance of the interior point algorithm. The CPU effort required for this benchmark solution is indicated in the figure by the horizontal lines. Further testing revealed that there was little reduction in CPU

effort achieved by further relaxation of the interior point tolerance, a finding that is easily
explained by examining the number of iterations required. The interior point method is
doing at most a few dozen iterations, and relaxing the convergence tolerance saves a few
of these; however, the POGS procedure required 1421, 6065, 28867 and 120002 iterations
respectively for the four ascending points appearing in the rightmost panel of the figure.
And the efficacy of the GPU notwithstanding, this takes some time.

This performance trade-off should not be entirely surprising since it is already apparent
in other gradient descent algorithms, and has been often remarked upon in applications of
the closely related EM algorithm. In some applications it can be easily disregarded since
decisions based on such data analysis only require a couple of digits accuracy. Nevertheless,
it is somewhat disconcerting in view of our usual obsessions with rates of convergence of
statistical procedures. There is an increasing tendency in statistical research to explicitly
consider computational effort as well as statistical performance in the evaluation of procedures. It would be nice to understand this better in the present context. Larger sample sizes
may not offer the precision we have come to expect, if we cannot reliably estimate models
with them. On a more positive note, it may also be that some form of Nesterov acceleration
can be used to improve the convergence speed of these proximal gradient methods.

Bibliography

I. Barrodale and F. Roberts. Solution of an overdetermined system of equations in the ℓ_1
norm. *Communications of the ACM*, 17:319–320, 1974.

P. Bloomfield and W. S. Steiger. *Least Absolute Deviations. Theory, Applications and
Algorithms*. Birkhäuser, Boston, 1983.

A. Charnes, W. W. Cooper, and R. O. Ferguson. Optimal estimation of executive compen-
sation by linear programming. *Management Science*, 1:138–151, 1955.

G. Dantzig. Linear programming. *Operations Research*, 50(1):42–47, 2002.

F. Y. Edgeworth. On a new method of reducing observations relating to several quantities.
Philosophical Magazine, 25:184–191, 1888.

R. W. Farebrother. Further details of contacts between Boscovich and Simpson in June
1760. *Biometrika*, 77:397–400, 1990.

A. V. Fiacco and G. P. McCormick. *Nonlinear Programming: Sequential Unconstrained
Minimization Techniques*. Wiley: New York, 1968.

C. Fougner. *POGS: Proximal Operator Graph Solver*, 2014. https://github.com/foges/
pogs.

C. Fougner and S. Boyd. Parameter selection and pre-conditioning for a graph form solver.
http://stanford.edu/~boyd/papers/pogs.html, 2015.

R. Frisch. La résolution des problèmes de programme linéaire par la mèthode du potential
logarithmique. *Cahiers du Seminaire d'Econometrie*, 4:7–20, 1956.

C. F. Gauss. *Theoria Motus Corporum Celestium*. Perthes et Besser, Hamburg, 1809.
Translated 1857 as *Theory of Motion of the Heavenly Bodies Moving about the Sun in
Conic Sections*, trans. C. H. Davis. Little, Brown, Boston. Reprinted Dover, New York,
1963.

C. F. Gauss. *Theoria Combinationis Observationum Erroribus Minimis Obnoxiae.* Dieterich, Göttingen, 1823. Translated as *Theory of the Combination of Observations Least Subject to Errors*, trans. G. W. Stewart. SIAM, Philadelphia, 1995.

C. Gutenbrunner and J. Jurečková. Regression quantile and regression rank score process in the linear model and derived statistics. *Annals of Statistics*, 20:305–330, 1992.

L. V. Kantorovich. Mathematical methods of organizing and planning production. *Management Science*, 6:366–422, 1960.

N. Karmarker. A new polynomial time algorithm for linear programming. *Combinatorica*, 4:373–395, 1984.

R. Koenker. Additive models for quantile regression: Model selection and confidence bandaids. *Brazilian Journal of Probability and Statistics*, 25:239–262, 2011.

R. Koenker. *Quantreg: An R package for quantile regression*, 2015. http://www.R-project.org.

R. Koenker and V. d'Orey. Computing regression quantiles. *Applied Statistics*, 36:383–393, 1987.

R. Koenker and P. Ng. SparseM: A sparse linear algebra package for R. *Journal of Statistical Software*, 8, 2003.

R. Koenker and P. Ng. Inequality constrained quantile regression. *Sankhyā*, 67:418–440, 2005.

I. J. Lustig, R. E. Marsden, and D. F. Shanno. Interior point methods for linear programming: computational state of the art with discussion. *ORSA J. on Computing*, 6:1–36, 1994.

S. Mehrotra. On the implementation of a primal-dual interior point method. *SIAM Journal of Optimization*, 2:575–601, 1992.

N. Parikh and S. Boyd. Proximal algorithms. *Foundations and Trends in Optimization*, 1: 123–221, 2013.

N. Parikh and S. Boyd. Block splitting for distributed optimization. *Mathematical Programming Computation*, 6:77–102, 2014.

S. Portnoy. Asymptotic behavior of the number of regression quantile breakpoints. *SIAM J. Science Statistical Computing,*, 12:867–883, 1989.

S. Portnoy and R. Koenker. The Gaussian hare and the Laplacian tortoise: Computability of squared-error versus absolute-error estimators, with discusssion. *Statistical Science*, 12:279–300, 1997.

P. J. Rousseeuw and M. Hubert. Regression depth. *Journal of the American Statistical Association*, 94:388–433, 1999.

S. Stigler. Boscovich, Simpson and a 1760 manuscript note on fitting a linear relation. *Biometrika*, 71:615–620, 1984.

6

Survival Analysis: A Quantile Perspective

Zhiliang Ying

Columbia University, New York, USA

Tony Sit

Chinese University of Hong Kong, Hong Kong, China

CONTENTS

This chapter discusses how quantile regression models can be extended so as to handle time-to-event data which are characterized by censoring. Censored quantile regression offers a valuable supplement to the Cox proportional hazards model for survival analysis, which has been an important topic in health sciences, industrial life testing and econometrical analysis. The chapter first presents basic concepts and models that are commonly adopted in survival analysis. It also summarizes recent developments in quantile regression for censored data and their relation to some existing survival models.

6.1 Introduction

The origins of survival analysis date back to early work in actuarial science centuries ago on modeling the mortality rate of a population via life tables. The need to study the reliability of military equipment during World War II also created a flourishing period for this research area. After the War, research in biomedicine and public health, especially cancer studies, generated a great deal of interest in the statistical community. Works by Kaplan and Meier (1958) and Cox (1972), among others, helped solidify survival analysis as a mainstay in the statistics literature. The areas of semiparametric modeling and nonparametric maximum likelihood estimation owe a great debt to these works.

Generally speaking, the statistical analysis of lifetime, survival time or failure time data deals with an outcome variable that is the time until the event of interest occurs. These events are generally regarded as "failures," and failure time data can be observed in a wide range of disciplines: from time of disease onset and time of death in medical studies to experiments in psychology on how long a subject takes to acquire skills for a certain task; from modeling the lifetime of a light bulb to modeling unemployment duration. Readers are referred to Cox and Oakes (1984), Kalbfleisch and Prentice (2003), and Lawless (2003) for more examples.

6.1.1 Notation

To facilitate our discussion, we first introduce some basic notation. Let T be a nonnegative random variable which represents the time to occurrence of an event, or in general, the failure time, of an individual from a population. We first consider the simplest setting in which observations are independent and identically distributed (i.i.d.) samples from this homogeneous population. We shall generalize this assumption in the later text so as to highlight the key concepts.

Typically, to describe the distributional features of T, we let $f(t)$ denote the probability density function of T. Its cumulative distribution function (cdf) is defined as

$$F(t) = \Pr(T \leqslant t) = \int_0^t f(x)dx.$$

The *survival function* is introduced to describe the probability of an individual surviving up to time t:

$$S(t) = \Pr(T \geqslant t) = \int_t^\infty f(x)dx = 1 - F(t). \tag{6.1}$$

From its definition, we can see that $S(t)$ is a monotone nonincreasing continuous function with $S(0) = 1$ and $S(\infty) = \lim_{t \to \infty} S(t) = 0$. We can also define the τth *quantile* of T as Q_τ where $\Pr(T \leqslant Q_\tau) = F(Q_\tau) = \tau$, $\tau \in (0,1)$. In other words, $Q_\tau = F^{-1}(\tau)$. In particular, $\tau = 0.5$ is the median of the failure time distribution.

While the survival function (6.1) quantifies the unconditional probability that the event of interest has not yet occurred by time t, it is also meaningful to define the conditional instantaneous probability of failure. Specifically, one can define the probability of an individual failing in the upcoming small time interval $[t, t + dt)$ among those who have not yet experienced the event of interest by t via the *hazard function* $\lambda(t)$:

$$\lambda(t) = \lim_{dt \downarrow 0} \frac{\Pr(t \leqslant T < t + dt \mid T \geqslant t)}{dt} = \frac{f(t)}{1 - F(t)} = \frac{f(t)}{S(t)}. \tag{6.2}$$

It is noteworthy that $\lambda(t)$ can be essentially any nonnegative function, although $S(t)$ is bounded between 0 and 1. From its definition (6.2), one can write

$$\lambda(t)dt \approx \Pr(t \leqslant T < t + dt \mid T \geqslant t),$$

which can be interpreted as the probability that an individual will experience the event of interest instantaneously given that she/he has survived up till t. In fact, in epidemiology, the hazard rate function was historically known as the *force of mortality*. By integrating $\lambda(t)$, one can derive the following relation between $S(t)$ and $\lambda(t)$:

$$\int_0^t \lambda(s)ds = \int_0^t \frac{f(s)}{1 - F(s)}ds = -\log\{1 - F(s)\}\Big|_0^t = -\log S(t), \qquad (6.3)$$

which leads to

$$S(t) = \exp\{-\Lambda(t)\},$$

where $\Lambda(t) = \int_0^t \lambda(s)ds$ is defined as the *cumulative hazard function*. By (6.2), we can write $f(t) = \lambda(t)\exp\{-\Lambda(t)\}$. It should also be noted that the above formulation can be extended to cases in which T does not have a density, that is, the distribution function F has jumps. We should assume continuity for a more general discussion. One can modify the concepts and models so as to introduce features such as jumps in the distribution function whenever necessary.

6.1.2 Censoring

One of the distinctive features of survival analysis is *censoring*. Essentially, censoring occurs when we only manage to collect partial information about an individual's survival time instead of knowing the exact time of her/his failure. This is common in clinical studies in which patients may drop out during the study period or have not experienced the event of interest before the study ends. Intuitively, the existence of censoring can potentially introduce bias into the inference for the survival distribution. Proper handling for this type of data is therefore required.

The most common type of censoring is so-called *right censoring*, which will be the focus here. An individual is called right censored if we only observe a lower bound for her/his failure time. This particular lower bound is called the *censoring time*, and is commonly denoted by C. The cases discussed earlier are typical examples of right censoring. Usually, we can only observe a minimum between the censoring time and the failure time, that is, we can only observe $\tilde{T} = T \wedge C$, where $a \wedge b = a$ if $a < b$ and b otherwise. There are other types of censoring depending on how the partial information is obtained. Examples include left censoring and interval censoring; for details, see Fleming and Harrington (1991) and references therein.

In addition to the aforementioned categorization of censoring mechanisms based on different cut-off points, we can distinguish different types of censoring schemes based on how the censoring time is generated in its relation to the failure time. The independent censoring assumption, which is commonly used, means that the two random variables, T and C, are independent. Although it appears to be a strong assumption, independent censoring may be reasonable when covariates are available and conditioned upon. A slightly weaker assumption is *noninformative censoring*, which specifies that the distribution of the survival time T does not contain information on the censoring times C. Mathematically, the censoring is noninformative if C is a stopping time whose distribution does not depend on any (non)parametric component of the failure time T distribution. Noninformative censoring is assumed in the justification of many statistical methods for right censored data using the counting processes or martingale approach. Further illustration can be found in Fleming and Harrington (1991) and Kalbfleisch and Prentice (2003).

It is natural for statisticians to model quantities of interest via regression models. In survival analysis, different models have been proposed that attempt to model the hazard rate function $\lambda(t)$ via subjects' p-dimensional covariates, which are denoted by Z. Random censoring can be restrictive in many cases. One typical example, in Altman (1991, Section 13.2.1), describes a study on how long subjects take before demonstrating any seasickness and vomiting. It is not difficult to see that a person who gets motion sickness is more likely to request a premature stop for the experiment, and hence the censoring time C is not independent of the failure time T. A less restrictive assumption on censoring is called *conditionally independent censoring* where, conditional on the covariates Z, C and T are statistically independent and hence

$$\lim_{dt \downarrow 0} \frac{\Pr\{T \in [t, t+dt] \mid T \geqslant t; Z\}}{dt} = \lim_{dt \downarrow 0} \frac{\Pr\{T \in [t, t+dt] \mid T \geqslant t, C \geqslant t; Z\}}{dt}.$$

In the next section we describe typical (non)parametric models used in survival modeling. Further elaboration on regression-type models is presented in Section 6.2.3.

6.2   Important models

This section discusses popular models used for the failure time distribution of a homogeneous population. We first introduce a number of common parametric models that are popular among researchers. They are adopted because of their distributional features, including the hazard rate and the mean, which can be easily characterized by a few parameters. Non-parametric estimators are also popular for estimating survival/hazard functions when the parametric models available are sometimes too restrictive. Two of the celebrated estimators, namely, the Kaplan–Meier estimator (Kaplan and Meier, 1958) and the Nelson–Aalen estimator (Aalen, 1978; Nelson, 1969, 1972), are introduced in Section 6.2.2. The above two classes of models are mainly used for modeling univariate failure times without taking other factors, such as group assignment or gender, into account. At the end of this section, we discuss a rich category of semiparametric regression models that enable statisticians to relate the survival time of a population to its corresponding covariate observations.

6.2.1   Parametric models

Among a wide array of parametric models that are used in general statistical analysis, a few distributions play a significant role in modeling survival data. We shall introduce the exponential, gamma, Weibull, and lognormal distributions, all of which are defined over the positive real line $\mathbb{R}^+$.

The exponential distribution is naturally adopted for modeling event times because of its connection with the Poisson distribution. The most distinctive feature of this distribution is its memorylessness property, which implies that the aging factor is not captured in this parametric model; the corresponding hazard function is also constant with respect to the time of failure. This implies that old equipment is as good as completely new equipment. To generalize this rather restrictive, if not entirely unrealistic, model feature, more flexible models including the gamma and Weibull distributions were developed. In particular, as we can see in Table 6.1, the value of α in the Weibull distribution controls the hazard rate. The distribution can model an increasing (decreasing) hazard if $\alpha > 1$ ($\alpha < 1$). When $\alpha = 1$, the distribution degenerates to an exponential distribution.

Sometimes, it is found that the log failure times are normally distributed, in which case the lognormal distribution is considered. Like the normal distribution, the lognormal distribution is specified by its mean and variance. Its probability density function, survival function and hazard function are also presented in Table 6.1.

TABLE 6.1

Distributional features of some common parametric models for survival data

Distribution	Probability density function $f(t)$	Survival function $S(t)$	Hazard function $\lambda(t)$
Exponential(λ)	$\lambda \exp\{-\lambda t\}$	$\exp\{-\lambda t\}$	λ
gamma(β, λ)	$\frac{\lambda \beta t^{\beta-1} \exp\{-\lambda t\}}{\Gamma(\beta)}$	$1 - \int_0^t u^{\beta-1} \exp\{-u\} du / \Gamma(\beta)$	$\frac{f(t)}{S(t)}$
Weibull(α, λ)	$\alpha \lambda t^{\alpha-1} \exp\{-\lambda t \alpha\}$	$\exp\{-\lambda t \alpha\}$	$\alpha \lambda t^{\alpha-1}$
lognormal(μ, σ)	$\frac{\exp\left\{-\frac{1}{2}\left(\frac{\log t - \mu}{\sigma}\right)^2\right\}}{t\sqrt{2\pi\sigma^2}}$	$1 - \Phi\left(\frac{\log x - \mu}{\sigma}\right)$	$\frac{f(t)}{S(t)}$

6.2.2 Nonparametric estimators

Nonparametric survival estimators provide an alternative solution for statisticians to model survival or hazard functions based on censored data in survival studies. Parallel to maximum likelihood estimators for various parametric models, a nonparametric estimator can be viewed as the maximizer for the empirical distribution associated with the observations. We shall consider n individuals each of whom has a counting process $N_i(t)$, $i = 1, \ldots, n$, registering the number of occurrences of the event of interest up to time t. The intensity process of $N_i(t)$ is given, for $i = 1, \ldots, n$, by

$$\lambda_i(t) = Y_i(t)\lambda(t), \tag{6.4}$$

where $Y_i(t) = I(\tilde{T}_i \geq t)$ is the *at-risk indicator* for individual i. The quantity $\sum_{i=1}^n Y_i(t)$ marks the number of subjects that are still at risk just prior to t. Another important quantity is the *censoring indicator*, $\Delta = I(T \leq C) = I(\tilde{T} = T)$, which takes the value 1 if the subject experiences a failure. In Section 6.2.2.1 we shall derive the celebrated Kaplan–Meier estimator for the empirical survival function $S(t)$. This is followed by discussion of the Nelson–Aalen estimator for the cumulative hazard function $\Lambda(t)$ in Section 6.2.2.2.

6.2.2.1 Kaplan–Meier estimator

When there is no censoring, to obtain the empirical survival function based on $(\tilde{T}_1, \ldots, \tilde{T}_n) = (T_1, \ldots, T_n)$, one can define

$$\hat{S}_n(t) = \frac{1}{n} \sum_{i=1}^n I(T_i \geq t)$$

as the nonparametric estimator for the true, unknown survival function S. In the analysis of survival data, however, due to censoring, we cannot directly assign a probability mass of n^{-1} to each of the n observations. We shall consider the case where there is right censoring.

Assume that we have observed $(\tilde{t}_i, \delta_i)$ for $i = 1, \ldots, n$. Among these n time-stamps, we let $0 = t_{(0)} < t_{(1)} < t_{(2)} < \ldots < t_{(k)} < t_{(k+1)} = \infty$ ($k \leq n$) represent the distinct observed failure times whose unknown survival function is denoted by S. For $j = 1, \ldots, k$, for simplicity, we assume there is exactly one event occurring at $t_{(j)}$. There are m_j items censored in $[t_{(j)}, t_{(j+1)})$. Defining $n_j = (1 + m_j) + \ldots + (1 + m_k) = \sum_{i=1}^n Y_i\{t_{(j)}\}$, we can write the likelihood of the data as

$$L = \prod_{j=0}^k \left[\{S(t_{(j)}^-) - S(t_{(j)})\} \prod_{l=1}^{m_j} S(t_{jl}) \right],$$

where t_{jl} denotes the censoring time of the lth individual who was censored during $[t_{(j)}, t_{(j+1)})$, $j = 1, \ldots, k$. We denote the optimizer of L as $\hat{S}$.

We can show that to maximize L, $\hat{S}(t)$ has to be discontinuous at the failure times $t_{(j)}$. Furthermore, we can see that $S(t_{jl}) \leqslant S(t_{(j)})$ for $j = 1, \ldots, k$. The corresponding maximum likelihood estimate $\hat{F}$ thus has the form $\hat{F}(t_{(j)}) = \prod_{i=1}^{j}(1 - \hat{\lambda}_i)$ and $\hat{S}(t_{(j)}^-) = \prod_{i=1}^{j-1}(1 - \hat{\lambda}_i)$, where $\hat{\lambda}_j$ denotes the hazard component at t_j, $j = 1, \ldots, k$. Thus, $\hat{F}$ can be expressed via the optimizers $\hat{\lambda}_j$, $j = 1, \ldots, k$, of the function

$$L(\lambda) = \prod_{j=1}^{k} \lambda_j (1 - \lambda_j)^{n_j - 1}.$$

By choosing $\hat{\lambda}_j = n_j^{-1}$, $j = 1, \ldots, k$, the Kaplan–Meier estimate (Kaplan and Meier, 1958) of the survival function is given by

$$\hat{S}(t) = \prod_{j:t_{(j)} \leqslant t} \frac{n_j - 1}{n_j} = \prod_{j:t_{(j)} \leqslant t} \left(1 - \frac{1}{n_j}\right). \tag{6.5}$$

It can be shown, if there is no censoring, that $\hat{S}(t)$ is equal to $1 - \hat{F}$, where $\hat{F}$ denotes the empirical cdf based on observations $t_1, \ldots, t_n$. For continuous-time or general distributions, we can proceed with a similar argument and consider it as a limit of the discrete-time case. $d\Lambda(t)$ will take the role of λ_j in the previous derivation and the corresponding product limit estimator is given by

$$\hat{S}(t) = \prod_{(0,t)} \{1 - d\hat{\Lambda}(u)\}.$$

To provide some intuition on the Kaplan–Meier estimator, we can express the survival function as

$$S(t) = \prod_{j=1}^{k'} S\{t_{(j)} \mid t_{(j-1)}\},$$

where $0 = t_{(0)} < t_{(1)} < \ldots < t_{(k')} = t$. We choose the time interval so small that there is at most one observed failure. Since $S(t_u \mid t_l) = S(t_u)/S(t_l)$ for $t_l < t_u$, when there is no event observed in $(t_{k-1}, t_k]$, $\hat{S}(t_k \mid t_{k-1}) = 1$; when there is an event observed at time $t_{(j)}$, where $t_{(j)} \in [t_{k-1}, t_k)$, it is sensible to estimate $\hat{S}(t_k \mid t_{k-1}) = (n_j - 1)n_j^{-1} = 1 - n_j^{-1}$, which yields the same result as shown in (6.5).

Efron (1967) provides an alternative approach to compute the Kaplan–Meier estimator. Since, in the absence of censoring, $\hat{S}(t)$ represents the proportion of individuals who can survive to at least t, the *redistribution to the right algorithm* makes use of this observation and proposes a method that can spread the contributions of censored subjects among the surviving population.

Specifically, a uniform probability mass of n^{-1} is assigned to each of the n time-stamps, including both failures and censored events. Whenever a censored observation is encountered while one scans from the earliest (leftmost) time-stamp to the last (rightmost) time-stamp, the mass assigned is redistributed to all the time-stamps on its right. An estimated value of the survival function $\hat{S}(t_j)$ is defined as the difference between $\hat{S}(t_{j-1})$ and the final weight for the time-stamp j, for $j = 1, \ldots, n$.

Estimation of the survival function of the survival time T also enables us to estimate the τth empirical quantile of T, denoted by $\hat{Q}_\tau$, for $\tau \in (0, 1)$. The most popular quantile value is the sample median $Q_{0.5}$, as it is a robust measure of the central tendency. Given $\hat{S}(t)$, $\hat{Q}_\tau$ is defined as the value which satisfies $\hat{S}(\hat{Q}_\tau) = 1 - \tau$. The corresponding confidence interval for $\hat{Q}_\tau$ can be obtained by observing the relation $\Pr(\ell_\alpha \leqslant \hat{Q}_\tau) = \Pr\{\hat{S}(\ell_\alpha) \geqslant 1 - \tau\}$, where ℓ_α denotes the lower α confidence limit for $\hat{Q}_\tau$. In fact, ℓ_α is the value such that the lower α confidence interval of $\hat{S}(\ell_\alpha)$ is exactly $1 - \tau$.

6.2.2.2 Nelson–Aalen estimator

In the derivation of (6.5), we can see that the estimator $\hat{\lambda}$ evaluated at each of the failure time points plays an important role in estimating the empirical survival function when there are censored data. Another important use of such an estimator is to offer a graphical check for the fit of parametric models. This estimator is called the Nelson–Aalen estimator, and was originally introduced by Nelson (1969, 1972), followed by Aalen (1978), who studied its asymptotic properties using martingale methods.

In fact, for $j = 1, \ldots, k$, by observing that $\hat{\lambda}_j = n_j^{-1}$ and recalling that the λ_j are conditional probabilities of failure (hazards) during $[t_{(j)}, t_{j+1})$, one can deduce that the nonparametric maximum likelihood for the cumulative hazard function Λ is

$$\hat{\Lambda}(t) = \sum_{j:t_j<t} \hat{\lambda}_j = \sum_{j:t_j<t} n_j^{-1}, \tag{6.6}$$

which is a right-continuous step function with increments of the empirical hazard estimates.

6.2.3 Regression models

Another important class of models in survival analysis is regression models. In various applications, it is common to observe samples of (censored) failure times whose values depend on explanatory variables (or covariates). Examples of these variables include treatment indicators (treatment versus placebo), individual attributes/measurements, and environmental conditions. Quantifying the relationship between these covariates and the lifetime is of interest. In this subsection, we introduce three semiparametric survival models that are popular among academic researchers and practitioners: the Cox proportional hazards model, the accelerated failure time model, and the Aalen additive hazard model.

6.2.3.1 Cox proportional hazards model

The multiplicative hazards models play a prominent role in survival analysis. These models model the effect of the covariates on the survival time on a multiplicative scale. Among all these models, the Cox proportional hazards model (Cox, 1972, 1975) has been the most popular and most applied survival regression model.

The Cox model stipulates that the hazard function of T is of the form

$$\lambda(t) = \lambda_0(t) \exp\{Z(t)^\top \beta_0\}, \tag{6.7}$$

where $Z(t) = (Z_1(t), \ldots, Z_p(t))^\top$ is a p-dimensional time-varying covariate. The parameter of interest is $\beta_0 = (\beta_{10}, \ldots, \beta_{p0})^\top$. An arbitrary unspecified baseline hazard function $\lambda_0(\cdot)$ is usually a nuisance parameter. Note that the model does not include an intercept term for it is subsumed in $\lambda_0(\cdot)$.

The formulation in (6.7) depicts a multiplicative effect of the covariates on the hazard function. The key model assumption imposed is that the relative risks are time-invariant. For instance, if we consider only one covariate which serves as a treatment indicator, so that $Z = 1$ for individuals from the treatment group and $Z = 0$ for the remaining control group subjects, the relative risk between these two groups is $\lambda(t \mid Z = 1)\{\lambda(t \mid Z = 0)\}^{-1} = \exp(\beta_{10})$.

Based on the random sample of $(\tilde{T}_i, Z_i, \Delta_i)$, $i = 1, \ldots, n$, we can estimate the regression parameter β_0 as the maximizer of Cox's partial likelihood function,

$$L(\beta) = \prod_t \prod_{i=1}^n \left[\frac{\exp\{Z_i(t)^\top \beta\}}{\sum_{\ell \in R(\tilde{T}_i)} Y_\ell(t) \exp\{Z_\ell(t)^\top \beta\}} \right]^{\Delta_i}, \tag{6.8}$$

where $R(t)$ is the set of items at risk of failure at time t^-, $R(t) = \{i : Y_i(t) = 1\}$. Under suitable regularity conditions, $n^{1/2}(\hat{\beta} - \beta_0)$ is asymptotically normal with mean zero; more details on the asymptotic properties of the estimator can be found in Andersen et al. (1993) and Andersen and Gill (1982).

Once the hazard function has been estimated, that is, both $\hat{\lambda}_0(\cdot)$ and $\hat{\beta}$ have been obtained, we can define the quantile correspondingly as the solution of

$$\tau = \Pr[T \leqslant Q_T\{\tau \mid Z(\cdot)\}] = 1 - \exp\left[-\int_0^{Q_T\{\tau \mid Z(\cdot)\}} \exp\{Z(t)^\top \hat{\beta}\} d\hat{\Lambda}_0(t)\right]. \qquad (6.9)$$

Since the right-hand side of (6.9) is a monotone increasing function, one may find the quantile value $Q_T(\tau \mid Z(\cdot))$ that solves (6.9). In particular, if $Z(\cdot) \equiv Z$, the above display can be simplified to

$$\hat{\Lambda}_0[Q_T\{\tau \mid Z(\cdot)\}] = -\log(1 - \tau)\exp(-Z^\top \beta).$$

6.2.3.2   Accelerated failure time model

The accelerated failure time (AFT) model considers a linear regression model for the log-transformed event time $\log T$ on a p-dimensional covariate $Z = (Z_1, \ldots, Z_p)^\top$ in the following fashion:

$$\log T = Z^\top \beta_0 + \epsilon, \qquad (6.10)$$

where $\beta_0 = (\beta_{10}, \ldots, \beta_{p0})^\top$ denotes a p-dimensional regression parameter and ϵ, the residual term, follows an unspecified distribution.

In contrast to the Cox proportional hazards model, the AFT model is easy to interpret for it directly captures the relation between the covariates and the log-transformed event time. If we define $\lambda_{\exp\{\epsilon\}}(t; \beta)$ as the hazard function associated with $\exp(\epsilon)$, we can express the hazard function of T as

$$\lambda(t) = \lambda_{\exp\{\epsilon\}}\{t \exp(-Z^\top \beta_0)\} \exp(-Z^\top \beta_0).$$

Model (6.10) resembles a typical linear regression model. To simplify our discussion, we first consider a simple linear regression model without censoring,

$$Y_i \triangleq \log T_i = \alpha + \beta_0 Z_i + \epsilon_i;$$

the least-squares estimates $\hat{\alpha}$ and $\hat{\beta}$ are defined as

$$(\hat{\alpha}, \hat{\beta}) = \operatorname{argmin}_{\alpha, \beta} \sum_{i=1}^n (y_i - \alpha - \beta z_i)^2 = n\int_{-\infty}^\infty e^2 dF_n(e),$$

where $F_n(\cdot)$ is the empirical cdf of $\mathbf{e} = (e_1, \ldots, e_n)^\top$ with $e_i \triangleq y_i - \alpha - \beta z_i$, $i = 1, \ldots, n$.

Miller's (1976) proposal for cases with censored observations is correspondingly to minimize

$$n\int_{-\infty}^\infty e^2 d\hat{F}(e) = \sum_{i=1}^n \hat{w}_i(\beta)(y_i - \alpha - \beta z_i)^2, \qquad (6.11)$$

where $\hat{F}(\cdot)$ denotes the product limit estimator based on $\{(e_i, \delta_i)\}_{i=1,\ldots,n}$. The weights $\{\hat{w}_i(\beta)\}_{i=1,\ldots,n}$ are the jump sizes of $\hat{F}(\cdot)$. Since, for each censored observation, there is no jump in $\hat{F}$, the weights enable (6.11) to be well defined even for censored (i.e. unobserved) y_i. As we can see from Efron's "redistribution to the right" concept, the weights, in fact, depend on the censored observations. The above discussion can easily be extended to the multiple regression setting.

One should note that the consistency of the estimates of $(\hat{\alpha}, \hat{\beta})$ is developed upon an assumption that the censoring distribution of the ith observation must be in the form of $G_{z_i}(x) = G_0(x - \beta z_i)$ for some distribution function G_0. This essentially means that the censoring distribution, after adjustment of individual means, is homogeneous among all the observations. This restriction can be rather stringent in many applications. See Miller (1981) for a comprehensive review.

Under this regression framework, Buckley and James (1979) defined the following *pseudo-random variables* in order to cope with the missing event times due to censoring:

$$Y_i^* = Y_i \delta_i + E(T_i \mid T_i > Y_i)(1 - \delta_i). \tag{6.12}$$

One can see that for observed failures with $\delta_i = 1$, the dependent variable remains Y_i; for censored time-stamps, the Buckley–James procedure assigns a reasonable estimate of these unknown failure times based on its conditional expectation given that the failure times must occur after the censored times. Indeed, with some manipulations, one can show that

$$
\begin{aligned}
E(Y_i^*) &= E(T_i) = \alpha + \beta Z_i, \\
\hat{E}(T_i \mid T_i > y_i) &= \hat{\beta} z_i + \frac{\sum_{\hat{e}_k > \hat{e}_i} \hat{w}_k(\hat{\beta}) \hat{e}_k}{1 - \hat{F}(\hat{e}_i)}, \quad \text{for } \{i : \delta_i = 0\},
\end{aligned}
$$

with $\hat{e}_i = y_i - \hat{\beta} z_i$ and $\hat{F}(\cdot)$ denoting the product-limit estimator. To facilitate computation, Buckley and James (1979) proposed a semiparametric algorithm which iterates between imputation of censored failure times and least-squares estimation. Unlike Miller's (1976) method, this approach does not require a specific form to be assumed for the censoring distribution G_i. In fact, the AFT model as well as its Buckley–James approach for censored data are highly relevant to quantile regression. We shall defer the discussion to Section 6.3.2.

To illustrate the connection between rank statistics and the AFT model, one can consider (6.10) and write the score function for β, given a particular weight function W, as

$$
\begin{aligned}
U(\beta) &= \sum_{i=1}^{n} \int_0^\infty W(t; \beta) Z_i \left\{ dN_i(t; \beta) - Y_i(t; \beta) \frac{\sum_{j=1}^{n} dN_j(t; \beta)}{\sum_{j=1}^{n} Y_j(t; \beta)} \right\} \\
&= \sum_{i=1}^{n} \int_0^\infty W(t; \beta) \left\{ Z_i - \bar{Z}(t; \beta) \right\} dN_i(t; \beta), \tag{6.13}
\end{aligned}
$$

where $N_i(t; \beta) = I(\epsilon_i \leq t) = I\{T_i \leq t \exp(-Z_i^\top \beta)\}$ denotes the counting process of ϵ_i; $Y_i(t; \beta) = I(\epsilon_i > t)$ is defined similarly. Furthermore, $\bar{Z}(t; \beta) = S^{(1)}(t; \beta)/S^{(0)}(t; \beta)$ with $S^{(\kappa)}(t; \beta) = n^{-1} \sum_{i=1}^{n} Y_i(t; \beta) Z_i^\kappa$, $\kappa = 0, 1$.

The above score function can be rewritten as a log-rank test statistic of the form

$$WLR = \sum_{i=1}^{n} W_i \{ Z_{(i)} - \bar{Z}_{(i)} \}, \tag{6.14}$$

where $\bar{Z}_{(i)} = n^{-1} \sum_{\ell \in \mathcal{R}_i} Z_\ell$, with $\mathcal{R}_i$ denoting the at-risk set at time $t_{(i)}$. It is easy to see that the log-rank test is a special case of (6.14) with specifically $W_i \equiv 1$. In fact, if we take $W_i = -\prod_{\ell=1}^{i} n_\ell (n_\ell + 1)^{-1}$, (6.14) becomes the generalized Wilcoxon test; if we assume $Q_i = n_i n^{-1}$, where $n_i = \|\mathcal{R}_i\|$, the test statistic corresponds to the Gehan generalized Wilcoxon test (see Gehan, 1965a,b).

The asymptotic properties of the estimator are more challenging to obtain due to the nonsmooth estimating equation $U(\cdot)$. Tsiatis (1990) and Ying (1993) have established the asymptotic normality of $n^{1/2}(\hat{\beta} - \beta_0)$ under regularity conditions. Tsiatis (1990) provides sufficient conditions for such asymptotic linearity, which were later relaxed by Ying (1993) without imposing the finite interval condition.

A number of proposals have been developed for estimating the asymptotic variance. Tsiatis (1990) applied a nonparametric density kernel to estimate the unknown hazard function of the survival time. Jin et al. (2003) and Zeng and Lin (2008), for example, propose resampling techniques in order to avoid estimation of $\lambda_{\exp(\epsilon)}(\cdot)$, which can be numerically unstable. More recently, Zeng and Lin (2007) consider a kernel-smoothed profile likelihood function for estimating the regression parameter.

6.2.3.3   Aalen additive hazard model

The Cox proportional hazards model introduced in Section 6.2.3.1 belongs to a family of hazard-based generalized linear regression models. Sometimes, however, proportionality in terms of hazard may not be reasonable as one often observes that relative risk decreases over time due to frailty. In view of this limitation, Aalen (1980) proposed an additive risk model which stipulates that

$$\lambda(t) = Z^{\top}\alpha(t), \tag{6.15}$$

where $\alpha(t) = (\alpha_1(t), \ldots, \alpha_p(t))^{\top}$ is an unknown vector of hazard functions with $\int_0^t |\alpha_j(s)|ds < \infty$, for $j = 1, \ldots, p$. The additive effects captured by this model can potentially be more informative than proportional effects, especially for cases where the baseline hazard is minute. Aalen (1980) proposed ordinary least-squares (OLS) estimators for the cumulative hazard function $A(t) = \int_0^t \alpha(s)ds$. Given a random sample of $(\tilde{T}_i, \Delta_i, Z_i)$, $i = 1, \ldots, n$, Aalen's (1980) OLS estimator is given by

$$\hat{A}_{\text{Aalen}}(t) = \sum_{\ell:t_\ell \leqslant t} (Z^{\top}Z)^{-1}Z^{\top}e_\ell, \tag{6.16}$$

where e_ℓ is a column vector with value 1 only in the position corresponding to the subject who encounters a failure at time t_ℓ, and zero elsewhere.

In fact, since $M(t) = N(t) - \Lambda(t)$, with $\Lambda(t)$ denoting the cumulative hazards, is an n-dimensional martingale, we can also view (6.16) from the counting process perspective:

$$dN(t) = \lambda(t)dt + dM(t) = Y(t)Z(t)^{\top}\alpha(t)dt + dM(t),$$

which yields the estimator

$$d\hat{A}_{\text{Aalen}}(t) = \{(Z^{\top}Z)^{-1}Z^{\top}\}dN(t).$$

Huffer and McKeague (1991) also established the asymptotic normality of the proposed estimator, in particular $n^{1/2}(\hat{A}_{\text{WLS}} - A)$ converges in distribution to a p-variate Gaussian martingale with predictable covariation process $\int_0^t V^{-1}ds$. See McKeague (1988) for further details of the definition of V.

While the Aalen additive hazard model (6.15) captures all effects of the covariates non-parametrically, it is of practical interest and benefit to further generalize the model and introduce a semiparametric component to it. McKeague and Sasieni (1994) extended (6.15) and proposed the following semiparametric additive intensity model:

$$\lambda(t) = Y(t)\{X(t)^{\top}\alpha(t) + Z(t)^{\top}\gamma\}, \tag{6.17}$$

where $\alpha(t)$ is a p-dimensional locally integrable function and γ is a q-dimensional regression coefficient vector. A special case of this model was studied by Lin and Ying (1994) where

$$\lambda(t) = Y(t)\{\alpha(t) + Z(t)^{\top}\gamma\},$$

whose nonparametric component only appears in the baseline $\alpha(\cdot)$. The (approximate) maximum likelihood estimators of the cumulative $A(t)$ and γ for (6.17) as well as their asymptotic properties can be found in McKeague and Sasieni (1994) as well as Lin and Ying (1994).

Before ending this section, we present an interesting connection between this model and the quantile (regression) model. Due to (6.15), we can write a more generic form of the hazard function as

$$\lambda(t \mid Z) = Z(t)^{\top}\alpha(t),$$

which, similar to our argument in (6.9), leads to

$$\log(1 - \tau) = -\int_0^{Q_T(\tau \mid Z(\cdot))} Z(s)^{\top}\, dA(s). \tag{6.18}$$

Specifically, if both Z and $\alpha(\cdot)$ are time-invariant, (6.18) can be simplified ato

$$Q_T(\tau \mid Z) = -\frac{\log(1 - \tau)}{Z^{\top}\alpha}.$$

The quantile obtained from this model is in fact inversely proportional to the mean hazard of the survival time. Recall that for an exponential random variable, its expectation is exactly equal to the inverse of its mean.

6.3 Quantile estimation based on censored data

For various medical follow-up studies, as we can see in Section 6.2, patients' survival properties are summarized by the Kaplan–Meier estimate of the survival distribution and/or semiparametric survival models. The median of the estimated survival distribution is sometimes a more useful summary statistic than the mean for it is more robust to heavy-tailedness/skewness of the true survival distribution. Furthermore, for events that occur only in a small portion of the population, only lower quantile estimates of the survival time (e.g. the lower quartile) can be practically meaningful. In this section, we shall first introduce methodologies for estimating the median survival time. Median estimation can be further generalized to quantile estimation. Section 6.3.2 outlines more recent developments in median/quantile regression. In closing, we shall elucidate the relationship among quantile regression, the additive hazards model, and other survival models.

6.3.1 Quantile estimation

This section is concerned with estimating the distribution of the median of the survival time. Assume that we have a random sample of n individuals whose censored event times and censoring status are denoted by $\{(\tilde{T}_i, \Delta_i)\} = \{(T_i \wedge C_i, \Delta_i)\}, i = 1, \ldots, n$. Further assume that $T_1, \ldots, T_n$ and $C_1, \ldots, C_n$ are i.i.d. with distribution functions F (with f as its density) and G (with g as its density), respectively. Under the random censorship assumption (i.e. that C_i is independent of T_i), the distribution of $(\tilde{T})$ satisfies

$$\Pr(\tilde{T} > t) = \{1 - F(t)\}\{1 - G(t)\}.$$

Let $0 < t_{(1)} < \ldots < t_{(n)}$. The nonparametric estimate of F is defined (see Section 6.2.2.1) for $t \leqslant \tilde{T}_{(n)}$ by

$$\hat{F}(t) = 1 - \prod_{i:\tilde{T}_{(i)} \leqslant t} \left(\frac{n - i}{n - i + 1}\right)^{\Delta_{(i)}}.$$

As discussed in Section 6.2.2.1, the median can be estimated from $\hat{F}$ by choosing $\hat{Q}_{0.5} = \inf\{t : \hat{F}(t) \geqslant 0.5\}$. Traditionally, the asymptotic variance of $\hat{Q}_{0.5}$ is given by

$$\text{Var}(\hat{Q}_{0.5}) = \frac{1}{n\{f(Q_{0.5})\}^2} \left[\{1 - F(Q_{0.5})\}^2 \int_0^{Q_{0.5}} \frac{dF}{(1-F)^2(1-G)} \right].$$

As discussed in Reid (1981), a more direct approach to discussing the properties of the sample median is to estimate its distribution. Assume that $n = 2k + 1$ is odd and all the observations are uncensored. Then we can write

$$\begin{aligned} \Pr(\hat{Q}_{0.5} = T_{(i)}) &= \Pr\{\hat{F}(T_{(i-1)}) < 1/2, \ \hat{F}(T_{(i)}) \geqslant 1/2\} \\ &= \Pr\{\hat{F}(T_{(i-1)}) < 1/2\} - \Pr\{\hat{F}(T_{(i)}) < 1/2\}. \end{aligned}$$

There are different ways to construct confidence intervals for the median. One is to numerically estimate the density function of the survival time based on the Kaplan–Meier estimate upon which the variance of the sample median can be computed. Another approach is to make use of the relationship

$$\Pr\left\{ \hat{F}^{-1}\left(0.5 - z_{\alpha/2}\hat{\sigma}_{0.5}\right) \leqslant Q_{0.5} \leqslant \hat{F}^{-1}\left(0.5 + z_{\alpha/2}\hat{\sigma}_{0.5}\right) \right\} \approx 1 - \alpha,$$

where z_α and $\hat{\sigma}_{0.5}$ denote the $1 - \alpha$ quantile of the standard normal distribution and an estimate of the asymptotic standard deviation of $\hat{F}(Q_{0.5})$, respectively, for $\alpha \in (0, 1)$. The third possible approach is to estimate the variance via bootstrap.

6.3.2 Median and quantile regression

Quantile regression is a valuable alternative to the celebrated Cox proportional hazards model. This approach, however, was a challenge for censored regression models until Portnoy (2003), who proposed a recursively reweighted estimator of the regression quantile process. This estimator can be viewed as a generalization of the Kaplan–Meier estimator.

Recall that quantile regression methods focus on analysis of the conditional quantile function. With a response of interest Y, given the covariate $Z = z$, one can define, for $\tau \in (0, 1)$, the τth conditional quantile of Y as

$$Q_Y(\tau \mid z) = \inf\{y : \Pr(Y \leqslant y \mid Z = z) = \tau\}. \tag{6.19}$$

Consider a general linear response $Y = Z^\top \beta + \epsilon$. To estimate these conditional quantiles based on a sample of n observations, Koenker and Bassett (1978) insightfully make use of the check function

$$\rho_\tau(u) = u\{\tau - I(u < 0)\},$$

to define the regression quantile estimator $\hat{\beta}$, which is the minimizer of

$$r(b) = \sum_{i=1}^n \rho_\tau\{Y_i - Z_i^\top \beta(\tau)\}. \tag{6.20}$$

The above optimization can be carried out easily based on Koenker and d'Orey (1987). With censoring, however, not all the response variables Y_i can be observed.

Recall from Section 6.2.3.2, that if we choose specifically $Z = (1, \tilde{Z}^\top)^\top$ in (6.10) and $\tau \in (0, 1)$, we can write a quantile regression model as

$$Q_{\log T}(\tau \mid Z) = Z^\top \beta(\tau), \quad \tau \in (0, 1), \tag{6.21}$$

where the regression coefficients $\beta(\tau)$ capture the covariate effects on the τth quantile of $\log(T)$. Note that when $\beta(\tau) = [Q_\epsilon(\tau), \beta^\top]^\top$, (6.21) reduces to the AFT model as described in (6.10). If we impose an assumption that $Q_\epsilon(\tau) = 0$, it will be consistent with Koenker and Bassett's (1978) formulation.

Linear model with homogeneous error

When the underlying model is governed by the AFT model with homogeneous error, one major challenge of extending the (6.20) framework to handle censored data is that subjects' failure times $Y = \log(T)$ can only be observed if they are not censored. Similar to the treatments for the AFT model presented in Buckley and James (1979) and Lai and Ying (1994), we can handle these unobserved event times via the (modified) Buckley–James projection. Observe that

$$\sum_{i=1}^{n} Z_i[I\{\log(T_i) - Z_i^\top \beta(\tau) \leqslant 0\} - \tau] = 0$$

is an estimating equation for $\beta(\tau)$, to handle censored $\log(T)$s, we can obtain their expected failure times given that they are lower-bounded by their corresponding censored times. Specifically, we can write

$$\sum_{i=1}^{n} Z_i \left[\Delta_i\{I(\log \tilde{T}_i \leqslant Z_i^\top \beta(\tau)) - \tau\} + (1 - \Delta_i)\frac{\int_{\tilde{T}_i}^{\infty}\{I(\log u \leqslant Z_i^\top \beta(\tau)) - \tau\}dF(u)}{1 - F(\tilde{T}_i)} \right],$$

$$\tag{6.22}$$

where F denotes the distribution function of T. By replacing F by its empirical estimate, say $\hat{F}$, we obtain the Buckley–James type estimator for the quantile regression coefficients. We have also conducted simulations which compare its performance with another model for quantile regression; more can be found in Section 6.3.3.

Linear model with independent censoring mechanism

The homogeneous error assumption can sometimes be rather restrictive in many applications. An alternative treatment that does not require the i.i.d. error is to correct the bias due to censoring via inverse probability weighting. This class of approaches is, however, founded upon the independent censoring assumption. For instance, Ying et al. (1995) proposed a quantile estimator for β under the model restriction of $\Pr(\epsilon \leqslant 0 \mid Z) = \pi \in (0, 1)$ and independent censoring. See Chapter 7 of this handbook for a more detailed discussion.

Redistribution of mass to the right approach

Another line of research includes that of Portnoy (2003) who mimics the "redistribution of mass to the right" idea adopted in the Kaplan–Meier estimator. This defines weights $w_i(\tau)$ that produce pseudo-observations for censored observations. A comprehensive summary and discussion on this development can be found in Chapter 7 of this handbook.

Quantile regression via counting process

The methodologies presented so far require either an i.i.d. error assumption or an independent censoring assumption. Besides, these methods only focus on finding a particular quantile regression parameter $\beta(\tau)$ – to estimate the covariate effects on different quantiles, separate regressions have to be run. For hazard-based regression models, such as the Cox model and Aalen's additive hazards regression model, martingale estimating equations can easily be constructed that only impose the minimum noninformative assumption on the censoring. A major breakthrough is due to Peng and Huang (2008), who recognize that the quantiles are on the survival time scale and can be converted into the hazard function so that natural martingale estimating equations emerge. Because the quantiles convert τ to the survival time scale, their approach naturally allows/requires simultaneous estimation of the regression quantiles as processes of τ whose asymptotic properties can be obtained more easily with little additional assumption either on the censoring or on homogeneity. More in-depth discussion and illustration of this novel methodology can be found in Chapter 7 of this handbook.

6.3.3   Discussion and miscellanea

In the previous section we introduced different approaches for handling censored quantile regression problems under various settings. Most of the contributions were developed upon the Koenker and Bassett (1978) estimating equation with individual handling for the unobservable failure times due to censoring. In addition to the inverse probability weighting and Buckley and James (1979) treatments, Peng and Huang (2008) and Huang (2010) leverage the counting process-based machinery developed for survival data to establish the corresponding inference. It has been documented (e.g. Miller and Halpern, 1982) that the Buckley–James approach provides better estimates in terms of efficiency and robustness for censored linear regression with homogeneous error terms. It is of interest to see how the Peng–Huang estimator compares with the Buckley–James approach.

We conducted numerical experiments to compare their performance. Following the setting adopted in Peng and Huang (2008), we considered the model

$$\log T = b_1 Z_1 + b_2 Z_2 + \epsilon,$$

where the error follows the extreme value distribution. $\mathbf{Z} = (1, Z_1, Z_2)^\top$ is generated from uniform$(0,1)$ and Bernoulli(0.5) random variables for Z_1 and Z_2, respectively. The censoring distribution is given by $C \sim \text{uniform}(0, c_u)I(Z_2 = 0) + \text{uniform}(0.1, c_u)I(Z_2 \neq 0)$ with various values of c_u so as to achieve different target censoring rates. The regression coefficients are set to $\mathbf{b}_0(\tau) = \{Q_\epsilon(\tau), 0.5, -0.5\}^\top$. Table 6.2 summarizes the biases and sample standard errors of the estimators. Both approaches lead to virtually unbiased estimates. The Buckley–James type estimator, however, appears to be more efficient for estimating one set of model parameters for a particular quantile value. This may be due to the fact that the Buckley–James type estimator requires a homogeneous error assumption, which is considered in this numerical exercise. In contrast, the Peng and Huang (2008) approach is more versatile in the sense that it performs well also for models with heterogeneous errors. Indeed, unlike other popular survival models, there has been little discussion on the semiparametric efficiency of quantile regression. Research is required in order to plug this particular gap.

Another semiparametric survival model that has a connection with Peng and Huang (2008)'s quantile regression model is Aalen additive hazard risk model presented in 6.2.3.3. As (6.15) shows, the hazard function considered in this model can capture the local effects of the covariates from the time perspective. Quantile regression models, meanwhile, quantifies the relationship between the covariates and the quantile of the survival time. Both models enjoy a similar level of flexibility although they impose an linearity assumption on different distribution features. Since there is a one-to-one relationship between the hazard, hence the survival, function of the event time and its quantile, it is expected that some goodness-of-fit procedures that can determine which model assumption is more suitable are of great academic and practical interest.

The relationship between the additive hazard and additive cumulative hazard model

TABLE 6.2

Comparison between Buckley–James type estimator (6.22) and Peng and Huang (2008) approaches

τ	n		Buckley–James type estimator		Peng and Huang (2008)	
			Bias	Ave Var	Bias	Ave Var
0.5	100	$\hat{\beta}^{(1)}$	−0.0043	0.1537	−0.0077	0.3102
		$\hat{\beta}^{(2)}$	0.0166	0.1826	0.0107	0.5370
		$\hat{\beta}^{(3)}$	−0.0077	0.2005	0.0025	0.2852
	200	$\hat{\beta}^{(1)}$	−0.0089	0.1377	0.0020	0.2441
		$\hat{\beta}^{(2)}$	0.0101	0.1391	0.0078	0.3809
		$\hat{\beta}^{(3)}$	0.0143	0.1898	−0.0046	0.2139

poses an intriguing question as to whether a quantile density model could serve as a counterpart to the corresponding quantile regression model based upon the link between the hazard and cumulative hazard functions. The concept of quantile density was first discussed in Tukey (1965), who called it the sparsity function, followed by Parzen (1979). In particular, if $F_T(\cdot)$, the distribution function of the failure time T, is continuous, we can write $F_T\{Q_T(\tau)\} = \tau$ for all $\tau \in (0,1)$. Taking the derivative on both sides, we have

$$f_T\{Q_T(\tau)\}Q'_T(\tau) = 1 \quad \Longleftrightarrow \quad Q'_T(\tau) = [f_T\{Q_T(\tau)\}]^{-1},$$

where $Q'_T(\tau)$ denotes the quantile density. There has been some work on modeling Q'_T, including Jones (1992) and Soni et al. (2012); see also Parzen (2004) for a comprehensive and concise summary on this topic. However, to the best of our knowledge, the extension of this model so that covariate effects can be incorporated remains an open question.

In many biomedical and other follow-up studies, it is common for statisticians to have to relate survival times with time-dependent covariates including, for instance, regular physiological measurements along the treatment horizon. It is of both academic and practical interest to see how the existing methods, especially that of Peng and Huang (2008), can be extended to handle time-dependent covariates. As presented in Section 6.2.3.3, the quantile regression model can be viewed as a dual with respect to typical survival models that model the hazard function of the survival time.

More recently, Gorfine et al. (2016) proposed a quantile regression model with time-varying covariates. It can be viewed as an extension of Robins and Tsiatis (1992) and Lin and Ying (1995) where a class of accelerated failure time models with time-dependent covariates in the presence of right censoring was studied. In particular, Gorfine et al. (2016) specify the baseline failure time T^* as

$$T^* = \int_0^T \exp\{\gamma_0^\top \tilde{Z}(t)\}\, dt,$$

where γ_0 denotes the model parameters that determine the increasing/decreasing hazard of the failure time. Accordingly, let $Z(t) = (1, \tilde{Z}(t)^\top)^\top$, $\bar{Z}(t) = \{Z(s) : 0 \leqslant s \leqslant t\}$ and $\bar{Z} = \{Z(s) : s \geqslant 0\}$. Then there exists a τth quantile regression coefficient $\beta_0(\tau)$ that satisfies

$$\Pr\left\{\int_0^T \exp\{\beta(\tau)^\top Z(t)\}\, dt \leqslant 1 \middle| \bar{Z}\right\} = \tau.$$

The regression parameters $\beta(\tau)$ can be estimated via inverse probability weighting to adjust for censoring by solving

$$\frac{1}{n}\sum_{i=1}^{n}\frac{\Delta_i Z_i(0)}{\hat{G}(\tilde{T}_i)}\left(I\left[\int_0^{\tilde{T}_i}\exp\{\beta(\tau)^\top Z_i(t)\,dt>1\}\right]-\tau\right)=0.$$

The above extension further generalizes the scope of quantile regression on survival data because in many practical cases, the covariates are time-varying. It is expected to be a valuable extension if this framework can be extended so that a less restrictive censoring mechanism can be imposed. One may borrow the ideas from various time-varying covariate survival models (see Bagdonavičius and Nikulin, 2001) to construct the corresponding counterpart for the quantile model.

Finally, due to technology advances, collection of high-dimensional covariate information has become technically and economically feasible. Topics in high-dimensional statistics and variable selection for semiparametric models have been extensively studied in recent years. An incomplete list of literature includes Lu and Zhang (2007), Li and Gu (2012), Li et al. (2014) and Xu et al. (2010) for transformation/Cox/AFT models. There have been valuable contributions to the quantile regression in high-dimensional settings. Wu and Liu (2009) showed the oracle properties of the penalised quantile regressions using smoothly clipped absolute deviation (SCAD) and adaptive lasso; Wang et al. (2013) developed an adaptive lasso-based variable selection method based on Portnoy's (2003) construction; more recently, Zheng et al. (2015), employed adaptive L_1 penalties for the ultra high-dimensional setting. However, solutions to ultra-high-dimensional variable selection problems for quantile regressions on censored data are still missing.

The quantile regression model has been popular since its introduction. Its extension to survival models brings in a whole new set of interesting problems. In addition to the references we have discussed, readers may also refer to the following (incomplete) list of literature that can offer a more comprehensive view of this area of active research: Yin et al. (2008), Wang and Wang (2009, 2014), Lin et al. (2012), Sun et al. (2012), Leng and Tong (2013, 2014), and He et al. (2013), among others.

Bibliography

O. O. Aalen. Nonparametric inference for a family of counting processes. *Annals of Statistics*, 6:701–26, 1978.

O. O. Aalen. A model for nonparametric regression analysis of counting processes. In W. Klonecki, A. Kozek, and J. Rosiński, editors, *Mathematical Statistics and Probability Theory: Proceedings, Sixth International Conference, Wisła (Poland), 1978*, pages 1–25. Springer, New York, 1980.

D. G. Altman. *Practical Statistics for Medical Research*. Chapman & Hall, London, 1991.

P. K. Andersen and R. D. Gill. Cox's regression model for counting processes: a large sample study. *Annals of Statistics*, 10:1100–20, 1982.

P. K. Andersen, Ø. Borgan, R. D. Gill, and N. Keiding. *Statistical Models Based on Counting Processes*. Springer, New York, 1993.

V. Bagdonavičius and M. Nikulin. *Accelerated Life Models: Modeling and Statistical Analysis*. CRC Press, Boca Ration, FL, 2001.

J. Buckley and I. James. Linear regression with censored data. *Biometrika*, 66:429–36, 1979.

D. R. Cox. Regression models and life tables. *Journal of the Royal Statistical Society, Series B*, 34:187–220, 1972.

D. R. Cox. Partial likelihood. *Biometrika*, 62:269–76, 1975.

D. R. Cox and D. Oakes. *Analysis of Survival Data*. Chapman & Hall, London, 1984.

B. Efron. The two sample problem with censored data. In L. LeCam and J. Neyman, editors, *Proceedings of the Fifth Berkeley Symposium on Mathematical Statistics and Probability*, volume IV, pages 831–53. University of California Press, Berkeley, 1967.

T. R. Fleming and D. P. Harrington. *Counting Processes and Survival Analysis*. Wiley, New York, 1991.

E. A. Gehan. A generalized Wilcoxon test for comparing arbitrarily singly-censored samples. *Biometrika*, 52:203–23, 1965a.

E. A. Gehan. A generalized two-sample Wilcoxon test for doubly censored data. *Biometrika*, 52:650–52, 1965b.

M. Gorfine, Y. Goldberg, and Y. Ritov. A quantile regression model for failure-time data with time-dependent covariates. *Biostatistics*, 18:132–46, 2016.

X. He, L. Wang, and H. G. Hong. Quantile-adaptive model-free variable screening for high-dimensional heterogeneous data. *Annals of Statistics*, 41:342–69, 2013.

Y. Huang. Quantile calculus and cnesored regression. *Annals of Statistics*, 38:1607–37, 2010.

F. W. Huffer and I. W. McKeague. Weighted least squares estimation for Aalen's additive risk model. *Journal of the American Statistical Association*, 86:114–29, 1991.

Z. Jin, D. Y. Lin, L. J. Wei, and Z. Ying. Rank-based inference for the accelerated failure time model. *Biometrika*, 90(2):341–353, 2003.

M. C. Jones. Estimating density, quantiles, quantile densities and density quantiles. *Annals of the Institute of Statistical Mathematics*, 44:721–7, 1992.

J. D. Kalbfleisch and R. L. Prentice. *The Statistical Analysis of Failure Time Data*. Wiley, Hoboken, NJ, 2nd edition, 2003.

E. L. Kaplan and P. Meier. Nonparametric estimation from incomplete observations. *Journal of the American Statistical Association*, 53(53):457–81, 1958.

R. Koenker and G. Bassett. Regression quantiles. *Econometrica*, 46:33–50, 1978.

R. Koenker and V. d'Orey. Computing regression quantiles. *Applied Statistics*, 36:383–393, 1987.

T. L. Lai and Z. Ying. A missing information principle and M-estimators in regression analysis with censored and truncated data. *Annals of Statistics*, 22:1222–55, 1994.

J. F. Lawless. *Statistical Models and Methods for Lifetime Data*. Wiley, Hoboken, NJ, 2003.

C. Leng and X. Tong. A quantile regression estimator for censored data. *Bernoulli*, 19:344–61, 2013.

C. Leng and X. W. Tong. Censored quantile regression via Box-Cox transformation under conditional independence. *Statisica Sinica*, 24:221–49, 2014.

J. Li and M. Gu. Adaptive LASSO for general transformation models with right censored data. *Computational Statistics and Data Analysis*, 56:2583–97, 2012.

J. Li, M. Gu, R. Zhang, and H. Lian. Variable selection for general transformation models with ranking data. *Statistics*, 48:81–100, 2014.

D. Y. Lin and Z. Ying. Semiparametric analysis of the additive risk model. *Biometrika*, 81: 61–71, 1994.

D.Y. Lin and Z. Ying. Semiparametric inference for the accelerated life model with time-dependent covariates. *Journal of Statistical Planning and Inference*, 44:47–63, 1995.

G. Lin, X. He, and S. Portnoy. Quantile regression with doubly censored data. *Computational Statistics and Data Analysis*, 56:797–812, 2012.

W. Lu and H. H. Zhang. Variable selection for poportional odds model. *Statistics in Medicine*, 26:3771–81, 2007.

I. W. McKeague. Asymptotic theory for weight least squares estimators in Aalen's additive risk model. In N. U. Prabhu, editor, *Statistical Inference from Stochastic Processes*, volume 80 of *Contemporary Mathematics*, pages 139–52. American Mathematical Society, Providence, RI, 1988.

I. W. McKeague and P. Sasieni. A partly parametric additive risk model. *Biometrika*, 81: 501–14, 1994.

R. G. Miller. Least squares regression with censored data. *Biometrika*, 63:449–64, 1976.

R. G. Miller. *Survival Analysis*. Wiley, New York, 1981.

R. G. Miller and J. Halpern. Regression with censored data. *Biometrika*, 69:521–31, 1982.

W. Nelson. Hazard plotting for incomplete failure data. *Journal of Quality Technology*, 1: 27–52, 1969.

W. Nelson. Theory and applications of hazard plotting for censored failure data. *Technometrics*, 14:945–65, 1972.

E. Parzen. Nonparameteric statisical data modeling (with discussion). *Journal of the American Statistical Association*, 74:105–31, 1979.

E. Parzen. Quantile probability and statistical data modeling. *Statistical Science*, 19:652–62, 2004.

L. Peng and Y. Huang. Survival analysis with quantile regression models. *Journal of the American Statistical Association*, 103:637–49, 2008.

S. Portnoy. Censored regression quantiles. *Journal of the American Statistical Association*, 98:1001–12, 2003.

N. Reid. Estimating the median survival time. *Biometrika*, 68:601–8, 1981.

J. Robins and A. A. Tsiatis. Semiparametric estimation of an accelerated failure time model with time-dependent covariates. *Biometrika*, 79:311–9, 1992.

P. Soni, I. Dewan, and K. Jain. Nonparametric estimation of quantile density function. *Computational Statistics and Data Analysis*, 56:3876–86, 2012.

Y. Sun, H. Wang, and P. B. Gilbert. Quantile regression for competing risks data with missing cause of failure. *Statisica Sinica*, 22:703–28, 2012.

A. A. Tsiatis. Estimating regression parameters using linear rank tests for censored data. *Annals of Statistics*, 18:354–72, 1990.

J. W. Tukey. Which part of the sample contains the information? *Proceedings of the National Academy of Sciences*, 53:127–34, 1965.

H. Wang and L. Wang. Locally weighted censored quantile regression. *Journal of the American Statistical Association*, 104:1117–28, 2009.

H. Wang and L. Wang. Quantile regression analysis of length-biased survival data. *Stat*, 3: 31–47, 2014.

H. Wang, J. Zhou, and Y. Li. Variable selection for censored quantile regression. *Statisica Sinica*, 23:145–67, 2013.

Y. Wu and Y. Liu. Variable selection in quantile regression. *Statisica Sinica*, 19:801–17, 2009.

J. Xu, C. Leng, and Z. Ying. Rank-based variable selection with censored data. *Statistics and Computing*, 20:165–76, 2010.

G. Yin, D. Zeng, and H. Li. Power-transformed linear quantile regression with censored data. *Journal of the American Statistical Association*, 103:1214–24, 2008.

Z. Ying. A large sample study of rank estimation for censored regression data. *Annals of Statistics*, 21:76–99, 1993.

Z. Ying, S. H. Jung, and L. J. Wei. Survival analysis with median regression models. *Journal of the American Statistical Association*, 90:178–84, 1995.

D. Zeng and D. Y. Lin. Efficient estimation for the accelerated failure time model. *Journal of the American Statistical Association*, 102:1387–96, 2007.

D. Zeng and D. Y. Lin. Efficient resampling methods for nonsmooth estimating functions. *Biostatistics*, 9:355–63, 2008.

Q. Zheng, L. Peng, and X. He. Globally adaptive quantile regression with ultra-high dimensional data. *Annals of Statistics*, 43:2225–2258, 2015.

7

Quantile Regression for Survival Analysis

Limin Peng

Department of Biostatistics and Bioinformatics, Emory University, USA

CONTENTS

7.1 Introduction

Quantile regression has a natural appeal in model flexibility and interpretability. It has received more and more attention in survival analysis because event times themselves are often of scientific interest, and quantiles are more flexible and robust quantitative tools for characterizing event times than mean-based devices. In addition, quantile regression allows the investigation of particular local features of the conditional distribution of an event time of interest. The usefulness of quantile regression for survival analysis has been demonstrated by many applications reported in the literature. For example, Koenker and Geling (2001) provided a detailed quantile regression analysis of a medfly longevity data set, which illustrates how to use quantile regression to assess and interpret covariate effects on different segments of the event time distribution. Such a capacity constitutes another major advantage of quantile regression over conventional survival models, such as the proportional hazards model and the accelerated failure time model, which implicitly exert pure location shift effects on survival times or monotone transformations thereof.

Let T denote the failure time of interest and $\tilde{\mathbf{Z}} \equiv (Z_1, \ldots, Z_p)^\mathsf{T}$ denote a $p \times 1$ vector of covariates. Define $\mathbf{Z} = (1, \tilde{\mathbf{Z}}^\mathsf{T})^\mathsf{T}$ and $Y = \log(T)$. For a random variable V, let $Q_Y(\tau|\mathbf{Z}) \equiv \inf\{t : \Pr(Y \leqslant t|\mathbf{Z}) \geqslant \tau\}$ denote the τth conditional quantiles of V given $\mathbf{Z}$, where $\tau \in (0,1)$. For the failure time T, a common quantile regression modeling strategy is to assume $Q_Y(\tau|\mathbf{Z})$ is linearly related to $\mathbf{Z}$. The resulting model takes the form

$$Q_Y(\tau|\mathbf{Z}) = \mathbf{Z}^\mathsf{T}\boldsymbol{\beta}_0(\tau), \quad \tau \in [\tau_L, \tau_U], \tag{7.1}$$

where $0 < \tau_L \leqslant \tau_U < 1$ and $\boldsymbol{\beta}_0(\tau)$ is a $(p+1) \times 1$ vector of unknown regression coefficients. A nonintercept coefficient in $\boldsymbol{\beta}_0(\tau)$ represents the change in the τth conditional quantile of $\log(T)$ given a one-unit change in the corresponding covariate.

The standard semiparametric AFT model,

$$\log(T) = \tilde{\mathbf{Z}}^\mathsf{T}\mathbf{b} + \varepsilon, \quad \varepsilon \perp\!\!\!\perp \tilde{\mathbf{Z}}, \tag{7.2}$$

is a special case of model (7.1) with $\boldsymbol{\beta}_0(\tau) = (Q_\varepsilon(\tau), \mathbf{b}^\mathsf{T})^\mathsf{T}$. Here $\perp\!\!\!\perp$ stands for statistical independence, and $Q_\varepsilon(\tau)$ denotes the τth quantile of ε. In contrast to model (7.1), the AFT model (7.2) requires the effects of covariates on $Q_Y(\tau|\mathbf{Z})$ to be constant for all $\tau \in [\tau_L, \tau_U]$ (i.e. location shift effects). The quantile regression model (7.1) generalizes model (7.2) in the sense that coefficients are formulated as functions of τ and hence covariate effects are allowed to vary across different quantile levels.

When $\tau_L = \tau_U$, model (7.1) is referred to as a locally linear quantile regression model because it only asserts "local" linearity between the conditional quantile of $\log(T)$ and $\mathbf{Z}$ at a single quantile level. When $\tau_L < \tau_U$, model (7.1) imposes a "global" quantile linearity for $\log(T)$ throughout a quantile index interval $[\tau_L, \tau_U]$, and thus is referred to as a globally linear quantile regression model. A globally linear quantile regression model implies more stringent data constraints than a locally linear quantile regression model, while offering the capacity to investigate more flexible patterns of covariate effects than those permitted by the classical AFT model.

In this chapter we first review quantile regression methods with randomly censored data in Section 7.2, and then discuss several more complex survival settings in Section 7.3. A discussion of an illustrative application in Section 7.4 concludes the chapter.

7.2 Quantile regression for randomly censored data

Let C denote time to censoring. Define $\tilde{T} = T \wedge C$ and $\Delta = I(T \leqslant C)$. The observed data under right censorship consists of n independent and identically distributed (i.i.d.) replicates of $(\tilde{T}, \Delta, \mathbf{Z})$, denoted by $(\tilde{T}_i, \Delta_i, \mathbf{Z}_i)$, $i = 1, \ldots, n$. In addition, we define $\tilde{Y} = \log(\tilde{T})$, $\tilde{Y}_i = \log(\tilde{T}_i)$, $U = \log(C)$, and $U_i = \log(C_i)$.

7.2.1 Random right censoring with C always known

Powell (1984, 1986) tackled quantile regression in a type I censoring case, where the censoring time C is fixed and prespecified. Specifically, utilizing the fact that $Q_{\tilde{Y}}(\tau|\mathbf{Z}) = \{\mathbf{Z}^\mathsf{T}\boldsymbol{\beta}_0(\tau)\} \wedge U$, Powell (1984, 1986) proposed estimating $\boldsymbol{\beta}_0(\tau)$ in model (7.1) by the minimizer of

$$r(\mathbf{b}, \tau) = \sum_{i=1}^{n} \rho_\tau\{\tilde{Y}_i - (\mathbf{Z}_i^\mathsf{T}\mathbf{b}) \wedge U_i\}$$

with respect to $\mathbf{b}$, where $\rho_\tau(x) = x\{\tau - I(x < 0)\}$. This estimation method is directly applicable to a more general case where C is always known but not necessarily fixed, and is independent of T given $\mathbf{Z}$. In the absence of right censoring (i.e. $C_i = \infty$), $r(\mathbf{b}, \tau)$ reduces to the check function of Koenker and Bassett (1978), the objective function for quantile regression with complete uncensored data.

Note that $r(\mathbf{b}, \tau)$ is not convex in $\mathbf{b}$ and thus may have multiple local minima. Further efforts have been made to improve the numerical performance of this approach by several authors, among them Fitzenberger (1997), Buchinsky and Hahn (1998), and Chernozhukov and Hong (2001). An implementation of Powell's method is available in the *crq* function in the contributed R package quantreg (Koenker, 2012).

7.2.2 Covariate-independent random right censoring

Under the assumption that T and C are independent and C is independent of $\mathbf{Z}$ (i.e. covariate-independent random censoring), a natural estimating equation for $\boldsymbol{\beta}_0(\tau)$, derived from Ying et al. (1995), is given by

$$n^{-1/2} \sum_{i=1}^{n} \mathbf{Z}_i \left[\frac{I\{\tilde{Y}_i - \mathbf{Z}^\mathsf{T}\boldsymbol{\beta}(\tau) > 0\}}{\widehat{G}\{\mathbf{Z}^\mathsf{T}\boldsymbol{\beta}(\tau)\}} - (1 - \tau) \right] = 0, \qquad (7.3)$$

where $\widehat{G}(\cdot)$ is the Kaplan–Meier estimate for $G(\cdot)$, the survival function of $\tilde{C}$.

Because the estimating function in (7.3) is not continuous in $\boldsymbol{\beta}(\tau)$, equation (7.3) usually does not have an exact zero-crossing. The estimator of $\boldsymbol{\beta}_0(\tau)$ is often alternatively obtained as a minimizer of the L_2 norm of the estimating function. Such an objective function is discontinuous and may have multiple minima. As suggested by Ying et al. (1995), grid search may be used for cases with low-dimensional $\mathbf{Z}$ and the simulated annealing algorithm (Lin and Geyer, 1992) may be used for high-dimensional cases.

For the same right censoring scenario, an alternative estimating equation for $\boldsymbol{\beta}_0(\tau)$ was exploited by Zhou (2006), which employed the inverse probability of censoring weighting (IPCW) technique (Robins and Rotnitzky, 1992). Specifically, an estimating equation for $\boldsymbol{\beta}_0(\tau)$ is given by

$$n^{-1/2} \sum_{i=1}^{n} \mathbf{Z}_i \left[\frac{I\{\tilde{Y}_i \leq \mathbf{Z}_i^\mathsf{T}\boldsymbol{\beta}(\tau), \Delta_i = 1\}}{\widehat{G}(\tilde{Y}_i)} - \tau \right] = 0. \qquad (7.4)$$

Since the estimating function in (7.4) is monotone (Fygenson and Ritov, 1994), though not continuous, the solution to equation (7.4) can be reformulated as the minimizer of the following L_1-type convex function with regard to $\mathbf{b}$:

$$\sum_{i=1}^{n} \left\{ I(\Delta_i = 1) \left| \frac{\tilde{Y}_i}{\widehat{G}(\tilde{Y}_i)} - \mathbf{b}^\mathsf{T} \frac{\mathbf{Z}_i}{\widehat{G}(\tilde{Y}_i)} \right| \right\} + \left| M - \mathbf{b}^\mathsf{T} \sum_{l=1}^{n} \left(-\frac{\mathbf{Z}_l I(\Delta_l = 1)}{\widehat{G}(\tilde{V}_l)} + 2\mathbf{Z}_l \tau \right) \right|,$$

where M is a sufficiently large positive number selected to bound

$$\mathbf{b}^\mathsf{T} \sum_{l=1}^{n} \left(-\frac{Z_l I(\Delta_l = 1)}{\widehat{G}(\tilde{Y}_l)} + 2Z_l \tau \right)$$

from the above for all $\mathbf{b}$ in the compact parameter space for $\boldsymbol{\beta}_0(\tau)$. This minimization problem can be readily solved by the *rq()* function in the contributed R package quantreg (Koenker, 2012).

7.2.3   Standard random right censoring

Standard random right censoring refers to a censoring scenario in which C is independent of T given $\mathbf{Z}$. It is a less restrictive censoring mechanism than those considered in Sections 7.2.1 and 7.2.2.

7.2.3.1   Approaches based on the principle of self-consistency

We first focus on the approaches that are oriented to a globally linear quantile regression model,

$$Q_Y(\tau|\mathbf{Z}) = \mathbf{Z}^\mathsf{T}\boldsymbol{\beta}_0(\tau), \quad \tau \in [0, \tau_U]. \tag{7.5}$$

This model is a special case of model (7.1) with $\tau_L = 0$.

Portnoy (2003) was the first to propose an estimation strategy for model (7.5) under the standard random right censoring assumption. The critical idea behind his approach is based on the interpretation of the Kaplan–Meier estimator as redistributing censoring probability to the right, as suggested by (Efron, 1967). The iterative self-consistent algorithm (Portnoy, 2003) was simplified into a grid-based sequential estimation procedure (Neocleous et al., 2006), which is implemented by the *crq* function in the contributed R package quantreg (Koenker, 2012). Further asymptotic analysis was developed in Portnoy and Lin (2010).

The grid-based procedure presented in Neocleous et al. (2006) defines a grid of τ-values, $\mathcal{G}$, as $0 = \tau_0 < \tau_1 < \ldots < \tau_M = \tau_U$. We will adopt the grid $\mathcal{G}$ throughout Section 7.2.3. Assuming no censoring occurs below the τ_1th conditional quantile of T, one can obtain an estimate for $\boldsymbol{\beta}_0(\tau_1)$ from applying uncensored quantile regression. Next, one can estimate $\boldsymbol{\beta}_0(\tau_{k+1})$ sequentially for $k = 1, 2, \ldots, M$ by minimizing

$$\sum_{i \in K^c} \rho_\tau(\tilde{Y}_i - \mathbf{Z}_i^\mathsf{T}\mathbf{b}) + \sum_{i \in K} \left\{ w_{k+1,i}\rho_\tau(\tilde{Y}_i - \mathbf{Z}_i^\mathsf{T}\mathbf{b}) + (1 - w_{k+1,i})\rho_\tau(Y^* - \mathbf{Z}_i^\mathsf{T}\mathbf{b}) \right\}, \tag{7.6}$$

where Y^* is an extremely large value and K denotes the set of indices of censored observations that have been previously crossed (i.e. $C_i \leq \mathbf{Z}_i^\mathsf{T}\hat{\boldsymbol{\beta}}(\tau)$). The weight $w_{k+1,i}$ is of the form $(\tau_{k+1} - \tau_l)/(1 - \tau_l)$, which approximates the conditional probability, $\Pr(C_i < T_i < \exp\{\mathbf{Z}_i\boldsymbol{\beta}_0(\tau_{k+1})\}|C_i < T_i, \mathbf{Z}_i)$ based on the estimates for $\boldsymbol{\beta}_0(\tau_1), \ldots, \boldsymbol{\beta}_0(\tau_k)$.

More recently, Peng (2012) proposed alternative formulations of the self-consistent approach based on stochastic integral equations. Define $N_i(t) = I(\tilde{Y}_i \leq t, \Delta_i = 1)$, $R_i(t) = I(\tilde{Y}_i \leq t, \Delta_i = 0)$, and $F_Y(t) \equiv \Pr(Y \leq t)$. First, consider Efron's (1967) self-consistent estimating equation for $F_Y(t)$ in the one-sample case:

$$F_Y(t) = n^{-1}\sum_{i=1}^n \left\{ N_i(t) + R_i(t)\frac{F_Y(t) - F_Y(\tilde{Y}_i)}{1 - F_Y(\tilde{Y}_i)} \right\}. \tag{7.7}$$

Expressing the right-hand side of (7.7) by a stochastic integral and applying (stochastic) integration by parts, one can rewrite equation (7.7) as

$$F_Y(t) = n^{-1}\sum_{i=1}^n \left[N_i(t) + R_i(t)\{1 - F_Y(t)\}\int_0^t \frac{R_i(u)}{\{1 - F_Y(u)\}^2}dF_Y(u) \right]. \tag{7.8}$$

With t replaced by $\mathbf{Z}_i^\mathsf{T}\boldsymbol{\beta}(\tau)$, equation (7.8) evolves into an estimating equation for $\boldsymbol{\beta}_0(\tau)$:

$$n^{1/2}\mathbf{S}_n^{(\mathrm{SC})}(\boldsymbol{\beta}, \tau) = 0, \tag{7.9}$$

where

$$\mathbf{S}_n^{(\mathrm{SC})}(\boldsymbol{\beta}, \tau) = n^{-1}\sum_{i=1}^n \mathbf{Z}_i \left[N_i\{\mathbf{Z}_i^\mathsf{T}\boldsymbol{\beta}(\tau)\} + R_i\{\mathbf{Z}_i^\mathsf{T}\boldsymbol{\beta}(\tau)\}(1 - \tau)\int_0^\tau \frac{R_i\{\mathbf{Z}_i^\mathsf{T}\boldsymbol{\beta}(u)\}}{(1 - u)^2}du - \tau \right].$$

Peng (2012) further justifies several asymptotically equivalent variants of estimating equation (7.9), one of which takes the form of $n^{1/2}\mathbf{S}_n^{(\mathrm{MSC})}(\boldsymbol{\beta}, \tau) = 0$, where

$$\mathbf{S}_n^{(\mathrm{MSC})}(\boldsymbol{\beta}, \tau) = n^{-1} \sum_{i=1}^{n} \mathbf{Z}_i \left[N_i\{\mathbf{Z}_i^{\mathsf{T}}\boldsymbol{\beta}(\tau)\} + R_i\{\mathbf{Z}_i^{\mathsf{T}}\boldsymbol{\beta}(\tau)\} \frac{\tau - \psi_i(\boldsymbol{\beta}, \tau)}{1 - \psi_i(\boldsymbol{\beta}, \tau)} - \tau \right].$$

Here $\psi_i(\boldsymbol{\beta}, \tau) = \sup\{A_i(\boldsymbol{\beta}, \tau)\} \cdot I(A_i(\boldsymbol{\beta}, \tau) \text{ is not empty}) + \tau I(A_i(\boldsymbol{\beta}, \tau) \text{ is empty})$ with $A_i(\boldsymbol{\beta}, \tau) = \{u : 0 \leqslant u < \tau, \mathbf{Z}_i^{\mathsf{T}}\boldsymbol{\beta}(u-) \leqslant \tilde{Y}_i \leqslant \mathbf{Z}_i^{\mathsf{T}}\boldsymbol{\beta}(u)\}$. The estimation based on $n^{1/2}\mathbf{S}_n^{(\mathrm{MSC})}(\boldsymbol{\beta}, \tau) = 0$ greatly resembles Neocleous et al.'s (2006) procedure presented earlier. The only distinction lies with the estimation of $\boldsymbol{\beta}_0(\tau_1)$. Specifically, $\boldsymbol{\beta}_0(\tau_1)$ is estimated by the minimizer of (7.6) with $w_{1,i} = 1$ according to Neocleous et al.'s (2006) procedure, while being estimated with $w_{1,i} = 0$ based on Peng's (2012) estimating equation. This indicates a close connection between the different formulations of self-consistent censored regression quantiles by Neocleous et al. (2006) and Peng (2012).

Large-sample studies for the estimator derived from equation (7.9), $\hat{\boldsymbol{\beta}}_{\mathrm{SC}}(\tau)$, are facilitated by the stochastic integral equation representation of (7.9). Specifically, under certain regularity conditions and given $\lim_{n\to\infty} \|\mathcal{G}\| = 0$, $\sup_{\tau \in [\nu, \tau_U]} \|\hat{\boldsymbol{\beta}}_{\mathrm{SC}}(\tau) - \boldsymbol{\beta}_0(\tau)\| \to_p 0$, where $0 < \nu < \tau_U$. If $n^{1/2} \lim_{n\to\infty} \|\mathcal{G}\| = 0$ is further satisfied, then $n^{1/2}\{\hat{\boldsymbol{\beta}}_{\mathrm{SC}}(\tau) - \boldsymbol{\beta}_0(\tau)\}$ converges weakly to a Gaussian process for $\tau \in [\nu, \tau_U]$.

7.2.3.2 Martingale-based approach

Under model (7.5), the same model considered in Section 7.2.3.1, Peng and Huang (2008) proposed to utilize the martingale structure of randomly right censored data to construct an estimating equation for model (7.5). Define $\Lambda_Y(t|\mathbf{Z}) = -\log\{1 - \Pr(Y \leqslant t|\mathbf{Z})\}$, $N(t) = I(\tilde{Y} \leqslant t, \Delta = 1)$, and $M(t) = N(t) - \Lambda_Y(t \wedge Y|\mathbf{Z})$. Let $N_i(t)$ and $M_i(t)$ be sample analogs of $N(t)$ and $M(t)$ respectively, $i = 1, \ldots, n$. Since $M_i(t)$ is the martingale process associated with the counting process $N_i(t)$, $E\{M_i(t)|\mathbf{Z}_i\} = 0$ for all $t > 0$. This implies

$$E\left\{ \sum_{i=1}^{n} \mathbf{Z}_i \left[N_i\{\mathbf{Z}_i^{\mathsf{T}}\boldsymbol{\beta}_0(\tau)\} - \Lambda_V\{\mathbf{Z}_i^{\mathsf{T}}\boldsymbol{\beta}_0(\tau) \wedge \tilde{Y}_i | \mathbf{Z}_i\} \right] \right\} = 0. \tag{7.10}$$

By the monotonicity of $\mathbf{Z}_i^{\mathsf{T}}\boldsymbol{\beta}_0(\tau)$ in $\tau \in (0, \tau_U)$ implied by model (7.5), we have

$$\Lambda_V\{\mathbf{Z}_i^{\mathsf{T}}\boldsymbol{\beta}_0(\tau) \wedge \tilde{Y}_i | \mathbf{Z}_i\} = \int_0^\tau I\{\tilde{Y}_i \geqslant \mathbf{Z}_i^{\mathsf{T}}\boldsymbol{\beta}_0(u)\} dH(u), \tag{7.11}$$

where $H(x) = -\log(1-x)$. Coupling the equalities (7.10) and (7.11) suggests the estimating equation

$$n^{1/2}\mathbf{S}_n^{(\mathrm{PH})}(\boldsymbol{\beta}, \tau) = 0, \tag{7.12}$$

where

$$\mathbf{S}_n^{(\mathrm{PH})}(\boldsymbol{\beta}, \tau) = n^{-1} \sum_{i=1}^{n} \mathbf{Z}_i \left[N_i\{\mathbf{Z}_i^{\mathsf{T}}\boldsymbol{\beta}(\tau)\} - \int_0^\tau I\{\tilde{Y}_i \geqslant \mathbf{Z}_i^{\mathsf{T}}\boldsymbol{\beta}(u)\} dH(u) \right].$$

An estimator of $\boldsymbol{\beta}_0(\tau)$, denoted by $\hat{\boldsymbol{\beta}}_{\mathrm{PH}}(\tau)$, can be obtained by approximating the stochastic solution to equation (7.12). Specifically, let $\hat{\boldsymbol{\beta}}_{\mathrm{PH}}(\tau)$ be a cadlag step function of τ that jumps only at the grid points of $\mathcal{G}$. The procedure to obtain $\hat{\boldsymbol{\beta}}_{\mathrm{PH}}(\tau)$ follows.

1. Set $\exp\{\mathbf{Z}_i^{\mathsf{T}}\hat{\boldsymbol{\beta}}_{\mathrm{PH}}(\tau_0)\} = 0$ for all i. Set $k = 0$.

2. Given $\exp\{\mathbf{Z}_i^{\mathsf{T}}\hat{\boldsymbol{\beta}}_{\mathrm{PH}}(\tau_l)\}$ for $l \leqslant k$, obtain $\hat{\boldsymbol{\beta}}_{\mathrm{PH}}(\tau_{k+1})$ as the minimizer of the L_1-type convex objective function

$$l_{k+1}(\mathbf{h}) = \sum_{i=1}^{n}\left|\Delta_i\tilde{Y}_i - \delta_i\mathbf{Z}_i^{\mathsf{T}}\mathbf{h}\right| + \left|Y^* - \sum_{l=1}^{n}(-\Delta_l\mathbf{Z}_l^{\mathsf{T}}\mathbf{h})\right|$$

$$+\left|Y^* - \sum_{r=1}^{n}\left[(2\mathbf{Z}_r^{\mathsf{T}}\mathbf{h})\sum_{l=0}^{k}I\{\tilde{Y}_r \geqslant \mathbf{Z}_r^{\mathsf{T}}\hat{\boldsymbol{\beta}}_{\mathrm{PH}}(\tau_l)\}\{H(\tau_{l+1}) - H(\tau_l)\}\right]\right|,$$

where Y^* is an extremely large value.

3. Replace k by $k + 1$ and repeat step 2 until $k = M$ or no feasible solution can be found for minimizing $l_k(\mathbf{h})$.

The *crq* function in the contributed R package quantreg (Koenker, 2012) provides an implementation of $\hat{\boldsymbol{\beta}}_{\mathrm{PH}}(\tau)$ based on an algorithm slightly different from the one presented above. More recently, Huang (2010) derived a grid-free estimation procedure for model (7.5) by using the concept of quantile calculus.

The uniform consistency and weak convergence of $\hat{\boldsymbol{\beta}}_{\mathrm{PH}}(\cdot)$ were established in Peng and Huang (2008). Moreover, $\hat{\boldsymbol{\beta}}_{\mathrm{PH}}(\cdot)$ was shown to be asymptotically equivalent to the self-consistent estimator $\hat{\boldsymbol{\beta}}_{\mathrm{SC}}(\cdot)$ (Peng, 2012). This theoretical result is consistent with the numerical results reported in Koenker (2008) and Peng (2012).

7.2.3.3 Locally weighted method

A locally weighted method was proposed by Wang and Wang (2009) to estimate a locally linear quantile regression model, which assumes that

$$Q_Y(\tau|\mathbf{Z}) = \mathbf{Z}^{\mathsf{T}}\boldsymbol{\beta}_0(\tau)$$

holds for a single τ. Similar to the strategy presented in Section 7.2.3.1, which uses the self-consistency principle, the fundamental idea of Wang and Wang (2009) is to redistribute the probability mass $\Pr(T_i > C_i|C_i, \mathbf{Z}_i)$ of the censored cases to the right through a local weighting scheme.

Suppose $F_0(t|\mathbf{z}) \equiv \Pr(T > t|\mathbf{Z} = \mathbf{z})$ is known. An estimator of $\boldsymbol{\beta}_0(\tau)$ can be obtained by minimizing the following objective function of $\boldsymbol{\beta}$:

$$n^{-1}\sum_{i=1}^{n}\left[w_i(F_0)\rho_\tau(\tilde{Y}_i - \mathbf{Z}_i^{\mathsf{T}}\boldsymbol{\beta}) + \{1 - w_i(F_0)\}\rho_\tau(Y^* - \mathbf{Z}_i^{\mathsf{T}}\boldsymbol{\beta})\right], \qquad (7.13)$$

where

$$w_i(F_0) = \begin{cases} 1 & \delta_i = 1, \text{ or } F_0(C_i|\mathbf{Z}_i) > \tau, \\ \frac{\tau - F_0(C_i|\mathbf{Z}_i)}{1 - F_0(C_i|\mathbf{Z}_i)}, & \delta_i = 0 \text{ and } F_0(C_i|\mathbf{Z}_i) < \tau. \end{cases}$$

Given $F_0(t|\mathbf{z})$ is usually unknown, one can proceed to estimate $F_0(\cdot|\mathbf{z})$ nonparametrically using the local Kaplan–Meier estimator,

$$\hat{F}(t|\mathbf{z}) = 1 - \prod_{j=1}^{n}\left\{1 - \frac{B_{nj}(\mathbf{z})}{\sum_{k=1}^{n}I(\tilde{Y}_k \geqslant \tilde{Y}_j)B_{nk}(\mathbf{z})}\right\}^{N_j(t)},$$

where $B_{nk}(\mathbf{z})$ is a sequence of nonnegative weights summing to 1, for example, Nadaraya–Watson type weight,

$$B_{nk}(\mathbf{x}) = K\left(\frac{\mathbf{z} - \mathbf{z}_k}{h_n}\right)\bigg/\sum_{i=1}^{n}K\left(\frac{\mathbf{z} - \mathbf{z}_i}{h_n}\right).$$

Here K is a density kernel function and h_n is the bandwidth, which is positive and converges to 0 as $n \to \infty$. A locally weighted censored regression estimator is given by the minimizer of the objective function (7.13) with F_0 replaced by $\hat{F}$. This estimator is shown to be consistent and asymptotically normal with $\sqrt{n}$ rate of convergence under regularity conditions.

7.2.4 Variance estimation and other inference

7.2.4.1 Variance estimation

The estimators of $\beta_0(\tau)$ discussed in Sections 7.2.1–7.2.3 generally have asymptotic variances that involve unknown density functions. As in uncensored cases, bootstrapping procedures have been successful for variance estimation under censored quantile regression. One may either use resampling methods that follow the idea of Parzen and Ying (1994) or apply simple bootstrapping procedures based on resampling with replacement.

Methods have also been developed for variance estimation without involving resampling. Under random right censoring with known censoring time or covariate-independent censoring, we can adapt Huang's (2002) technique to avoid density estimation. Specifically, let $\hat{\beta}(\tau)$ be general notation for an estimator of $\beta_0(\tau)$, and $\mathbf{S}_n(\beta(\tau), \tau)$ denote the estimating function associated with $\hat{\beta}(\tau)$, for example, the left-hand side of (7.3) and (7.4). It can be shown that $\mathbf{S}_n\{\beta_0(\tau), \tau\}$ converges to a multivariate normal distribution with variance matrix $\mathbf{\Sigma}(\tau)$. Suppose one can obtain a consistent estimator of $\mathbf{\Sigma}(\tau)$, denoted by $\hat{\mathbf{\Sigma}}(\tau)$. The following are the main steps to estimate the asymptotic variance of $\hat{\beta}(\tau)$:

A.1 Find a symmetric and nonsingular $(p + 1) \times (p + 1)$ matrix $\mathbf{E}_n(\tau) \equiv \{\mathbf{e}_{n,1}(\tau), \dots, \mathbf{e}_{n,p+1}(\tau)\}$ such that $\hat{\mathbf{\Sigma}}(\tau) = \{\mathbf{E}_n(\tau)\}^2$.

A.2 Calculate

$$\mathbf{D}_n(\tau) = \left(\mathbf{S}_n^{-1}\{\mathbf{e}_{n,1}(\tau), \tau\} - \hat{\beta}(\tau), \dots, \mathbf{S}_n^{-1}\{\mathbf{e}_{n,p+1}(\tau), \tau\} - \hat{\beta}(\tau) \right),$$

where $\mathbf{S}_n^{-1}(\mathbf{e}, \tau)$ is defined as the solution to $\mathbf{S}_n(\mathbf{b}, \tau) - \mathbf{e} = 0$.

A.3 Estimate the asymptotic variance matrix of $n^{1/2}\{\hat{\beta}(\tau) - \beta_0(\tau)\}$ by $n\{\mathbf{D}_n(\tau)\}^{\otimes 2}$.

Under standard random censoring, Sun et al. (2016) provided a sample-based variance estimation procedure for model (7.5). The key idea is to find consistent estimates for $\mathbf{B}(\beta_0(\tau))$ and $\mathbf{J}(\beta_0(\tau))$, two pivotal quantities in the asymptotic variance formula for $\hat{\beta}_{\mathrm{PH}}(\tau)$, and then get the plug-in variance estimate. Define $\mathbf{L}_n(\mathbf{b}) = n^{-1/2} \sum_{i=1}^n \mathbf{Z}_i N_i(e^{\mathbf{Z}_i^\mathsf{T}\mathbf{b}})$, $\tilde{\mathbf{L}}_n(\mathbf{b}) = n^{-1/2} \sum_{i=1}^n \mathbf{Z}_i Y_i(e^{\mathbf{Z}_i^\mathsf{T}\mathbf{b}})$, $\boldsymbol{\iota}_j(u) = \mathbf{Z}_j N_j(e^{\mathbf{Z}_j^\mathsf{T}\hat{\beta}(u)})$, and $\mathbf{\Omega}_n(u) = n^{-1} \sum_{j=1}^n \{\boldsymbol{\iota}_j(u)\}^{\otimes 2}$. The procedure for estimating $\mathbf{B}(\beta_0(u))$ and $\mathbf{J}(\beta_0(u))$ follows.

B.1 Find a symmetric and nonsingular $(p + 1) \times (p + 1)$ matrix $\mathbf{E}_n(u) \equiv \{\mathbf{e}_{n,1}(u), \dots, \mathbf{e}_{n,p+1}(u)\}$ such that $\mathbf{\Omega}_n(u) = \{\mathbf{E}_n(u)\}^2$.

B.2 Solve the equation

$$\mathbf{L}_n(\mathbf{b}) = \mathbf{L}_n(\hat{\beta}(u)) + \mathbf{e}_{n,j}(u), \tag{7.14}$$

for $\mathbf{b}$, and denote the solution by $\mathbf{b}_{n,j}(u)$ $(j = 1, \dots, p + 1)$.

B.3 Calculate

$$\mathbf{D}_n(u) \equiv \left(\mathbf{b}_{n,1}(u) - \hat{\beta}(u), \dots, \mathbf{b}_{n,p+1}(u) - \hat{\beta}(u) \right)$$

and

$$\tilde{\mathbf{E}}_n(u) \equiv \left(\tilde{\mathbf{L}}_n(\mathbf{b}_{n,1}(u)) - \tilde{\mathbf{L}}_n(\hat{\beta}(u)), \dots, \tilde{\mathbf{L}}_n(\mathbf{b}_{n,p+1}(u)) - \tilde{\mathbf{L}}_n(\hat{\beta}(u)) \right).$$

B.4 Compute $n^{-1/2}\mathbf{E}_n(u)\mathbf{D}_n(u)^{-1}$ and $n^{-1/2}\tilde{\mathbf{E}}_n(u)\mathbf{D}_n(u)^{-1}$, which provide consistent estimates for $\mathbf{B}(\boldsymbol{\beta}_0(u))$ and $\mathbf{J}(\boldsymbol{\beta}_0(u))$, respectively.

The working estimating equation in step B.2, $\mathbf{L}_n(\mathbf{b}) = \mathbf{L}_n(\hat{\boldsymbol{\beta}}(u)) + \mathbf{e}_{n,j}(u)$, is monotone and can be solved via L_1-minimization. The procedures outlined by B.1–B.4 can be stably implemented.

7.2.4.2   Second-stage inference

Second-stage inference for the globally linear quantile regression model (7.5) permits exploration of varying covariate effect patterns across different quantile levels. Questions of interest often include: (1) how to summarize $\hat{\boldsymbol{\beta}}(\tau)$ for a range of τ; (2) how to determine whether some covariates have constant or other types of parametric effects so that a simpler model may be considered.

A general formulation of problem (1) corresponds to estimating a functional of $\boldsymbol{\beta}_0(\cdot)$, say $\boldsymbol{\Psi}(\boldsymbol{\beta}_0)$. A natural estimator for $\boldsymbol{\Psi}(\boldsymbol{\beta})$ is given by $\boldsymbol{\Psi}(\hat{\boldsymbol{\beta}}_0)$. Such an estimator may be justified by the functional delta method provided that $\boldsymbol{\Psi}$ is compactly differentiable at $\boldsymbol{\beta}_0$ (Andersen et al., 1998).

Question (2) can be formulated as a hypothesis testing problem. For example, if the interest is in determining whether a covariate effect is constant or not, the null hypothesis may take the form $H_{0,j}$: $\beta_0^{(j)}(\tau) = \rho_0$, $\tau \in [\tau_L, \tau_U]$, where the superscript $^{(j)}$ indicates the jth component of a vector, and ρ_0 is an unspecified constant, $j = 2, \ldots, p+1$. Examples of the test procedure for H_0 with survival data are presented in Peng and Huang (2008), Peng and Fine (2009), Li and Peng (2011), and Ji et al. (2012). Not ethat accepting $H_{0,j}$ for all $j \in \{2, \ldots, p+1\}$ may indicate the adequacy of an AFT model. Therefore, a procedure for testing the goodness-of-fit of an AFT model can be developed through the second-stage inference. A more general formulation of the second-stage inference framework for addressing problem (2) can follow the work of Koenker and Xiao (2002) on quantile regression process inference developed for uncensored data. Specifical treatments or adaptations that accommodate survival data features are expected for the research endeavors in this direction.

7.2.4.3   Model checking

Model checking is often of practical importance. When interest lies only in checking the linearity between covariates and conditional quantile at a single quantile level, one can adapt the approaches developed for uncensored data, for example, the work by Zheng (2000), Horowitz and Spokoiny (2002), and He and Zhu (2003). When the focus is on testing a global linear relationship between conditional quantiles and covariates, a natural approach is to use a stochastic process which has mean zero under the assumed model. For example, Peng and Huang (2008) proposed a diagnostic process for model (7.5) which takes the form $K_n(\tau) = n^{-1/2} \sum_{i=1}^n q(\mathbf{Z}_i) M_i(\tau; \hat{\boldsymbol{\beta}})$, where $q(\cdot)$ is a known bounded function and

$$M_i(\tau; \boldsymbol{\beta}) = N_i\big(\exp\{\mathbf{Z}_i^\mathsf{T}\boldsymbol{\beta}(\tau)\}\big) - \int_0^\tau \tau I\{\tilde{Y}_i \geqslant \mathbf{Z}_i^\mathsf{T}\boldsymbol{\beta}(u)\}dH(u).$$

It can be shown that $K(\tau)$ converges weakly to a zero-mean Gaussian process, whose distribution can be approximated by that of

$$K^*(\tau) = n^{-1/2} \sum_{i=1}^n q(\mathbf{Z}_i) M_i(\tau; \hat{\boldsymbol{\beta}})(1 - \zeta_i) + n^{-1/2} \sum_{i=1}^n q(\mathbf{Z}_i)\{M_i(\tau; \boldsymbol{\beta}^*) - M_i(\tau; \hat{\boldsymbol{\beta}})\}.$$

Here $\boldsymbol{\beta}^*(\tau)$ denotes the resampling estimator obtained by perturbing the estimating equation (7.12) by $\{\zeta_i\}_{i=1}^n$, independent variates from a known nonnegative distribution with mean 1 and variance 1. An unusual pattern of $K(\cdot)$ compared to that of $K^*(\cdot)$ would suggest a lack of fit of model (7.5). A formal lack-of-fit test is given by the supremum statistic, $\sup_{\tau \in [l,u]} |K(\tau)|$, where $0 < l < u < \tau_U$. The p-value may be approximated by the empirical proportion of $\sup_{\tau \in [l,u]} |K^*(\tau)|$ exceeding $\sup_{\tau \in [l,u]} |K(\tau)|$.

7.3 Quantile regression in other survival settings

In practice, survival data often involve more complex censoring mechanisms than those treated above. Truncation may also present, for example, in many observational studies. Recent developments have extended quantile regression methods that are more complex than the random right censoring case. For example, Ji et al. (2012) proposed a modification of Peng and Huang's (2008) martingale-based approach for survival data subject to known random left censoring and/or random left truncation in addition to random right censoring. Wu and Yin (2013) extended Peng and Huang's (2008) approach to deal with survival data with cure fraction. Wang and Wang (2014) proposed an adaptation of Zhou's (2006) method for quantile regression analysis of length-biased survival data. Huang and Peng (2009), Luo et al. (2013), and Sun et al. (2016) studied quantile regression and its extension for analyzing recurrent events data. Quantile regression methods for competing risks and semi-competing risks data were developed, for example, by Peng and Fine (2009), Li and Peng (2011), and Sun et al. (2012). Recently, Yang et al. (2016) proposed a unified algorithm for performing quantile regression with different types of censoring, such as right censoring, double censoring, and interval censoring. Their empirical results demonstrated impressive efficiency gains and good robustness to initial estimator selection, greatly encouraging further formal investigations. In this subsection, we briefly outline the extensions of Peng and Huang's (2008) methods for left censored or truncated survival data and for survival data with a cure fraction (Ji et al., 2012; Wu and Yin, 2013). The methods for competing risks and semi-competing risks data are presented in detail in a separate chapter.

7.3.1 Known random left censoring and/or left truncation

The extension of Peng and Huang (2008) to accommodate known left censoring and/or left truncation (Ji et al., 2012) is briefly described below. We follow the notation in Section 7.2. In addition, we let L denote left censoring time, always observed, and A denote left truncation time. Define $X = L \vee (T \wedge C)$ and δ as the censoring indicator which equals 1 if $L < T \leqslant C$, 2 if $T \leqslant L$, and 3 if $T > C$, where $\vee$ is the maximum operator. When X is subject to left truncation by A, the observed data include n i.i.d. replicates of $(X', L', A', \delta', \mathbf{Z})$, denoted by $\{(X'_i, L'_i, A'_i, \delta'_i, \mathbf{Z}_i)\}_{i=1}^n$, where $\{X', L', A', \delta', \mathbf{Z}\}$ follows the conditional distribution of $\{X, L, A, \delta, \mathbf{Z}\}$ given $X \geqslant A$. It is assumed that (L, C, A) is independent of T given $\mathbf{Z}$. Such data can be referred to as doubly censored data with left truncation. With $L = 0$, the data reduce to the usual left truncated right censored data.

For doubly censored data with left truncation, an estimating equation for model (7.5) can be constructed based on the same idea for equation (7.12), which is to utilize the martingale structure associated with the observed survival data. We define a counting process as $N'(t) = I(\log X' \leqslant t, \delta' = 1)$, and an at-risk process as $R'(t) = I(\log(L' \vee A') < t \leqslant \log X')$. It can be shown that $M'(t) = N'(t) - \int_0^t R'(s) d\Lambda_Y(s|\mathbf{Z})$ is a martingale process. This fact suggests an estimating equation for $\boldsymbol{\beta}_0(\cdot)$,

$$n^{1/2} \mathbf{S}'_n(\boldsymbol{\beta}, \tau) = 0, \tag{7.15}$$

where

$$\mathbf{S}'_n(\boldsymbol{\beta}, \tau) = n^{-1} \sum_{i=1}^{n} \mathbf{Z}_i \left(N'_i \{ \mathbf{Z}_i^{\mathsf{T}} \boldsymbol{\beta}(\tau) \} - \int_0^{\tau} I\{ \log(L'_i \vee A'_i) < \mathbf{Z}_i^{\mathsf{T}} \boldsymbol{\beta}(u) \leq \log(X'_i) \} dH(u) \right).$$

An estimator of $\boldsymbol{\beta}_0(\tau)$ can be obtained based on equation (7.15) via an algorithm similar to that for $\hat{\boldsymbol{\beta}}_{\mathrm{PH}}(\tau)$.

7.3.2   Censored data with a survival cure fraction

In the presence of a substantial fraction of long-term survivors (i.e. cured subjects) who are either cured or immune to the event of interest, the survival times tend to be highly right-skewed. When interest lies in the survival of uncured subjects, adjustments to modeling and estimation are necessary.

Wu and Yin (2013) considered a mixture cure rate model which assumes $T = \eta T^* + (1 - \eta)\infty$, where $T^* < \infty$ denotes the survival time of a susceptible subject, and η is either 0 or 1, indicating whether the subject is susceptible or not. A logistic regression was used to model susceptibility,

$$\mathrm{Pr}(\eta = 1 | \mathbf{W}, \mathbf{Z}) = \pi(\boldsymbol{\gamma}^{\mathsf{T}} \mathbf{W}) \equiv \frac{\exp(\boldsymbol{\gamma}_0^{\mathsf{T}} \mathbf{W})}{1 + \exp(\boldsymbol{\gamma}^{\mathsf{T}} \mathbf{W})},$$

where $\mathbf{W}$ is a q-vector of covariates related to η. For T^*, a quantile regression model is assumed and takes the form

$$Q_{T^*}(\tau | \mathbf{Z}) = \exp\{ \mathbf{Z}^{\mathsf{T}} \boldsymbol{\beta}_0(\tau) \}, \quad \tau \in (0, \tau_U).$$

When $\boldsymbol{\gamma}_0$ is known, $\boldsymbol{\beta}_0(\tau)$ can be estimated by solving $n^{1/2} \mathbf{S}_n(\boldsymbol{\beta}, \tau; \boldsymbol{\gamma}_0) = 0$, where

$$\mathbf{S}_n(\boldsymbol{\beta}, \tau; \boldsymbol{\gamma}) = n^{-1} \sum_{i=1}^{n} \mathbf{Z}_i \left[N_i \{ \mathbf{Z}_i^{\mathsf{T}} \boldsymbol{\beta}(\tau) \} - \int_0^{\tau} I\{ \tilde{Y}_i \geq \mathbf{Z}_i^{\mathsf{T}} \boldsymbol{\beta}(u) \} H_{\boldsymbol{\gamma}}(du | \mathbf{W}) \right] = 0,$$

where $H_{\boldsymbol{\gamma}}(u | \mathbf{W}) = -\log\{ 1 - \pi(\boldsymbol{\gamma}^{\mathsf{T}} \mathbf{W}) u \}$.

Since $\boldsymbol{\gamma}_0$ is typically unknown, Wu and Yin (2013) proposed an estimating equation for $\boldsymbol{\gamma}_0$, $n^{1/2} \mathbf{R}_n(\boldsymbol{\gamma}; \boldsymbol{\beta}(\cdot)) = 0$, where $\mathbf{R}_n(\boldsymbol{\gamma}; \boldsymbol{\beta}(\cdot))$ equals

$$n^{-1} \sum_{i=1}^{n} \int_0^{\tau_U} \frac{\mathbf{W}_i \{ 1 - \pi(\boldsymbol{\gamma}^{\mathsf{T}} \mathbf{W}_i) \}}{1 - \pi(\boldsymbol{\gamma}^{\mathsf{T}} \mathbf{W}_i) u} \left[dN_i(\exp\{ \mathbf{Z}_i^{\mathsf{T}} \boldsymbol{\beta}(u) \}) - I[X_i \geq \exp\{ \mathbf{Z}_i^{\mathsf{T}} \boldsymbol{\beta}(u) \}] H_{\boldsymbol{\gamma}}(du | \mathbf{W}_i) \right].$$

Wu and Yin (2013) developed an iterative procedure to solve equations $n^{1/2} \mathbf{S}_n(\boldsymbol{\beta}, \tau; \boldsymbol{\gamma}_0) = 0$ and $n^{1/2} \mathbf{R}_n(\boldsymbol{\gamma}; \boldsymbol{\beta}(\cdot)) = 0$ simultaneously. To avoid the difficulty from the entanglement of estimating $\boldsymbol{\beta}_0(\tau)$ and estimating $\boldsymbol{\gamma}_0$, Wu and Yin (2013) further suggest estimating $\boldsymbol{\gamma}_0$ separately based on a nonparametric estimator of $F_{T^*}(t | \mathbf{z})$, the conditional distribution function of T^* given $\mathbf{Z} = \mathbf{z}$. The resulting estimator of $\boldsymbol{\gamma}_0$ is then plugged into $n^{1/2} \mathbf{S}_n(\boldsymbol{\beta}, \tau; \boldsymbol{\gamma}_0) = 0$ to get a final estimate for $\boldsymbol{\beta}_0(\tau)$.

7.4   An illustration of quantile regression for survival analysis

To illustrate the use quantile regression for survival analysis, we consider a data set from a dialysis study that investigated predictors of mortality risk in a cohort of 191 incident dialysis patients (Kutner et al., 2002). Analysis covariates included patient's age (AGE), the indicator of fish consumption over the first year of dialysis (FISHH), the indicator of baseline HD dialysis modality (BHDPD), the indicator of moderate to severe symptoms of restless leg syndrome (BLEGS), the indicator of eduction equal to or higher than college (HIEDU), and the indicator of being black (BLACK). We first fit the data with AFT model (7.2), where T stands for time to death. In Table 7.1, we present the estimation results based on the log-rank estimator, Gehan's estimator, and the least-squares estimator. All covariates except for BLEGS are consistently shown to have significant effects on survival by all three different estimators. For the coefficient of BLEGS, Gehan's estimator and the least-square estimator yield significant p-values while the log-rank estimator does not. In addition, a Cox proportional hazards analysis of this data set indicates that BLEGS is not a significant predictor for dialysis mortality risk. These traditional survival analyses pose a question regarding the association between restless leg symptoms and dialysis survival.

TABLE 7.1

Results from fitting an AFT model to the dialysis data set. Coef: coefficient estimate; SE: standard error

	Gehan's estimator			Log-rank estimator			Least-squares estimator		
	Coef	SE	p-value	Coef	SE	p-value	Coef	SE	p-value
AGE	−0.031	0.005	< 0.001	−0.035	0.004	<0.001	0.033	0.006	< 0.001
FISHH	0.402	0.139	0.004	0.485	0.128	<0.001	0.507	0.163	0.002
BHDPD	−0.505	0.156	0.001	−0.473	0.136	< 0.001	−0.509	0.164	0.002
BLEGS	−0.340	0.166	0.040	−0.173	0.145	0.232	−0.412	0.176	0.019
HIEDU	−0.352	0.133	0.008	−0.364	0.161	0.024	0.335	0.139	0.016
BLACK	0.640	0.144	<0.001	0.591	0.138	< 0.001	0.643	0.153	< 0.001

We next conduct quantile regression based on model (7.5) for the same data set. Figure 7.1 displays Peng and Huang's (2008) estimator of $\beta_0(\tau)$ along with 95% pointwise confidence intervals. In Figure 7.1, we observe that the coefficient for BLEGS diminishes gradually with τ, whereas estimates for the other coefficients seem to be fairly constant. We apply the second-stage inference to formally investigate the constancy of each coefficient. The results confirm our observation from Figure 7.1, suggesting a varying effect of BLEGS and constant effects of the other covariates. This suggests that BLEGS may affect the survival experience of dialysis patients with short survival times but may have little impact on that of long-term survivors. In addition, the evidence for the nonconstancy of BLEGS coefficient indicates the lack of fit of the AFT model for the dialysis data.

We further estimate the average quantile effects defined as $\int_l^u \beta_0^{(i)}(u)du$ $(i = 2,\ldots,7)$. The results are given in Table 7.2. We observe that the estimated average effects are similar to the AFT coefficients obtained by Gehan's estimator and the least-squares estimator, but have relatively larger discrepancies with the log-rank estimates. This may indirectly reflect the presence of some non-constant covariate effect. When quantile regression model (7.5) holds without any varying covariate effects, we would expect to see more consistent results between Gehan's estimator, which emphasizes early survival, and the log-rank estimator, which treats early and late survival information equally.

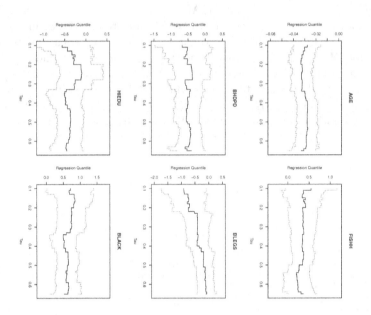

FIGURE 7.1
Peng and Huang's estimator (solid lines) and 95% pointwise confidence intervals (dotted lines) of regression quantiles in the dialysis example.

TABLE 7.2
Estimation of average covariate effects based on quantile regression. AveEff: Estimated average effect; SE: standard error

	AveEff	SE	p value
AGE	−0.030	0.003	< 0.001
FISHH	0.327	0.116	0.005
BHDPD	−0.489	0.162	0.003
BLEGS	−0.369	0.161	0.022
HIEDU	−0.350	0.137	0.011
BLACK	0.654	0.144	< 0.001

Bibliography

P. K. Andersen, Ø. Borgan, R. D. Gill, and N Keiding. *Statistical Models Based on Counting Processes*. New York: Springer-Verlag, 2nd edition, 1998.

M. Buchinsky and J. Hahn. A alternative estimator for censored quantile regression. *Econometrica*, 66:653–671, 1998.

V. Chernozhukov and H. Hong. Three-step censored quantile regression and extramarital affairs. *Journal of the American Statistical Association*, 97:872–882, 2001.

B. Efron. The two sample problem with censored data. In L. LeCam and J. Neyman, editors, *Proceedings of the Fifth Berkeley Symposium on Mathematical Statistics and Probability*, volume IV, pages 831–53. University of California Press, Berkeley, 1967.

B. Fitzenberger. A guide to censored quantile regressions. In G.S. Maddala and C.R. Rao, editors, *Handbooks of Statistics, Volume 15: Robust Inference*, pages 405–437. North-Holland, Amsterdam, 1997.

M. Fygenson and Y. Ritov. Monotone estimating equations for censored data. *Annals of Statistics*, 22:732–746, 1994.

X. He and L.-X. Zhu. A lack-of-fit test for quantile regression. *Journal of the American Statistical Association*, 98:1013–1022, 2003.

J. L. Horowitz and V. G. Spokoiny. An adaptive, rate-optimal test of linearity for median regression models. *Journal of the American Statistical Association*, 97:822–835, 2002.

Y. Huang. Calibration regression of censored lifetime medical cost. *Journal of the American Statistical Association*, 98:318–327, 2002.

Y. Huang. Quantile calculus and censored regression. *Annals of Statistics*, 38(3):1607–1637, 2010.

Y. Huang and L. Peng. Accelerated recurrence time models. *Scandivanian Journal of Statistics*, 36:636–648, 2009.

S. Ji, L. Peng, Y. Cheng, and H. Lai. Quantile regression for doubly censored data. *Biometrics*, 68:101–112, 2012.

R. Koenker. Censored quantile regression redux. *Journal of Statistical Software*, 27: http://www.jstatsoft.com, 2008.

R. Koenker. *Quantile Regression*. R package version 4.81, http://cran.r-project.org/web/packages/quantreg/quantreg.pdf, 2012.

R. Koenker and G. Bassett. Regression quantiles. *Econometrica*, 46:33–50, 1978.

R. Koenker and O. Geling. Reapprasing medfly longevity: A quantile regression survival analysis. *Journal of the American Statistical Association*, 96:458–468, 2001.

R. Koenker and Z. Xiao. Inference on the quantile regression process. *Econometrica*, 70: 1583–1612, 2002.

N. G. Kutner, P. W. Clow, R. Zhang, and X. Aviles. Association of fish intake and survival in a cohort of incident dialysis patients. *American Journal of Kidney Diseases*, 39:1018–1024, 2002.

R. Li and L. Peng. Quantile regression for left-truncated semi-competing risks data. *Biometrics*, 67:701–710, 2011.

D. Y. Lin and C. J. Geyer. Computational methods for semi-parametric linear regression with censored data. *Journal of Computational and Graphical Statistics*, 1:77–90, 1992.

X. Luo, C.-Y. Huang, and L. Wang. Quantile regression for recurrent gap time data. *Biometrics*, 69:375385, 2013.

T. Neocleous, K. Vanden Branden, and S. Portnoy. Correction to censored regression quantiles by S. Portnoy, 98 (2003), 1001-1012. *Journal of the American Statistical Association*, 101:860–861, 2006.

L. J. Parzen, M. I. Wei and Z. Ying. A resampling method based on pivotal estimating functions. *Biometrika*, 81:341–350, 1994.

L. Peng. A note on self-consistent estimation of censored regression quantiles. *Journal of Multivariate Analysis*, 105:368–379, 2012.

L. Peng and J. P. Fine. Competing risks quantile regression. *Journal of the American Statistical Association*, 104:1440–1453, 2009.

L. Peng and Y. Huang. Survival analysis with quantile regression models. *Journal of the American Statistical Association*, 103:637–649, 2008.

S. Portnoy. Censored regression quantiles. *Journal of the American Statistical Association*, 98:1001–1012, 2003.

S. Portnoy and G. Lin. Asymptotics for censored regression quantiles. *Journal of Nonparametric Statistics*, 22:115–130, 2010.

J. Powell. Least absolute deviations estimation for the censored regression model. *Journal of Econometrics*, 25:303–325, 1984.

J. Powell. Censored regression quantiles. *Journal of Econometrics*, 32:143–155, 1986.

J. M. Robins and A. Rotnitzky. Recovery of information and adjustment for dependent censoring using surrogate markers. In N. Jewell, K. Dietz, and V. Farewell, editors, *AIDS Epidemiology-Methodological Issues*, pages 24–33. Birkhäuser, Boston, 1992.

X. Sun, L. Peng, Y. Huang, and H.-C. Lai. Generalizing quantile regression for counting processes with applications to recurrent events. *Journal of the American Statistical Association*, 111:145–156, 2016.

Y. Sun, H.-J Wang, and J. Gilbert. Quantile regression for competing risks data with missing cause of failure. *Statistica Sinica*, 22, 2012.

H. Wang and L. Wang. Locally weighted censored quantile regression. *Journal of the American Statistical Association*, 104:1117–1128, 2009.

H. J. Wang and L. Wang. Quantile regression analysis of length-biased survival data. *Stat*, 3:31–47, 2014.

Y. Wu and G. Yin. Cure rate quantile regression for censored data with a survival fraction. *Journal of the American Statistical Association*, 108:1517–1531, 2013.

X. Yang, N. N. Narisetty, and X. He. A new approach to censored quantile regression estimation. Preprint, 2016.

Z. Ying, S. H. Jung, and L. J. Wei. Survival analysis with median regression models. *Journal of the American Statistical Association*, 90:178–184, 1995.

J. X. Zheng. A consistent test of conditional parametric distributions. *Econometric Theory*, 16:667–691, 2000.

L. Zhou. A simple censored median regression estimator. *Statistica Sinica*, 16:1043–1058, 2006.

8

Survival Analysis with Competing Risks and Semi-competing Risks Data

Ruosha Li

University of Texas Health Science Center, Houston, Texas, USA

Limin Peng

Emory University, Atlanta, Georgia, USA

CONTENTS

8.1 Competing risks data

8.1.1 Introduction

Competing risks data arise when study subjects are subject to multiple types of failure events. The competing risks events are *terminating*, in that the onset of one event precludes the observation of other events. Moreover, the event times are correlated, imposing dependent censoring on each other. Denote the latent failure times to different events by $T_1, T_2, \ldots, T_k$, where $k \geqslant 2$ denotes the number of failure types. One can only observe the time to the first failure $T = \min(T_1, T_2, \ldots, T_k)$ and the type of the first failure ε, where $\varepsilon \in \{1, 2, \ldots, k\}$.

In practice, researchers are often interested in analyzing a specific type of event separately from other competing risks events. An example is the E1178 Cummings et al. (1993) breast cancer study, where patients were followed until breast cancer recurrence or nonrecurrence-related death, whichever occurred first. The two competing endpoints represent different mechanisms of disease progression, and it is of interest to analyze the endpoints separately to gain specific insights. Without loss of generality, we let $\varepsilon = 1$ denote the event type to be analyzed.

One approach to describing a specific event of interest is to adopt a *crude quantity* which accounts for the presence of other competing events. An example of the crude quantity is the cumulative incidence function (CIF). Specifically, the conditional CIF for event I given covariate vector $\mathbf{Z}$ is

$$\bar{F}_1(t|\mathbf{Z}) = P\{T_1 \leqslant t, T_1 \leqslant \min(T_2, \ldots, T_k)|\mathbf{Z}\} = P(T \leqslant t, \varepsilon = 1|\mathbf{Z}). \tag{8.1}$$

The CIF is practically meaningful and characterizes the probability that event I occurs before time t, in the presence of other competing events that may preclude the observation of event I. It is worth noting that $\lim_{t \to \infty} \bar{F}_1(t|\mathbf{Z})$ does not equal to 1 but equals $P(\varepsilon = 1|\mathbf{Z})$.

The CIF entails the *cumulative incidence quantile function*, defined as

$$\bar{Q}_1(\tau|\mathbf{Z}) = \inf\{t : \bar{F}_1(t|\mathbf{Z}) \geqslant \tau\},$$

for $0 < \tau < P(\varepsilon = 1|\mathbf{Z})$, where τ denotes the quantile level. The cumulative incidence quantile has a natural interpretation, namely the earliest time at which the conditional probability of event I exceeds τ, in the presence of other competing events.

Another way to describe the event of interest is to employ a *net quantity*, which hypothesizes the removal of other competing events. For example, the *marginal distribution function* $F_1(t|\mathbf{Z}) = P(T_1 \leqslant t|\mathbf{Z})$ correspond to the probability of a type I event occurring before time t in the absence of other competing events. Accordingly, a marginal quantile

$$Q_1(\tau|\mathbf{Z}) = \inf\{t : F_1(t|\mathbf{Z}) \geqslant \tau\}$$

can be defined as the inverse function of $F_1(t|\mathbf{Z})$.

Net quantities are of interest in many practical scenarios as well. For example, in many studies of organ failure, researchers are interested in studying the risk factors for death. However, patients may receive organ transplantation prior to death and thereby be dependently censored. In this case, death and organ transplantation form the competing risks data. A net quantity for the death event corresponds to a hypothetical scenario where patients never receive organ transplantation. Therefore, it can facilitate insights into the natural history of death by removing the impact of organ transplantation.

In Section 8.1.2 we introduce quantile regression methods that target the cumulative incidence quantile, $\bar{Q}_1(\tau|\mathbf{Z})$. The methods are illustrated using the breast cancer data set in Section 8.1.3. We discuss quantile regression methods for the marginal quantile, $Q_1(\tau|\mathbf{Z})$, in Section 8.1.4.

8.1.2   Cumulative incidence quantile regression

Peng and Fine (2009) proposed a competing risks quantile regression model, by formulating covariate effects on the conditional cumulative incidence quantile as

$$\bar{Q}_1(\tau|\mathbf{Z}) = g\{\mathbf{Z}^T \beta_0(\tau)\}, \ \tau \in [\tau_L, \tau_U]. \tag{8.2}$$

The covariate vector $\mathbf{Z}$ is of dimension $(p+1) \times 1$ and has 1 as its first element. The monotone link function $g(\cdot)$ is prespecified. For example, we can let $g(x)$ be $\exp(x)$ or the identity link. The unknown regression coefficient $\beta_0(\tau)$ is a $(p+1) \times 1$ function of τ and represents covariate effects on the τth cumulative incidence quantile, where $0 < \tau_L \leqslant \tau \leqslant \tau_U < 1$ and $[\tau_L, \tau_U]$ is a prespecified quantile range of interest.

In practice, the failure time $T = \min(T_1, \ldots, T_k)$ is often subject to independent censoring by another event time C. For example, C may represent time to administrative censoring in many clinical settings. Write $\tilde{T} = T \wedge C$, $\delta = I(T \leqslant C)$, and $\tilde{\delta} = \delta\varepsilon$, where $\wedge$ is the minimum operator. The observed data involve n independent and identically distributed (i.i.d.) replicates of $(\tilde{T}, \tilde{\delta}, \mathbf{Z})$, expressed as $(\tilde{T}_i, \tilde{\delta}_i, \mathbf{Z}_i)$ for $i = 1, \ldots, n$.

For a specific $\tau \in [\tau_L, \tau_U]$, Peng and Fine (2009) proposed estimating $\beta_0(\tau)$ using the estimating equation

$$\mathbf{S}_n(\mathbf{b}, \tau) = n^{-1/2} \sum_{i=1}^{n} \mathbf{Z}_i \left(\frac{I\{\tilde{T}_i \leqslant g(\mathbf{Z}_i^T \mathbf{b})\} I(\tilde{\delta}_i = 1)}{\hat{G}(\tilde{T}_i | \mathbf{Z}_i)} - \tau \right). \tag{8.3}$$

Here, $\hat{G}(t|\mathbf{Z})$ is a reasonable estimator for $P(C \geqslant t|\mathbf{Z})$. Because C is independently censored by T, $\hat{G}(t|\mathbf{Z})$ can be constructed using standard nonparametric or semiparametric methods for independently right censored data. The estimating function in (8.3) is not smooth in $\mathbf{b}$, and an exact zero-crossing of $\mathbf{S}_n(\mathbf{b}, \tau)$ may not exist. Instead, the problem can be reformulated as finding the minimizer of an L_1 type convex function, defined as

$$U_n(\mathbf{b}, \tau) = \sum_{i=1}^{n} I(\tilde{\delta}_i = 1) \left| \frac{g^{-1}(\tilde{T}_i)}{\hat{G}(\tilde{T}_i | \mathbf{Z}_i)} - \mathbf{b}^T \frac{\mathbf{Z}_i}{\hat{G}(\tilde{T}_i | \mathbf{Z}_i)} \right| + \left| M - \mathbf{b}^T \sum_{l=1}^{n} \frac{-\mathbf{Z}_l I(\tilde{\delta}_l = 1)}{\hat{G}(\tilde{T}_l | \mathbf{Z}_i)} \right|$$

$$+ \left| M - \mathbf{b}^T \sum_{k=1}^{n} (2\mathbf{Z}_k \tau) \right|,$$

where M is a very large positive constant, such as $M = 10^6$. The minimization of $U_n(\mathbf{b}, \tau)$ can be implemented easily in standard L_1-minimization software, such as the $rq()$ function in the R package quantreg. To estimate $\beta_0(\tau)$ over $[\tau_L, \tau_U]$, one can impose a fine grid on this quantile interval and solve $\mathbf{S}_n(\mathbf{b}, \tau) = 0$ at each gridpoint. Following this, $\hat{\beta}(\tau)$ can be defined as a cadlag function on $[\tau_L, \tau_U]$ that jumps only at the gridpoints.

The estimator $\hat{\beta}(\tau)$ has desirable asymptotic properties. First, $\hat{\beta}(\tau)$ is uniformly consistent for the true regression coefficient $\beta_0(\tau)$ for $\tau \in [\tau_L, \tau_U]$. Moreover, $\sqrt{n}\{\hat{\beta}(\tau) - \beta_0(\tau)\}$ converges weakly to a mean zero Gaussian process with covariance matrix

$$\mathbf{\Phi}(\tau', \tau) = \mathbf{A}\{\beta_0(\tau')\}^{-1} \mathbf{\Sigma}(\tau', \tau) \mathbf{A}\{\beta_0(\tau)\}^{-1}. \tag{8.4}$$

Here $\mathbf{\Sigma}(\tau', \tau)$ is the limiting covariance matrix of $\mathbf{S}_n\{\beta_0(\tau), \tau\}$, and $\mathbf{A}(\mathbf{b}) = \partial E[\mathbf{Z}\bar{F}_1\{g(\mathbf{Z}^T\mathbf{b})|\mathbf{Z}\}]/\partial \mathbf{b}$. While $\mathbf{\Sigma}(\tau', \tau)$ can be consistently estimated, estimation of $\mathbf{A}(\mathbf{b})$ is difficult due to its unknown density component.

Covariance estimation can be conducted in a fast and efficient manner. Specifically, the following three-step procedure allows one to estimate $\mathbf{\Phi}(\tau', \tau)$ without resampling.

1. Construct a consistent estimator for $\mathbf{\Sigma}(\tau', \tau)$ and refer to it as $\hat{\mathbf{\Sigma}}(\tau', \tau)$. Find a symmetric and nonsingular $(p + 1) \times (p + 1)$ matrix $\mathbf{E}_n(\tau) \equiv \{\mathbf{e}_{n,1}(\tau), \dots, \mathbf{e}_{n,p+1}(\tau)\}$ such that $\hat{\mathbf{\Sigma}}(\tau, \tau) = \{\mathbf{E}_n(\tau)\}^2$.

2. Calculate

$$\mathbf{D}_n(\tau) = \left(\mathbf{S}_n^{-1}\{\mathbf{e}_{n,1}(\tau), \tau\} - \hat{\beta}(\tau), \dots, \mathbf{S}_n^{-1}\{\mathbf{e}_{n,p+1}(\tau), \tau\} - \hat{\beta}(\tau) \right),$$

where $\mathbf{S}_n^{-1}(\mathbf{e}, \tau)$ is defined as the solution to $\mathbf{S}_n(\mathbf{b}, \tau) - \mathbf{e} = 0$.

3. Estimate the asymptotic covariance matrix $\mathbf{\Phi}(\tau', \tau)$ by

$$\hat{\mathbf{\Phi}}(\tau', \tau) = n \mathbf{D}_n(\tau') \mathbf{E}_n(\tau')^{-1} \hat{\mathbf{\Sigma}}(\tau', \tau) \mathbf{E}_n(\tau)^{-1} \mathbf{D}_n(\tau)^T.$$

In the special case with $\tau = \tau'$, an estimator for the asymptotic variance matrix of $\sqrt{n}\{\hat{\beta}(\tau) - \beta_0(\tau)\}$ is given by $n\{\mathbf{D}_n(\tau)\}^{\otimes 2}$.

The key rationale for this procedure is that the functional linearity of $\mathbf{S}_n(\mathbf{b}, \tau)$ in $\mathbf{b}$ around $\boldsymbol{\beta}_0(\tau)$ allows us to estimate its derivative $\mathbf{A}\{\boldsymbol{\beta}_0(\tau)\}$ through evaluating the difference in $\mathbf{S}_n(\mathbf{b}, \tau)$ after imposing a small perturbance on $\mathbf{b} = \boldsymbol{\beta}_0(\tau)$. More specifically, $\sqrt{n}[\mathbf{S}_n^{-1}\{\mathbf{e}_{n,j}(\tau), \tau\} - \widehat{\boldsymbol{\beta}}(\tau)]$ can be shown to be asymptotically equivalent to $\mathbf{A}\{\boldsymbol{\beta}_0(\tau)\}^{-1}\mathbf{e}_{n,j}(\tau)$, where $j = 1, 2, \ldots, p+1$. Therefore, $\sqrt{n}\mathbf{D}_n(\tau) \approx \mathbf{A}\{\boldsymbol{\beta}_0(\tau)\}^{-1}\mathbf{E}_n(\tau)$, and matrix $\mathbf{A}\{\boldsymbol{\beta}_0(\tau)\}$ in (8.4) can be consistently estimated by $n^{-1/2}\mathbf{E}_n(\tau)\{\mathbf{D}_n(\tau)\}^{-1}$. This procedure facilitates standard error estimates and Wald-type hypothesis testing for $\boldsymbol{\beta}_0(\tau)$ at any specific $\tau \in [\tau_L, \tau_U]$.

Moreover, it is often practically important to draw overall conclusions on $\boldsymbol{\beta}_0(\tau)$, by accounting for a range of $\tau \in [\tau_L, \tau_U]$ simultaneously. This can be achieved through appropriate second-stage inference procedures. For example, one sensible approach to summarizing the effect of $Z^{(j)}$ ($j = 2, 3, \ldots, p+1$) on $\bar{Q}_1(\tau | \mathbf{Z})$ for a range of τ is to use the trimmed mean effect, defined by $\eta_0^{(j)} = \{\int_{\tau_L}^{\tau_U} \beta_0^{(j)}(\tau)d\tau\}/(\tau_U - \tau_L)$. Here and in the sequel, $u^{(j)}$ denotes the jth component of a vector $\mathbf{u}$. A Wald-type test can be used to examine whether the trimmed mean effect of a given covariate $Z^{(j)}$ is significantly different from 0, namely $H_{0,j} : \eta_0^{(j)} = 0$. Another important question is to assess whether the effect of a covariate on $\bar{Q}_1(\tau | \mathbf{Z})$ is constant over τ. To address this, we can formulate the null hypothesis as $\widetilde{H}_{0,j} : \beta_0^{(j)}(\tau) \equiv \rho_0$, where ρ_0 is an unspecified constant. A test statistic can be formulated as $\mathcal{T} = n^{1/2} \int_{\tau_L}^{\tau_U} \Xi(\tau)\{\widehat{\beta}^{(j)}(\tau) - \hat{\eta}^{(j)}\}d\tau$, where $\Xi(\tau)$ is a nonconstant weight function satisfying $\int_{\tau_L}^{\tau_U} \Xi(\tau)d\tau = 1$.

Competing risk quantile regression methods have been extended to handle the situation where the cause of failure ε_i is subject to missingness. In some clinical settings, the actual cause of failure may be unknown for some subjects. Such competing risks data can be expressed as $(\widetilde{T}_i, \mathbf{Z}_i, \delta_i, R_i, R_i\delta_i\varepsilon_i, \delta_i\mathbf{V}_i)$, where $\delta_i = I(T_i \leqslant C_i)$. The complete case indicator is $R_i \in \{0, 1\}$, and it equals 1 either if $\delta_i = 0$ or if $\delta_i = 1$ and ε_i is observed. The vector $\mathbf{V}_i$ contains auxiliary variables for estimating the missing failure type. We write $\mathbf{W}_i = (\widetilde{T}_i, \mathbf{Z}_i, \mathbf{V}_i)$ and $\mathbf{Q}_i = (\mathbf{W}_i, \delta_i)$ for brevity.

In the presence of missing cause of failure, Sun et al. (2012) studied the cumulative incidence quantile regression for model (8.2) using augmented inverse probability weighting methods. The missing at random (MAR) assumption was posed for the failure time, that is,

$$r(\mathbf{W}_i) \equiv P(R_i = 1 | \varepsilon_i, \delta_i = 1, \mathbf{W}_i) = P(R_i = 1 | \delta_i = 1, \mathbf{W}_i).$$

Write $\pi(\mathbf{Q}_i) = \delta_i r(\mathbf{W}_i) + (1 - \delta_i)$. Suppose that $r(\mathbf{W}_i)$ can be modeled via a proper parametric model. Then one can obtain $\hat{\pi}(\mathbf{Q}_i)$ by plugging in the model-based estimate of $r(\mathbf{W}_i)$. Similarly, write $\rho(\mathbf{W}_i) = P(\varepsilon_i = 1 | \delta_i = 1, \mathbf{W}_i)$. Another parametric model can be used to obtain $\hat{\rho}(\mathbf{W}_i)$, the estimator for $\rho(\mathbf{W}_i)$.

Under the MAR assumption, Sun et al. (2012) defined an estimating function $\mathbf{S}_n(\mathbf{b}, \tau)$ as

$$n^{-1/2} \sum_{i=1}^n \mathbf{Z}_i \left[\left(\frac{R_i}{\hat{\pi}(Q_i)} \frac{I\{\widetilde{T}_i \leqslant g(\mathbf{Z}_i^T \mathbf{b})\}\delta_i I(\varepsilon_i = 1)}{\hat{G}(\widetilde{T}_i | \mathbf{Z}_i)} + \left\{ 1 - \frac{R_i}{\hat{\pi}(Q_i)} \right\} \frac{I\{\widetilde{T}_i \leqslant g(\mathbf{Z}_i^T \mathbf{b})\}\delta_i \hat{\rho}(\mathbf{W}_i)}{\hat{G}(\widetilde{T}_i | \mathbf{Z}_i)} \right) - \tau. \right]$$

The estimating function is doubly robust (Robins et al., 1994), in that the resulting estimator $\hat{\boldsymbol{\beta}}(\tau)$ is consistent when at least one of $\hat{r}(\cdot)$ or $\hat{\rho}(\cdot)$ is consistent for their respective estimands.

8.1.3   Data analysis example

Peng and Fine (2009) illustrated the cumulative incidence quantile regression methods using the data set from a breast cancer study. In the E1178 breast cancer trial, the two competing endpoints are breast cancer recurrence (BCR) and nonrecurrence-related death (NRD). The study enrolled 167 eligible patients, who were randomized to receive either the tamoxifen therapy or placebo. During the follow-up period, 101 patients experienced BCR and 42 patients died without BCR. The observed risk factors include age at randomization, tumor size, and the number of positive nodes.

For the purpose of illustration, we display the analysis results for the BCR endpoint in Figure 8.1. The covariate vector $\mathbf{Z}$ included treatment (tamoxifen versus placebo), age, tumor size, and log-transformed number of nodes. In this analysis, $[\tau_L, \tau_U] = [0.1, 0.45]$ and $g(\cdot)$ is the identity link. In Figure 8.1, the bold solid lines and the dotted lines correspond

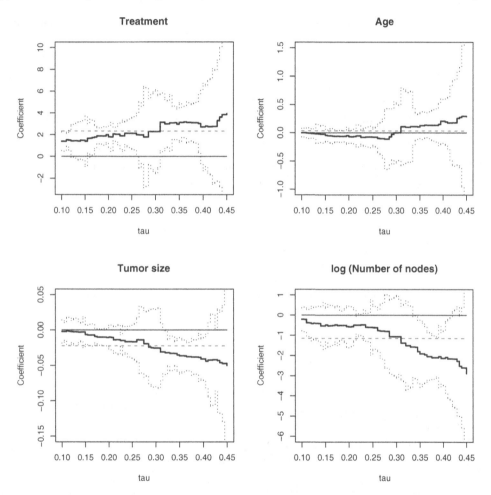

FIGURE 8.1

E1178 trial example: estimated regression coefficients for the breast cancer recurrence endpoint. Bold solid lines represent coefficient estimates; dotted lines represent 95% pointwise confidence intervals; dashed lines represent estimates for trimmed mean covariate effects.

to the estimated regression coefficients and the 95% pointwise confidence intervals (CI), respectively. Here, a positive regression coefficient corresponds to slower progression to BCR. The dashed lines show the estimated trimmed mean effect for $\tau \in [0.1, 0.45]$. For example, the estimated regression coefficient for the treatment variable is positive across

τ, and the lower bound of the CI tends to stay above 0. Therefore, tamoxifen treatment is associated with larger cumulative incidence quantiles for the BCR endpoint. On average, treatment prolongs the cumulative incidence quantile of BCR by approximately 2 years, according to the trimmed mean effect estimate. The hypothesis tests for $H_{0,j}$ yield p-values 0.02, 0.79, 0.04, and 0.020 respectively for treatment, age, tumor size, and log(number of nodes). These results suggest that the tamoxifen treatment postpones BCR when compared to placebo. Moreover, smaller tumors and fewer positive nodes were associated with slower progression to BCR. The effect of age on BCR is not significant.

Another interesting observation from Figure 8.1 is that the magnitudes of the tumor size effect and the log(number of nodes) effect appear to increase with τ. The hypothesis tests for nonconstant effect give p-values of 0.36, 0.33, 0.04 and 0.04 respectively for the coefficients of treatment, age, tumor size, and log(number of nodes). These results suggest that increase in either tumor size or log(number of nodes) may significantly reduce time to BCR for patients with moderate or low risks of BCR, but the impacts may diminish among patients subject to high risk of BCR. Such a heterogeneous effect pattern may also be uncovered by other methods, such as using a traditional survival regression model with the interaction between the risk factors of interest and covariates indicating risk categories. The competing risks quantile regression modeling adopted in this example provides a useful alternative perspective to capture heterogeneous effect patterns without explicitly classifying subjects into different risk groups, and hence may provide a more robust picture of the relationship between the covariates and the survival outcome of interest.

8.1.4   Marginal quantile regression

With competing risks data, analysis of the net quantities is complicated by the curse of identifiability. Specifically, the marginal model for a specific event of interest is not nonparametrically identifiable (Tsiatis, 1978). The identifiability issue can be remedied by posing adequate assumptions on the dependence structure between the event of interest and the competing risks events. In practice, conducting sensitivity analysis for the net quantities under various plausible assumptions has been advocated. The sensitivity analysis can provide bounds for the estimands of interest.

Consider the situation with $k = 2$ here without loss of generality. Let C be an independent censoring time. Let $T = T_1 \wedge T_2$, $\widetilde{T} = T \wedge C$, and $\delta = I(T \leqslant C)$. Also define $\widetilde{\delta} = \delta\varepsilon$, which takes values in $\{0, 1, 2\}$. Let $\mathbf{Z}$ be a $(p+1) \times 1$ covariate vector that has 1 as first element. Therefore, the observed data include n i.i.d. replicates of $\{\widetilde{T}, \widetilde{\delta}, \mathbf{Z}\}$, denoted by $\{\widetilde{T}_i, \widetilde{\delta}_i, \mathbf{Z}_i\}$ for $i = 1, 2, \ldots, n$.

Let $Q_j(\tau|\mathbf{Z})$ denote the marginal quantile function of T_j, $j = 1, 2$. A flexible approach to modeling T_1 and T_2 is to assume marginal quantile regression models (Ji et al., 2012)

$$Q_1(\tau|\mathbf{Z}) = g\{\mathbf{Z}^T \boldsymbol{\beta}_0(\tau)\}, \quad \tau \in (0, 1), \tag{8.5}$$

$$Q_2(\tau|\mathbf{Z}) = h\{\mathbf{Z}^T \boldsymbol{\alpha}_0(\tau)\}, \quad \tau \in (0, 1), \tag{8.6}$$

where the unknown regression coefficients $\boldsymbol{\beta}_0(\tau)$ and $\boldsymbol{\alpha}_0(\tau)$ represents unknown covariate effects on the τth marginal quantile of T_1 and T_2, respectively. The $g(\cdot)$ and $h(\cdot)$ are prespecified monotone link functions, such as the $\exp(\cdot)$ function or the identity link function. While model (8.5) corresponds to T_1 and is of primary interest, model (8.6) corresponds to the competing event and is necessary for identifying model (8.5). In the following, we let $g(\cdot) = h(\cdot)$ without loss of generality.

To tackle the identifiability issue, the dependence structure between T_1 and T_2 needs to be specified. To this end, we can adopt a copula model that relates the joint survival function of (T_1, T_2) to the marginal survival functions as

$$P(T_1 > t_1, T_2 > t_2 | \mathbf{Z}) = H\{S_1(t_1 | \mathbf{Z}), S_2(t_2 | \mathbf{Z})\}. \tag{8.7}$$

Here, $H(\cdot, \cdot)$ is a known copula function, and $S_j(t | \mathbf{Z}) = 1 - P(T_j \leqslant t | \mathbf{Z})$ represents the marginal survival function for $j = 1, 2$. For example, $H(\cdot, \cdot)$ can be chosen from a variety of parametric classes, such as the Clayton copula $H(u, v) = \{u^{-r_c} + v^{-r_c} - 1\}^{-1/r_c}$, $r_c > 0$ (Clayton, 1978), and the Frank copula $H(u, v) = \log_{r_f}\left\{1 + \frac{(r_f^u - 1)(r_f^v - 1)}{r_f - 1}\right\}$, $r_f > 0$ and $r_f \neq 1$ (Genest, 1987), where r_c and r_f are known copula parameters. In practice, the copula parameter may be chosen according to prior knowledge on the strength of the association between T_1 and T_2. Alternatively, one can perform a sensitivity analysis to obtain bounds of $\beta_0(\tau)$ and hence $Q_1(\tau | \mathbf{Z})$ by perturbing r in a plausible range.

Estimation of the regression coefficients is facilitated by a counting process approach using martingales associated with the cause-specific hazard function. Let $N_1(t) = I(\tilde{T} \leqslant t, \tilde{\delta} = 1)$ denote the counting process for T_1, and let $Y(u) = I(\tilde{T} \geqslant u)$ be the at-risk indicator. Let $\lambda_1^*(t | \mathbf{Z}) = \lim_{\Delta \to 0} P\{t \leqslant T < t + \Delta, \varepsilon = 1 | T \geqslant t; \mathbf{Z}\}/\Delta$ be the conditional cause-specific hazard function for event 1, respectively. Thus $M_1(t) = N_1(t) - \int_0^t Y(s)\lambda_1^*(s | \mathbf{Z}) ds$ is a martingale. Therefore, we see that

$$E\{N_1(t) - \int_0^t Y(s)\lambda_1^*(s | \mathbf{Z}) ds\} = 0, \quad \forall t \geqslant 0. \tag{8.8}$$

Noting that $\lambda_1^*(t | \mathbf{Z}) = -\partial \log[H\{S_1(t_1 | \mathbf{Z}), S_2(t_2 | \mathbf{Z})\}]/\partial t_1|_{t_1 = t_2 = t}$ and employing the variable transformation $s = Q_1(u | \mathbf{Z})$ inside the integral, we see that

$$\int_0^t Y(s)\lambda_1^*(s | \mathbf{Z}) ds = \int_0^{F_1\{t | \mathbf{Z}\}} Y\{Q_1(u | \mathbf{Z})\}\phi_1(1 - u, S_2\{Q_1(u | \mathbf{Z}) | \mathbf{Z}\}) du, \tag{8.9}$$

where $\phi_1(v_1, v_2) = \partial \log\{H(v_1, v_2)\}/\partial v_1$. Moreover, under models (8.5) and (8.6), $S_2\{Q_1(u | \mathbf{Z}) | \mathbf{Z}\}$ can be expressed as

$$S_2\{Q_1(u | \mathbf{Z}) | \mathbf{Z}\} = 1 - \int_0^1 I\{\mathbf{Z}^T \boldsymbol{\alpha}_0(v) \leqslant \mathbf{Z}^T \boldsymbol{\beta}_0(u)\} dv. \tag{8.10}$$

Combining (8.5), (8.8), (8.9), and (8.10) yields

$$E\left[n^{-1/2} \sum_{i=1}^n \mathbf{Z}_i\left\{N_{1i}[g\{\mathbf{Z}_i^T \boldsymbol{\beta}_0(\tau)\}] - \int_0^\tau Y[g\{\mathbf{Z}_i^T \boldsymbol{\beta}_0(u)\}]\right.\right.$$
$$\left.\left. \times \phi_1\left(1 - u, 1 - \int_0^1 I\{\mathbf{Z}_i^T \boldsymbol{\alpha}_0(v) \leqslant \mathbf{Z}_i^T \boldsymbol{\beta}_0(u)\} dv\right) du\right\}\right] = 0, \tag{8.11}$$

where $N_{1i}(t)$ is the sample analog of $N_1(t)$.

By reversing the role of T_1 and T_2, an equation parallel to (8.11) can be derived for $\boldsymbol{\alpha}_0(\cdot)$. Following this approach, Ji et al. (2012) propose estimating $\boldsymbol{\beta}_0(\tau)$ and $\boldsymbol{\alpha}_0(\tau)$ from the estimating equations

$$\mathbf{S}_n^{*(k)}(\boldsymbol{\beta}, \boldsymbol{\alpha}, \tau) = 0, \quad k = 1, 2, \tag{8.12}$$

where

$$\mathbf{S}_n^{*(1)}(\boldsymbol{\beta}, \boldsymbol{\alpha}, \tau) = n^{-1/2} \sum_{i=1}^n \mathbf{Z}_i \Big\{ N_{1i}[g\{\mathbf{Z}_i^T \boldsymbol{\beta}(\tau)\}] I\{g^{-1}(\widetilde{T}_i) \leqslant \mathbf{Z}_i^T \boldsymbol{\alpha}(\tau_{U,2})\}$$

$$- \int_0^\tau Y_i[g\{\mathbf{Z}_i^T \boldsymbol{\beta}(u)\}] I\{\mathbf{Z}_i^T \boldsymbol{\beta}(u) \leqslant \mathbf{Z}_i^T \boldsymbol{\alpha}(\tau_{U,2})\}$$

$$\times \phi_1\Big(1 - u, 1 - \int_0^{\tau_{U,2}} I\{\mathbf{Z}_i^T \boldsymbol{\alpha}(v) \leqslant \mathbf{Z}_i^T \boldsymbol{\beta}(u)\} \, dv\Big) \, du \Big\},$$

$$\mathbf{S}_n^{*(2)}(\boldsymbol{\beta}, \boldsymbol{\alpha}, \tau) = n^{-1/2} \sum_{i=1}^n \mathbf{Z}_i \Big\{ N_{2i}[g\{\mathbf{Z}_i^T \boldsymbol{\alpha}(\tau)\}] I\{g^{-1}(\widetilde{T}_i) \leqslant \mathbf{Z}_i^T \boldsymbol{\beta}(\tau_{U,1})\}$$

$$- \int_0^\tau Y_i[g\{\mathbf{Z}_i^T \boldsymbol{\alpha}(u)\}] I\{\mathbf{Z}_i^T \boldsymbol{\alpha}(u) \leqslant \mathbf{Z}_i^T \boldsymbol{\beta}(\tau_{U,1})\}$$

$$\times \phi_2\Big(1 - \int_0^{\tau_{U,1}} I\{\mathbf{Z}_i^T \boldsymbol{\beta}(v) \leqslant \mathbf{Z}_i^T \boldsymbol{\alpha}(u)\} \, dv, 1 - u\Big) \, du \Big\}.$$

The terms in (8.12) can be shown to have expectation 0 at the true parameters. The indicator functions $I\{g^{-1}(\widetilde{T}_i) \leqslant \mathbf{Z}_i^T \boldsymbol{\alpha}(\tau_{U,2})\}$ and $I\{\mathbf{Z}_i^T \boldsymbol{\beta}(u) \leqslant \mathbf{Z}_i^T \boldsymbol{\alpha}(\tau_{U,2})\}$ are inserted to take into consideration that $\boldsymbol{\alpha}_0(u)$ may not be identifiable for upper quantile levels due to censoring. The indicators $I\{g^{-1}(\widetilde{T}_i) \leqslant \mathbf{Z}_i^T \boldsymbol{\beta}(\tau_{U,1})\}$ and $I\{\mathbf{Z}_i^T \boldsymbol{\alpha}(u) \leqslant \mathbf{Z}_i^T \boldsymbol{\beta}(\tau_{U,1})\}$ in $\mathbf{S}_n^{*(2)}(\boldsymbol{\beta}, \boldsymbol{\alpha}, \tau)$ serve the same purpose. With these additional indicator functions, the proposed estimating equations only involve the estimation of $\{\boldsymbol{\beta}_0(\tau), \ \tau \in (0, \tau_{U,1})\}$ and $\{\boldsymbol{\alpha}_0(\tau), \ \tau \in (0, \tau_{U,2})\}$, and thus do not demand the identifiability of $\boldsymbol{\beta}_0(\tau)$ and $\boldsymbol{\alpha}_0(\tau)$ at quantile levels close to 1. The bounds $\tau_{U,1}$ and $\tau_{U,2}$ need to satisfy certain identifiability conditions, and some empirical rules can be employed to select them in practice.

The proposed estimating equations (8.12) can be solved via an iterative algorithm, starting from a reasonable initial estimator for $\boldsymbol{\beta}_0(\tau)$ and $\boldsymbol{\alpha}_0(\tau)$. When the censoring rate on T_2 is very high, the stability of the algorithm may be challenged. A good remedy is to adopt an accelerated failure time (AFT) model for T_2, which requires that $\alpha_0^{(j)}(\tau)$ be constant over τ for $j = 2, 3, \ldots, p + 1$. See Ji et al. (2012) for further details.

8.2   Semi-competing risks data

8.2.1   Introduction

Semi-competing risks data involve a nonterminating event and a terminating event. The terminating event may dependently censor the nonterminating event but remains observable if the nonterminating event occurs first (Fine et al., 2001). In this section, we let T_1 denote time to the nonterminating event and T_2 denote time to the terminating event.

A commonly encountered scenario of semi-competing risks data involves time to a landmark event of disease T_1 and time to death T_2. For example, in the Danish diabetes registry study (Kofoed-Enevoldsen et al., 1987), an important event is the onset of diabetic nephropathy (DN), a clinical syndrome indicative of kidney damage. Time to DN is subject to dependent censoring due to death, because patients may die without experiencing DN first. Similar scenarios that involve an illness event and a death event arise frequently in biomedical studies.

Another common scenario of semi-competing risks data involves a nonterminating event of interest and disease-related patient dropout. In many clinical studies, patients may withdraw from the study due to adverse events and worsening disease status. The dropout event may be associated with the event of interest and should not be treated as independent censoring. Instead, the problem can be formulated as a semi-competing risks problem, with T_1 corresponding to time to the nonterminating event of interest and T_2 corresponding to time to dropout.

With semi-competing risks data, analysis of T_2 can follow standard censored regression methods, because T_2 is only subject to independent censoring by C. However, analysis of T_1 is complicated by the dependent censoring by T_2. As with competing risks data, there are two types of statistical quantities for describing T_1. Crude quantities, such as the cumulative incidence quantile function, reflect the progression of the nonterminating event in the presence of the terminating event. By comparison, net quantities, such as the marginal quantile function, correspond to the removal of the terminating event. We discuss quantile regression methods that target the cumulative incidence quantiles in Section 8.2.2 and the marginal quantiles in Section 8.2.3.

8.2.2 Cumulative incidence quantile regression

The cumulative incidence quantile regression model in (8.2), specified as

$$\bar{Q}_1(\tau|\mathbf{Z}) = g\{\mathbf{Z}^T \beta_0(\tau)\}, \quad \tau \in [\tau_L, \tau_U],$$

can facilitate meaningful insights into the nonterminating event time T_1. For example, when T_1 and T_2 correspond to time to illness and time to death respectively, $\bar{Q}_1(\tau|\mathbf{Z})$ represents the smallest time t when the conditional probability of the illness occurring by time t exceeds τ, accounting for the fact that death may preclude the observation of the illness event.

If the terminating event and the non-terminating event in semi-competing risks data are only subject to right censoring, the observed data can be expressed as $\{\widetilde{T}, \widetilde{\delta}, \eta, \widetilde{D}, \mathbf{Z}\}$, where $\widetilde{T} = T_1 \wedge T_2 \wedge C$, $\widetilde{D} = T_2 \wedge C$, $\widetilde{\delta} = I(T_1 \leqslant \widetilde{D})$, $\eta = I(T_2 \leqslant C)$, and $\mathbf{Z}$ is the covariate vector, which includes 1 as the first component. To conduct cumulative incidence quantile regression for $\bar{Q}_1(\tau|\mathbf{Z})$, we only need to use the information on $\{\widetilde{T}, \widetilde{\delta}, \mathbf{Z}\}$. Therefore, the regression methods in Section 8.1.2 can be readily adopted for estimation and inferences.

In many observational studies, the semi-competing risks data are subject to an additional complication due to left truncation to the terminating event. For example, in the Danish diabetes registry study, there was a time-lag between diabetes diagnosis and study enrollment. As a result, only the patients who remained alive until study enrollment were available for analysis. Let L denote the truncation time, corresponding to time to study enrollment in the Danish diabetes registry. In the presence of the left truncation to T_2, observed data include n i.i.d. replicates of $(\widetilde{T}^*, \widetilde{D}^*, \widetilde{\delta}^*, \eta^*, L^*, \mathbf{Z}^*)$, which follows the conditional distribution of $(\widetilde{T}, \widetilde{D}, \widetilde{\delta}, \eta, L, \mathbf{Z})$ given $L < Y$. We need to account for the left truncation appropriately to avoid the potential selection bias.

Li and Peng (2011) proposed cumulative incidence quantile regression methods that account for the complication of left truncation. Their method is motivated by the equation

$$E\left\{\frac{I(\widetilde{T}_i^* \leqslant t, \widetilde{\delta}_i^* = 1, \eta_i^* = 1)}{W(\widetilde{D}_i^*, \mathbf{Z}_i^*)}\middle|\mathbf{Z}_i^*\right\} = \bar{F}_1(t|\mathbf{Z}_i^*), \tag{8.13}$$

where $W(y, \mathbf{z}) = G(y, \mathbf{z})/\alpha(\mathbf{z})$, $G(y, \mathbf{z}) = P(L < y \leqslant C|\mathbf{z})$, and $\alpha(\mathbf{z}) = P(L < T_2|\mathbf{Z} = \mathbf{z})$. Suppose that an adequate estimator $\widehat{W}(y, \mathbf{z})$ is available. Then the coefficient vector $\beta_0(\tau)$

can be estimated through the following estimating equation:

$$\mathbf{S}_n(\mathbf{b},\tau) \equiv n^{-1/2} \sum_{i=1}^{n} \mathbf{Z}_i^* \left[\frac{I\{\widetilde{T}_i^* \leqslant g(\mathbf{Z}_i^{*T}\mathbf{b}), \widetilde{\delta}_i^* = 1, \eta_i^* = 1\}}{\hat{W}(\widetilde{D}_i^*, \mathbf{Z}_i^*)} - \tau \right] = 0,$$

The inverse weights $W(y,\mathbf{z})$ can be estimated under suitable assumptions. For example, the assumption that (L,C) is independent of $(T_1,T_2,\mathbf{Z})$ is realistic in many settings with administrative right censoring and left truncation. In this case, $G(y,\mathbf{z})$ does not depend on $\mathbf{z}$ and can be simplified to $G(y)$. $\alpha(\mathbf{z})$ represents the probability that a subject with covariate vector $\mathbf{z}$ is not truncated. Under realistic assumptions, Li and Peng (2011) proposed consistent estimators, $\hat{\alpha}(\mathbf{z})$ and $\hat{G}(y)$, which yield an adequate estimator for the inverse weights $W(y,\mathbf{z})$.

The solution to $\mathbf{S}_n(\mathbf{b},\tau) = 0$, referred to as $\hat{\beta}(\tau)$, can be shown to converge to $\beta_0(\tau)$ uniformly for $\tau \in [\tau_L, \tau_U]$. Moreover, we can extend the three-step procedure in Section 8.1.2 to obtain a resampling-free covariance estimator. Second-stage inference procedures can be adopted to evaluate the overall significance and constancy of $\beta_0(\tau)$.

8.2.3 Marginal quantile regression

To estimate the marginal quantile function for the nonterminating event time T_1, we can assume that T_1 and T_2 marginally follow the quantile regression models in (8.5) and (8.6), where $\beta_0(\tau)$ is the regression coefficient of primary interest. To tackle lack of nonparametric identifiability of the marginal quantities for T_1, we assume that the association between T_1 and T_2 follow a prespecified copula model with *unspecified* copula parameter. Specifically, we assume that

$$\Pr(T_1 > s, T_2 > t \mid \mathbf{Z}) = H\{S_1(s|\mathbf{Z}), S_2(t|\mathbf{Z}), g(\bar{\mathbf{Z}}^T \mathbf{r}_0)\}, \tag{8.14}$$

where $S_j(\cdot|\mathbf{Z})$ denotes the conditional marginal survival function for T_j, $j = 1,2$, and $\bar{\mathbf{Z}}$ is a subvector of $\mathbf{Z}$, or $\mathbf{Z}$ itself, that also contains 1 as first element. The $H(\cdot)$ function is a copula function with known form, such as the Clayton copula or the Frank copula, and its association parameter is linked to covariates through $g(\bar{\mathbf{Z}}^T \mathbf{r}_0)$. Thus, the unknown association parameter $\mathbf{r}_0$ explains the possibly covariate-dependent association between T_1 and T_2. When $\bar{\mathbf{Z}} = 1$, the association between T_1 and T_2 is assumed to be homogeneous for all covariate groups. It is worth noting that the association model in (8.14) poses much weaker assumptions than the association model posed for competing risks data. This is because $\{\tilde{D},\eta\}$, the additional information in semi-competing risks data when compared to competing risks data, allows us to identify the strength of association between T_1 and T_2.

The proposed method in Li and Peng (2015) for estimating $\beta_0(\tau)$ and $\mathbf{r}_0$ is based on the following two equalities:

$$\Pr(\widetilde{T} > t \mid \widetilde{D} > t, \mathbf{Z}) = \frac{\Pr(T_1 > t, T_2 > t \mid \mathbf{Z})}{\Pr(T_2 > t \mid \mathbf{Z})} = K_A\{F_1(t \mid \mathbf{Z}), F_2(t \mid \mathbf{Z}), g(\bar{\mathbf{Z}}^T \mathbf{r}_0)\},$$

$$\Pr(\widetilde{T} \leqslant s \mid \widetilde{D} > t, \mathbf{Z}) = \frac{\Pr(T_1 \leqslant s, T_2 > t \mid \mathbf{Z})}{\Pr(T_2 > t \mid \mathbf{Z})} = K_B\{F_1(s \mid \mathbf{Z}), F_2(t \mid \mathbf{Z}), g(\bar{\mathbf{Z}}^T \mathbf{r}_0)\},$$

where $s \leqslant t$, $K_A(u,v,\theta) = H(1-u,1-v,\theta)/(1-v)$, and $K_B(u,v,\theta) = \{1 - v - H(1-u, 1-v, \theta)\}/(1-v)$. Moreover, we notice that $\alpha_0(u)$ can be related to $F_2(\cdot|\mathbf{Z})$ by

$$F_2(t \mid \mathbf{Z}) = \int_0^1 I\{F_2(t \mid \mathbf{Z}) \geqslant u\}du = \int_0^1 I[t \geqslant \exp\{\mathbf{Z}^T \alpha_0(u)\}]du.$$

Here and in the sequel, we let $g(x) = h(x) = \exp(x)$ without loss of generality. Note that T_2 is only subject to independent censoring by C. Therefore, a suitable censored quantile regression method, such as that proposed in Peng and Huang (2008), would provide an adequate estimator for $\boldsymbol{\alpha}_0(u)$, namely $\hat{\boldsymbol{\alpha}}(u)$ for $u \in (0, \tau_{U,2}]$. Here, $\tau_{U,2}$ is an upper bound in $(0,1)$ subject to regularity conditions.

Motivated by these facts, Li and Peng (2015) consider estimating $\boldsymbol{\beta}_0(\tau)$ and $\mathbf{r}_0$ using the following estimating functions:

$$\mathbf{S}_n(\boldsymbol{\beta}, \hat{\boldsymbol{\alpha}}, \mathbf{r}, \tau) = 0, \quad \int_{\tau_a}^{\tau_b} \mathbf{W}_n(\boldsymbol{\beta}, \hat{\boldsymbol{\alpha}}, \mathbf{r}, \tau) d\tau = 0, \qquad (8.15)$$

where

$$\mathbf{S}_n(\boldsymbol{\beta}, \boldsymbol{\alpha}, \mathbf{r}, \tau) = n^{-1/2} \sum_{i=1}^n \mathbf{Z}_i I\{\mathbf{Z}_i^T \boldsymbol{\beta}(\tau) \leq \mathbf{Z}_i^T \boldsymbol{\alpha}(\tau_{U,2})\} P_i(\boldsymbol{\beta}, \boldsymbol{\alpha}, \mathbf{r}, \tau),$$

$$\mathbf{W}_n(\boldsymbol{\beta}, \boldsymbol{\alpha}, \mathbf{r}, \tau) = n^{-1/2} \sum_{i=1}^n \bar{\mathbf{Z}}_i Q_i(\boldsymbol{\beta}, \boldsymbol{\alpha}, \mathbf{r}, \tau),$$

and

$$P_i(\boldsymbol{\beta}, \boldsymbol{\alpha}, \mathbf{r}, \tau) = I\{\log \tilde{T}_i > \mathbf{Z}_i^T \boldsymbol{\beta}(\tau)\} - I\{\log \tilde{D}_i > \mathbf{Z}_i^T \boldsymbol{\beta}(\tau)\}$$
$$\times K_A \left\{ \tau, \int_0^{\tau_{U,2}} I\{\mathbf{Z}_i^T \boldsymbol{\beta}(\tau) \geq \mathbf{Z}_i^T \boldsymbol{\alpha}(u)\} du, g(\bar{\mathbf{Z}}_i^T \mathbf{r}) \right\},$$

$$Q_i(\boldsymbol{\beta}, \boldsymbol{\alpha}, \mathbf{r}, \tau) = \int_{t \in (0,\infty)} I\{\mathbf{Z}_i^T \boldsymbol{\beta}(\tau) \leq \log t \leq \mathbf{Z}_i^T \boldsymbol{\alpha}(\tau_{U2}) \wedge \log \tilde{D}_i\} \times \left(I\{\log \tilde{T}_i \leq \mathbf{Z}_i^T \boldsymbol{\beta}(\tau)\} \right.$$
$$\left. - K_B \left[\tau, \int_0^{\tau_{U2}} I\{\log t \geq \mathbf{Z}_i^T \boldsymbol{\alpha}(u)\} du, g(\bar{\mathbf{Z}}_i^T \mathbf{r}) \right] \right) dt,$$

which involves $\boldsymbol{\alpha}(u)$ for u up to $\tau_{U,2}$ instead of 1.

An iterative procedure can be used to solve the estimating equations in (8.15) as follows.

Step A Estimate $\{\hat{\boldsymbol{\alpha}}(\tau), \tau \in (0, \tau_{U,2}]\}$ using the method proposed in Peng and Huang (2008).

Step B Set $k = 0$ and choose an initial value for $\hat{\boldsymbol{\beta}}$ and $\hat{\mathbf{r}}$, denoted by $\hat{\boldsymbol{\beta}}^{[k]}$ and $\hat{\mathbf{r}}^{[k]}$.

Step C Update $\hat{\boldsymbol{\beta}}^{[k]}$ with $\hat{\boldsymbol{\beta}}^{[k+1]} = \hat{\boldsymbol{\beta}}(\hat{\mathbf{r}}^{[k]})$.

Step D Solve $\hat{\mathbf{r}}^{[k]}$ from $\mathbf{W}_n(\hat{\boldsymbol{\beta}}^{[k]}, \hat{\boldsymbol{\alpha}}, \mathbf{r}) = 0$. Then increase k by 1 and go to step C until certain convergence criteria are satisfied.

In practice, $\hat{\boldsymbol{\beta}}^{[0]}$ may be chosen as the naive estimate obtained by treating $(\tilde{T}, \tilde{\delta})$ as independently censored survival data and employing Peng and Huang's (2008) method, and $\hat{\mathbf{r}}^{(0)}$ can be chosen as the $\mathbf{r}$ corresponding to the Kendall's tau coefficient between $\tilde{T}_i$ and $\tilde{D}_i$, $i = 1, 2, \ldots, n$. The resulting $\hat{\boldsymbol{\beta}}(\tau)$ and $\hat{\mathbf{r}}$ possess nice asymptotic properties. Specifically, $\hat{\boldsymbol{\beta}}(\tau)$ converges uniformly to $\boldsymbol{\beta}_0(\tau)$, and $\sqrt{n}\{\hat{\boldsymbol{\beta}}(\tau) - \boldsymbol{\beta}_0(\tau)\}$ converges weakly to a zero-mean Gaussian process. Similarly, $\hat{\mathbf{r}}$ converges in probability to $\mathbf{r}_0$, and $\sqrt{n}\{\hat{\mathbf{r}} - \mathbf{r}_0\}$ asymptotically follows a zero-mean normal distribution. Bootstrap methods can be used to do covariance estimation and inference.

8.3 Summary and open problems

In this chapter, we have introduced quantile regression methods for competing risks and semi-competing risks data. As a crude quantity, the cumulative incidence quantiles can be modeled directly, without additional modeling of the association between the event times. When the scientific interest centers on the marginal quantities, however, associational assumptions are necessary to tackle the curse of non-identifiability. In this case, posing the copula model assumption is a sensible and convenient solution. The methods introduced here can render robust and comprehensive analysis for competing risks and semi-competing risks data.

There remain several interesting topics in this area. The methods we present here are all semiparametric. Considering nonparametric quantile regression models can provide enhanced flexibility and thus may be desirable in some practical settings. Moreover, it is of interest to extend the methods to accommodate additional real-life complications, such as left censoring and interval censoring to the event times. It would be useful to incorporate existing survival analysis techniques such as Ji et al. (2012) and Huang (1996) into the competing risks quantile regression setting to handle these survival data features. Future research is required to develop detailed methods.

Bibliography

D. G. Clayton. A model for association in bivariate life tables and its application in epidemiological studies of familial tendency in chronic disease incidence. *Biometrika*, 65(1): 141–151, 1978.

F. J. Cummings, R. Gray, D. C. Tormey, T. E. Davis, H. Volk, J. Harris, G. Falkson, and J. M. Bennett. Adjuvant tamoxifen versus placebo in elderly women with node-positive breast cancer: Long-term follow-up and causes of death. *Journal of Clinical Oncology*, 11 (1):29–35, 1993.

J. P. Fine, H. Jiang, and R. Chappell. On semi-competing risks data. *Biometrika*, 88(4): 907–919, 2001.

C. Genest. Frank's family of bivariate distributions. *Biometrika*, 74(3):549–555, 1987.

J. Huang. Efficient estimation for the proportional hazards model with interval censoring. *Annals of Statistics*, 24(2):540–568, 1996.

S. Ji, L. Peng, Y. Cheng, and H. Lai. Quantile regression for doubly censored data. *Biometrics*, 68(1):101–112, 2012.

A. Kofoed-Enevoldsen, K. Borch-Johnsen, S. Kreiner, J. Nerup, and T. Deckert. Declining incidence of persistent proteinuria in type I (insulin-dependent) diabetic patients in Denmark. *Diabetes*, 36(2):205–209, 1987.

R. Li and L. Peng. Quantile regression for left-truncated semicompeting risks data. *Biometrics*, 67(3):701–710, 2011.

R. Li and L. Peng. Quantile regression adjusting for dependent censoring from semicompeting risks. *Journal of the Royal Statistical Society, Series B*, 77(1):107–130, 2015.

L. Peng and J. P. Fine. Competing risks quantile regression. *Journal of the American Statistical Association*, 104(488):1440–1453, 2009.

L. Peng and Y. Huang. Survival analysis with quantile regression models. *Journal of the American Statistical Association*, 103(482):637–649, 2008.

J. M. Robins, A. Rotnitzky, and L. P. Zhao. Estimation of regression coefficients when some regressors are not always observed. *Journal of the American Statistical Association*, 89 (427):846–866, 1994.

Y. Sun, H. J. Wang, and P. B. Gilbert. Quantile regression for competing risks data with missing cause of failure. *Statistica Sinica*, 22(2):703–728, 2012.

A. A. Tsiatis. An example of non-identifiability in competing risks. *Scandinavian Actuarial Journal*, 1978(4):235–239, 1978.

9.

Instrumental Variable Quantile Regression

Victor Chernozhukov

MIT, Cambridge, Massachusetts, USA

Christian Hansen

University of Chicago, Chicago, Illinois, USA

Kaspar Wüthrich

University of California, San Diego, California, USA

CONTENTS

This chapter reviews the instrumental variable quantile regression model of Chernozhukov and Hansen (2005). We discuss the key conditions used for identification of structural quantile effects within this model which include the availability of instruments and a restriction on the ranks of structural disturbances. We outline several approaches to obtaining point estimates and performing statistical inference for model parameters. Finally, we point to possible directions for future research.

9.1   Introduction

Empirical analyses often focus on understanding the structural (causal) relationship between an outcome, Y, and variables of interest, D. In many cases, interest is not only in how D affects measures of the center of the distribution Y but also in other features of the distribution. For example, in understanding the effect of a government-subsidized saving program, one might be more interested in the effect of the program on the lower tail of the savings distribution conditional on individual characteristics than in the effect of the program on the mean of the savings distribution. Quantile regression, as introduced by Koenker and Bassett (1978), offers one useful way to estimate such effects and to summarize the impact of changes in D on the conditional distribution of Y.

Of course, variables of interest are often endogenous or self-selected in observational data. For example, individuals choose whether to participate in government-subsidized savings plans. Similarly, in trying to understand the demand relationship between quantity and price, one must face the fact that prices and quantities are jointly determined. Endogeneity of covariates renders conventional quantile regression inconsistent for estimating the causal effects of variables on the quantiles of outcomes of interest. Instrumental variables (IV) provide a powerful tool for learning about structural effects in the presence of endogenous right-hand-side variables, and we focus this review on a generalization of the classical linear IV model to accommodate estimating structural quantile treatment effects (QTE) in the presence of endogenous covariates.

We specifically focus on the instrumental variable quantile regression model developed in Chernozhukov and Hansen (2005). Chernozhukov and Hansen (2005) provide conditions under which structural QTE are nonparametrically identified through the use of instrumental variables. The key identifying assumption is a condition that restricts how structural errors, which we will refer to as rank variables, vary across different potential states of the endogenous variables. The simplest, though strongest, version of this condition is rank invariance, which requires that individual ranks are invariant to the potential states of the endogenous variable. Rank invariance is implied by many classical structural models which posit a single source of unobserved heterogeneity, and the framework developed in Chernozhukov and Hansen (2005) is indeed a natural generalization of the classical structural simultaneous equation model, corresponding to a structural simultaneous equation model with nonadditive errors.

There are alternative sets of modeling assumptions that one could employ to build a quantile model with endogeneity. Abadie et al. (2002) offer an approach within the local average treatment effect framework of Imbens and Angrist (1994). This approach differs from the framework discussed in this review in a few key respects. The Abadie et al. (2002) framework does not restrict the behavior of rank variables across potential treatment states and thus allows for essentially unrestricted heterogeneity of effects at the cost of only identifying QTE for the subpopulation of compliers. To achieve identification without restricting structural errors, Abadie et al. (2002) restrict attention to a setting with a binary endogenous treatment variable and impose a monotonicity restriction on the relation between the instrument and treatment. When the endogenous variable of interest is continuous, triangular models as in Imbens and Newey (2009) provide another alternative framework for identifying and estimating QTE. As in Abadie et al. (2002), the Imbens and Newey (2009) framework does not restrict the evolution of ranks across treatment states. Instead, it relies on monotonicity of the selection mechanism in a scalar disturbance. We refer interested readers to Chapter 10 of this handbook, which discusses the Abadie et al. (2002) approach in detail and contains further comparative discussion of the two modeling frameworks, and to Wüthrich (2014), which establishes a connection between the estimands of both models under the assumptions of the Abadie et al. (2002) framework. Section 9.2.5 discusses the approach by Imbens and Newey (2009) and compares it to the framework discussed in this chapter.

We devote the remainder of this review to providing an overview of the model of Chernozhukov and Hansen (2005) along with outlining approaches to estimating parameters and performing inference within this model.

9.2 Model overview

9.2.1 The instrumental variable quantile regression model

The instrumental variable quantile regression (IVQR) model is developed within the conventional potential outcome framework. Potential real-valued outcomes, which vary among observational units, are indexed against potential treatment states $d \in \mathcal{D}$ and denoted Y_d. The potential outcomes $\{Y_d\}$ are latent because, given the observed treatment D, the observed outcome for each observational unit is only one component

$$Y := Y_D$$

of the potential outcomes vector $\{Y_d\}$. Note that throughout this review we use capital letters to denote random variables and lower-case letters to denote the potential values the random variables may take. We also do not explicitly state technical measurability assumptions as these can be deduced from the context.

The objective of causal or structural analysis is to learn about features of the distributions of potential outcomes Y_d. Of primary interest to us are the τth quantiles of potential outcomes under various potential treatment states d, conditional on observed characteristics $X = x$, denoted by

$$Q_{Y_d}(\tau|x) = q(\tau, d, x).$$

We note that, after conditioning on observed characteristics $X = x$, each potential outcome Y_d can be related to its quantile function $q(\tau, d, x)$ as

$$Y_d = q(U_d, d, x), \quad \text{where } U_d \sim U(0,1) \tag{9.1}$$

is the structural error term and (9.1) follows from the Fisher–Skorohod representation of random variables.

Given the conditional quantiles of the potential outcomes, we are interested in QTE which are given by the difference in τth quantiles of two different conditional potential outcomes Y_{d_1} and Y_{d_0}:

$$q(\tau, d_1, x) - q(\tau, d_0, x).$$

These QTE may then be used to summarize the impact of variables of interest D on the quantiles of potential outcomes as suggested in Doksum (1974) and Lehmann (1975).

It is important to note that the structural error U_d in (9.1) is responsible for heterogeneity of potential outcomes among individuals with the same observed characteristics x. This error term determines the relative ranking of observationally equivalent individuals in the distribution of potential outcomes given the individuals' observed characteristics, and thus we refer to U_d as the rank variable. Because U_d drives differences between observationally equivalent individuals, one may think of U_d as representing some unobserved characteristic, such as ability or proneness, where we adopt the term "proneness" from Doksum (1974) who uses it as in "prone to learn fast" or "prone to grow taller." This interpretation of the structural error makes quantile analysis an interesting tool for describing and learning the structure of heterogeneous treatment effects while accounting for unobserved heterogeneity; see Doksum (1974), Heckman et al. (1997), and Koenker (2005). For example, consider a returns-to-training model, where the Y_d are potential earnings under different training levels d, and $q(\tau, d, x)$ is the conditional earnings function which describes how an individual with training d, characteristics x, and latent "ability" τ is rewarded by the labor market. The earnings function may differ for different levels of τ, implying heterogeneous effects of training on earnings of people who have different levels of "ability." For example, it may be that the largest returns to training accrue to those in the upper tail of the conditional distribution, that is, to the "high-ability" workers.

In observational data, the realized treatment D is often selected in relation to potential outcomes, inducing endogeneity. This endogeneity makes the conventional quantile regression of Y on D and X, which relies upon the restriction

$$P[Y \leqslant Q_Y(\tau|D, X)|D, X] = \tau \text{ a.s.,}$$

inappropriate for measuring the structural quantile function $q(\tau, d, x)$ and thus for learning about QTE. Indeed, the conditional quantile function, $Q_Y(\tau|d, x)$, solving these equations will generally differ from the structural quantile function of latent potential outcomes, $q(\tau, d, x)$, under endogeneity. The IVQR model presented below provides conditions under which we can identify and estimate the quantiles of the latent potential outcomes through the use of instruments Z that affect D but are independent of potential outcomes by making use of the nonlinear quantile-type conditional moment restrictions

$$P[Y \leqslant q(\tau, D, X)|X, Z] = \tau \text{ a.s.}$$

Formally, the IVQR model consists of five key conditions (some are representations).

Assumption 1 (IVQR model) *Consider a common probability space (Ω, F, P) and the set of potential outcome variables $(Y_d, d \in \mathcal{D})$, endogenous variables D, exogenous covariates X, and instrumental variables Z. The following conditions hold jointly with probability 1:*

- A1 *(Potential outcomes). Conditional on X and for each d, $Y_d = q(U_d, d, X)$, where $\tau \mapsto q(\tau, d, X)$ is nondecreasing on $[0, 1]$ and left-continuous and $U_d \sim U(0, 1)$.*

- A2 *(Independence). Conditional on X and for each d, U_d is independent of instrumental variables Z.*

- A3 *(Selection). $D := \delta(Z, X, \nu)$ for some unknown function δ and random vector ν.*

- A4 *(Rank similarity). Conditional on (X, Z, ν), $\{U_d\}$ are identically distributed.*

- A5 *(Observables) The observed random vector consists of $Y := Y_D$, D, X and Z.*

The following theorem summarizes the main econometric implications of the model.

Theorem 1 (Main implications of the IVQR model) *Suppose conditions A1–A5 hold. (i) Then we have for $U := U_D$, with probability 1,*

$$Y = q(U, D, X), \quad U \sim U(0, 1)|X, Z. \tag{9.2}$$

(ii) If (9.2) holds and $\tau \mapsto q(\tau, d, X)$ is strictly increasing for each d, then for each $\tau \in (0, 1)$, a.s.

$$P[Y \leqslant q(\tau, D, X)|X, Z] = \tau. \tag{9.3}$$

(iii) If (9.2) holds, then for any closed subset I of $[0,1]$, a.s.

$$P(U \in I) \leqslant P[Y \in q(I,D,X)|X,Z],\tag{9.4}$$

where $q(I,d,x)$ is the image of I under the mapping $\tau \mapsto q(\tau,d,x)$.

The first result states that the main consequence of A1–A5 is a simultaneous equation model (9.2) with nonseparable error U that is independent of Z, X, and normalized so that $U \sim U(0,1)$. The second result considers econometric implications when $\tau \mapsto q(\tau, D, X)$ is strictly increasing, which requires that Y is nonatomic conditional on X and Z. In this case, we obtain the conditional moment restriction (9.3). This implication follows from the first result and the fact that

$$\{Y \leqslant q(\tau, D, X)\} \text{ is equivalent to } \{U \leqslant \tau\}$$

when $q(\tau, D, X)$ is strictly increasing in τ. The final result deals with the case where Y may have atoms conditional on X and Z, for example when Y is a count or discrete response variable. The first two results were obtained in Chernozhukov and Hansen (2005), and the third result is in the spirit of results given in Chesher et al. (2013), Chesher (2005), and Chesher and Smolinski (2010).

The model and the results of Theorem 1 are useful for two reasons. First, Theorem 1 serves as a means of identifying QTE in a reasonably general heterogeneous effects model. Second, by demonstrating that the IVQR model leads to the conditional moment restrictions (9.3) and (9.4), Theorem 1 provides an economic and causal foundation for estimation based on these restrictions.

Equations (9.3) and (9.4) implicitly define the identification region for the structural quantile function $(\tau, d, x) \mapsto q(\tau, d, x)$. The identification region for the case of strictly increasing $\tau \mapsto q(\tau, d, x)$ can be stated as the set $\mathcal{M}$ of functions $(\tau, d, x) \mapsto m(\tau, d, x)$ that satisfy the following relations, for all $\tau \in (0,1]$:

$$P[Y < m(\tau, D, X)|X, Z] = \tau \text{ a.s.}\tag{9.5}$$

This representation of the identification region $\mathcal{M}$ is implicit. Without imposing additional conditions, statistical inference about $q \in \mathcal{M}$ from (9.5) can be performed using weak-identification robust inference as described in Chernozhukov and Hansen (2008), Jun (2008), Santos (2012), or Chernozhukov et al. (2009). Section 9.2.2 discusses conditions under which point identification is obtained; and we mainly focus on the point-identified case in discussing estimation and inference in this review.

The identification region for the case of weakly increasing $\tau \mapsto q(\tau, d, x)$ can be stated as the set $\mathcal{M}$ of functions $(\tau, d, x) \mapsto m(\tau, d, x)$ that satisfy the following relations: for any closed subset I of $(0,1]$,

$$P(U \in I) \leqslant P[Y \in m(I, D, X)|X, Z] \text{ a.s.,}$$

where $m(I, D, X)$ is the image of I under the mapping $\tau \mapsto m(\tau, D, X)$. The inference problem here falls in the class of conditional moment inequalities, and approaches such as those described in Andrews and Shi (2013) or Chernozhukov et al. (2013b) can be used.

9.2.2 Conditions for point identification

Here we briefly discuss the key conditions under which the moment equations (9.3) point-identify the structural quantile function $q(\tau, d, x)$. We focus on the simplest case where $D \in \{0, 1\}$ and $Z \in \{0, 1\}$ and refer to Chernozhukov and Hansen (2013) for more details and extensions to multi-valued and continuous D and to Chernozhukov and Hansen (2006) for a discussion of identification in linear-in-parameters models. The following analysis is conditional on $X = x$, but we suppress this dependence for ease of notation.

It follows from Theorem 1 that there is at least one function $q(\tau, d)$ that solves $P[Y \leqslant q(\tau, D)|Z] = \tau$ a.s. The function $q(\tau, d)$ can be equivalently represented by a vector of its values $q = (q(\tau, 0), q(\tau, 1))'$. Therefore, for vectors of the form $y = (y_0, y_1)'$, we have a vector of moment equations

$$\Pi(y) := (P[Y \leqslant y_D|Z = 0] - \tau, P[Y \leqslant y_D|Z = 1] - \tau)',$$

where $y_D := (1 - D) \cdot y_0 + D \cdot y_1$. We say that $q(\tau, d)$ is identified in some parameter space, $\mathcal{L}$, if $y = q$ is the only solution to $\Pi(y) = 0$ among all $y \in \mathcal{L}$. Define the Jacobian $\partial \Pi(y)$ of $\Pi(y)$ with respect to $y = (y_0, y_1)'$ as

$$
\partial \Pi(y) := \begin{bmatrix} f_Y(y_0|D = 0, Z = 0) P[D = 0|Z = 0] & f_Y(y_1|D = 1, Z = 0) P[D = 1|Z = 0] \\ f_Y(y_0|D = 0, Z = 1) P[D = 0|Z = 1] & f_Y(y_1|D = 1, Z = 1) P[D = 1|Z = 1] \end{bmatrix}
$$

$$
:= \begin{bmatrix} f_{Y,D}(y_0, 0|Z = 0) & f_{Y,D}(y_1, 1|Z = 0) \\ f_{Y,D}(y_0, 0|Z = 1) & f_{Y,D}(y_1, 1|Z = 1) \end{bmatrix} \tag{9.6}
$$

The key condition for point identification is full rank of $\partial \Pi(y)$ at $y = q$. This local identification condition can be extended to a global condition; see Chernozhukov and Hansen (2005, 2013).

Full rank of $\partial \Pi(y)$ requires the impact of Z on the joint distribution of (Y, D) to be rich enough. To illustrate, note that full rank of $\partial \Pi(y)$ is equivalent to $\det(\partial \Pi(y)) \neq 0$, which implies that

$$\frac{f_{Y,D}(y_1, 1|Z = 1)}{f_{Y,D}(y_0, 0|Z = 1)} > \frac{f_{Y,D}(y_1, 1|Z = 0)}{f_{Y,D}(y_0, 0|Z = 0)} \tag{9.7}$$

(or the same condition with $>$ replaced by $<$). Inequality (9.7) may be interpreted as a *monotone likelihood ratio condition*. That is, the instrument Z should have a monotonic impact on the likelihood ratio in (9.7), which is generally stronger than the usual condition that D is correlated with Z. Nevertheless, the full-rank condition will be trivially satisfied in many useful contexts. For instance, if the instrument satisfies one-sided noncompliance (e.g., those not offered the treatment cannot receive that treatment), $P[D = 1|Z = 0] = 0$, so that the right-hand side of (9.7) equals 0, which makes (9.7) hold trivially.

9.2.3   Discussion of the IVQR model

Condition A1 imposes monotonicity on the structural function of interest which makes its relation to $q(\tau, d, x)$ apparent. Condition A2 states that potential outcomes are independent of Z, given X, which is a conventional independence restriction employed in nonlinear IV models. Condition A3 provides a convenient representation of a treatment selection mechanism, stated for the purpose of discussion. In A3, the unobserved random vector ν is responsible for the difference in treatment choices D across observationally identical individuals. Dependence between ν and $\{U_d\}$ is the source of endogeneity that makes the conventional exogeneity assumption $U \sim U(0, 1)|X, D$ break down. This failure leads to inconsistency of exogenous quantile methods for estimating the structural quantile function. Within the model outlined above, this breakdown is resolved through the use of instrumental variables.

The independence imposed in conditions A2 and A3 is weaker than the assumption that both the disturbances $\{U_d\}$ in the outcome equation and the disturbances ν in the selection equation are *jointly* independent of the instrument Z which is maintained, for example, in Abadie et al. (2002). The assumption that structural errors $\{U_d\}$ and first-stage unobservables ν are jointly independent of instruments may be violated in practical examples. For example, this condition would not hold when the instrument is measured with error as discussed in Hausman (1977), or when the instrument is not assigned exogenously relative to the selection equation as in Imbens and Angrist (1994, Example 2).

Condition A4 is the key restriction of the IVQR model. This assumption restricts the variation in ranks across potential outcomes and is key to identifying the structural quantile function $q(\tau, d, x)$ and the associated QTE. The simplest, though strongest, version of this condition is rank invariance, which means that ranks U_d do not vary with potential treatment states d:

$$U_d = U, \quad \text{for each } d \in \mathcal{D}. \tag{9.8}$$

Rank invariance is a strong condition that has been used in many interesting models without endogeneity such as Doksum (1974), Heckman et al. (1997), and Koenker and Geling (2001). Rank invariance implies that a common unobserved factor U, such as innate ability, determines the ranking of a given person across treatment states. For example, under rank invariance, people who are strong (highly ranked) earners without a training program ($d = 0$) remain strong earners having done the training ($d = 1$). Indeed, the earnings of a person with characteristics x and rank $U = \tau$ in training state 0 is $Y_0 = q(\tau, 0, x)$ and in state 1 is $Y_1 = q(\tau, 1, x)$; that is, the individual's rank, τ, in the earnings distribution is exactly the same whether or not the person receives training. Finally, note that condition A3 is a pure representation under rank invariance as nothing restricts the unobserved component ν in this case.

While convenient, rank invariance seems too strong a condition for many applications as discussed, for example, in Heckman et al. (1997). Rank invariance maintains that an individual's rank in the outcome distribution under every possible state of the endogenous variables is exactly the same. Thus, the potential outcomes $\{Y_d\}$ are jointly degenerate, which allows identification of individual treatment effects even though no individual is ever observed in more than one state of the endogenous variable. Rank invariance also rules out the possibility that there may be many unobserved factors that determine individual ranks which may be differentially relevant under different states of the endogenous variables.

Rank similarity condition A4 relaxes these undesirable features of rank invariance by allowing the rank variables $\{U_d\}$ to change across d in a way that reflects unobserved, asystematic variation in ranks across states of the endogenous variables while also providing sufficient structure to allow identification of QTE via the moment restrictions in Theorem 1. More specifically, condition A4 relaxes exact rank invariance by allowing "slippages," in the terminology of Heckman et al. (1997), in an individuals's rank away from some common level U. Conditional on U, which may enter disturbance ν in the selection equation, and any other components of ν from the selection condition A3, rank similarity yields that the slippages of ranks away from common level U under different potential states of the endogenous variable, $U_d - U$, are identically distributed across $d \in \mathcal{D}$. In this formulation, we implicitly assume that any selection of the state of the endogenous variables occurs without knowing the exact potential outcomes. That is, selection may depend on U and even the distribution of slippages, but does not depend on the exact slippage $U_d - U$. This assumption is consistent with many empirical situations where the exact latent outcomes are not known before receipt of treatment. We also note that conditioning on appropriate covariates X may be important to achieve rank similarity. Finally, we note that rank similarity has testable implications. Dong and Shen (2015) and Frandsen and Lefgren (2015) exploit these conditions to develop tests of unconditional rank similarity, and their approaches could be extended to test some forms of conditional rank similarity.

9.2.4 Examples

We present two examples that highlight the nature of the model, its strengths, and its limitations.

Example 1 (Demand with nonseparable error). The following is a generalization of the classic supply–demand example taken from Chernozhukov and Hansen (2006). Consider the model

$$Y_p = q\,(U, p)\,, \\ \tilde{Y}_p = \rho\,(\mathcal{U}, p, z)\,, \\ P \in \{p : \rho\,(U, p, Z) = q\,(\mathcal{U}, p)\}\,, \tag{9.9}$$

where the functions q and ρ are increasing in their first argument. The function $p \mapsto Y_p$ is the random demand function, and $p \mapsto \tilde{Y}_p$ is the random supply function. Additionally, q and ρ may depend on covariates X, but this dependence is suppressed.

The random variable U is the level of demand and describes the demand curve at different states of the world. Demand is maximal when $U = 1$ and minimal when $U = 0$, holding p fixed. Note that we imposed rank invariance (9.8), as is typical in classic supply–demand models, by making U invariant to p.

Model (9.9) incorporates traditional additive error models for demand which have $Y_p = q(p) + \epsilon$ where $\epsilon = Q_\epsilon(U)$. The model is much more general in that the price can affect the entire distribution of the demand curve, while in traditional models it only affects the location of the distribution of the demand curve.

The τ-quantile of the demand curve $p \mapsto Y_p$ is given by $p \mapsto q(\tau, p)$. Thus, the curve $p \mapsto Y_p$ lies below the curve $p \mapsto q(\tau, p)$ with probability τ. Therefore, the various quantiles of the potential outcomes play an important role in describing the distribution and heterogeneity of the stochastic demand curve. The QTE may be characterized by $\partial q(\tau, p)/\partial p$ or by an elasticity $\partial \ln q(\tau, p)/\partial \ln p$. For example, consider the model $q(\tau, p) = \exp\left(\beta(\tau) + \alpha(\tau)\ln p\right)$ which corresponds to a Cobb–Douglas model for demand with nonseparable error $Y_p = \exp(\beta(U) + \alpha(U)\ln p)$. The log transformation gives $\ln Y_p = \beta(U) + \alpha(U)\ln p$, and the QTE for the log-demand equation is given by the elasticity of the original τ-demand curve $\alpha(\tau) = \partial Q_{\ln Y_p}(\tau)/\partial \ln p = \partial \ln q(\tau, p)/\partial \ln p$.

The elasticity $\alpha(U)$ is random and depends on the state of the demand U and may vary considerably with U. For example, this variation could arise when the number of buyers varies and aggregation induces a nonconstant elasticity across the demand levels. Chernozhukov and Hansen (2008) estimate a simple demand model based on data from a New York fish market that were first collected and used by Graddy (1995). They find point estimates of the demand elasticity, $\alpha(\tau)$, that vary quite substantially from -2 for low quantiles to -0.5 for high quantiles of the demand curve.

The third condition in (9.9), $P \in \{p : \rho\,(U, p, Z) = q\,(\mathcal{U}, p)\}$, is the equilibrium condition that generates endogeneity; the selection of the clearing price P by the market depends on the potential demand and supply outcomes. As a result, we have a representation that is consistent with condition A3, $P = \delta(Z, \nu)$, where ν consists of U and $\mathcal{U}$ and may include "sunspot" variables if the equilibrium price is not unique. Thus, what we observe can be written as

$$Y := q(U, P), \quad P := \delta(Z, \nu), \quad U \text{ is independent of } Z. \tag{9.10}$$

Identification of the τth quantile of the demand function, $p \mapsto q(p, \tau)$ is obtained through the use of instrumental variables Z, like weather conditions or factor prices, that shift the supply curve and do not affect the level of the demand curve, U, so that independence assumption A2 is met. Furthermore, the IVQR model allows arbitrary correlation between

Z and ν. This property is important as it allows, for example, Z to be measured with error or to be exogenous relative to the demand equation but endogenous relative to the supply equation.

Example 2 (Savings). Chernozhukov and Hansen (2004) use the framework of the IVQR model to examine the effects of participating in a 401(k) plan on an individual's accumulated wealth. Since wealth is continuous, wealth, Y_d, in the participation state $d \in \{0, 1\}$ can be represented as

$$Y_d = q(U_d, d, X), \quad U_d \sim U(0, 1),$$

where $\tau \mapsto q(\tau, d, X)$ is the conditional quantile function of Y_d and U_d is an unobserved random variable. U_d is an unobservable that drives differences in accumulated wealth conditional on X under participation state d. Thus, one might think of U_d as the preference for saving and interpret the quantile index τ as indexing rank in the preference for saving distribution. One could also model the individual as selecting the 401(k) participation state to maximize expected utility:

$$
\begin{aligned}
D &= \arg\max_{d \in \mathcal{D}} E\left[W\{Y_d, d\} \big| X, Z, \nu \right] \\
&= \arg\max_{d \in \mathcal{D}} E\left[W\{q(U_d, d, X), d\} \big| X, Z, \nu \right],
\end{aligned}
\tag{9.11}
$$

where $W\{Y_d, d\}$ is the random indirect utility derived under participation state d. Of course, utility may depend on both observables in X as well as realized and unrealized unobservables. Only dependence on Y_d and d is highlighted. As a result, the participation decision is represented by

$$D = \delta(Z, X, \nu),$$

where Z and X are observed, ν is an unobserved information component that may be related to ranks U_d and includes other unobserved variables that affect the participation state, and the function δ is unknown. This model fits into the IVQR model with the independence condition A2 requiring that U_d is independent of Z, conditional on X.

Under rank invariance (9.8), the preference for saving vector U_d may be collapsed to a single random variable $U = U_0 = U_1$. In this case, a single preference for saving is responsible for an individual's ranking across both treatment states. The more general rank similarity condition A4 relaxes the exact invariance of ranks U_d across d by allowing noisy, asystematic variations of U_d across d, conditional on (ν, X, Z). This relaxation allows for variation in rank across the treatment states, requiring only an "expectational rank invariance." Similarity implies that, given the information in (ν, X, Z) employed to make the selection of treatment D, the expectation of any function of rank U_d does not vary across the treatment states. That is, *ex ante*, conditional on (ν, X, Z), the ranks may be considered to be the same across potential treatments, but the realized, *ex post*, rank may be different across treatment states.

From an econometric perspective, the similarity assumption is nothing but a restriction on the unobserved heterogeneity component which precludes systematic variation of U_d across the treatment states. To be more concrete, consider the following simple example where

$$U_d = F_{\nu + \eta_d}(\nu + \eta_d),$$

where $F_{\nu + \eta_d}(\cdot)$ is the distribution function of $\nu + \eta_d$ and $\{\eta_d\}$ are mutually i.i.d. conditional on ν, X, and Z. The variable ν represents an individual's "mean" saving preference, while η_d is a noisy adjustment. Clearly similarity holds in this case, $U_d \stackrel{d}{=} U_{d'}$ given ν, X, and Z. This more general assumption leaves the individual optimization problem (9.11) unaffected, while allowing variation in an individual's rank across different potential outcomes.

While we feel that rank similarity may be a reasonable assumption in many contexts, imposing rank similarity is not innocuous. In the context of 401(k) participation, matching practices of employers could jeopardize the validity of the similarity assumption. To be more concrete, let $U_d = F_{\nu+\eta_d}(\nu + \eta_d)$ as before but let $\eta_d = dM$ for random variable M that depends on the match rate and is independent of ν, X, and Z. Then conditional on $\nu = v$, X, and Z, $U_0 = F_\nu(v)$ is degenerate but $U_1 = F_{\nu+M}(v + M)$ is not. Therefore, U_1 is not equal to U_0 in distribution. Similarity may still hold in the presence of the employer match if the rank, U_d, in the asset distribution is insensitive to the match rate. The rank may be insensitive if, for example, individuals follow simple rules of thumb such as target saving when they make their savings decisions. Also, if the variation of match rates is small relative to the variation of individual heterogeneity or if the covariates capture most of the variation in match rates, then similarity may be satisfied approximately.

9.2.5 Comparison to other approaches

There are, of course, other assumptions that one could employ to build a quantile model with endogeneity. In this subsection, we briefly compare the IVQR framework to triangular models as in Imbens and Newey (2009); see Chesher (2003), Koenker and Ma (2006), Lee (2007), and Chernozhukov et al. (2015a) for related models and results. We also note that triangular models are related to the Rosenblatt transform; see, for example, Chapter 12 of this handbook. A comparison between the IVQR model and the popular Abadie et al. (2002) approach is provided in Chapter 10 of this handbook.

The triangular model takes the form of a triangular system of equations

$$Y = g(D, \epsilon),$$
$$D = h(Z, \eta),$$

where Y is the outcome, D is a continuous scalar endogenous variable, ϵ is a vector of disturbances, Z is a vector of instruments with a continuous component, η is a scalar reduced-form error, and we ignore other covariates X for simplicity. It is important to note that the triangular system generally rules out simultaneous equations which typically have that the reduced form relating D to Z depends on a vector of disturbances. For example, in a supply and demand system, the reduced form for both price and quantity will generally depend on the unobservables from both the supply equation and the demand equation; see Example 1 in Section 9.2.4.

Outside of η being a scalar, the key conditions that allow identification of quantile effects in the triangular system are: (a) the function $\eta \mapsto h(Z, \eta)$ is strictly increasing in η; and (b) D and ϵ are independent conditional on V for some observable or estimable V. The variable V is thus the "control function" conditional on which changes in D may be taken as causal. Imbens and Newey (2009) use $V = F_{D|Z}(d, z) = F_\eta(\eta)$ as a control variable and show that this variable satisfies condition (b) under the additional condition that (ϵ, η) is independent of Z. Identification then proceeds as follows. Under the assumed monotonicity of $h(Z, \eta)$ in η, $D = h(Z, \eta)$ can be used to identify V. Using V obtained in this first step, one may then construct the distribution of $Y|D, V$. Integrating over the distribution of V and using iterated expectations, one has

$$\int F_{Y|D,V}(y|d, v) F_V(dv) = \int \mathbf{1}(g(d, \epsilon) \leqslant y) F_\epsilon(d\epsilon)$$
$$= \Pr(g(d, \epsilon) \leqslant y) := G(y, d)$$

and the structural quantile function Y_d can be obtained as $G^{-1}(\tau, d)$.

It should be emphasized that the triangular model is neither more nor less general than the IVQR model reviewed here. The key difference between the approaches is that the IVQR model uses an essentially unrestricted selection equation (ν may be vector-valued) but requires monotonicity and a scalar disturbance (U) in the structural equation. The triangular system, on the other hand, relies on monotonicity of the selection mechanism in a scalar disturbance (η) but does not restrict the unobserved heterogeneity in the outcome equation (ϵ may be a vector of disturbances). In addition, the triangular system, as developed in Imbens and Newey (2009), requires a more stringent independence condition in that the instruments Z needs to be independent of both the structural disturbances, ϵ, and the reduced-form disturbance, η. That the approaches impose structure on different parts of the model makes them complementary with a researcher's choice between the two being dictated by whether it is more natural to impose restrictions on the structural function or the reduced form in a given application.

Finally, we note that the triangular model and the IVQR model can be made compatible by imposing the conditions from the triangular model on the selection equation and the conditions from the IVQR model on the structural model. Torgovitsky (2015) studies identification when both sets of conditions are imposed and shows that the requirements on the instruments may be substantially relaxed relative to the IVQR model or Imbens and Newey (2009) in this case.

9.3 Basic estimation and inference approaches

In this section, we present various approaches to estimating and doing inference for the parameters of the IVQR model in the leading case where $\tau \mapsto q(\tau, d, X)$ is strictly increasing. We focus on linear-in-parameters structural quantile models at a single quantile of interest τ:

$$q(\tau, d, x) = d'\alpha_0(\tau) + x'\beta_0(\tau). \qquad (9.12)$$

In (9.12), $\alpha_0(\tau)$ captures the causal effect of the endogenous variables D on the τth quantile of the conditional distribution of potential outcomes Y_d given $X = x$. Similarly, $\beta_0(\tau)$ provides the causal effect of controls X on the τth quantile of the conditional potential outcome distributions. We note that D may also contain interactions of endogenous variables and covariates. Because $\alpha_0(\tau)$ is the chief object of interest in many studies, we focus most of our discussion on estimating and doing inference for $\alpha_0(\tau)$, treating $\beta_0(\tau)$ as a nuisance parameter. Note that in what follows we will often suppress the dependence of $\alpha_0(\tau)$ and $\beta_0(\tau)$ on the quantile level τ.

In interpreting the parameters in (9.12), it is important to note that the quantile index, τ, refers to the quantile of potential outcome Y_d given that exogenous variables are set to $X = x$, and not to the unconditional quantile of Y_d. For example, suppose that one of the control variables in the savings example in Section 9.2.4 is income. An individual at the 10th percentile of the distribution of Y_d given an income of \$200,000, which is far above the median income, may not necessarily be in the lower tail of the unconditional distribution of Y_d as even a relatively low saver with a high level of income may still save substantially more than the median saver in the overall population, that is, without conditioning on income; see Frölich and Melly (2013) for a further discussion of this point. In some applications, features of the conditional distribution are not the chief objects of interest, and researchers are interested in effects of treatments on unconditional quantiles. Unconditional QTE can be obtained from the conditional quantile functions in three steps. First, obtain the conditional potential outcome distribution functions, $F_{Y_d}(y|x)$, as

$$F_{Y_d}(y|x) = \int_0^1 \mathbb{1}\left(d'\alpha_0(\tau) + x'\beta_0(\tau) \leqslant y\right) d\tau,$$

where $\mathbb{1}(\cdot)$ is the indicator function that returns 1 when the expression in parentheses is true and 0 otherwise. Second, the unconditional potential outcome distributions, $F_{Y_d}(y)$, are obtained by integrating $F_{Y_d}(y|x)$ with respect to the marginal distribution of covariates, $F_X(x)$:

$$F_{Y_d}(y) = \int F_{Y_d}(y|x)\,dF_X(x).$$

Finally, the unconditional τ-QTE is given by $F_{Y_{d_1}}^{-1}(\tau) - F_{Y_{d_0}}^{-1}(\tau)$. This discussion suggests that, given estimators of the parameters $\alpha_0(\tau)$ and $\beta_0(\tau)$ and the distribution of covariates $F_X(x)$, unconditional QTE can be estimated based on the plug-in principle; see, for instance, Machado and Mata (2005), Melly (2005), or Chernozhukov et al. (2013a).

Model (9.12) provides a simple and widely used baseline for discussion of estimation and inference. Extending the discussion to allow for nonlinear parametric specifications of the potential outcome quantile functions or to estimation at a small number of quantile indices that are widely spaced is straightforward. In some applications, we may be interested in understanding QTE across a range of quantile indices, say $\tau \in [\delta, 1 - \delta]$ for some $\delta > 0$. Chernozhukov and Hansen (2006) explicitly consider this case and provide uniform convergence results which allow for inference about a variety of hypotheses surrounding the behavior of QTE viewed as a function of τ such as tests of monotonicity of treatment effects or tests that treatment effects are uniformly 0 across a range of τ. Finally, we note that Chernozhukov et al. (2007), Horowitz and Lee (2007), Chen and Pouzo (2009, 2012), and Gagliardini and Scaillet (2012) consider fully nonparametric approaches to estimating structural quantile models.

9.3.1 Generalized methods of moments and related approaches

The most direct way to estimate the parameters of the linear IVQR model is to note that the main implication of the model, equation (9.3), implies unconditional moment conditions

$$\mathrm{E}\left[\left(\tau - \mathbb{1}\left(Y - D'\alpha_0 - X'\beta_0 \leqslant 0\right)\right)\Psi\right] = 0, \tag{9.13}$$

where $\Psi := \Psi(X, Z)$ is a vector of functions of the instruments and endogenous variables.[1] Supposing that α_0 is an $s \times 1$ vector and β_0 is a $k \times 1$ vector, a minimal necessary condition for identifying the model parameters will be $\dim(\Psi) = r \geqslant k + s$.

For $\theta := (\alpha, \beta)$ and $V := (Y, D, X, Z)$, let

$$g_\tau(V, \theta) = \left(\tau - \mathbb{1}\left(Y - D'\alpha - X'\beta \leqslant 0\right)\right)\Psi.$$

With a given set of instruments, Ψ, and observables $\{V_i\}_{i=1}^N = \{Y_i, D_i, X_i, Z_i\}_{i=1}^N$, one may then form the sample analog of the right-hand side of equation (9.13),

$$\hat{g}_N(\theta) = \frac{1}{N}\sum_{i=1}^N g_\tau(V_i, \theta), \tag{9.14}$$

---

[1]A natural choice of instruments would be $\Psi = (Z', X')'$, though the instruments and generalized methods of moments weighting matrix could be chosen to produce a pointwise efficient procedure following Chamberlain (1987).

and estimate $\theta_0 = (\alpha_0', \beta_0')'$ by the generalized method of moments (GMM) as

$$\widehat{\theta} = (\widehat{\alpha}', \widehat{\beta}')' = \arg\min_{\theta \in \Theta} m_N(\theta) \qquad (9.15)$$

for Θ the parameter space for θ and

$$m_N(\theta) := N\widehat{g}_N(\theta)'\Omega_N\widehat{g}_N(\theta),$$

where Ω_N is the GMM weighting matrix which will typically be set to

$$\Omega_N = \left(\tau(1-\tau)\frac{1}{N}\sum_{i=1}^{N}\Psi_i\Psi_i'\right)^{-1}.$$

Maintaining sufficient conditions for point identification as in Chernozhukov and Hansen (2005, 2006, 2013) and assuming that a suitable solution to the GMM optimization problem (9.15) can be found, asymptotic properties of $\widehat{\theta}(\tau)$ would then follow from standard results for GMM with nonsmooth moment conditions as in Newey and McFadden (1994); see Abadie (1995) and Chernozhukov and Hong (2003). We note that if the GMM problem (9.13) is overidentified, overidentification-type tests can be used to assess the joint validity of the underlying assumptions.

The chief difficulty in implementing estimation based on (9.15) is that the function being minimized is both nonsmooth and nonconvex in general. We also note that in many applications, s will be small, often 1, but k may be quite large. Solving (9.15) then involves optimizing a nonsmooth, nonconvex function over $s + k$ arguments, where $s + k$ may be quite large. Directly solving this problem thus poses a substantial computational challenge and has led to the adoption of different approaches to estimating the parameters of the IVQR model.

Within the conventional GMM framework, one option is to take the quasi-Bayesian approach of Chernozhukov and Hong (2003); see also Chapter 4 of this handbook for a review of subsequent work on related methods. The Chernozhukov and Hong (2003) approach uses the GMM criterion function to form a "quasi-likelihood,"

$$L_N(\theta) = \exp\left(-\frac{1}{2}N\widehat{g}_N(\theta)'\Omega_N\widehat{g}_N(\theta)\right),$$

which, when coupled with a prior density $\pi(\theta)$ over model parameters θ, defines a "quasi-posterior" density for θ:

$$\pi_N(\theta) = \frac{L_N(\theta)\pi(\theta)}{\int L_N(\widetilde{\theta})d\pi(\widetilde{\theta})} \propto L_N(\theta)\pi(\theta).$$

Rather than try to solve the optimization problem (9.15), one can then use Markov chain Monte Carlo (MCMC) sampling to attempt to explore the implied quasi-posterior distribution. Chernozhukov and Hong (2003) show that measures of central tendency from the quasi-posterior, such as the quasi-posterior mean,

$$\widehat{\theta} = (\widehat{\alpha}', \widehat{\beta}')' = \int \theta d\pi_N(\theta),$$

and quasi-posterior median are consistent for model parameters with the same asymptotic distribution as the solution to (9.15). Chernozhukov and Hong (2003) also demonstrate that valid frequentist confidence intervals may be obtained by taking quasi-posterior quantiles. For example, a frequentist 95% confidence interval may be constructed by taking the 2.5th and 97.5th quantiles of the quasi-posterior distribution. This approach bypasses the need to optimize a nonconvex and nonsmooth criterion at the cost of needing to design a sampler that adequately explores the quasi-posterior in a reasonable amount of computation time.

A second option is to directly smooth the GMM criterion function as in Kaplan and Sun (2017), building upon ideas in Amemiya (1982) and Horowitz (1998). Specifically, one modifies the moment condition (9.14) to

$$\hat{g}_N^{h_N}(\theta) = \frac{1}{N} \sum_{i=1}^N \left(\tau - G_{h_N} \left(Y_i - D_i'\alpha - X_i'\beta \right) \right) \Psi_i \qquad (9.16)$$

by smoothing the indicator function, where $G_h(\cdot)$ denotes a smoothing function with smoothing parameter h. $G_h(\cdot)$ can be defined as the survival function associated with any kernel function $K_h(\cdot)$, that is, $G_h(u) = \int_u^\infty K_h(v)dv$, that satisfies regularity conditions provided in Kaplan and Sun (2017). One can then proceed to estimate model parameters by replacing $\hat{g}_N(\theta)$ in (9.15) with $\hat{g}_N^{h_N}(\theta)$ and applying any optimizer which is appropriate for smooth, nonconvex optimization problems or the quasi-Bayesian approach described above. Solving the smoothed problem can offer some computational gains relative to attempting to solve the original problem, though nonconvexities remain after smoothing. The resulting estimator is first-order equivalent to the GMM estimator for the original problem. The estimator can, however, enjoy higher-order improved performance. Kaplan and Sun (2017) provide a plug-in approach to choosing the smoothing parameter h_N and also demonstrate that the estimated parameters obtained from solving the smoothed problem may perform better in small samples than those from solving the unsmoothed problem or the inverse quantile regression discussed in Section 9.3.2.

9.3.2   Inverse quantile regression

Rather than work directly with moment condition (9.13), Chernozhukov and Hansen (2006, 2008) take a different approach which they label "inverse quantile regression" (IQR). IQR is based on the observation that (9.3), coupled with the linear quantile model (9.12), implies that the τth quantile of $Y - D'\alpha_0$ conditional on covariates X and instruments Z is equal to $X'\beta_0(\tau)$:

$$Q_{Y-D'\alpha_0}(\tau|X, Z) = X'\beta_0 + Z'\gamma_0, \quad \text{with } \gamma_0 \equiv 0. \qquad (9.17)$$

That is, at the true value of the coefficient vector on the endogenous variables α_0, the conventional linear τ-quantile regression of $Y - D'\alpha_0$ on X and Z would yield coefficients on the instruments of exactly 0 in the population. This observation then suggests an estimation approach based on concentrating X out of the problem using conventional quantile regression, which is convex and can be solved very quickly, and then solving a lower-dimensional nonconvex optimization problem over only the dimension of D to find $\hat{\alpha}$.

Specifically, the IQR procedure works as follows. Let a denote an arbitrary hypothesized value for α_0. Using the hypothesized value a, estimate coefficients $\beta(a)$ and $\gamma(a)$ from the model $Q_{Y-D'a}(\tau|X, Z) = X'\beta(a) + Z'\gamma(a)$ by running the ordinary linear τ-quantile regression of $Y - D'a$ on X and Z. Let $\hat{\beta}(a)$ and $\hat{\gamma}(a)$ denote the resulting estimators of $\beta(a)$ and $\gamma(a)$. Also, let $\hat{\Omega}_N(a)$ denote the estimated covariance matrix of $\sqrt{N}(\hat{\gamma}(a) - \gamma(a))$, and note that this covariance matrix is available in any common implementation of the ordinary quantile regression. We can then define the IQR estimator of α_0 as

$$\hat{\alpha} = \arg\min_{a \in \mathcal{A}} W_N(a), \qquad (9.18)$$

where $\mathcal{A}$ is the parameter space for α and

$$W_N(a) := N\hat{\gamma}(a)'\hat{\Omega}_N(a)^{-1}\hat{\gamma}(a). \qquad (9.19)$$

Given $\hat{\alpha}$, we can then estimate β_0 as $\hat{\beta}(\hat{\alpha})$.

In terms of point estimation, the main virtue of IQR is that, by concentrating out the coefficients on exogenous variables X, it produces a nonconvex optimization problem over only the parameters α. In many applications, the dimension of D is small, so one can approach the nonconvex optimization problem using highly robust optimization procedures that deal effectively with objectives with many local optima. Chernozhukov and Hansen (2006) recommend using a grid search to solve (9.18), though other approaches are certainly available. Using a grid search is particularly appealing when coupled with weak-identification robust inference as discussed in Section 9.3.3.

Chernozhukov and Hansen (2006) analyze the properties of $(\hat{\alpha}(\tau)', \hat{\beta}(\tau)')'$ under assumptions that guarantee strong identification. They verify asymptotic normality of the estimator, provide a consistent estimator of the asymptotic variance, and show how instruments and observation weights can be chosen to produce an efficient estimator of the coefficients for a single quantile, following Chamberlain (1987). Chernozhukov and Hansen (2006) also analyze the behavior of the process $(\hat{\alpha}(\tau)', \hat{\beta}(\tau)')'$ not just at a point but viewed as a function of τ, providing uniform convergence results and discussing in detail applications of these convergence results to testing hypotheses about the behavior of $(\alpha_0(\tau)', \beta_0(\tau)')'$ across the index τ.

9.3.2.1 A useful interpretation of IQR as a GMM estimator.

It is useful to interpret IQR as first-order equivalent to a particular GMM estimator, where we first profile out the coefficients on exogenous variables.

To this end, let us define

$$g_\tau(V, \alpha; \beta, \delta) = \left(\tau - \mathbb{1}\left(Y \leqslant D'\alpha + X'\beta\right)\right) \Psi(\alpha, \delta(\alpha)), \tag{9.20}$$

with "instrument"

$$\Psi(\alpha, \delta(\alpha)) := (Z - \delta(\alpha)X). \tag{9.21}$$

In (9.21),

$$\delta(\alpha) = M(\alpha)J^{-1}(\alpha),$$

where δ is a matrix parameter,

$$M(\alpha) = \mathrm{E}\left[ZX'f_\varepsilon(0 \mid X, Z)\right], \quad J(\alpha) = \mathrm{E}\left[XX'f_\varepsilon(0 \mid X, Z)\right],$$

and $f_\varepsilon(0|X, Z)$ is the conditional density of $\varepsilon = Y - D'\alpha - X'\beta(\alpha)$ where $\beta(\alpha)$ is defined by

$$\mathrm{E}\left[(\tau - \mathbb{1}(Y \leqslant D'\alpha + X'\beta(\alpha))X\right] = 0.$$

To proceed with estimation, for a hypothesized value a, we first profile out the coefficients on the exogenous variables as in IQR,

$$\hat{\beta}(a) = \arg\min_{b \in \mathcal{B}} \frac{1}{N} \sum_{i=1}^{N} \rho_\tau \left(Y_i - D_i'a - X_i'b\right) \tag{9.22}$$

where $\mathcal{B}$ is the parameter space for β. We may then plug the solution of (9.22) into (9.20) to form

$$\hat{g}_N(a) = \frac{1}{N} \sum_{i=1}^{N} g(V_i, a, \hat{\beta}(a), \hat{\delta}(a)), \tag{9.23}$$

where
$$\hat{\delta}(a) = \widehat{M}(a)\hat{J}^{-1}(a),$$

for

$$\widehat{M}(a) = \frac{1}{Nh_N} \sum_{i=1}^{N} Z_i X_i' K_{h_N} \left(Y_i - D_i' a - X_i' \widehat{\beta}(a) \right),$$

$$\widehat{J}(a) = \frac{1}{Nh_N} \sum_{i=1}^{N} X_i X_i' K_{h_N} \left(Y_i - D_i' a - X_i' \widehat{\beta}(a) \right),$$

and $K_{h_N}(\cdot)$ a kernel function with bandwidth h_N. Then we consider the GMM estimator based on the concentrated moments (9.23):

$$\hat{\alpha}(\tau) = \arg\min_{a \in \mathcal{A}} m_N(a),$$

for

$$m_N(a) := N\hat{g}_N(a)'\widehat{\Sigma}(a,a)^{-1}\hat{g}_N(a). \tag{9.24}$$

$\widehat{\Sigma}(a,a)$ in $m_N(a)$ is an estimator of the covariance function of the sample concentrated moment functions (9.23) such as

$$\widehat{\Sigma}(a_1, a_2) = \frac{1}{N} \sum_{i=1}^{N} g\left(V_i, a_1, \hat{\beta}(a_1) \right) g\left(V_i, a_2, \hat{\beta}(a_2) \right)'. \tag{9.25}$$

The estimator $\hat{\alpha}$ is first-order equivalent to the estimator $\tilde{\alpha}$ which employs the moment function

$$g_\tau^*(V, \alpha) = (\tau - \mathbb{1}(Y \leqslant D'\alpha + X'\beta_0))\Psi(\alpha_0, \delta(\alpha_0)).$$

That is, the sample objective function for $\tilde{\alpha}$ uses

$$\hat{g}_N(\alpha) = \frac{1}{N} \sum_{i=1}^{N} g_\tau^*(V, \alpha), \tag{9.26}$$

where

$$\widehat{\Sigma}(\alpha, \alpha) = \mathrm{E}\left[g_\tau^*(V_i, \alpha_0) g_\tau^*(V_i, \alpha_0)' \right].$$

This equivalence holds because the moments possess the Neyman orthogonality property that we discuss later. Moreover, by examining the first-order properties of the IQR estimator we can conclude that $\tilde{\alpha}$ and IQR are first-order equivalent.

9.3.3   Weak identification robust inference

The good behavior of asymptotic approximation results for the point estimators provided in Sections 9.3.1 and 9.3.2 relies on strong identification of the model parameters as discussed in Section 9.2.2. Because checking these conditions may be difficult, it is useful to have inference procedures that are robust to weak or non-identification.

Chernozhukov and Hansen (2008) present a simple weak-identification robust inference procedure that results naturally from the IQR estimator. The basic idea underlying this procedure is precisely relation (9.17), which states that the instruments Z should have no explanatory power in the conventional τ-quantile regression of $Y - D'\alpha_0$ on X and Z at the true value of the structural parameter α_0. Thus, a valid test of the hypothesis that $\alpha_0 = a$ for some hypothesized a can be obtained by considering a test of the hypothesis that $\gamma(a) = 0$ for $\gamma(a)$ denoting the population value of the τ-quantile regression coefficients defined in Section 9.3.2. Also, note that $W_N(a)$ in (9.19) is simply the standard Wald statistic for testing $\gamma(a) = 0$ and that $W_N(\alpha_0)$ converges in distribution to $\chi^2_{\dim(Z)}$ regardless of the strength of identification of α_0; see Chernozhukov and Hansen (2008) for details.[2] It then follows that a valid $100(1-p)\%$ confidence region for α_0 may be constructed as the set

$$\{a \in \mathcal{A} : W_N(a) \leqslant c_{1-p}\} \qquad (9.27)$$

where c_{1-p} is such that $P[\chi^2_{\dim(Z)} > c_{1-p}] = p$, and the set may be approximated numerically by considering as in the grid $\{a_j, j = 1, \ldots, J\}$. Thus, a natural byproduct of solving (9.18) through a grid search is a confidence set for the structural parameter α_0 that is valid regardless of the strength of identification of the parameter. We note that this procedure could also be adapted for use with the orthogonal scores defined in Section 9.4.1 to provide weak-identification robust inference in settings with high-dimensional X or other settings where robustness to estimation of the nuisance parameter β_0 is a major concern.

The approach of Chernozhukov and Hansen (2008) outlined above is in the spirit of the weak-identification robust procedure of Anderson and Rubin (1949). The procedure is relatively simple to implement, but suffers from the same well-known lack of power as other Anderson–Rubin-type statistics in overidentified models under strong identification. To overcome this potential inefficiency, Jun (2008) proposes a different statistic analogous to the proposal of Kleibergen (2005) which is locally efficient under strong identification but may suffer from substantial declines in power against alternatives that are distant from the true parameter value. In the following, we discuss the related approach of Andrews and Mikusheva (2016) which extends the conditional likelihood ratio approach of Moreira (2003) to general nonlinear settings. This approach retains efficiency under strong identification but also maintains good power against distant alternatives.

The Andrews and Mikusheva (2016) approach employs a quasi-likelihood ratio (QLR) statistic,

$$QLR_N(a) = m_N(a) - \inf_{a \in \mathcal{A}} m_N(a), \qquad (9.28)$$

where $m_N(a)$ is the GMM objective function (9.24). Under weak identification, the distribution of $QLR_N(a)$ is nonstandard and depends on a nuisance function that is not consistently estimable. Andrews and Mikusheva (2016) provide a sufficient statistic (in LeCam's Gaussian limit experiment)

$$S(a) = \sqrt{N}\left(\hat{g}_N(a) - \hat{\Sigma}(a, \alpha_0)\hat{\Sigma}(\alpha_0, \alpha_0)^{-1}\hat{g}_N(\alpha_0)\right)$$

for this functional nuisance parameter, where $g_N(a)$ and $\hat{\Sigma}(a_1, a_2)$ are defined in (9.23) and (9.25). Andrews and Mikusheva (2016) also outline a procedure to simulate the distribution of $QLR(a)$ conditional on $S(a)$ that proceeds as follows. First, draw $\zeta_b^* \sim N(0, \hat{\Sigma}(\alpha_0, \alpha_0))$ for $b = 1, \ldots, B$, for a large number B. For each ζ_b^*, the QLR statistic for that draw is then calculated as

$$QLR_{N,b}^*(a) = m_{N,b}^*(a) - \inf_{a \in \mathcal{A}} m_{N,b}^*(a),$$

where

$$m_{N,b}^*(a) = N\hat{g}_{N,b}^*(a)'\hat{\Sigma}(a, a)^{-1}\hat{g}_{N,b}^*(a)$$

---

[2]The same statement would also hold for the GMM objective function based on (9.23) discussed in Section 9.3.2.1.

for

$$\hat{g}_{N,b}^{*}(a) = S(a) + \hat{\Sigma}(a, \alpha_0)\hat{\Sigma}(\alpha_0, \alpha_0)^{-1}\zeta_b^{*}.$$

The simulated distribution then provides an appropriate critical value, $c_{1-p}(S(a))$, for performing a valid p-level test of the null hypothesis that $\alpha_0 = a$ by rejecting when $QLR_N(a) > c_{1-p}(S(a))$. It then follows that a valid $100(1-p)\%$ confidence region for α_0 is given by

$$\{a \in \mathcal{A} : QLR_N(a) \leqslant c_{1-p}(S(a))\}.$$

9.3.4 Finite-sample inference

The inference procedures reviewed so far in this section all rely on asymptotic approximations. Chernozhukov et al. (2009) provide a finite-sample inference approach which can also be used if the validity of the assumptions necessary to justify these approximations is questionable and is valid in setups with weak or set identification.

Their approach makes use of the fact under the assumptions of the IVQR model, the event $\{Y \leqslant q(\tau, D, X)\}$ conditional on (Z, X) is distributed exactly as a Bernoulli(τ) random variable regardless of the sample size. This random variable depends only on τ, which is known, and so is pivotal in finite samples. For the GMM objective function $m_N(\theta_0)$ defined in (9.15), this implies that $m_N(\theta_0) \overset{d}{=} \tilde{m}_N$ conditional on $\{X_i, Z_i\}_{i=1}^{N}$, where

$$\tilde{m}_N := \left(\frac{1}{\sqrt{N}} \sum_{i=1}^{N} (\tau - B_i) \cdot \Psi_i\right)' \Omega_N \left(\frac{1}{\sqrt{N}} \sum_{i=1}^{N} (\tau - B_i) \cdot \Psi_i\right)$$

and $\{B_i\}_{i=1}^{N}$ are i.i.d. Bernoulli random variables that are independent of $\{X_i, Z_i\}_{i=1}^{N}$ and have $\mathrm{E}[B_i] = \tau$. This result provides the finite-sample distribution of the GMM function $m_N(\theta)$ at $\theta = \theta_0$, which does not depend on any unknown parameters. Given the finite-sample distribution of $m_N(\theta_0)$, a p-level test of the null hypothesis that $\theta = \theta_0$ is given by the rule that rejects the null if $m_N(\theta) > c_{1-p}$, where the critical value c_{1-p} is the $(1-p)$th quantile of $\tilde{m}_N(\tau)$. It then follows that a valid $100(1-p)\%$ joint confidence set for θ is given by

$$\{\theta \in \Theta : m_N(\theta) \leqslant c_{1-p}\}.$$

We note that inference is simultaneous on all components of θ and that for joint inference the approach is not conservative. Inference about subcomponents of θ such as α may be made by projections and may be conservative.

The chief difficulty with the finite-sample approach is computational. Implementing the approach requires inversion of the function $m_N(\theta)$, which may be quite difficult if the number of parameters is large. To alleviate this problem, Chernozhukov et al. (2009) develop suitable MCMC algorithms.

9.4 Advanced inference with high-dimensional X

9.4.1 Neyman orthogonal scores

Here we deal with the case where we have high-dimensional covariates. Such cases are common in current high-dimensional data sets where one may see very many potential control variables. High-dimensional covariates also arises in semiparametric problems; for example, we may be interested in a partially linear structural quantile model

$$q(\tau, d, w) = \alpha_0(\tau)d + g(\tau, w)$$

where W is a low-dimensional set of variables and we approximate $g(\tau, w) \approx x'\beta_0(\tau)$ using a collection of approximating functions $x = h(w)$. In settings with high-dimensional X, estimation of $\beta_0(\tau)$ may contaminate estimation of the parameters of interest, $\alpha_0(\tau)$, leading to a breakdown of estimation and inference based directly on (9.13). The potential for contamination is especially acute in high-dimensional settings where some form of regularization will be used to make informative estimation feasible but may arise more generally.

Due to the potentially poor finite-sample performance of estimators based directly on (9.13), one might prefer to base estimation and inference on "orthogonal" moment conditions that are relatively insensitive to estimation of the nuisance parameters β_0. Specifically, we may prefer to base estimation and inference for α_0 on moment functions

$$g(V, \alpha; \eta), \quad \text{where } V = (Y, D, Z, X)$$

and η denotes nuisance parameters with true values η_0 that include β_0 as a subcomponent, that identify α_0 via

$$\mathrm{E}[g(V, \alpha_0; \eta_0)] = 0 \tag{9.29}$$

and obey the Neyman orthogonality condition

$$\partial_\eta \mathrm{E}[g(V, \alpha_0; \eta)]\Big|_{\eta=\eta_0} = 0, \tag{9.30}$$

where ∂_η denotes a functional derivative operator. Equation (9.30) is the key orthogonality condition that ensures that the moment conditions defining α_0 are locally insensitive to perturbations in the nuisance parameters. This property results in the first-order properties of estimation and inference of α_0 based on sample analogs to (9.29) being insensitive to estimation of nuisance functions as long as sufficiently high-quality estimators of the nuisance functions are available. The idea of using orthogonal estimating equations goes back at least to Neyman (1959, 1979) where they were used in construction of Neyman's celebrated $C(\alpha)$ statistic. The use of moment conditions satisfying the orthogonality condition (9.30) is crucial for establishing good properties of semiparametric estimators in modern, high-dimensional estimation settings when regularized estimation or other machine learning tools are used in estimation of nuisance functions; see, for example, Belloni et al. (2017) and Chernozhukov et al. (2015b, 2016).

The orthogonal moment functions for the IVQR setting are given by

$$g_\tau(V, \alpha, \eta) = (\tau - \mathbb{1}(Y \leqslant D'\alpha + X'\beta))\Psi(\alpha, \delta(\alpha)),$$

where $\Psi(\alpha, \delta(\alpha))$ and $\delta(\alpha)$ are defined in Section 9.3.2.1. The nuisance parameter and its true value are then given by

$$\eta := (\beta, \delta(\alpha)) \quad \text{and} \quad \eta_0 := (\beta_0, \delta(\alpha_0)).$$

Observe that the Neyman orthogonality condition holds for these moment conditions because, under appropriate smoothness conditions,

$$\partial_\beta \mathrm{E}[g(V, \alpha_0; \eta)]\Big|_{\eta=\eta_0} = M(\alpha_0) - M(\alpha_0)J^{-1}(\alpha_0)J(\alpha_0) = 0,$$

$$\partial_\delta \mathrm{E}[g(V, \alpha_0; \eta)]\Big|_{\eta=\eta_0} = \mathrm{E}\left[(\tau - \mathbb{1}(Y \leqslant D'\alpha_0 + X'\beta_0))X\right] = 0.$$

9.4.2   Estimation and inference using orthogonal scores

We start, similarly to the IQR estimator, by first profiling out the coefficients on exogenous variables using an ℓ_1-penalized quantile regression estimator to define

$$\hat{\beta}(a) = \arg\min_{b\in\mathcal{B}} \frac{1}{n} \sum_{i=1}^{N} \rho_\tau(Y_i - D_i'a - X_i'b) + \lambda \sum_{j=1}^{\dim(b)} \psi_j|b_j|, \tag{9.31}$$

for a hypothesized value a. We then estimate

$$\widehat{M}(a) = \frac{1}{Nh_N} \sum_{i=1}^{N} Z_i X_i' K_{h_N}\left(Y_i - D_i'a - X_i'\hat{\beta}(a)\right),$$

$$\hat{J}(a) = \frac{1}{Nh_N} \sum_{i=1}^{N} X_i X_i' K_{h_N}\left(Y_i - D_i'a - X_i'\hat{\beta}(a)\right),$$

for $K_{h_N}(\cdot)$ a kernel function with bandwidth h_N as before. Since $\hat{J}(a)$ is high-dimensional and need not be invertible, we may estimate row components $\delta_j(a)$ of matrix $\delta(a)$ by solving the ℓ_1-regularized problem

$$\hat{\delta}_j(a) = \arg\min_{\delta} \frac{1}{2}\delta'\hat{J}(a)\delta - \hat{M}_j(a)\delta + \vartheta\|\delta\|_1,$$

where $\hat{M}_j(a)$ is the jth row of $\hat{M}(a)$, interpreted as a row vector itself, and ϑ is a penalty level. The solution $\hat{\delta}_j(a)$ obeys the Karush–Kuhn–Tucker condition

$$\|\hat{\delta}_j(a)'\hat{J}(a) - \hat{M}_j(a)\|_\infty \leqslant \vartheta, \ \forall j, \tag{9.32}$$

so we may think of $\hat{\delta}_j(a)$ as a regularized estimator of $M_j(a)J^{-1}(a)$.

Alternatively we can use the regularized estimator via the Dantzig form of the lasso (see Candes and Tao, 2007) by minimizing a norm of $\hat{\delta}(a)$ subject to the above constraints (9.32).

We may then plug in the solution of (9.31) to form a concentrated sample moment function analogous to (9.14) given by

$$\hat{g}_N(a) = \frac{1}{N} \sum_{i=1}^{N} \left(\tau - \mathbb{1}\left(Y_i - D_i'a - X_i'\hat{\beta}(a) \leqslant 0\right)\right) \Psi(a, \hat{\delta}(a)). \tag{9.33}$$

These concentrated moments can be used to set up the continuously updated GMM estimator,

$$\hat{\alpha} = \arg\min_{a\in\mathcal{A}} N\hat{g}_N(a)'\widehat{\Sigma}(a,a)^{-1}\hat{g}_N(a),$$

where again $\widehat{\Sigma}(a,a)$ is an estimator of the covariance function of the sample concentrated moment functions (9.33). The estimator $\hat{\alpha}$ would then follow standard properties of the infeasible GMM estimator that replaced the estimators $\hat{\beta}(a)$ and $\hat{\delta}(a)$ with their true values β_0 and $\delta(\alpha_0)$ as long as instruments are low-dimensional and identification is strong. If the set of instruments were also high-dimensional, further regularization would be called for to make reliable estimation and inference feasible.

We can also directly use the concentrated moments to set up standard Anderson–Rubin-type inference for α_0 under weak or partial identification as in Section 9.3.3. Similarly, we could base inference from more refined approaches, such as Andrews and Mikusheva (2016), on the concentrated moments. Indeed, we can use these concentrated moments to form a QLR statistic,

$$QLR_N(a) = N\widehat{g}_N(a)'\widehat{\Sigma}(a,a)^{-1}\widehat{g}_N(a) - \inf_{a \in \mathcal{A}} N\widehat{g}_N(a)'\widehat{\Sigma}(a,a)^{-1}\widehat{g}_N(a). \qquad (9.34)$$

Because of the orthogonality property, estimation of the nuisance parameters does not affect the first-order behavior of the empirical moments, so inference based on (9.34) falls back exactly into the setting of Andrews and Mikusheva (2016). One could then employ their approach to compute the critical values for $QLR_N(a)$ conditional on a sufficient statistic, $c_{1-p}(QLR_N(a))$. It then follows that a valid $100(1 - p)\%$ confidence region for α_0 may be constructed by considering as in the grid $\{a_j, j = 1, \ldots, J\}$ exactly as in approximating (9.27).

9.5 Conclusion

In this chapter, we have reviewed the structural IVQR model developed in Chernozhukov and Hansen (2005) which can be used to estimate causal quantile effects in the presence of endogeneity. The model makes use of instrumental variables that satisfy conventional independence and relevance conditions from the nonlinear IV literature. Specifically, instruments are assumed to be independent of unobservables associated to potential outcomes but related to endogenous right-hand-side variables in the model. The presence of instruments alone is insufficient to identify QTE, and the IVQR model imposes an additional condition on structural unobservables, termed rank similarity, that restricts the distribution of unobservables in potential outcomes across different potential states of the endogenous variables. Under these conditions, an IV-style moment condition can be derived which then provides a basis for identification and estimation of QTE. We provided two concrete examples of economic models that fall within the IVQR framework.

We then reviewed leading approaches to estimating model parameters and performing inference for QTE within the IVQR model based on the moment conditions implied by the model. Estimation and inference are complicated by the nonsmooth and nonconvex nature of the IVQR moment conditions. We discuss estimation and inference approaches that attempt to alleviate this issue. We also review approaches to inference which remain valid under weak identification or even nonidentification.

There are, of course, many open areas for research in quantile models with endogeneity. As discussed in Section 9.2.5, Abadie et al. (2002) and Imbens and Newey (2009) offer alternative approaches to identifying QTE by imposing alternate sets of assumptions to those used in the IVQR model. These approaches and the IVQR model are nonnested and further understanding their connections may be interesting. Wüthrich (2014) provides a contribution in this direction by showing the connection between the estimands of both models within the structure of the Abadie et al. (2002) framework. It would also be interesting to analyze the properties of the IVQR estimands when some of the underlying assumptions are violated. Toward this end, Wüthrich (2014) provides a characterization of QTE estimands based on the IVQR model with binary treatments in the absence of rank similarity. Another topic that may deserve further consideration is the systematic analysis of estimation and inference based on the orthogonal moment equations sketched in Section 9.4.1, especially in high-dimensional settings. We also note that the IVQR model may be useful for uncovering structural objects even if quantile effects are not the chief objects of interest; see, for example, Berry and Haile (2014). It may be interesting to further explore application of the IVQR model and related estimation methods in structural economic applications. Finally, a potentially interesting but more unexplored area may be to think about quantile-like quantities for multivariate outcomes with endogenous covariates.

Bibliography

A. Abadie. Changes in Spanish labor income structure during the 1980s: A quantile regression approach. CEMFI Working Paper No. 9521, October 1995.

A. Abadie, J. Angrist, and G. Imbens. Instrumental variables estimates of the effect of subsidized training on the quantiles of trainee earnings. *Econometrica*, 70(1):91–117, 2002.

T. Amemiya. Two stage least absolute deviations estimators. *Econometrica*, 50:689–711, 1982.

T. W. Anderson and H. Rubin. Estimation of the parameters of a single equation in a complete system of stochastic equations. *Annals of Mathematical Statistics*, 20:46–63, 1949.

D. W. K. Andrews and X. Shi. Inference based on conditional moment inequalities. *Econometrica*, 81(2):609–666, 2013.

I. Andrews and A. Mikusheva. Conditional inference with a functional nuisance parameter. *Econometrica*, 84(4):1571–1612, 2016.

A. Belloni, V. Chernozhukov, I. Fernández-Val, and C. Hansen. Program evaluation and causal inference with high-dimensional data. *Econometrica*, 85:233–298, 2017.

S. T. Berry and P. A. Haile. Identification in differentiated products markets using market level data. *Econometrica*, 82(5):1749–1797, 2014.

E. Candes and T. Tao. The dantzig selector: Statistical estimation when p is much larger than n. *Annals of Statistics*, 35:2313–2351, 2007.

G. Chamberlain. Asymptotic efficiency in estimation with conditional moment restrictions. *Journal of Econometrics*, 34(3):305–334, 1987.

X. Chen and D. Pouzo. Efficient estimation of semiparametric conditional moment models with possibly nonsmooth residuals. *Journal of Econometrics*, 152(1):46–60, 2009.

X. Chen and D. Pouzo. Estimation of nonparametric conditional moment models with possibly nonsmooth moments. *Econometrica*, 80(1):277–322, 2012.

V. Chernozhukov and C. Hansen. The effects of 401(k) participation on the wealth distribution: An instrumental quantile regression analysis. *Review of Economics and Statistics*, 86(3):735–751, 2004.

V. Chernozhukov and C. Hansen. An IV model of quantile treatment effects. *Econometrica*, 73(1):245–262, 2005.

V. Chernozhukov and C. Hansen. Instrumental quantile regression inference for structural and treatment effect models. *Journal of Econometrics*, 132(2):491–525, 2006.

V. Chernozhukov and C. Hansen. Instrumental variable quantile regression: A robust inference approach. *Journal of Econometrics*, 142(1):379–398, 2008.

V. Chernozhukov and C. Hansen. Quantile models with endogeneity. *Annual Review of Economics*, 5:57–81, 2013.

V. Chernozhukov and H. Hong. An MCMC approach to classical estimation. *Journal of Econometrics*, 115(2):293–346, 2003.

V. Chernozhukov, G. W. Imbens, and W. K. Newey. Instrumental variable estimation of nonseparable models. *Journal of Econometrics*, 139(1):4–14, 2007.

V. Chernozhukov, C. Hansen, and M. Jansson. Finite sample inference for quantile regression models. *Journal of Econometrics*, 152(2):93–103, 2009.

V. Chernozhukov, I. Fernández-Val, and B. Melly. Inference on counterfactual distributions. *Econometrica*, 81(6):pp. 2205–2268, 2013a.

V. Chernozhukov, S. Lee, and A. Rosen. Intersection bounds: Estimation and inference. *Econometrica*, 81(2):667–737, 2013b.

V. Chernozhukov, I. Fernández-Val, and A. E. Kowalski. Quantile regression with censoring and endogeneity. *Journal of Econometrics*, 186(1):201–221, 2015a.

V. Chernozhukov, C. Hansen, and M. Spindler. Valid post-selection and post-regularization inference: An elementary, general approach. *Annual Review of Economics*, 7:649–688, 2015b.

V. Chernozhukov, D. Chetverikov, M. Demirer, E. Duflo, C. Hansen, and W. Newey. Double machine learning for treatment and causal parameters. arXiv:1608.00060, 2016.

A. Chesher. Identification in nonseparable models. *Econometrica*, 71(5):1405–1441, 2003.

A. Chesher. Nonparametric identification under discrete variation. *Econometrica*, 73(5): 1525–1550, 2005.

A. Chesher and K. Smolinski. Sharp identified sets for discrete variable IV models. CeMMAP Working Paper CWP11/10, 2010. URL http://www.cemmap.ac.uk/wps/cwp1110.pdf.

A. Chesher, A. Rosen, and K. Smolinski. An instrumental variable model of multiple discrete choice. *Quantitative Economics*, 4(2):157–196, 2013.

K. Doksum. Empirical probability plots and statistical inference for nonlinear models in the two-sample case. *Annals of Statistics*, 2:267–277, 1974.

Y. Dong and S. Shen. Testing for rank invariance or similarity in program evaluation. Working paper, 2015. URL http://www.yingyingdong.com/Research/Rank_Invariance_Test_R1_Maintext.pdf.

B. R. Frandsen and L. J. Lefgren. Testing rank similarity. Working paper, 2015. URL https://economics.byu.edu/frandsen/Documents/testingranksimilarity20151111.pdf.

M. Frölich and B. Melly. Unconditional quantile treatment effects under endogeneity. *Journal of Business and Economic Statistics*, 31(3):346–357, 2013.

P. Gagliardini and O. Scaillet. Nonparametric instrumental variable estimation of structural quantile effects. *Econometrica*, 80(4):1533–1562, 2012.

K. Graddy. Testing for imperfect competition at the Fulton fish market. *Rand Journal of Economics*, 26(1):75–92, 1995.

J. A. Hausman. Errors in variables in simultaneous equation models. *Journal of Econometrics*, 5(3):389–401, 1977.

J. J. Heckman, J. Smith, and N. Clements. Making the most out of programme evaluations and social experiments: Accounting for heterogeneity in programme impacts. *Review of Economic Studies*, 64(4):487–535, 1997.

J. L. Horowitz. Bootstrap methods for median regression models. *Econometrica*, 66(6): 1327–1351, 1998.

J. L. Horowitz and S. Lee. Nonparametric instrumental variables estimation of a quantile regression model. *Econometrica*, 75(4):1191–1208, 2007.

G. W. Imbens and J. D. Angrist. Identification and estimation of local average treatment effects. *Econometrica*, 62(2):467–475, 1994.

G. W. Imbens and W. K. Newey. Identification and estimation of triangular simultaneous equations models without additivity. *Econometrica*, 77(5):1481–1512, 2009.

S. J. Jun. Weak identification robust tests in an instrumental quantile model. *Journal of Econometrics*, 144:118–138, 2008.

D. M. Kaplan and Y. Sun. Smoothed estimating equations for instrumental variables quantile regression. *Econometric Theory*, 33(1):105–157, 2017.

F. Kleibergen. Testing parameters in GMM without assuming that they are identified. *Econometrica*, 73(4):1103–1124, 2005.

R. Koenker. *Quantile Regression*. Cambridge University Press, Cambridge, 2005.

R. Koenker and G. Bassett. Regression quantiles. *Econometrica*, 46:33–50, 1978.

R. Koenker and O. Geling. Reappraising medfly longevity: A quantile regression survival analysis. *Journal of the American Statistical Association*, 96:458–468, 2001.

R. Koenker and L. Ma. Quantile regression methods for recursive structural equation models. *Journal of Econometrics*, 134(2):471–506, 2006.

S. Lee. Endogeneity in quantile regression models: A control function approach. *Journal of Econometrics*, 141(2):1131–1158, 2007.

E. L. Lehmann. *Nonparametrics: statistical methods based on ranks*. Holden-Day, San Francisco, 1975.

J. A. F. Machado and J. Mata. Counterfactual decomposition of changes in wage distributions using quantile regression. *Journal of Applied Econometrics*, 20(4):445–465, 2005.

B. Melly. Decomposition of differences in distribution using quantile regression. *Labour Economics*, 12(4):577–590, 2005.

M. J. Moreira. A conditional likelihood ratio test for structural models. *Econometrica*, 71: 1027–1048, 2003.

W. K. Newey and D. McFadden. Large sample estimation and hypothesis testing. In R.F. Engle and D. McFadden, editors, *Handbook of Econometrics, Vol. IV*, pages 2111–2245. North-Holland, Amsterdam, 1994.

J. Neyman. Optimal asymptotic tests of composite statistical hypotheses. In U. Grenander, editor, *Probability and Statistics: The Harald Cramér Volume*. Wiley, New York, 1959.

J. Neyman. $c(\alpha)$ tests and their use. *Sankhyā*, 41:1–21, 1979.

A. Santos. Inference in nonparametric instrumental variables with partial identification. *Econometrica*, 80(1):213–275, 2012.

A. Torgovitsky. Identification of nonseparable models using instruments with small support. *Econometrica*, 83(3):1185–1197, 2015.

K. Wüthrich. A comparison of two quantile models with endogeneity. Working Paper, Universität Bern, Department of Economics, 2014. URL `http://www.vwl.unibe.ch/papers/dp/dp1408.pdf`.

10

Local Quantile Treatment Effects

Blaise Melly

University of Bern, Bern, Switzerland

Kaspar Wüthrich

University of California, San Diego, California, USA

CONTENTS

This chapter reviews instrumental variable models of quantile treatment effects. We focus on models that achieve identification through a monotonicity assumption in the treatment choice equation. We discuss the key conditions, the role of control variables as well as the estimands in detail and review the literature on estimation and inference. Then we consider extensions to multiple and continuous instruments, to the regression discontinuity design, and discuss the testability of the assumptions. Finally, we compare this approach to the alternative instrumental variable approach reviewed in Chapter 9 of this handbook. Two open research problems are highlighted in the conclusion.

10.1   Introduction

Policy-makers are often interested in the causal effect of a binary treatment D on an outcome variable Y. This effect is frequently summarized by the average treatment effect $E[Y_1] - E[Y_0]$, where Y_1 and Y_0 are the treatment and control potential outcomes that were introduced by Neyman in 1923 (see Spława-Neyman, 1990) and popularized by Rubin (1974). This single measure, however, gives only a partial view of the effects of the treatment. The treatment may not affect the mean of the outcome distribution but may increase its variance or change its shape. From a policy perspective, an intervention that helps to raise the lower tail of an income distribution is often more appreciated than an intervention that shifts the median, even if the average treatment effects of both interventions are identical. In another application, in clinical trials, a drug may increase the long-term survival probability without affecting short-term survival.

We naturally have more information about the treatment effect if we know the entire distributions of the potential outcomes, $F_{Y_0}(y)$ and $F_{Y_1}(y)$, or equivalently their inverse, the quantile functions $Q_{Y_0}(\tau)$ and $Q_{Y_1}(\tau)$, where $\tau \in (0,1)$ denotes the quantile index. An intuitive way to summarize the treatment effect is to report the quantile treatment effect (QTE) function

$$\Delta(\tau) = Q_{Y_1}(\tau) - Q_{Y_0}(\tau).$$

QTEs allow numerous interesting hypotheses to be assessed. For instance, if the treatment has a pure location shift effect, then QTEs will be constant as a function of the quantile. The location–scale shift model implies a monotone QTE function. If the distribution of Y_1 has first-order stochastic dominance over the distribution of Y_0, then the QTE function will be above the zero line. Thus, the QTE function represents an intuitive way to report the effect of a treatment on the marginal distribution of the outcome variable.

It would be even more informative to know the entire joint distribution of Y_1 and Y_0. Heckman et al. (1997) discuss several of the additional evaluation questions that could be analyzed if we knew this joint distribution. For instance, we could calculate the distribution of the individual treatment effects, $F_{Y_1-Y_0}(y)$, and in particular recover the proportion of units who benefit from the treatment. Alas, this joint distribution is not point-identified even in the ideal situation where the treatment has been randomized. The classical inequalities of Hoeffding and Fréchet provide bounds on joint distributions when the marginals are known. Makarov (1982) derives the sharp bounds on the distribution of the individual effects and Fan and Park (2010) provide estimation and inference procedures for these bounds.

Rank preservation is a common assumption that allows the joint distribution to be recovered from the marginals. If we assume that each individual maintains his rank in the distribution of the outcome regardless of his treatment status, then the τ QTE is the treatment effect for individuals at the τ quantile of the potential outcome distributions. Doksum (1974) and Lehmann (1975) were the first to suggest this measure of the treatment effect and to discuss its properties. The idea is that each subject possesses an underlying "proneness" or ability – proneness to die early, to learn fast, to grow taller, depending on the application – which does not change with the treatment. In some applications, rank preservation is natural because it seems unlikely that the treatment makes weak subjects robust and strong subjects weak. If we consider an application where the outcome is the wage and the treatment is the sector of employment, it is implausible that the best professors of philosophy are also the best bricklayers. The methods reviewed in this chapter do not rely on the rank invariance assumption. We therefore interpret QTE as the difference between the same quantile of two distributions. However, the rank preservation assumption can always be added on the top to enrich the interpretation of the results.

In a randomized controlled trial with perfect compliance, the sample quantiles in the treated and control groups are consistent for $Q_{Y_1}(\tau)$ and $Q_{Y_0}(\tau)$ and, consequently, their difference is consistent for $\Delta(\tau)$. In practice, $\Delta(\tau)$ can be estimated using a simple quantile regression of Y on D and a constant. However, it is often impossible to impose perfect compliance: some subjects may be assigned randomly to the treatment but may refuse to take it, others may be assigned to the control group but may find a way to get the treatment anyway. As a result of this self-selection, the observed treatment variable is endogenous, which renders standard quantile regression inconsistent, just as is the case for least-squares methods. On the other hand, since the assignment has been randomized, it is still possible to identify the causal effect of the assignment on the outcome distribution. In the context of a clinical trial, this is called the intention-to-treat (ITT) effect. While the ITT effect might be an interesting parameter, especially when the only possible intervention is to assign the treatment, it does not provide an estimate of the treatment effect.

Instrumental variable methods provide a powerful tool to address this problem. There are several approaches to instrumental variable identification and estimation of QTE. In this chapter we focus on models that achieve identification through a monotonicity assumption in the treatment choice equation. This approach was first developed by Imbens and Angrist (1994) for average effects and then extended by Imbens and Rubin (1997), Abadie (2002), and Abadie et al. (2002) to distributional effects. Identification in these models does not rely on rank preservation or other similar assumptions, which is in sharp contrast to the instrumental variable approach reviewed in Chapter 9 of this handbook. We compare both approaches in Section 10.5. We also limit our survey to models for binary endogenous variables because almost no results have been obtained for multi-valued or continuous treatments within the framework that we consider.

In the simplest setup, both the treatment and the assignment (instrument) are binary. With imperfect compliance, there are some individuals who do not react to a change in the assignment; either they always refuse the treatment or they always find a way to get the treatment. Since we allow the treatment effect to be arbitrarily heterogeneous, we cannot identify the treatment effects for these units. There is no information in the data for them about one of both potential outcomes. This implies that it is also impossible to identify the treatment effect for the whole population (QTE) or, if some units are always treated independently of the assignment, for the treated subpopulation (QTE on the treated). We are only able to identify effects for the individuals who respond to a change in the value of the instrument. These individuals are called compliers because they comply with the assignment.

Our main estimand is, therefore, the QTE for the compliers, which we call the local quantile treatment effect (LQTE) because it corresponds to the QTE for a subpopulation. This terminology is in analogy to the local average treatment effect (LATE) of Imbens and Angrist (1994) and is *not* related to nonparametric methods that are local with respect to the covariates. Whether or not the subpopulation of compliers is of interest depends heavily on the empirical context; see, for instance, the controversial discussion in Imbens (2010), Deaton (2010), and Heckman and Urzúa (2010). While other populations would naturally be of interest, we would like to emphasize again that there is no information in the data about their treatment effects. The LQTE is the QTE for the the largest population for which the effect is identified. As will be obvious in Section 10.5, the approaches that recover treatment effects for larger populations achieve identification by means of extrapolation from the compliers.

The methods surveyed in this chapter have already been applied numerous times in very different setups. For instance, Abadie et al. (2002) estimate the effect of Job Training Partnership Act training programs on the earnings distribution of previously unemployed individuals. Eren and Ozbeklik (2014) similarly study the effects of the Job Corps programs. They both identify the causal effects of the programs by exploiting a randomized experiment with imperfect compliance. The methods are, however, not limited to the analysis of experimental data. Many variables of interest cannot be randomized in economics and other social sciences. For this reason, researchers use "natural experiments" to identify treatment effects. Ananat and Michaels (2008) note that the probability of a divorce is higher when the first-born child is a female. Using this source of variation, they find that a divorce has little effect on women's mean household income but it increases women's odds of having very high or very low income. Cawley and Meyerhoefer (2012) estimate the impact of obesity on the distribution of medical costs, instrumenting the respondent's weight with the weight of a biological relative. Frölich and Melly (2013) use twin birth as an instrument for having several children and estimate its effect on the household income.

The basic framework was already developed in the 1990s; Angrist and Pischke (2008), Imbens (2014) and Imbens and Rubin (2015) provide interesting surveys of this approach to instrumental variables. In Section 10.2, after briefly summarizing the basics, we focus on the particularities of the quantile estimands and their implications for the identification of the effects in setups without and with covariates. Section 10.3 briefly reviews the literature on estimation and inference. In Section 10.4 we consider two extensions within the same framework – nonbinary instruments and the regression discontinuity design – as well as the testability of the identifying assumptions. Section 10.5 compares the model in this chapter to the instrumental variable quantile regression model introduced by Chernozhukov and Hansen (2005) and reviewed in Chapter 9 of this handbook. Finally, Section 10.6 briefly summarizes the findings and highlights two important open problems in this literature.

10.2   Framework, estimands and identification

10.2.1   Without covariates

We develop our presentation in the context of a randomized trial with noncompliance. The units of observation are assigned to a treatment but this assignment cannot be perfectly enforced. We consider the simplest case where both the assignment Z and the treatment D are binary. In many applications it is reasonable to assume that there is no interference between units. This assumption, first introduced by Cox (1958), is called the stable unit of treatment assignment (SUTVA) by Rubin (1980).

Assumption 2 (SUTVA) *For any unit the value of the treatment when exposed to the assignment z and of the outcome when exposed to the assignment z and the treatment d is the same regardless of the treatments and assignments that other units receive.*

The SUTVA excludes any kind of interaction between units. A classical violation of this assumption is a setting where the treatment is a vaccine that immunizes the unit against a contagious disease. The effect of this vaccine will necessarily be a function of the number of persons who have already been vaccinated. Peer effects and general equilibrium effects are also excluded; for example, the SUTVA is violated if the wage effect of getting a college degree depends on the proportion of the labor force having the same degree **and the intervention is large enough to affect this proportion**. Thus, like all partial equilibrium approaches, it is well suited only for small-scale interventions.

We use the potential outcomes notation. D_z denotes the potential treatment status when the assignment is set exogenously to $z \in \{0, 1\}$. The observed treatment is related to the potential treatments by $D = D_1 Z + D_0 (1 - Z)$. In the case of perfect compliance, $D_z = z$ such that $D = Z$ and the treatment itself has been randomized. With imperfect compliance this equality breaks down and we have to take into account the fact that the treatment has been self-selected and is, therefore, endogenous. Similarly, we define four potential outcomes Y_{zd} for each combination of z and $d \in \{0, 1\}$. Note that without Assumption 2 the potential outcomes of each unit would depend on the assignments and treatments of all units in the population. It is impossible to identify any treatment effect without imposing some restriction on the dependence between units.

With a randomized assignment it is easy to identify the average causal effect of the assignment on the outcome, $E[Y_{1D_1} - Y_{0D_0}]$, as well as its effect on the distribution and quantile functions of the outcome. If compliance is low, the ITT will underestimate the absolute value of the effect of the treatment. In the following, we will take the extent of noncompliance into account in order to estimate the effect of the treatment.

With D and Z being binary, we can partition the population into four types $\mathcal{T}$ defined by D_1 and D_0:

D_1	D_0	Type
1	1	Always-takers ($\mathcal{T} = a$)
1	0	Compliers ($\mathcal{T} = c$)
0	1	Defiers ($\mathcal{T} = d$)
0	0	Never-takers ($\mathcal{T} = n$)

The never-takers and the always-takers do not react to the assignment and do not contribute to the ITT effect. We have no source of random variation for these types and we will not be able to identify any treatment effects for them. Both the compliers and the defiers react to a change in the assignment, but they do so in opposite directions. Therefore, the ITT effect is a weighted average of the individual treatment effects with positive weights for the compliers, negative weights for the defiers and zero weights for both other types (e.g. Angrist et al., 1996). Given that we observe each unit only in one state of the world, we are unable to determine the type of each unit to separate compliers from defiers. There are at least two ways to solve this mixture problem: either we restrict the outcome to be the same for the compliers and the defiers (homogeneity assumption for the treatment effect) or we assume that there are no defiers (homogeneity assumption for the assignment effect). The approach to instrumental variables that we review in this chapter follows the second assumption. De Chaisemartin (2014) shows that these two types of assumption can be combined and would lead to a similar interpretation of the same estimators.

We impose the following assumptions:

Assumption 3

1. *Independent instrument:* $(Y_{11}, Y_{10}, Y_{01}, Y_{00}, D_1, D_0) \perp\!\!\!\perp Z$.

2. *Exclusion restriction:* $P(Y_{1d} = Y_{0d}) = 1$ *for* $d \in \{0, 1\}$.

3. *Relevant instrument:* $\Pr(\mathcal{T} = c) > 0$ *and* $\Pr(Z = 1) \in (0, 1)$.

4. *Monotonicity:* $\Pr(\mathcal{T} = d) = 0$.

Assumption 3.1 is an unconfounded instrument restriction. It is mechanically satisfied if the assignment has been randomized. Full independence is required for the identification of the whole LQTE process but a slighter weaker local quantile independence condition is enough to identify a single LQTE; see Assumption 5 in Section 10.4.3. The exclusion restriction (Assumption 3.2) requires that the assignment Z must have no direct effect on the potential outcomes to be a valid instrumental variable. It is important to note that this exclusion restriction is unrelated to – and, in particular, not implied by – the randomization of the instrument (Assumption 3.1). In medical studies it is well known that there is a psychological effect of being assigned to a treatment. For this reason, even the control group receives a placebo treatment and the exclusion restriction is then likely to be satisfied. There are cases where it is more difficult to find a placebo treatment. For instance, if unemployed individuals receive an invitation to attend a CV-writing course, they may decide not to attend the course but start reading books to improve their application documents. This would violate the exclusion restriction. Under Assumptions 2 and 3.2 we can define potential outcomes in terms of D alone: $Y_0 = Y_{00} = Y_{10}$ and $Y_1 = Y_{01} = Y_{11}$. Assumption 3.3 requires that at least some individuals react to changes in the value of the instrument. The strength of the instrument can be measured by $\Pr(\mathcal{T} = c)$. These first three assumptions are common to all instrumental variable models. Assumption 3.4, which is often referred to as monotonicity, is specific to the approach we review in this chapter. It requires that D_z weakly increases with z for all individuals. Note that it is mechanically satisfied if there is perfect one-sided compliance, which is quite common. If those assigned to receive the control treatment can be denied access to the active treatment, then there are no defiers and no always-takers. In addition to automatically satisfying the monotonicity assumption, this implies that the treated are all compliers such that the LQTE is equal to the quantile treatment effect on the treated.

Imbens and Angrist (1994) show that, under Assumptions 2 and 3, the Wald (1940) estimator is consistent for the average treatment effect for the compliers, usually referred to as the LATE:

$$E\left(Y_1 - Y_0 | \mathcal{T} = c\right) = \frac{E\left[Y | Z = 1\right] - E\left[Y | Z = 0\right]}{E\left[D | Z = 1\right] - E\left[D | Z = 0\right]}.$$

The result for the average of the dependent variable naturally also apply to transformations of the dependent variable such as the distribution function ($F_Y\left(y\right) = E\left[1\left(Y \leqslant y\right)\right]$). This directly provides results for the distributional treatment effect for the compliers, $F_{Y_1}\left(y | \mathcal{T} = c\right) - F_{Y_0}\left(y | \mathcal{T} = c\right)$. Obtaining the effect on the distribution is enough to test some hypotheses such as the absence of any effect or first-order stochastic dominance. On the other hand, it does not allow us to test the hypothesis that the treatment effect exerts a pure location shift or a location–scale shift of the distribution of the outcome. More generally, the interpretation of distributional treatment effects is not straightforward. We may find that the effect is positive over a part of the support of Y and negative over another part, but we cannot identify the relevance of these two parts without knowing separately the distributions of Y_1 and Y_0. For instance, it may be that the treatment effect is positive over 99% of the quantiles and negative only over 1%; or it may be the opposite.

The quantile treatment effects for the compliers, $Q_{Y_1}\left(\tau | \mathcal{T} = c\right) - Q_{Y_0}\left(\tau | \mathcal{T} = c\right)$, overcome these limitations. They have an intuitive and straightforward interpretation. Their average is the traditional LATE estimand. Their unit of measurement is the same as the unit of the outcome itself, which allows for testing the location shift and location–scale shift hypothesis. The rest of this subsection shows how we can identify the LQTE.

Imbens and Rubin (1997) show that the distributions of Y_1 and Y_0 are identified for the compliers by working directly with the densities. Abadie (2002) gives more convenient representations for the cumulative distribution functions (cdfs):

$$F_{Y_1}\left(y | \mathcal{T} = c\right) = \frac{E\left[1\left(Y \leqslant y\right) D | Z = 1\right] - E\left[1\left(Y \leqslant y\right) D | Z = 0\right]}{E\left[D | Z = 1\right] - E\left[D | Z = 0\right]}. \quad (10.1)$$

The reason is that

$$E\left[1\left(Y \leqslant y\right) D | Z = 1\right] = F_{Y_1}\left(y | \mathcal{T} = a\right)\Pr\left(\mathcal{T} = a\right) + F_{Y_1}\left(y | \mathcal{T} = c\right)\Pr\left(\mathcal{T} = c\right)$$

and

$$E\left[1\left(Y \leqslant y\right) D | Z = 0\right] = F_{Y_1}\left(y | \mathcal{T} = a\right)\Pr\left(\mathcal{T} = a\right),$$

such that the numerator simplifies to $F_{Y_1}\left(y | \mathcal{T} = c\right)\Pr\left(\mathcal{T} = c\right)$. Similarly, the denominator simplifies to $\Pr\left(\mathcal{T} = c\right)$. A similar reasoning can be used to identify the distribution of the control potential outcome:

$$F_{Y_0}\left(y | \mathcal{T} = c\right) = \frac{E\left[1\left(Y \leqslant y\right)\left(1 - D\right) | Z = 1\right] - E\left[1\left(Y \leqslant y\right)\left(1 - D\right) | Z = 0\right]}{E\left[1 - D | Z = 1\right] - E\left[1 - D | Z = 0\right]}. \quad (10.2)$$

These two distributions can then be inverted to obtain the quantile functions:

$$Q_{Y_1}(\tau|\mathcal{T} = c) = \inf\{y : F_{Y_1}(y|\mathcal{T} = c) \geqslant \tau\}$$

and

$$Q_{Y_0}(\tau|\mathcal{T} = c) = \inf\{y : F_{Y_0}(y|\mathcal{T} = c) \geqslant \tau\}.$$

Consequently, the LQTEs are identified as

$$\Delta(\tau|\mathcal{T} = c) = Q_{Y_1}(\tau|\mathcal{T} = c) - Q_{Y_0}(\tau|\mathcal{T} = c).$$

10.2.2   In the presence of covariates: conditional LQTE

In almost all applications we also observe a vector of covariates X. We may want to include them in the estimation for two reasons. First, the validity of the independence assumption (Assumption 3.1) may be plausible only after conditioning on covariates. This is the case for stratified randomized experiment when the assignment probabilities are different across the strata. This is also the case in observational studies when the instrument has not been randomized. For instance, Frölich and Melly (2013) use twin birth as an instrument for having several children. Since it is well known that the probability of a twin birth is a function of the race of the parents and increases with the age of the mother, they control for these characteristics to satisfy the exclusion restriction. Even if this assumption is valid unconditionally (e.g. because the instrument has been unconditionally randomized), we will see in Section 10.3 that we may wish to include covariates in the estimation for efficiency reasons. Therefore, we consider the case where the instrument is valid after conditioning on the covariates, which also covers randomized instruments as a special case.

Assumption 4 *For all $x \in \text{supp}(X)$:*

1. *Independent instrument:* $(Y_{11}, Y_{10}, Y_{01}, Y_{00}, D_1, D_0) \perp\!\!\!\perp Z|X = x$.
2. *Exclusion restriction:* $P(Y_{1d} = Y_{0d}|X = x) = 1$ *for* $d \in \{0, 1\}$.
3. *Relevant instrument:* $\Pr(\mathcal{T} = c|X = x) > 0$ *and* $\Pr(Z = 1|X = x) \in (0, 1)$.
4. *Monotonicity:* $\Pr(\mathcal{T} = d|X = x) = 0$.

Assumption 4 is simply the conditional version of Assumption 3. Note that Assumption 4.3 requires the support of X to be identical in the $Z = 0$ and $Z = 1$ subpopulations. If the support condition is not met initially, we need to define the parameters relative to the common support.

The conditional distributions for the compliers are identified using the same approach as above; for example, for Y_1,

$$F_{Y_1}(y|X = x, \mathcal{T} = c) = \frac{E[1(Y \leqslant y)D|X = x, Z = 1] - E[1(Y \leqslant y)D|X = x, Z = 0]}{E[D|X = x, Z = 1] - E[D|X = x, Z = 0]}.$$

$$(10.3)$$

The nonparametric estimation of these conditional distributions will naturally suffer from the curse of dimensionality. For this reason, Abadie et al. (2002) impose a linear restriction for the compliers:

$$Q_{Y_0}(\tau|X = x, \mathcal{T} = c) = X'\beta(\tau)$$

and

$$Q_{Y_1}(\tau|X = x, \mathcal{T} = c) = X'\beta(\tau) + \Delta(\tau|\mathcal{T} = c).$$

The particular form of conditional quantile functions does not matter for what follows. In particular, we can easily introduce interactions between the treatment and the covariates, which allows for heterogeneous treatment effects with respect to the observables.

Equation (10.3) shows that $\Delta(\tau|\mathcal{T}=c)$ and $\beta(\tau)$ are identified but does not suggest a convenient parametric estimator. Abadie et al. (2002), applying a result in Abadie (2003), give the following weighted quantile regression representation:

$$(\beta(\tau), \Delta(\tau|\mathcal{T}=c)) = \underset{\tilde{\beta}, \tilde{\Delta}}{\arg\min} E\left[W^{\text{AAI}} \cdot \rho_\tau \left(Y - X\tilde{\beta} - D\tilde{\Delta}\right)\right] \qquad (10.4)$$

$$W^{\text{AAI}} = 1 - \frac{D(1-Z)}{1 - E(Z|X)} - \frac{(1-D)Z}{E(Z|X)}. \qquad (10.5)$$

The proof follows from the fact that

$$E\left[W^{\text{AAI}}|Y, X, \mathcal{T}=a\right] = 1 - \frac{1 - E[Z|Y, X, D_0 = D_1 = 1]}{1 - E(Z|X)} = 0,$$

$$E\left[W^{\text{AAI}}|Y, X, \mathcal{T}=n\right] = 1 - \frac{E[Z|Y, X, D_0 = D_1 = 0]}{E(Z|X)} = 0,$$

$$E\left[W^{\text{AAI}}|Y, X, \mathcal{T}=c\right] = 1.$$

In other words, on average, the weights W^{AAI} find the conditional compliers and annihilate the always- and never-takers. This suggests a weighted quantile regression estimator. However, the sample analog of (10.4) is not globally convex because the weights are positive and negative. We discuss the solution suggested by Abadie et al. (2002) below in Section 10.3.

10.2.3   In the presence of covariates: unconditional LQTE

In Section 10.2.2 we discussed *conditional* quantile treatment effects, that is, the quantile is defined within the population with the same X. For instance, Abadie et al. (2002) are interested in the effect of a training program on earnings. Assuming rank invariance, the conditional LQTE for $\tau = 0.1$ is the effect for the individuals who are at the 0.1 quantile of the earnings distribution given their age, education level, race, etc. This includes white workers in their peak earnings years with a college degree who earn well above the median of the whole population. This is not necessarily the estimand that the policy-maker is interested in; she may be interested in the effects for unconditionally poor households. Similarly, in other applications, she may be interested in unconditionally low-birthweight, or unconditionally low-achieving students. This distinction does not really play a role for averages because the unconditional mean is simply the average of the conditional means, but it matters for quantiles because the law of iterated expectation does not apply. While we may be interested in unconditional LQTE, we may need to include control variables in order to satisfy Assumption 4, in particular the (conditional) independence assumption, or we may wish to add other covariates for efficiency reasons. In other words, the set of covariates that we wish to include depends on both the estimand that we are interested in and on the exclusion restriction that we have to satisfy. There may be a tension between these objectives.

It is possible to separate the definition of the estimand from the assumptions needed for a causal interpretation. A first representation of the unconditional cdf of the potential outcomes for the compliers is given by

$$
F_{Y_1}(y|\mathcal{T} = c) = \int F_{Y_1}(y|X = x, \mathcal{T} = c)dF_X(x|\mathcal{T} = c)
$$

$$
= \int F_{Y_1}(y|X = x, \mathcal{T} = c)\frac{\Pr(\mathcal{T} = c|X = x)}{\Pr(\mathcal{T} = c)}dF_X(x)
$$

$$
= \frac{\int (E\left[1(Y \leqslant y)D|X = x, Z = 1\right] - E\left[1(Y \leqslant y)D|X = x, Z = 0\right])dF_X(x)}{\int (E\left[D|X = x, Z = 1\right] - E\left[D|X = x, Z = 0\right])dF_X(x)}, \quad (10.6)
$$

where the first equality holds by the law of total expectation, the second by Bayes' law and the third by the representation of the conditional distribution in (10.3) and the fact that $\Pr(\mathcal{T} = c|X = x) = E[D|X = x, Z = 1] - E[D|X = x, Z = 0]$ as seen above. A similar result applies to $F_{Y_0}(y|\mathcal{T} = c)$.

Parts (b) and (c) of Theorem 3.1 in Abadie (2003) applied to the cdf provide a weighted representation of $F_{Y_1}(y|\mathcal{T} = c)$ and $F_{Y_0}(y|\mathcal{T} = c)$ (see Frölich and Melly, 2013):

$$
F_{Y_1}(y|\mathcal{T} = c) = \frac{E\left[1(Y < y)DW^{\mathrm{FM}}\right]}{E\left[DW^{\mathrm{FM}}\right]},
$$

$$
F_{Y_0}(y|\mathcal{T} = c) = \frac{E\left[1(Y < y)(1 - D)W^{\mathrm{FM}}\right]}{E\left[DW^{\mathrm{FM}}\right]}, \quad (10.7)
$$

where

$$
W^{\mathrm{FM}} = \frac{Z - \Pr(Z = 1|X)}{\Pr(Z = 1|X)(1 - \Pr(Z = 1|X))}(2D - 1).
$$

This result can be obtained from (10.6) via iterated expectations arguments. These weights are different from the W^{AAI} weights defined in Section 2.3. They find the conditional compliers in the following sense:

$$
E\left[W^{\mathrm{FM}}|X, Y, \mathcal{T} = a\right] = E\left[W^{\mathrm{FM}}|X, Y, \mathcal{T} = n\right] = 0,
$$

$$
E\left[W^{\mathrm{FM}}|X, Y, \mathcal{T} = c\right] = 2.
$$

In addition, they balance covariates between treated compliers and nontreated compliers in the following sense:

$$
E\left[W^{\mathrm{FM}}|X, Y, \mathcal{T} = c, D = 1\right] = \frac{1}{\Pr(Z = 1|X)}
$$

$$
= \frac{1}{\Pr(Z = 1|X, \mathcal{T} = c)}
$$

$$
= \frac{1}{\Pr(D = 1|X, \mathcal{T} = c)},
$$

$$
E\left[W^{\mathrm{FM}}|X, Y, \mathcal{T} = c, D = 0\right] = \frac{1}{1 - \Pr(Z = 1|X)}
$$

$$
= \frac{1}{1 - \Pr(D = 1|X, \mathcal{T} = c)}.
$$

This means that the weights W^{FM} not only find the conditional compliers but also reweight them such that the treated and control compliers have the same covariates distributions. They share the first property with the weights W^{AAI} of Abadie et al. (2002) and the second property with the inverse probability weights used when the treatment is exogenous conditional on the covariates. W^{FM} indeed nests inverse probability weights as a strict special case when the treatment is used as its own instrument, which is justified under exogeneity. Firpo (2007) discusses the estimation of unconditional QTEs following this approach.

Frölich and Melly (2013) suggest considering unconditional quantile treatment effects

$$\Delta\left(\tau|\mathcal{T}=c\right)=Q_{Y_1}\left(\tau|\mathcal{T}=c\right)-Q_{Y_0}\left(\tau|\mathcal{T}=c\right)$$

while keeping Assumption 4 (i.e. we need covariates for the identification). Both unconditional quantile functions can be obtained by inverting the unconditional cdfs obtained in (10.6) or (10.7). Instead, it is also possible to directly obtain the LQTE by a weighted quantile regression

$$\left(Q_{Y_0}\left(\tau|\mathcal{T}=c\right),\Delta\left(\tau|\mathcal{T}=c\right)\right)=\underset{\tilde{Q}_{Y_0},\tilde{\Delta}}{\arg\min}E\left[W^{\text{FM}}\cdot\rho_\tau(Y-\tilde{Q}_{Y_0}-\tilde{\Delta}D)\right]. \tag{10.8}$$

This optimization problem is nonconvex but can be solved relatively easily by rewriting it as two one-dimensional problems and noting that the objective function changes only when the fitted values cross observed outcomes.

10.3   Estimation and inference

In the absence of covariates, the sample analogs of (10.1) and (10.2) provide natural estimators of the cdfs of the potential outcomes. The estimated cdfs will necessarily be non-monotone, but they can be monotonized, for instance, using the rearrangement method of Chernozhukov et al. (2010). This allows the cdfs to be inverted and to obtain the quantile functions and subsequently the LQTEs.

In the presence of covariates, Abadie et al. (2002) suggest estimating conditional LQTE based on the weighted quantile regression representation (10.4). Estimation based on (10.4) may be a difficult task because the weights (10.5) are positive and negative, implying that the objective function has many local minima. Therefore, Abadie et al. (2002) suggest replacing the weights (10.5) with their projection on (Y,D,X), which can be shown to be always positive. Their estimation strategy consists of three steps: (i) estimation of the instrument propensity score $E\left(Z|X\right)$ using nonparametric power series; (ii) estimation of the positive weights using nonparametric power series of the estimated weights on (Y,D,X); and (iii) weighted quantile regression using the estimated positive weights. The resulting estimator is $\sqrt{n}$ consistent and asymptotically normal. However, Hong and Nekipelov (2010) show that the estimator suggested in Abadie et al. (2002) does not attain the semiparametric efficiency bound. The alternative and efficient estimator that they suggest uses density weighting for the compliers, which is similar to the efficient weighted least-squares estimator in the presence of heteroskedasticity. This approach is nevertheless **not often used in practice** because (i) it requires estimation of the conditional density of Y for the compliers, and (ii) the weighted estimator is more difficult to interpret if the conditional model is misspecified.

For unconditional effects, analog estimators based on all three representations (10.6), (10.7), and (10.8) have been suggested. While parametric restrictions (e.g. linearity) are necessary to achieve the $\sqrt{n}$ consistency of the conditional LQTE estimators, unconditional LQTE can be estimated at the $\sqrt{n}$ rate without any parametric restrictions because the unconditional distributions are averages of conditional distributions.

Belloni et al. (2017) provide estimators based on (10.6) for environments with many control variables, either because many variables are available in the raw data set or because we want to include interactions and other transformations of the control variables. They suggest LASSO-type methods that automatically select the relevant ones. Assuming that reduced-form relationships are approximately sparse, they show that valid inference can be performed after data-driven selection of control variables. Moreover, they derive the limiting laws of the estimators of the whole quantile treatment effect process. This allows **for the** construction of confidence band for the LQTE function over a quantile continuum, to test functional hypotheses and for dominance relations between the potential outcomes. Hsu et al. (2015) suggest a weighted cdf estimator based on (10.7) and derive the asymptotic distribution for the whole LQTE process. Finally, Frölich and Melly (2013) analyze a weighted quantile regression estimator based on (10.8). Their estimator is $\sqrt{n}$-consistent, asymptotically normal, and achieves the the semiparametric efficiency bound. In addition, Frölich and Melly (2013) show that adding covariates increases the precision of the estimator of the unconditional effect if these additional covariates (i) do not affect the instrument propensity score and (ii) affect the outcome. These conditions are often satisfied, for instance, when the instrument has been randomized. On the other hand, including covariates that affect the instrument propensity score and do not affect the outcome will increase the variance of the estimator. Note that such a comparison of variances does not make sense for the conditional LQTE because the definition of the estimand is a function of the covariates included.

Note that all the asymptotic results mentioned in this section rely on the continuity of the dependent variable. This assumption is not an identifying condition. On the contrary, the LQTE framework accommodates discrete outcomes and outcomes with mass points very naturally. It is also possible to show that the analog estimators of the cdfs of the potential outcomes are asymptotically normally distributed even for discrete outcomes. On the other hand, continuity of the dependent variable is a condition for obtaining well-behaved asymptotic distributions for the quantiles and LQTE estimators.

10.4   Extensions

10.4.1   Regression discontinuity design

A fuzzy regression discontinuity design (RDD) exploits a discontinuity in the probability of treatment when a running variable R exceeds a threshold r_0. This section also covers the sharp RDD as a special case where the probability jumps from 0 to 1 at the threshold. If the distribution of the potential outcomes is continuous in R, then the discontinuity becomes a valid instrumental variable for the treatment. In this context, compliers are the units that switch their treatment status at the discontinuity, and monotonicity means that all units that switch treatment at the discontinuity do so in the same direction. This design fits in the framework outlined in Section 10.2, with the exception that the identification is local at $R = r_0$, which requires nonparametric estimation.

The local version of (10.1) is given by

$$
F_{Y_1}\left(y|\mathcal{T} = c, R = r_0\right) =
$$
$$
\lim_{\varepsilon \to 0} \frac{E\left[1\left(Y \leqslant y\right) D | r_0 < R < r_0 + \varepsilon\right] - E\left[1\left(Y \leqslant y\right) D | r_0 - \varepsilon < R < r_0\right]}{E\left[D | r_0 < R < r_0 + \varepsilon\right] - E\left[D | r_0 - \varepsilon < R < r_0\right]}, \tag{10.9}
$$

and similarly for the control potential outcome. Note that the distributions are identified only for the local compliers, that is, the compliers whose running variable is arbitrarily close to r_0. A natural estimand to consider is the QTE for the local compliers:

$$\Delta(\tau|\mathcal{T} = c, R = r_0) \equiv Q_{Y_1}(\tau|\mathcal{T} = c, R = r_0) - Q_{Y_0}(\tau|\mathcal{T} = c, R = r_0).$$

The representation of $F_{Y_1}(y|\mathcal{T} = c, R = r_0)$ in (10.9) is a function of four conditional means at boundary points (from the left and right of r_0). Frandsen et al. (2012) **estimate these means using local linear techniques because they are** not subject to bias from ignoring the running variable and automatically corrects for boundary effects. The cdfs are subsequently inverted to obtain an estimator of $\Delta(\tau|\mathcal{T} = c, R = r_0)$. They prove uniform consistency and asymptotic Gaussianity of the estimators for the whole QTE process for the local compliers. Of course, this nonparametric estimator only converges at the one-dimensional nonparametric rate.

10.4.2 Multi-valued and continuous instruments

In the previous two sections we focused on the case with a binary instrument. If the instrument is multi-valued (or there are several instruments), then it is obviously possible to identify an LQTE with respect to any pair of distinct values of Z, satisfying Assumption 3. Instead of estimating many pairwise effects, one may prefer to estimate the LQTE for the largest complying subpopulation. This is simply obtained by considering the value of Z that minimizes the treatment probability and the value that maximizes the treatment probability (given a monotonicity assumption defined with respect to $\Pr(D = 1|Z)$).

When the instrument is continuous, it is possible to identify a continuum of treatment effects. Heckman and Vytlacil (2005) developed this approach for average treatment effects and called the resulting parameters marginal treatment effects. They show that the LATE can be represented as a weighted average of marginal treatment effects. Inversely, the marginal treatment effects can be considered as the limit form of the LATE parameter. Carneiro and Lee (2009) extend these ideas to the estimation of the quantile analogs, the marginal quantile treatment effects (MQTEs). To simplify the notation we do not incorporate covariates in this section. Suppose that individuals choose their treatment status according to the equation

$$D = 1\{\Pr(D = 1|Z) \geqslant \eta\}, \tag{10.10}$$

where $\eta|Z \sim U(0,1)$ is a scalar error term. Z is a continuous instrument that is independent of Y_0 and Y_1. For binary Z, Vytlacil (2002) shows that (10.10) is equivalent to Assumption 3.4. Hence, this model can be seen as a generalization of the LQTE framework to general instruments. In this model, the main estimands of interest are the MQTEs

$$\Delta(\tau|p) \equiv Q_{Y_1}(\tau|\eta = p) - Q_{Y_0}(\tau|\eta = p).$$

Carneiro and Lee (2009) show that $\Delta(\tau|v)$ can only be identified for values of p in the common support of $Pr(D = 1|Z)$ for observations in $D = 1$ and $D = 0$.

They note that

$$F_Y(y|\Pr(D = 1|Z) = p, D = 1) \cdot p = F_{Y_1}(y|\Pr(D = 1|Z) = p, D = 1) \cdot p$$
$$= F_{Y_1}(y|\eta \leqslant p) \cdot p$$
$$= \int_0^p F_{Y_1}(y|\eta = h)\, dh.$$

By taking the derivative with respect to p on both sides, this implies that

$$F_{Y_1}(y|\eta = p) = F_Y(y|\Pr(D = 1|Z) = p, D = 1) + \frac{\partial F_Y(y|\Pr(D = 1|Z) = p, D = 1)}{\partial p}.$$

A similar result applies to the cdf of Y_0. Both cdfs can be inverted to obtain the quantile functions.

Yu (2014) suggests semiparametric estimators of the MQTEs based on this representation. He allows for the presence of control variables and considers both conditional and unconditional MQTEs. Finally, he derives the corresponding weak limits and shows the validity of the bootstrap for inference.

10.4.3 Testing instrument validity

Assumption 3 imposes testable implications on the joint distribution of (Y, D, Z). Under this set of assumptions, $F_Y(y|D = 1, Z = 1)$ is a mixture of $F_{Y_1}(y|\mathcal{T} = c)$ and $F_{Y_1}(y|\mathcal{T} = a)$ with identified mixing probabilities. One of the mixing distributions is also separately identified because $F_{Y_1}(y|\mathcal{T} = a) = F_Y(y|D = 1, Z = 0)$. Imbens and Rubin (1997) note that a violation of Assumption 3 may cause the resulting density function for the compliers, $f_{Y_1}(y|\mathcal{T} = c)$, to be negative. An equivalent result holds for the distribution of the control outcome.

More formally, Assumption 3 implies that, for every Borel set B in $\text{supp}(Y)$,

$$P(Y \in B, D = 1|Z = 1) - P(Y \in B, D = 1|Z = 0) = P(Y_1 \in B, D_1 > D_0),$$
$$P(Y \in B, D = 0|Z = 0) - P(Y \in B, D = 0|Z = 1) = P(Y_0 \in B, D_1 > D_0).$$

Because the right-hand sides are nonnegative by the definition of probabilities, we obtain the following testable restriction (Balke and Pearl, 1997; Heckman and Vytlacil, 2005):

$$P(Y \in B, D = 1|Z = 1) - P(Y \in B, D = 1|Z = 0) \geqslant 0, \tag{10.11}$$
$$P(Y \in B, D = 0|Z = 0) - P(Y \in B, D = 0|Z = 1) \geqslant 0. \tag{10.12}$$

Kitagawa (2015) shows that this testable restriction possesses two important features. First, it is optimal in the sense that any other feature of the observable data distribution cannot contribute to further screening out violations of Assumption 3. Second, validity of Assumption 3 is a refutable but nonverifiable hypothesis. In particular, it is possible to construct a joint probability law (Y_1, Y_0, D_1, D_0, Z) that satisfies (10.11) and (10.12) but violates Assumption 3. Consequently, accepting the null hypothesis never allows us to confirm Assumption 3 no matter how large the sample is.

To implement this testing idea, Kitagawa (2015) proposes a variance-weighted Kolmogorov–Smirnov test statistic based on the empirical distribution and a bootstrap algorithm to obtain critical values. He also provides an extension of the test to settings with conditioning covariates. Mourifié and Wan (2014) show that an alternative formulation of (10.11) and (10.12) fits into the intersection bounds framework of Chernozhukov et al. (2013), which provides an alternative test of the model.

This test exploits Assumption 3.1. Huber and Mellace (2015) note that this independence assumption is too strong if one is only interested in the average effect. Therefore, they propose an alternative test for a slightly weaker mean independence assumption. The fact that the observed distribution $F_Y(y|D = 1, Z = 1)$ is a mixture of the distributions of Y_1 for the always-takers and compliers allows $E[Y_1|\mathcal{T} = a]$ to be bounded using a result in Horowitz and Manski (1995): in one extreme scenario the always-takers are all at the bottom of the distribution, and in the other extreme scenario they are all at the top of the distribution. The test suggested by Huber and Mellace (2015) involves checking that $E[Y_1|\mathcal{T} = a]$, which is point-identified by $E[Y|D = 1, Z = 0]$, lies within the bounds obtained from the mixture. Similarly, $E[Y_1|\mathcal{T} = n]$ must lie within its bounds.

The same line of reasoning applies to quantile effects. Full independence is needed for the identification of the whole LQTE process. If one is interested in a single LQTE, then Assumption 3.1 may be replaced by the following:

Assumption 5 $(D_1, D_0) \perp\!\!\!\perp Z$ *and*

$$F_{Y_d}\left(Q_{Y_d}\left(\tau|\mathcal{T}=c\right)|\mathcal{T}=t, Z=0\right) = F_{Y_d}\left(Q_{Y_d}\left(\tau|\mathcal{T}=c\right)|\mathcal{T}=t, Z=1\right),$$

for $d \in \{0,1\}$ and $t \in \{c, a, n\}$.

Compared with Assumption 3.1, full independence between the instrument and the potential outcomes has been replaced by a local quantile independence restriction. Chesher (2003), for instance, emphasizes that this local condition is weaker than the global one. The proofs of the results in Section 10.2 clearly go through with Assumption 5 replacing Assumption 3.1. Under this alternative assumption, Kitagawa's first testable restriction (10.11) is valid only for two sets B, $(-\infty, Q_{Y_1}(\tau|\mathcal{T}=c)]$ and $(Q_{Y_1}(\tau|\mathcal{T}=c), \infty)$, and the second restriction (10.12) only for two other sets, $(-\infty, Q_{Y_0}(\tau|\mathcal{T}=c)]$ and $(Q_{Y_0}(\tau|\mathcal{T}=c), \infty)$. Interestingly, these four restrictions correspond to the moment inequalities of Huber and Mellace (2015) but applied to $1\left(Y_1 \leqslant Q_{Y_1}(\tau|\mathcal{T}=c)\right)$ and $1\left(Y_0 \leqslant Q_{Y_0}(\tau|\mathcal{T}=c)\right)$ instead of Y_1 and Y_0.

To summarize, Assumption 5 is sufficient for the consistency of the LQTE estimator at a single quantile, it is still testable but it is more difficult to reject. Note, however, that the goal of an instrument validity test is not necessarily to test the weakest set of assumptions under which the estimator is consistent but may be to assess the plausibility of the model that motivates the estimator. For instance, if an economic model delivers an exclusion restriction or the instrument has been randomized, then it makes sense to test the model that includes the full independence assumption. As discussed above, even asymptotically it is difficult to reject an invalid instrument; it is therefore judicious to test all the implications of the model to increase the power of the test

10.5   Comparison to the instrumental variable quantile regression model

In this section, we compare the framework reviewed in this chapter with another popular approach to instrumental variable estimation that accommodates binary treatments and binary instruments, the instrumental variable quantile regression (IVQR) model introduced by Chernozhukov and Hansen (2005) and reviewed in Chapter 9 of this handbook.

It is instructive to depart from the potential outcomes notation and to consider the general two-equation structural model given by

$$Y = q(D, \varepsilon), \qquad\qquad (10.13)$$
$$D = h(Z, \eta), \qquad\qquad (10.14)$$

where $q(\cdot)$ and $h(\cdot)$ are general nonseparable functions and ε and η are unobservable components that can be scalars or vectors. We refer to equation (10.13) as the *outcome equation* and to equation (10.14) as the *selection equation*. Note that this system of structural equations is recursive or triangular. To simplify the exposition, we omit covariates throughout this section and refer the interested reader to the original references for further details.

The equation-based model can be related to potential outcomes by

$$Y_d = q(d, \varepsilon_d),$$
$$D_z = h(z, \eta_z).$$

This formulation plays a key role in understanding different approaches to instrumental variable estimation. Vytlacil (2002) shows that Assumption 4 is equivalent to the latent index selection model

$$D_z = 1\{v(z) \geqslant \eta\}, \tag{10.15}$$

where $v(\cdot)$ is a nontrivial function of Z and $(Y_1, Y_0, \eta) \perp\!\!\!\perp Z$. Thus, Assumption 3 restricts the unobserved heterogeneity in the selection equation to being scalar, that is, $\eta_1 = \eta_0 = \eta$, while leaving the heterogeneity in the outcome equation unrestricted. As will be discussed below, restrictions on the dimensionality of the unobservables play a key role in nonseparable instrumental variable models.

A natural alternative is to consider a model that restricts the unobservables in the outcome equation, while leaving the selection equation unconstrained. This leads to the IVQR model. By the Skorohod representation of random variables, potential outcomes can be related to their structural quantile functions by $Y_d = q(d, \varepsilon_d)$, where $q(\cdot)$ is the structural quantile function of Y_d and $\varepsilon_d \sim U(0, 1)$. ε_d can be interpreted as a rank variable because it determines individual ranks in the distribution of Y_d.

The key assumption underlying the IVQR model is the rank preservation assumption. Formally, rank preservation requires that, conditional on Z, $\varepsilon_1 = \varepsilon_0$. Given the interpretation of ε_d as a rank variable, this assumption thus restricts the ranks to be invariant accross potential outcome distributions, whence the name of this assumption. Chernozhukov and Hansen (2005) show that rank preservation can be weakened to rank similarity. Formally, rank similarity requires that, conditional on Z and the disturbance in the selection equation, ε_1 and ε_0 are identically distributed. Rank similarity thus allows for random slippages from an individual's rank level ε. Chernozhukov and Hansen (2005) show that under the aforementioned assumptions and a full-rank condition on the Jacobian of the moment condition

$$P(Y \leqslant q(D, \tau)|Z) = P(Y < q(D, \tau)|Z) = \tau, \tag{10.16}$$

the treatment effect $q(D, \tau)$ is identified for the whole population for all $\tau \in (0, 1)$,

It is interesting to compare the identification strategy of the IVQR model to the instrumental variable framework reviewed in this chapter. Restricting the dimensionality of unobservables in the outcome equation yields point identification of the effect for the whole population. However, restricting the dimensionality of the unobservables in the selection equation only point-identifies QTEs for the compliers.

On the surface, the IVQR model does not seem to be connected to the approach reviewed in this chapter – the two estimands differ (i.e. the QTE and the LQTE, respectively) and the underlying assumptions are nonnested, noncontradictory, and concern different aspects of the models (i.e. the outcome equation and the selection equation, respectively). For these reasons, Chernozhukov and Hansen (2013) describe both models as complements and, for example, Chernozhukov and Hansen (2004) use comparisons of both models as specification checks for the underlying assumptions.

Wüthrich (2016) shows that there is actually a close connection between the estimands of both models. Under Assumptions 2 and 4, the **IVQR** model captures the LQTE at transformed quantile levels:

$$\Delta^{\mathrm{IVQR}}(\tau) = \Delta(\tau'|\mathcal{T} = c), \tag{10.17}$$

where

$$\tau' = F_{Y_0}(q(0, \tau)|\mathcal{T} = c) = F_{Y_1}(q(1, \tau)|\mathcal{T} = c)$$

This result has interesting implications for the connection between both models. First, if the LQTE estimands are constant across quantiles, the estimates of both models converge to the same true effect. Second, if the LQTEs are positive (or negative) at all quantiles, then the sign of the quantile estimands will be the same in both models. Finally, monotonicity of the LQTE function (which is implied, for example, by a location–scale shift model for the compliers) implies monotonicity of the IVQR estimands.

It is important to note that (10.17) does not rely on the rank preservation assumption and thus also provides a characterization of the IVQR estimands absent the rank preservation assumption. To this end, (10.17) implies that the IVQR estimands are quite robust: they preserve sign and monotonicity of $\Delta(\tau|\mathcal{T} = c)$ whenever these properties are invariant across quantiles. Furthermore, the results show that the estimates based on the IVQR model are not arbitrary under misspecification but correspond to well-defined (functions of) causal effects for the compliers.

The results in Wüthrich (2016) confirm that with unrestricted treatment effect heterogeneity all the information about the treatment effects has to come from the compliers. Moreover, they show how the IVQR model extrapolates from the compliers to the whole population. This motivates the use of the IVQR as an approach to extrapolation in the LQTE framework.

10.6   Conclusion and open problems

In this chapter we have reviewed instrumental variable methods to estimate QTEs. In addition to the traditional exclusion and relevance conditions for the instrument, the models considered impose that the treatment either weakly increases or weakly decreases with the instrument for all units in the population. This monotonicity assumption is sometimes satisfied by construction (e.g. one-sided perfect compliance) but there are naturally also cases where it is a strong assumption; see, for instance, the examples discussed by de Chaisemartin (2014). If it can be made, the entire distributions of the control and treated outcomes are identified for the units which react to the instrument without imposing any restrictions on treatment effect heterogeneity. Estimation and inference methods have been developed for the standard setup with a binary instrument, a binary treatment and a continuous outcome. Stata and R code is available to implement most of these methods. For instance, Frölich and Melly (2010) provide a Stata package to estimate conditional and unconditional LQTEs based the approaches by Abadie et al. (2002) and Frölich and Melly (2013), respectively. We have also summarized extensions to multi-valued and continuous instruments, which are now well understood.

From our point of view, the most pressing open research questions are inference methods for discrete outcomes and identification of the effects of nonbinary treatments. Since the monotonicity assumption does not restrict the outcomes at all, identification follows for discrete outcomes in exactly the same way as it does for continuous outcomes. The LQTE framework therefore accommodates discrete outcomes and outcomes with mass points very naturally. This is in sharp contrast to the instrumental variable quantile regression model (Chernozhukov and Hansen, 2005), where continuity is essential for point identification. It is also possible to show that the analog estimators of the cdfs of the potential outcomes are asymptotically normally distributed even for discrete outcomes. However, the existing literature still assumes continuity to provide inference tools based on the asymptotic Gaussianity of the quantile estimators. Inference procedures that accommodate discrete outcomes would be useful and deserve closer attention from future research. Chernozhukov et al. (2016) make initial steps in this direction and suggest procedures that could also be applied to the LQTE model.

On the other hand, the LQTE framework does not easily extend to nonbinary treatments. When an independent instrument is nonbinary, any binary transformation of this instrument will also satisfy the independence assumption and the results for binary instrument can be used. But when the endogenous treatment is nonbinary, then the instrument does not necessarily satisfy the exclusion restriction for any binary transformation of the treatment. In addition, the number of types of compliers increases exponentially in the number of points in the support of the treatment. For average effects, Angrist and Imbens (1995) show that a weighted average of local effects is identified. This is unfortunately not the case for quantile effects. This raises new challenges that have not yet been overcome.

Instead of restricting heterogeneity in the treatment choice equation, an alternative approach reviewed in Chapter 9 of this handbook restricts heterogeneity in the outcome equation by imposing a stochastic rank preservation condition. As we have shown in Section 10.5, this assumption allows the treatment effects to be extrapolated from the compliers (or, more generally, from the population for which the effects are identified) to the whole population. This implies that there is a close connection between these models in the sense that the QTEs identified by one model correspond to the treatment effect identified by the other model at another quantile.

This relationship holds for the standard setup. Since the IVQR model imposes assumptions on the outcome equation but not on the selection equation, which is the opposite of the LQTE model, it will accommodate well the opposite types of generalizations. For instance, it applies without modification to multi-valued and continuous treatments if the instrument is rich enough to identify all the parameters. Torgovitsky (2015) and D'Haultfoeuille and Février (2015) show that combining the rank invariance and the monotonicity assumption allows the large support condition for the instrument to be suppressed. In contrast, point identification breaks down when the outcome is not continuous (see Chesher, 2010). Intuitively, the one-to-one mapping from the outcome to the unobserved error term is lost when there are mass points.

Bibliography

A. Abadie. Bootstrap tests for distributional treatment effects in instrumental variable models. *Journal of the American Statistical Association*, 97:284–292, 2002.

A. Abadie. Semiparametric instrumental variable estimation of treatment response models. *Journal of Econometrics*, 113:231–263, 2003.

A. Abadie, J. Angrist, and G. W. Imbens. Instrumental variables estimates of the effect of subsidized training on the quantiles of trainee earnings. *Econometrica*, 70:91–117, 2002.

E. O. Ananat and G. Michaels. The effect of marital breakup on the income distribution of women with children. *Journal of Human Resources*, 43(3):611–629, 2008.

J. Angrist and G. W. Imbens. Two-stage least squares estimation of average causal effects in models with variable treatment intensity. *Journal of the American Statistical Association*, 90:431–442, 1995.

J. Angrist and J.-S. Pischke. *Mostly Harmless Econometrics: An Empiricist's Companion.* Princeton University Press, Princeton, NJ, 2008.

J. Angrist, G. W. Imbens, and D. B. Rubin. Identification of causal effects using instrumental variables. *Journal of the American Statistical Association*, 91(434):444–455, 1996.

A. Balke and J. Pearl. Bounds on treatment effects from studies with imperfect compliance. *Journal of the American Statistical Association*, 92(439):1171–1176, 1997.

A. Belloni, V. Chernozhukov, I. Fernández-Val, and C. Hansen. Program evaluation and causal inference with high-dimensional data. *Econometrica*, 85(1):233–298, 2017.

P. Carneiro and S. Lee. Estimating distributions of potential outcomes using local instrumental variables with an application to changes in college enrollment and wage inequality. *Journal of Econometrics*, 149(2):191–208, 2009.

J. Cawley and C. Meyerhoefer. The medical care costs of obesity: An instrumental variables approach. *Journal of Health Economics*, 31(1):219–230, 2012.

V. Chernozhukov and C. Hansen. The effects of 401(k) participation on the wealth distribution: An instrumental quantile regression analysis. *Review of Economics and Statistics*, 86(3):735–751, 2004.

V. Chernozhukov and C. Hansen. An IV model of quantile treatment effects. *Econometrica*, 73:245–261, 2005.

V. Chernozhukov and C. Hansen. Quantile models with endogeneity. *Annual Review of Economics*, 5(1):57–81, 2013.

V. Chernozhukov, I. Fernández-Val, and A. Galichon. Quantile and probability curves without crossing. *Econometrica*, 78(3):1093–1125, 2010.

V. Chernozhukov, S. Lee, and A. Rosen. Intersection bounds: Estimation and inference. *Econometrica*, 81:667–737, 2013.

V. Chernozhukov, I. Fernández-Val, B. Melly, and K. Wüthrich. Generic inference on quantile and quantile effect functions for discrete outcomes. arXiv:1608.05142, 2016.

A. Chesher. Identification in nonseparable models. *Econometrica*, 71:1405–1441, 2003.

A. Chesher. Instrumental variable models for discrete outcomes. *Econometrica*, 78(2): 575–601, 2010.

D. R. Cox. *Planning of Experiments*. Wiley, New York, 1958.

C. de Chaisemartin. Tolerating defiance? Local average treatment effects without monotonicity. Warwick Economics Research Paper Series 1020, 2014.

A. Deaton. Instruments, randomization, and learning about development. *Journal of Economic Literature*, 48(2):424–455, 2010.

X. D'Haultfoeuille and P. Février. Identification of nonseparable triangular models with discrete instruments. *Econometrica*, 83(3):1199–1210, 2015.

K. Doksum. Empirical probability plots and statistical inference for nonlinear models in the two-sample case. *Annals of Statistics*, 2:267–277, 1974.

O. Eren and S. Ozbeklik. Who benefits from job corps? A distributional analysis of an active labor market program. *Journal of Applied Econometrics*, 29(4):586–611, 2014.

Y. Fan and S. S. Park. Sharp bounds on the distribution of treatment effects and their statistical inference. *Econometric Theory*, 26(03):931–951, 2010.

S. Firpo. Efficient semiparametric estimation of quantile treatment effects. *Econometrica*, 75:259–276, 2007.

B. R. Frandsen, M. Frölich, and B. Melly. Quantile treatment effects in the regression discontinuity design. *Journal of Econometrics*, 168(2):382–395, 2012.

M. Frölich and B. Melly. Estimation of quantile treatment effects with Stata. *Stata Journal*, 10(3):423, 2010.

M. Frölich and B. Melly. Unconditional quantile treatment effects under endogeneity. *Journal of Business & Economic Statistics*, 31(3):346–357, 2013.

J. J. Heckman and S. Urzúa. Comparing IV with structural models: What simple IV can and cannot identify. *Journal of Econometrics*, 156(1):27–37, 2010.

J. J. Heckman and E. Vytlacil. Structural equations, treatment effects, and econometric policy evaluation 1. *Econometrica*, 73:669–738, 2005.

J. J. Heckman, J. Smith, and N. Clements. Making the most out of programme evaluations and social experiments: Accounting for heterogeneity in programme impacts. *Review of Economic Studies*, 64:487–535, 1997.

H. Hong and D. Nekipelov. Semiparametric efficiency in nonlinear late models. *Quantitative Economics*, 1(2):279–304, 2010.

J. Horowitz and C. Manski. Identification and robustness with contaminated and corrupted data. *Econometrica*, 63:281–302, 1995.

Y.-C. Hsu, T.-C. Lai, and R. P. Lieli. Estimation and inference for distribution functions and quantile functions in endogenous treatment effect models. *IEAS Working Paper*, 15-A003, 2015.

M. Huber and G. Mellace. Testing instrument validity for LATE identification based on inequality moment constraints. *Review of Economics and Statistics*, 97(2):398–411, 2015.

G. W. Imbens. Better LATE than nothing: Some comments on Deaton (2009) and Heckman and Urzua (2009). *Journal of Economic Literature*, 48(2):399–423, 2010.

G. W. Imbens. Instrumental variables: An econometrician's perspective. *Statistical Science*, 29(3):323–358, 2014.

G. W. Imbens and J. Angrist. Identification and estimation of local average treatment effects. *Econometrica*, 62:467–475, 1994.

G. W. Imbens and D. B. Rubin. Estimating outcome distributions for compliers in instrumental variables models. *Review of Economic Studies*, 64:555–574, 1997.

G. W. Imbens and D. B. Rubin. *Causal Inference in Statistics, Social, and Biomedical Sciences*. Cambridge University Press, New York, 2015.

T. Kitagawa. A test for instrument validity. *Econometrica*, 83(5):2043–2063, 2015.

E. L. Lehmann. *Nonparametrics: Statistical Methods Based on Ranks*. Holden-Day, San Francisco, 1975.

G. D. Makarov. Estimates for the distribution function of a sum of two random variables when the marginal distributions are fixed. *Theory of Probability and Its Applications*, 26 (4):803–806, 1982.

I. Mourifié and Y. Wan.   Testing LATE assumptions.   *Available at SSRN: http://dx.doi.org/10.2139/ssrn.2429664*, 2014.

D. B. Rubin. Estimating causal effects of treatments in randomized and nonrandomized studies. *Journal of Educational Psychology*, 66:688–701, 1974.

D. B. Rubin. Randomization analysis of experimental data: The Fisher randomization test comment. *Journal of the American Statistical Association*, 75(371):591–593, 1980.

J. Spława-Neyman. On the application of probability theory to agricultural experiments. Essay on principles. Section 9. *Statistical Science*, 5:465–472, 1990.

A. Torgovitsky. Identification of nonseparable models using instruments with small support. *Econometrica*, 83(3):1185–1197, 2015.

E. Vytlacil. Independence, monotonicity, and latent index models: An equivalence result. *Econometrica*, 70:331–341, 2002.

A. Wald. The fitting of straight lines if both variables are subject to error. *Annals of Mathematical Statistics*, 11(3):284–300, 1940.

K. Wüthrich. A comparison of two quantile models with endogeneity. Universität Bern, Departement Volkswirtschaft, 2016.

P. Yu. Marginal quantile treatment effect. Department of Economics, University of Auckland, 2014.

11

Quantile Regression with Measurement Errors and Missing Data

Ying Wei

Columbia University, New York, USA

CONTENTS

11.1 Introduction

In many applications, data are imperfectly collected. Variables are often measured with error and data are missing for various reasons. When the covariates of interest, denoted here by $\mathbf{x}$, are not directly observable and are, instead, measured with error, it is well known that such errors can lead to substantial attenuation of the estimated effects (Carroll et al., 2006). Likewise, ignoring missing observations in the data can lead to efficiency loss or biased estimation (Little and Rubin, 2014). While there is an abundant literature on measurement errors and missing data, there have been little attention devoted to quantile methods directly, primarily due to the lack of parametric likelihood in quantile regression. In recent years, several methods have been developed specifically for quantile regression. In what follows, we review some methods dealing with measurement errors and missing data in quantile regression models.

11.2 Quantile regression with measurement errors

11.2.1 Linear quantile regression with measurement errors

We first consider the linear quantile regression model

$$Q_y(\tau|\mathbf{x}) = \mathbf{x}^\top \boldsymbol{\beta}_{0,\tau}, \tag{11.1}$$

where y is the response and $\mathbf{x} = (x_1, \ldots, x_p)^\top$ is a p-dimensional covariate. Suppose $\{y_i, \mathbf{x}_i\}$ is a random sample from model (11.1) with sample size n. Then an estimating equation for $\boldsymbol{\beta}_{0,\tau}$ can be written as

$$n^{-1} \sum_{i=1}^{n} \Psi_\tau(y_i - \mathbf{x}_i^\top \boldsymbol{\beta}_\tau) \mathbf{x}_i \approx 0, \tag{11.2}$$

where $\Psi_\tau(u) = \tau - I\{u < 0\}$, $I\{\cdot\}$ denotes the indicator function, and $\boldsymbol{\beta}_\tau \in \mathbf{R}^p$ is a p-dimensional unknown coefficient vector.

When $\mathbf{x}$ is measured with error and, instead, only a surrogate $\mathbf{w}_i$ is observed, naively replacing $\mathbf{x}_i$ by the observed $\mathbf{w}_i$ will result in substantial bias (Wei and Carroll, 2009), even under the surrogacy condition $f(y|\mathbf{x}, \mathbf{w}) = f(y|\mathbf{x})$. The surrogacy condition means the contaminated $\mathbf{w}$ does not provide additional information about the response y if the true covariate $\mathbf{x}$ is known and is generally imposed in the measurement error literature. All the methods discussed in this chapter also assume the surrogacy condition.

Regression calibration (Carroll and Stefanski, 1990; Rosner et al., 1989) is often a simple solution to correct measurement error bias for mean regressions. It replaces $\mathbf{x}_i$ by its conditional mean $E(\mathbf{x}_i|\mathbf{w}_i)$, which is often estimated from a set of replicas. This approach is only valid when the estimating functions are linear in $\mathbf{x}_i$, which is not the case for quantile regression. Other likelihood-based approaches cannot be applied directly, since quantile regression does not assume any prespecified parametric error distributions. To circumvent this difficulty, Wei and Carroll (2009) proposed an approach involving semiparametric joint estimating equations to estimate the conditional quantiles with error in covariates.

11.2.1.1 Semiparametric joint estimating equations

When the covariates $\mathbf{x}_i$ are measured with error, the original sample estimation equation (11.2) is biased. To correct error-induced bias, the estimating equations must be reconstructed by taking the measurement errors into account. The corrected estimating equations take the form

$$\mathbf{S}_n^0(\boldsymbol{\beta}_\tau) = n^{-1} \sum_{i=1}^{n} \int_{\mathbf{x}} \Psi_\tau(y_i - \mathbf{x}^\top \boldsymbol{\beta}_\tau) \mathbf{x} \cdot f(\mathbf{x}|y_i, \mathbf{w}_i) d\mathbf{x} = 0, \tag{11.3}$$

where $f(\mathbf{x}|y_i, \mathbf{w}_i)$ is the conditional density of $\mathbf{x}$, given the observed $(y_i, \mathbf{w}_i)$. The integration in (11.3) makes the function $\mathbf{S}_n^0$ continuous in its argument. The summand of (11.3) is $E_\mathbf{x}\{\Psi_\tau(y - \mathbf{x}^\top \boldsymbol{\beta}_\tau)\mathbf{x}|y, \mathbf{w}\}$, the conditional mean of the original score function, given the observed y and $\mathbf{w}$. Letting $\Psi_{\text{new}}(y, \mathbf{w}, \boldsymbol{\beta}_\tau) = \int_{\mathbf{x}} \Psi_\tau(y - \mathbf{x}^\top \boldsymbol{\beta}_\tau)\mathbf{x} \cdot f(\mathbf{x}|y, \mathbf{w}) d\mathbf{x}$, it is easy to show that $E_y[\Psi_{\text{new}}(y, \mathbf{w}, \boldsymbol{\beta}_{0,\tau})|\mathbf{w}] \equiv 0$ for all $\mathbf{w}$. Therefore, $\Psi_{\text{new}}(y, \mathbf{w}, \boldsymbol{\beta}_\tau)$ is an unbiased estimating function, that is, it has mean zero. To use the new estimating equation (11.3), the conditional density $f(\mathbf{x}|y_i, \mathbf{w}_i)$ must be estimated. Under the surrogacy condition,

$$f(\mathbf{x}|y_i, \mathbf{w}_i) = \frac{f(y_i|\mathbf{x})f(\mathbf{x}|\mathbf{w}_i)}{\int_{\mathbf{x}} f(y_i|\mathbf{x})f(\mathbf{x}|\mathbf{w}_i)d\mathbf{x}}. \tag{11.4}$$

We operate under the framework where the density component $f(y|\mathbf{x})$ does not have parametric form under model (11.2).

Wei and Carroll (2009) assumed that the linear quantile model (11.2) holds at all quantile levels $\tau \in (0,1)$, that is, $Q_y(\tau|\mathbf{x}) = \mathbf{x}^\top \boldsymbol{\beta}_0(\tau)$, where $\{\boldsymbol{\beta}(\tau), \tau\} = (\beta_1(\tau), \dots, \beta_p(\tau))^\top \times \tau \in \mathbf{R}^p \times (0,1)$ is a p-dimensional quantile coefficient process on the interval $(0,1)$. With this joint modeling, the density $f(y|\mathbf{x})$ can be obtained as

$$f(y|\mathbf{x}) = \lim_{\delta \to 0} \frac{\delta}{\mathbf{x}^\top[\boldsymbol{\beta}_0(\tau_y + \delta) - \boldsymbol{\beta}_0(\tau_y)]}, \tag{11.5}$$

where $\tau_y = \{\tau \in (0,1) : \mathbf{x}^\top \boldsymbol{\beta}_0(\tau) = y\}$.

Under the joint modeling approach, the estimating equations (11.3) need to be solved *jointly* for all the τ values, even if one is interested in only one particular quantile level τ. Consequently, the estimation equation (11.3) at a single quantile needs to be extended to the semiparametric joint estimating equation

$$n^{-1} \sum_{i=1}^{n} \int_{\mathbf{x}} [\tau - I\{y_i - \mathbf{x}^\top \boldsymbol{\beta}(\tau) < 0\}] \cdot \mathbf{x} \cdot f\{\mathbf{x}|y_i, \mathbf{w}_i; \boldsymbol{\beta}(\tau)\}d\mathbf{x} = 0, \tag{11.6}$$

where we use $f\{\mathbf{x}|y, \mathbf{w}; \boldsymbol{\beta}(\tau)\}$ to indicate its dependence on the entire unknown quantile process $\boldsymbol{\beta}(\tau)$ and hence the parametric space $\{\boldsymbol{\beta}(\tau), \tau\} \in \mathbf{R}^p \times (0,1)$ is of infinite dimension. One can approximate the infinite-dimensional (11.6) by estimating the quantile model on a fine grid of quantile levels $\varepsilon = \tau_1 < \tau_2 < \cdots < \tau_{k_n} = 1 - \varepsilon$ and reduce (11.6) to

$$S_n(\boldsymbol{\theta}) = n^{-1} \sum_{i=1}^{n} \int_{\mathbf{x}} \Psi(y_i - \mathbf{x}^\top \boldsymbol{\theta}) \otimes \mathbf{x} \cdot f(\mathbf{x}|y_i, \mathbf{w}_i; \boldsymbol{\theta})d\mathbf{x} = 0, \tag{11.7}$$

where $\Psi(y_i - \mathbf{x}^\top \boldsymbol{\theta}) = \{\Psi_{\tau_1}(y_i - \mathbf{x}^\top \boldsymbol{\beta}_{\tau_1}), \dots, \Psi_{\tau_{k_n}}(y_i - \mathbf{x}^\top \boldsymbol{\beta}_{\tau_{k_n}})\}^\top$ is a k_n-dimensional vector, $\otimes$ stands for the Kronecker product, and $\boldsymbol{\theta} = (\boldsymbol{\beta}_{\tau_1}^\top, \boldsymbol{\beta}_{\tau_2}^\top, \dots, \boldsymbol{\beta}_{\tau_{k_n}}^\top)^\top$ is the set of quantile coefficients.

An iterative EM-like estimation algorithm

The solution of (11.7) can be obtained from a nonparametric analog of the EM algorithm, which iteratively updates the conditional distribution $f(\mathbf{x}|y_i, \mathbf{w}_i, \boldsymbol{\theta})$ and quantile coefficients $\boldsymbol{\theta}$, without assuming a specific likelihood functions, as in classical EM algorithms. Let ν be the indicator of iteration steps. The main steps of the algorithm are then the following:

Step 1. Set the initial values of $\boldsymbol{\theta}$ by simply regressing y_i on $\mathbf{w}_i$.

Step 2. Update the distribution $f^{(\nu)}(\mathbf{x}|y_i, \mathbf{w}_i)$ based on $\widehat{\boldsymbol{\theta}}^{(\nu-1)} = (\widehat{\boldsymbol{\beta}}_{\tau_1}^{(\nu-1)}, \dots, \widehat{\boldsymbol{\beta}}_{\tau_{k_n}}^{(\nu-1)})$, that is,

$$f^{(\nu)}(\mathbf{x}|y_i, \mathbf{w}_i) = \frac{f(y_i|\mathbf{x}, \widehat{\boldsymbol{\theta}}^{(\nu-1)})f(\mathbf{x}|\mathbf{w}_i)}{\int_{\mathbf{x}} f(y_i|\mathbf{x}, \widehat{\boldsymbol{\theta}}^{(\nu-1)})f(\mathbf{x}|\mathbf{w}_i)dx},$$

where

$$f(y_i|\mathbf{x}, \widehat{\boldsymbol{\theta}}^{(\nu-1)}) = \sum_{k=1}^{k_n} \left\{ \frac{\tau_k - \tau_{k-1}}{\mathbf{x}^\top (\widehat{\boldsymbol{\beta}}_{\tau_k}^{(\nu-1)} - \widehat{\boldsymbol{\beta}}_{\tau_{k-1}}^{(\nu-1)})} \right.$$

$$\left. \times I(\mathbf{x}^\top \widehat{\boldsymbol{\beta}}_{\tau_{k-1}}^{(\nu-1)} \leqslant y_i < \mathbf{x}^\top \widehat{\boldsymbol{\beta}}_{\tau_k}^{(\nu-1)}) \right\}. \tag{11.8}$$

Step 3. Estimate $\widehat{\boldsymbol{\theta}}^{(\nu)} = (\widehat{\boldsymbol{\beta}}_{\tau_1}^{(\nu)}, \dots, \widehat{\boldsymbol{\beta}}_{\tau_{k_n}}^{(\nu)})$ based on the new estimating function $\Psi_{\text{new}}(y_i, \mathbf{w}_i; \boldsymbol{\beta}_\tau)$ evaluated at $f^{(\nu)}(\mathbf{x}|y_i, \mathbf{w}_i)$. Approximating the integral in $\Psi_{\text{new}}(y_i, \mathbf{w}_i; \boldsymbol{\beta}_\tau)$ numerically, one can rewrite the estimating equations as

$$\sum_{i=1}^n \sum_{j=1}^m \Psi_{\tau_k}(y_i - \widetilde{\mathbf{x}}_{i,j}^\top \boldsymbol{\beta}_{\tau_k}) \widetilde{\mathbf{x}}_{i,j} f^{(\nu)}(\widetilde{x}_{i,j}|y_i, \mathbf{w}_i) = 0, \quad k = 1, \dots, k_n,$$

where $\widetilde{\mathbf{x}}_i = (\widetilde{\mathbf{x}}_{i,1}, \widetilde{\mathbf{x}}_{i,2}, \dots, \widetilde{\mathbf{x}}_{i,m})$ is a fine grid of possible $\mathbf{x}_i$ values, akin to a set of abscissas in Gaussian quadrature. Solving the estimating equations is essentially a weighted quantile regression with response y_i over the covariates $\widetilde{x}_{i,j}$ with weights $f^{(\nu)}(\mathbf{x}_{i,j}|y_i, \mathbf{w}_i)$.

Step 4. Iterate Steps 2 and 3 until the algorithm converges.

As in any measurement error problem, we need to understand how the observed covariates $\mathbf{w}$ relate to the true covariates $\mathbf{x}$. Depending on the individual application, many measurement error models have been proposed to estimate $f(\mathbf{x}|\mathbf{w})$. Among them, the most commonly used model is the classical additive measurement error model that assumes $\mathbf{w}_i = \mathbf{x}_i + \mathbf{u}_i$, where the measurement error $\mathbf{u}_i$ is independent of $\mathbf{x}_i$ and has mean zero and covariance matrix Σ_u^2. Under the classical measurement error model, the conditional distribution $f(\mathbf{x}|\mathbf{w})$ can be estimated parametrically or nonparametrically. In a parametric context, one assumes the distributions for both $\mathbf{u}$ and $\mathbf{x}$ follow parametric likelihood with parameters ζ_u and ζ_x, respectively. The parameter ζ_u is often estimated by a set of replication data, where we observe $\mathbf{w}_{ij}$, $j = 1, \dots, m_i$ for some $\mathbf{x}_i$ and then estimate ζ_x from the entire data. Gaussian likelihood is a standard choice. When parametric likelihood is implausible, $f(\mathbf{x}|\mathbf{w})$ can also be estimated nonparametrically using deconvolution methods, where less is assumed about the distribution of $\mathbf{x}$ (Staudenmayer et al., 2008).

Suppose that $\widehat{\boldsymbol{\theta}}_n$ is the estimated coefficients following the EM algorithm above with a well-estimated $f(\mathbf{x}|\mathbf{w})$. Then $\widehat{\boldsymbol{\theta}}_n$ is consistent and asymptotically normally distributed,

$$\sqrt{n}(\widehat{\boldsymbol{\theta}}_n - \boldsymbol{\theta}_0) \to N(0, \Sigma),$$

as $n \to \infty$, where $\Sigma = D^{-1} V D^{-1}$, $D = \lim_{n \to \infty} \frac{\partial}{\partial \boldsymbol{\theta}_0} S_n(\boldsymbol{\theta}_0)$, and $V = \lim_{n \to \infty} \text{var}\{S_n(\boldsymbol{\theta}_0)\}$. Moreover, if we define $\widehat{\boldsymbol{\beta}}_n(\tau)$ by a natural linear spline expansion $\widehat{\boldsymbol{\theta}}_n$ (i.e. $\widehat{\boldsymbol{\beta}}(\tau)$ is a p-continuous, piecewise linear function on $[0,1]$ that satisfies $\widehat{\boldsymbol{\beta}}(\tau_k) = \widehat{\boldsymbol{\beta}}_{\tau_k}$ and is subject to the constraints $\widehat{\boldsymbol{\beta}}'(0) = \widehat{\boldsymbol{\beta}}'(1) = \mathbf{0}$) then, with a sufficient number of knots $k_n \to \infty$ and $k_n n^{-1} \to 0$ and under other regulation conditions, Wei and Carroll (2009) showed that $\widehat{\boldsymbol{\beta}}_n(\tau)$ uniformly converges to $\boldsymbol{\beta}_0(\tau)$, that is,

$$\sup_{\tau \in [1/(k_n+1),\, k_n/(k_n+1)]} \|\widehat{\boldsymbol{\beta}}_n(\tau) - \boldsymbol{\beta}_0(\tau)\| = o_p(1).$$

11.2.1.2 Other methods for linear quantile regression with measurement errors

Any analysis of measurement error problems involves two sets of models: one is the model of interest, which describes the association between y and $\mathbf{x}$; and the other is the measurement error model, which associates the observed covariates $\mathbf{w}$ to the underlying true covariates $\mathbf{x}$. Several statistical methods have recently been proposed that are specifically tailored for certain measurement error models to simplify computation and improve estimation efficiency.

Orthogonal quantile regression

He and Liang (2000) considered an orthogonal regression approach for a special class of linear quantile model with measurement errors. They assumed that the random sample $(y_i, \mathbf{x}_i, \mathbf{w}_i)$ is observed from a linear quantile model $y_i = \alpha_\tau + \mathbf{x}_i^\top \boldsymbol{\beta}_\tau + \epsilon_i$, where $Q_{\epsilon_i}(\tau) = 0$, and a classical additive measurement error model $\mathbf{w}_i = \mathbf{x}_i + \mathbf{u}_i$. They further assumed that the joint distribution of $(\epsilon_i, \mathbf{u}_i)$ is spherically symmetric with a finite first moment. In this setting, it was shown that the estimated coefficients from an orthogonal quantile regression,

$$(\widehat{\alpha}_\tau, \widehat{\boldsymbol{\beta}}_\tau) = \arg\min_{\alpha, \boldsymbol{\beta}} \sum_{i=1}^{n} \rho_\tau \left(\frac{y_i - \alpha - \mathbf{w}_i^\top \boldsymbol{\beta}}{\sqrt{1 + \|\boldsymbol{\beta}\|^2}} \right), \tag{11.9}$$

are consistent. In (11.9), $\rho_\tau(\cdot)$ is the original quantile regression loss function and $\frac{y_i - \alpha - \mathbf{w}_i^\top \boldsymbol{\beta}}{\sqrt{1 + \|\boldsymbol{\beta}\|^2}}$ is the orthogonal residual rather than the vertical distance in the regression space. Under additional regularity conditions, the $(\widehat{\alpha}_\tau, \widehat{\boldsymbol{\beta}}_\tau)$ values are asymptotically normally distributed. Specifically,

$$\sqrt{n}(\widehat{\alpha}_\tau - \alpha_\tau) \to N(0, \tau(1 - \tau)f^{-2}(q_\tau)),$$

where q_τ is the unique solution to $E\rho_\tau(\epsilon_i - q) = 0$, and

$$\sqrt{n}(\widehat{\boldsymbol{\beta}}_\tau - \boldsymbol{\beta}_\tau) \to N(0, \Sigma_\beta),$$

where $\Sigma_\beta = f^{-2}(q_\tau)(1 + \|\boldsymbol{\beta}\|^2)\Sigma_x^{-1}Q\Sigma_x^{-1}$ with $\xi = (\epsilon - \mathbf{u}^\top \boldsymbol{\beta})/\sqrt{1 + \|\boldsymbol{\beta}\|^2} - q_\tau$, $\Sigma_x = E(\mathbf{x}^\top \mathbf{x})$ and $Q = \tau(1 - \tau)\Sigma_x + \mathrm{Cov}\{\Psi_\tau(\xi)(\mathbf{u} + \xi\boldsymbol{\beta}/(\sqrt{1 + \|\boldsymbol{\beta}\|^2}))\}$.

Smoothed and corrected estimation/loss function

The measurement error-induced bias could be corrected by solving the corrected estimating equations, as in (11.3), or, equivalently, by minimizing a corrected quantile regression loss function. However, either approach is computationally intensive, partly because the quantile regression loss function is not differentiable at zero. Wang et al. (2012) and Wu et al. (2015) proposed smoothing out the corrected loss function. The fundamental rationale of the smoothing scheme can be described as follows.

Recall that the true parameter $\boldsymbol{\beta}_0$ minimizes $E[\rho_\tau(y - \mathbf{x}^\top \boldsymbol{\beta}_0)]$, where $\rho_\tau(u) = u(\tau - I\{u < 0\})$ is the original quantile regression loss function. We approximate the non-differentiable indicator function $I\{u < 0\}$ by a smooth function $K(u, h)$ with a positive bandwidth h. The function $K(u, h)$ converges to $I\{u < 0\}$ as $h \to 0$ and, consequently, the smoothed loss function $\widetilde{\rho}_\tau(u, h) = u(\tau - K(u, h))$ converges to $\rho_\tau(u)$ as $h \to 0$. If the covariates $\mathbf{x}_i$ are error free, minimizing the smoothed loss function $\sum_{i=1}^{n} \widetilde{\rho}_\tau(y_i - \mathbf{x}_i^\top \boldsymbol{\beta}, h)$ with $h \to 0$ leads to consistent estimation of $\boldsymbol{\beta}$. When $\mathbf{x}_i$ is measured with error and only its surrogate $\mathbf{w}_i$ is available, the strategy is to find a function $\rho_\tau^*(y_i - \mathbf{w}_i^\top \boldsymbol{\beta}, h)$ such that $E_{\mathbf{w}_i}[\rho_\tau^*(y_i - \mathbf{w}_i^\top \boldsymbol{\beta}, h)|y_i, \mathbf{x}_i] = \widetilde{\rho}_\tau(y_i - \mathbf{x}_i^\top \boldsymbol{\beta}, h)$. This way, minimizing

$$\sum_{i=1}^{n} \rho_\tau^*(y_i - \mathbf{w}_i^\top \boldsymbol{\beta}, h)$$

also produces consistent $\boldsymbol{\beta}$ estimation, provided that h converges to zero at an appropriate rate. The function $\rho_\tau^*(u, h)$ depends on the distribution of $(\mathbf{x}_i, \mathbf{w}_i)$ in choosing the right form of the function K and constructing ρ_τ^* accordingly.

Both Wang et al. (2012) and Wu et al. (2015) worked with two families of measurement error models. One is the normal additive measurement error model, where $\mathbf{w}_i = \mathbf{x}_i + \mathbf{u}_i$ and the measurement error $\mathbf{u}_i \sim N(0, \Sigma_u)$ is a random normal vector that is independent of $\mathbf{x}_i$ and y_i. The other is an additive Laplace measurement error model, where the measurement error $\mathbf{u}_i$ follows a multivariate Laplace distribution (Kotz et al., 2001) with mean zero and variance Σ_u.

Normal measurement errors. When $(\mathbf{x}_i, \mathbf{w}_i)$ are jointly normal,

$$y_i - \mathbf{w}_i^\top \boldsymbol{\beta} \sim N(y_i - \mathbf{x}_i^\top \boldsymbol{\beta}, \boldsymbol{\beta}^\top \Sigma_u \boldsymbol{\beta}).$$

According to Stefanski and Cook (1995), for any normal vector $\epsilon \sim N(\mu_\epsilon, \sigma_\epsilon)$ and a smooth function $g(\cdot)$, $E_\epsilon[E_u[g(\epsilon + i\sigma_\epsilon u)|\epsilon]] = g(\mu_\epsilon)$, where $i = \sqrt{-1}$ and $u \sim N(0, 1)$. Following this result, Wang et al. (2012) suggested smoothing $\rho_\tau(u)$ by

$$\widetilde{\rho_\tau}(u, h) = u\{\tau - 1/2 - SI(u/h)\},$$

where $SI(u/h) = (1/\pi) \int_0^{u/h} \sin(t)/t \, dt\}$ is the sine integral function that ranges between $-1/2$ and $1/2$, and $1/2 + SI(u/h)$ converges to the indicator function $I\{u < 0\}$. The smoothed and corrected loss function is then constructed as

$$\rho_\tau^*(y_i - \mathbf{w}_i^\top \boldsymbol{\beta}, h) = A(y_i - \mathbf{w}_i^\top \boldsymbol{\beta}, \boldsymbol{\beta}^\top \Sigma_u \boldsymbol{\beta}, h),$$

where

$$
\begin{aligned}
A(\epsilon, \sigma_\epsilon, h) &= E_u[\widetilde{\rho_\tau}(\epsilon + i\sigma_\epsilon u, h)|\epsilon] \\
&= \epsilon\left(\tau - \frac{1}{2}\right) + \frac{1}{\pi} \int_0^{1/h} \left\{ \frac{\epsilon \sin(y\epsilon)}{y} - \sigma_\epsilon \cos(y\epsilon) \right\} e^{y^2 \sigma_\epsilon^2/2} dy.
\end{aligned}
$$

Let $\widehat{\boldsymbol{\beta}}(h) = \arg\min \sum_{i=1}^n \rho_\tau^*(y_i - \mathbf{w}_i^\top \boldsymbol{\beta}, h)$. Then $\widehat{\boldsymbol{\beta}}(h) \to \boldsymbol{\beta}_0$ almost as surely as $n \to \infty$, $h \to 0$, and $h = C(\log n)^{-\delta}$ for $\delta < 1/2$ and some positive constant C. A similar smoothed and corrected score function was studied by Wu et al. (2015) in the context of censored quantile regression.

Laplace measurement errors. In the Laplace measurement error model,

$$y - \mathbf{w}^\top \boldsymbol{\beta} | y, \mathbf{x} \sim L(y - \mathbf{x}^\top \boldsymbol{\beta}, \boldsymbol{\beta}^\top \Sigma_u \boldsymbol{\beta}),$$

where L is the Laplace (double exponential) distribution. For any Laplace random variable $\epsilon \sim L(\mu_\epsilon, \sigma_\epsilon)$ and a continuous function $g(\cdot)$, $E\{g(\epsilon) - (1/2)\sigma_\epsilon^2 g''(\epsilon)\} = g(\mu_\epsilon)$, where $g''(\cdot)$ is the second derivative of $g(\cdot)$. Following this result, Wang et al. (2012) suggested smoothing $\rho_\tau(u)$ by

$$\widetilde{\rho_\tau}(u, h) = u\left\{\tau - 1 + \int_{u < \epsilon/h} K(u) du\right\},$$

where K is a kernel function, and constructing the smoothed and corrected loss function accordingly, as

$$\rho_\tau^*(y_i - \mathbf{w}_i^\top \boldsymbol{\beta}, h) = \widetilde{\rho_\tau}(y_i - \mathbf{w}_i^\top \boldsymbol{\beta}, h) - \frac{\boldsymbol{\beta}^\top \Sigma_u \boldsymbol{\beta}}{2} \frac{\partial^2 \widetilde{\rho_\tau}(\epsilon, h)}{\partial \epsilon} \bigg|_{\epsilon = y_i - \mathbf{w}_i^\top \boldsymbol{\beta}}.$$

Let $\widehat{\boldsymbol{\beta}}(h) = \arg\min \sum_{i=1}^n \rho_\tau^*(y_i - \mathbf{w}_i^\top \boldsymbol{\beta}, h)$. Then $\widehat{\boldsymbol{\beta}}(h) \to \boldsymbol{\beta}_0$ almost surely as $n \to \infty$, $h \to 0$, and $(nh)^{-1/2} \log n \to 0$. Similar smoothed and corrected score functions were studied by Wu et al. (2015) in the context of censored quantile regression.

11.2.2 Nonparametric and semiparametric quantile regression model with measurement errors

In the proceeding section, we have assumed that model (11.1) holds for all τ values, that is, all the conditional quantiles are linear in $\mathbf{x}_i$. The assumption of linearity could be relaxed to a more general family of nonparametric or semiparametric models, that is,

$$y_i = g_\tau(x_i) + \mathbf{z}_i^\top \boldsymbol{\beta}_{0,\tau} + \epsilon_i(\tau), \qquad (11.10)$$

where x_i is a scalar covariate, $g_\tau(\cdot)$ is an unknown smooth function, and $\mathbf{z}_i$ is a vector of linear covariates. A model of this form is also called a partially linear model. It has been well studied and widely used due to its flexibility (e.g. Lin and Ying, 2001; Fan and Li, 2004). Estimating (11.10) is a great deal more challenging, however, due to the curse of dimensionality, especially in the presence of measurement error.

Schennach (2008) suggested an instrumental variable (IV) approach to estimate the non-parametric quantile function $(Q_y(\tau|x) = g_\tau(x))$ with measurement errors. The IV approach assumes that an instrumental vector $\mathbf{v}$ exists that is independent of the response y and satisfies

$$x = h(\mathbf{v}) + \xi,$$

where $h(\cdot)$ is some function and ξ is the prediction error of the instruments, with ξ independent of $\mathbf{v}$ and $E(\xi) = 0$. In the additive measurement error model $w = x + u$ and $E(u) = 0$, the true covariate x and its surrogate value w have the same conditional mean, given the instruments $\mathbf{v}$, that is, $E(x|\mathbf{v}) = E(w|\mathbf{v})$. Hence $h(\mathbf{v})$ can be estimated by simply regressing w_i with the instruments $\mathbf{v}_i$.

Recall that $S(u) = I\{u < 0\} - \tau$ is the quantile loss function, such that $g_\tau(x)$ is the solution to the equation $E[S(y - q)|x] = 0$ for any x. Due to measurement error, directly regressing y_i on x_i is not feasible. Schennach (2008) essentially approximated $E[S(y-q)|x]$ by functions of the observed $y_i, \mathbf{v}_i, w_i$. Through the convolution operator, the author related the observable $E[S(y - q)|h(\mathbf{v}) = v]$ to the function of interest $E[S(y - q)|x]$ by

$$E[S(y - q)|h(\mathbf{v}) = v] \;=\; \int E[S(y - q)|x = v - z] f_\xi(-z)dz.$$

A closed-form expression for $E[S(y-q)|x]$ can then be established through Fourier transform and deconvolution techniques and can be estimated from the observed y_i, $\mathbf{v}_i$, and w_i.

The EM algorithm outlined in Section 11.2.1.1 can also be extended to estimate model (11.10). To estimate $g_\tau(x)$, we can approximate it by a normalized B-spline, as does de Boor (2001), that is, $g_\tau(x) \approx \boldsymbol{\pi}(x)^\top \boldsymbol{\alpha}_\tau$, where $\boldsymbol{\pi}(x)$ is a set of k_n basis functions given a set of knots and the order of the spline. Given the spline approximation, the model is linear in form and the conditional quantile of y_i can then be estimated by $\boldsymbol{\pi}(x_i)^\top \widehat{\boldsymbol{\alpha}}_\tau + \mathbf{z}_i^\top \widehat{\boldsymbol{\beta}}_\tau$, where $(\widehat{\boldsymbol{\alpha}}_\tau^\top, \widehat{\boldsymbol{\beta}}_\tau^\top)^\top$ is the solution of the estimating equation

$$\sum_{i=1}^n \Psi_\tau(y_i - \boldsymbol{\pi}(x_i)^\top \boldsymbol{\alpha}_\tau + \mathbf{z}_i^\top \boldsymbol{\beta}_\tau) \cdot (\boldsymbol{\pi}(x_i)^\top, \mathbf{z}_i^\top)^\top = 0.$$

If the true x_i is measured with error, naively replacing x_i by the observed w_i will result in substantial bias in both $g_\tau(x)$ and $\boldsymbol{\beta}_\tau$. Following a similar idea as in the linear quantile model, we define a new estimating function,

$$\begin{aligned}
&\Psi_{\text{new}}(\tau; y_i, w_i, \mathbf{z}_i; \boldsymbol{\alpha}, \boldsymbol{\beta}) \\
&= \int_x (\tau - I\{y_i - \boldsymbol{\pi}(x)^\top \boldsymbol{\alpha}_\tau - \mathbf{z}_i^\top \boldsymbol{\beta}_\tau\}) \cdot (\boldsymbol{\pi}(x)^\top, \mathbf{z}_i^\top)^\top \cdot f(x|y_i, \mathbf{w}_i, \mathbf{z}_i; \boldsymbol{\alpha}, \boldsymbol{\beta}) \, dx,
\end{aligned}$$

based on which we can construct the joint estimating equation

$$\mathbf{S}_n^0(\tau; \boldsymbol{\alpha}, \boldsymbol{\beta}) = n^{-1} \sum_{i=1}^{n} \Psi_{\text{new}}(\tau; y_i, w_i, \mathbf{z}_i; \boldsymbol{\alpha}, \boldsymbol{\beta}) = 0.$$

Unlike the estimating equations for the linear quantile model, the estimating equations for the semiparametric model are not unbiased due to the spline approximation; however, they are asymptotically unbiased as the number of knots k_n goes to infinity and $k_n n^{-1}$ goes to zero.

In practice, the density $f(x|y_i, \mathbf{w}_i, \mathbf{z}_i; \boldsymbol{\alpha}, \boldsymbol{\beta})$ needs to be estimated. The estimation algorithms outlined for the linear model can also be adapted for the semiparametric model with certain modifications. Specifically, (1) model (11.10) will be estimated on a grid of τ values and the coefficient process $\boldsymbol{\alpha}(\tau)$ and $\boldsymbol{\beta}(\tau)$ approximated by the linear spline expansion; and (2) we can decompose the conditional density $f(x|y_i, w_i, \mathbf{z}_i)$ by

$$f(x|y_i, w_i, \mathbf{z}_i) = \frac{f(y_i|x, \mathbf{z}_i)f(x|w_i, \mathbf{z}_i)}{\int_x f(y_i|x, \mathbf{z}_i)f(x|w_i, \mathbf{z}_i)dx}. \tag{11.11}$$

The first density $f(y_i|x, \mathbf{z}_i)$ can be estimated by taking the derivative over the estimated quantile function $\boldsymbol{\pi}(x_i)^\top \boldsymbol{\alpha}(\tau) + \mathbf{z}_i^\top \boldsymbol{\beta}(\tau)$, while the second density $f(x|w_i, \mathbf{z}_i)$ can be estimated via the parametric approach discussed above. After taking the modifications (1) and (2) into account, the modified iterative algorithm can be shown to converge and lead to a consistent estimation.

11.3   Quantile regression with missing data

In many applications, observations are missing. Ignoring the missing data (using only complete cases for analysis) will undermine study efficiency and sometimes introduce substantial bias. There is a vast literature on handling missing data. Classical options include conditional mean imputation (Afifi and Elashoff, 1969a,b), reweighing methods, likelihood-based approaches including multiple imputation (MI; Rubin, 1996, 2004) and Bayesian methods (Rubin and Schafer, 1990). Little and Rubin (2014) reviewed these approaches. Statistical methods combining quantile regression with missing data were not well developed until recent years.

As for the mean regression calibration approach designed to correct measurement errors, simple conditional mean imputation is only valid when the estimation equations are linear, but that is not the case for quantile regression. The reweighting method restricts the analysis to complete observations but views these as a nonrepresentative sample. To correct the resulting bias, the method reweights each completely observed observation by the reciprocal of its probability of being observed. This inverse probability weighting (IPW) approach (Seaman and White, 2013) requires a model for the probability of data being missing that is simple to implement and generally applicable to any regression model, including quantile regression. Lipsitz et al. (1997) considered an inverse probability approach for longitudinal data with dropout. The disadvantage of IPW is its inflated variances. Alternative approaches with possibly better efficiency are generally likelihood-based. Again, since quantile regression does not impose any specific parametric form for the conditional distributions, those likelihood methods cannot be directly applied but can be implemented through semiparametric joint quantile modeling.

Most of the statistical methods handling missing data are built upon a missing at random

(MAR) assumption, which asserts that the probability of missing an observation in a variable X is unrelated to the actually missing value, after controlling for all the other variables in the analysis. For example, in a survey, some respondents do not report their incomes. The probability of missing income data depends on the respondent's age, race, and marital status. However, given this information, the probability of missing income does not depend on the respondent's actual income level. When data are missing not at random (MNAR), the analysis requires more intensive modeling and good external knowledge of the sources of the omission. In fact, censored data are a special but important case of MNAR data, and we refer the reader to Chapters 6 and 7 for related statistical methods. Although MAR is often a plausible assumption, many studies, such as that of Collins et al. (2001), have found that violations of MAR often have only a minor impact on estimates and standard errors. The statistical methods presented in this section all rely on the MAR assumption.

We also assume throughout the section that n_1 is the number of complete observations and n_0 is the remaining sample size associated with missing observations. To avoid trivial situations, we assume $0 < \lim_{n \to \infty} n_0/n_1 = \lambda < \infty$. Therefore, the proportion of missing observations is nonnegligible and nondominating.

11.3.1 Statistical methods handling missing covariates in quantile regression

We first consider the case where only some of the covariates are missing. The quantile model of interest can be written as

$$Q_{y_i}(\tau; \mathbf{x}_i, \mathbf{z}_i) = \mathbf{x}_i^\top \boldsymbol{\beta}_{0,\tau} + \mathbf{z}_i^\top \boldsymbol{\gamma}_{0,\tau}, \quad \text{for } i = 1, 2, \ldots, n, \tag{11.12}$$

where $Q_{y_i}(\tau; \mathbf{x}_i, \mathbf{z}_i)$ is the τth conditional quantile of the response variable y_i, given $\mathbf{x}_i$ and $\mathbf{z}_i$. Assume that the covariate $\mathbf{z}_i$ is q-dimensional, including a vector of 1s, and $\mathbf{x}_i$ is p-dimensional. Furthermore, we assume that the covariate $\mathbf{z}_i$ and response y_i are completely observed but some values of $\mathbf{x}_i$ are missing. We further denote δ_i as the binary indicator of whether $\mathbf{x}_i$ was fully observed.

Whether the impact of the missing $\mathbf{x}_i$ values is *ignorable* depends on whether the omission is related to the outcome y. Note that $S(\boldsymbol{\beta}_\tau, \boldsymbol{\gamma}_\tau, y, \mathbf{x}, \mathbf{z}) = \Psi_\tau(y - \mathbf{x}^\top \boldsymbol{\beta}_\tau - \mathbf{z}^\top \boldsymbol{\gamma}_\tau)(\mathbf{x}^\top, \mathbf{z}^\top)^\top$ is the original estimating function for regression quantiles. Its conditional mean for the true parameter $E_y\{S(\boldsymbol{\beta}_{0,\tau}, \boldsymbol{\gamma}_{0,\tau}, y, \mathbf{x}, \mathbf{z})|\mathbf{x}, \mathbf{z}\} = 0$. If one ignores the missing $\mathbf{x}_i$ and only uses the complete cases for regression, the estimation equations are equivalent to

$$\sum_{i=1}^n \delta_i S(\boldsymbol{\beta}_\tau, \boldsymbol{\gamma}_\tau, y, \mathbf{x}, \mathbf{z}) = 0. \tag{11.13}$$

When the omission of $\mathbf{x}_i$ is unrelated to the outcome y after conditioning on $\mathbf{x}_i$ and $\mathbf{z}_i$, the estimating equations (11.13) remain unbiased, since $E\{\delta_i S(\boldsymbol{\beta}_\tau, \boldsymbol{\gamma}_\tau, y, \mathbf{x}, \mathbf{z})|\mathbf{x}_i, \mathbf{z}_i\} = E\{\delta_i|\mathbf{x}_i, \mathbf{z}_i\}E\{S(\boldsymbol{\beta}_\tau, \boldsymbol{\gamma}_\tau, y, \mathbf{x}, \mathbf{z})|\mathbf{x}_i, \mathbf{z}_i\} = 0$. In other words, a complete-case analysis would still be valid but suboptimal in estimating efficiency, since it does not utilize all the data. When the missingness of $\mathbf{x}_i$ is associated with outcome y, however, the estimating equations (11.13) using the complete cases could be seriously biased.

11.3.1.1 Multiple imputation algorithm

One way to improve the estimation efficiency of the complete-case analysis is to utilize multiple imputation. By imputing missing $\mathbf{x}_i$ values, one can include all the observed data in the estimation. The key step of MI is to simulate the missing $\mathbf{x}$ from the conditional density $f(\mathbf{x}|y, \mathbf{z})$. The density $f(\mathbf{x}|y_i, \mathbf{z}_i)$ can be decomposed by $f(y|\mathbf{x}, \mathbf{z})$ and $f(\mathbf{x}|\mathbf{z})$. Under the quantile regression framework, $f(y|\mathbf{x}, \mathbf{z})$ is unspecified but could be derived from joint quantile modeling, as in Section 11.2.1. In a simpler case, where the omission of $\mathbf{x}$ is unrelated to y, an MI algorithm could be outlined as follows:

1. *Quantile regression with the complete data only.* Run a quantile regression at a fine grid of quantile levels, $0 < \tau_1 < \cdots < \tau_k < \cdots < \tau_{k_n} < 1$, where k_n is the number of quantile levels, and denote the resulting coefficients by $\widehat{\boldsymbol{\beta}}_{n_1, \tau_k}$.

2. *Estimation of the conditional density* $f(y|\mathbf{x}, \mathbf{z})$. Under the assumption that the linear quantile model (11.12) holds for all quantile levels τ, the conditional density $f(y|\mathbf{x}, \mathbf{z})$ can be well approximated by

$$
\widehat{f}\{y|\mathbf{x}, \mathbf{z}, \widehat{\boldsymbol{\beta}}_{n_1}(\tau)\}
$$
$$
= \sum_{k=1}^{K_n} \left\{ \frac{\tau_{k+1} - \tau_k}{(\mathbf{x}^\top, \mathbf{z}^\top)\widehat{\boldsymbol{\beta}}_{n_1, \tau_{k+1}} - (\mathbf{x}^\top, \mathbf{z}^\top)\widehat{\boldsymbol{\beta}}_{n_1, \tau_k}} \right.
$$
$$
\left. \times I\{(\mathbf{x}^\top, \mathbf{z}^\top)\widehat{\boldsymbol{\beta}}_{n_1, \tau_k} \leqslant y < (\mathbf{x}^\top, \mathbf{z}^\top)\widehat{\boldsymbol{\beta}}_{n_1, \tau_{k+1}}\} \right\}. \tag{11.14}
$$

As in the measurement error model, the estimated conditional desnity $\widehat{f}\{y|\mathbf{x}, \mathbf{z}, \widehat{\boldsymbol{\beta}}_{n_1}(\tau)\}$ is a consistent approximation to the true density with evenly spaced τ_k and with the number of quantile levels $k_n \to \infty$ and $k_n n^{-1} \to 0$.

3. *Estimation of the conditional density* $f(\mathbf{x}|\mathbf{z})$. One models $\mathbf{x}$ given $\mathbf{z}$ parametrically as $f(\mathbf{x}|\mathbf{z}, \eta)$, and estimates η using the complete data. We denote the estimate by $\widehat{\eta}$, and the estimated conditional density of $\mathbf{x}$ given $\mathbf{z}$ by $f(\mathbf{x}|\mathbf{z}, \widehat{\eta})$.

4. *Simulate the missing* $\mathbf{x}_i$ *from the estimated* $f(\mathbf{x}|y, \mathbf{z})$. With estimated $f(\mathbf{x}|\mathbf{z})$ and $f(y|\mathbf{x}, \mathbf{z})$, the estimated conditional density function is

$$
\widehat{f}(\mathbf{x}|y_j, \mathbf{z}_j) \propto \widehat{f}\{y_j|\mathbf{x}, \mathbf{z}_j, \widehat{\boldsymbol{\beta}}_{n_1}(\tau)\} f(\mathbf{x}|\mathbf{z}_j, \widehat{\eta}).
$$

For each j, $j = 1, \ldots, n_0$, we simulate missing $\mathbf{x}_j$ from the estimated density $\widehat{f}(\mathbf{x}|y_j, \mathbf{z}_j)$ and denote by $\widetilde{\mathbf{x}}_{j(\ell)}$ the ℓth imputed $\mathbf{x}$ associated with $(y_j, \mathbf{z}_j)$. Consequently, $\widetilde{\mathbf{x}}_{j(\ell)} \sim \widehat{f}(\mathbf{x}|y_j, \mathbf{z}_j)$.

5. *Re-estimation together with the imputed data.* We assemble a new objective function including the completely observed data and the ℓth imputed data set,

$$
S_{n(\ell)}(\boldsymbol{\beta}) = \sum_{i=1}^{n_1} \rho_\tau\{y_i - (\mathbf{x}_i^\top, \mathbf{z}_i^\top)\boldsymbol{\beta}\} + \sum_{j=1}^{n_0} \rho_\tau\{y_j - (\widetilde{\mathbf{x}}_{j(\ell)}^\top, \mathbf{z}_j^\top)\boldsymbol{\beta}\}, \tag{11.15}
$$

and define $\widehat{\boldsymbol{\beta}}_{*(\ell)} = \arg\min_{\boldsymbol{\beta}} S_{n(\ell)}(\boldsymbol{\beta})$ as the estimated coefficient using the ℓth assembled complete data. We repeat this imputation estimation step m times and the MI estimator is $\widetilde{\boldsymbol{\beta}}_{n, \tau} = m^{-1}\sum_{\ell=1}^m \widehat{\boldsymbol{\beta}}_{*(\ell)}$.

Wei et al. (2012) show that, under regulation conditions and for $K_n \to \infty$ and $K_n n^{-1} \to 0$, the MI estimator $\widetilde{\boldsymbol{\beta}}_{n, \tau}$ is consistent and asymptotically normally distributed, that is,

$$
\sqrt{n}(\widetilde{\boldsymbol{\beta}}_{n, \tau} - \boldsymbol{\beta}_{0, \tau}) \to N(\mathbf{0}, \boldsymbol{\Psi}_\tau^{-1} \boldsymbol{\Sigma} \boldsymbol{\Psi}_\tau^{-1}), \tag{11.16}
$$

where

$$\boldsymbol{\Psi}_\tau = \frac{\partial}{\partial \boldsymbol{\beta}_{0,\tau}} E\left[\varphi_\tau\{y_i - (\mathbf{x}_i^\top, \mathbf{z}_i^\top)\boldsymbol{\beta}_{0,\tau}\}(\mathbf{x}_i^\top, \mathbf{z}_i^\top)^\top\right],$$

$$\Sigma = (\lambda + 1)^{-1}\mathbf{V}_1 + (1 + 1/\lambda)^{-1}\left[m^{-1}\mathbf{V}_0 + \{(m-1)/m\}\mathbf{U}_0\right],$$

$$\mathbf{V}_1 = \text{var}[\varphi_\tau\{y_i - (\mathbf{x}_i^\top \mathbf{z}_i^\top)\boldsymbol{\beta}_{0,\tau}\}(\mathbf{x}_i^\top, \mathbf{z}_i^\top)^\top],$$

$$\mathbf{V}_0 = \lim_{n \to \infty} \text{var}[\varphi_\tau\{y_j - (\tilde{\mathbf{x}}_{j(\ell)}^\top, \mathbf{z}_j^\top)\boldsymbol{\beta}_{0,\tau}\}(\tilde{\mathbf{x}}_{j(\ell)}^\top, \mathbf{z}_j^\top)^\top],$$

$$\mathbf{U}_0 = \lim_{n \to \infty} \text{cov}[\varphi_\tau\{y_j - (\tilde{\mathbf{x}}_{j(\ell)}^\top, \mathbf{z}_j^\top)\boldsymbol{\beta}_{0,\tau}\}(\tilde{\mathbf{x}}_{j(\ell)}^\top, \mathbf{z}_j^\top), \varphi_\tau\{y_j - (\tilde{\mathbf{x}}_{j(\ell')}^\top, \mathbf{z}_j^\top)\boldsymbol{\beta}_{0,\tau}\}(\tilde{\mathbf{x}}_{j(\ell')}^\top, \mathbf{z}_j^\top)^\top].$$

Note that the asymptotic variance of the estimator using the completely observed data only is $\boldsymbol{\Psi}_\tau^{-1}\mathbf{V}_1\boldsymbol{\Psi}_\tau^{-1}$. The MI estimator gains estimation efficiency with a larger effective sample size n, but the MI estimator also contains additional sources of variability, including the sampling variability from MI, the inherited variability from using the complete-data estimated parameters, and their correlations. Hence, it is possible that the MI estimators are less efficient than the complete-data estimator. Such phenomena are common for MI estimators (Tsiatis, 2007). In practice, if less than 10% of observations are missing, imputation may not be worthwhile and one could assess the estimation variabilities of both the MI estimator and complete-data estimator to decide which estimator to use.

11.3.1.2 Modified MI algorithms

MI-IPW algorithm

The MI algorithm outlined in Section 11.3.1.1 assumes that missing $\mathbf{x}_i$ are unrelated to the outcome y_i. Under this relatively strong assumption, the regressions in Steps 1 and 3, which only use completely observed cases, yield unbiased estimates. When the omission does depend on y_i, however, the estimates in both steps will be biased. One way to correct the bias is to incorporate IPW in Steps 1 and 3. The resulting MI-IPW algorithm follows the same structure but replaces Steps 1 and 3 as follows.

1*. Estimate the conditional quantile process using IPW with complete data only. Conduct weighted quantile regressions on quantile levels $0 < \tau_1 < \cdots < \tau_k < \cdots < \tau_{K_n} < 1$,

$$\hat{\boldsymbol{\beta}}_{n_1,\tau_k} = \arg\min_{\boldsymbol{\beta}} \sum_{i=1}^{n_1} w_i \rho_\tau\{y_i - (\mathbf{x}_i^\top, \mathbf{z}_i^\top)\boldsymbol{\beta}\},$$

where $\rho_\tau(r) = r\{\tau - I(r < 0)\}$ is an asymmetric L_1 loss function, and the weights are

$$w_i = \frac{1/\text{Prob}(\delta_i = 1|y_i, \mathbf{z}_i)}{\sum_{j=1}^{n_1} 1/\text{Prob}(\delta_j = 1|y_j, \mathbf{z}_j)}.$$

The missing probability $\text{Prob}(\delta_i = 1|y_i, \mathbf{z}_i)$ can be estimated from logistically regressing δ_i against y_i and $\mathbf{z}_i$.

3* Use the IPW approach to estimate the parameter η and the conditional density $f(\mathbf{x}|\mathbf{z}, \eta)$.

Robust shrinkage MI algorithm

The MI estimator $\tilde{\boldsymbol{\beta}}_{n,\tau}$ is more efficient than the complete-case estimates. However, imputation itself could induce additional bias when the parametric likelihood for $\mathbf{x}$ given $\mathbf{z}$ is misspecified. On the other hand, although less efficient, the complete-case estimator $\hat{\boldsymbol{\beta}}_{n_1,\tau}$

is unbiased. (In the case where the omission is associated with y, $\widehat{\beta}_{n_1,\tau}$ is estimated using the IPW approach.) One can use a shrinkage estimator (Chen et al., 2008) to balance the two estimates and make it more robust against misspecified likelihood.

Let $\widehat{\theta}_\tau = \widehat{\beta}_{n_1,\tau} - \widetilde{\beta}_{n,\tau}$ be componentwise differences of the MI and complete-data estimators, with elements $(\widehat{\theta}_{1,\tau}, \ldots, \widehat{\theta}_{p,\tau})^\top$. Let V be the covariance matrix of $\widehat{\theta}_\tau$ with diagonal elements $(v_{11}, \ldots, v_{pp})$, which can be estimated from the sample as outlined in Wei et al. (2012, Appendix 2). The shrinkage MI estimator is then constructed as

$$\widehat{\beta}_{n,\tau}^{(s)} = \widehat{\beta}_{n_1,\tau} + K(\widetilde{\beta}_{n,\tau} - \widehat{\beta}_{n_1,\tau}), \tag{11.17}$$

where K is a diagonal matrix with jth diagonal elements $= v_{jj}/(v_{jj} + \widehat{\theta}_{j,\tau}^2)$. Recall that the asymptotic variances are $O_p(n^{-1})$. The idea behind this method is that, if there is no bias, then $\widehat{\theta}_{j,\tau}^2 = O_p(n^{-1})$ and the shrinkage factor $K \approx I$, so that the MI estimator will have the greatest weights. Conversely, if there is bias, $\widehat{\theta}_{j,\tau}^2 = O(1)$ and the elements of $K \to 0$, so that the complete-data estimator will have greater weight.

Fractional imputation algorithm

In the MI algorithm, one has to evaluate and simulate the imputation density $f(x|y,z)$ nonparametrically for each missing value. That step is computationally burdensome. As increasing numbers of statistical applications involve complex and massive data, computational efficiency becomes a crucial feature of any statistical method. An alternative imputation algorithm that extends the fractional imputation of Kim (2011) can be considered to reduce the computation burden.

A set of unbiased estimating equations for regression quantiles with covariates missing at random can be constructed as

$$\sum_{i=1}^n \delta_i S(\mathbf{x}_i, y_i, \mathbf{z}_i, \boldsymbol{\beta}) + \sum_{i=1}^n (1 - \delta_i) E_\mathbf{x}\{S(\mathbf{x}, y_i, \mathbf{z}_i \boldsymbol{\beta})|y_i, \mathbf{z}_i\}, \tag{11.18}$$

where $S(\cdot)$ is the original estimating function as defined in (11.13). In some sense, the MI approach essentially approximates the conditional expectation in (11.18) by a Monte Carlo integration,

$$E_\mathbf{x}\{S(\mathbf{x}, y_i, \mathbf{z}_i, \boldsymbol{\beta})|y_i, \mathbf{z}_i\} \approx \frac{1}{m} \sum_{\ell=1}^m \{S(\widetilde{\mathbf{x}}_{i(\ell)}, y_i, \mathbf{z}_i \boldsymbol{\beta}),$$

where $\widetilde{\mathbf{x}}_{i(\ell)}$ is a random draw from $f(\mathbf{x}|y_i, \mathbf{z}_i)$.

Note that

$$f(\mathbf{x}|y, \mathbf{z}) = \frac{f(y_i|x, z_i)f(x|z_i)}{\int_x f(y_i|x, z_i)f(x|z_i)dx}.$$

One can then rewrite the conditional expectation as

$$E_\mathbf{x}\{S(\mathbf{x}, y_i, \mathbf{z}_i, \boldsymbol{\beta})|y_i, \mathbf{z}_i\} = \frac{\int_\mathbf{x} S(\mathbf{x}, y_i, \mathbf{z}_i, \boldsymbol{\beta})f(y_i|\mathbf{x}, \mathbf{z}_i)f(\mathbf{x}|\mathbf{z}_i)dx}{\int_\mathbf{x} f(y_i|\mathbf{x}, \mathbf{z}_i)f(\mathbf{x}|\mathbf{z}_i)dx}. \tag{11.19}$$

Therefore, an alternative way to approximate $E_\mathbf{x}\{S(\mathbf{x}, y_i, \mathbf{z}_i, \boldsymbol{\beta})|y_i, \mathbf{z}_i\}$ is

$$E_\mathbf{x}\{S(\mathbf{x}, y_i, \mathbf{z}_i, \boldsymbol{\beta})|y_i, \mathbf{z}_i\} \approx \frac{\frac{1}{m}\sum_{\ell=1}^m S(\mathbf{x}, y_i, \mathbf{z}_i, \boldsymbol{\beta})f(y_i|\widetilde{\mathbf{x}}_{i(\ell)}^*, \mathbf{z}_i)}{\frac{1}{m}\sum_{\ell=1}^m f(y_i|\widetilde{\mathbf{x}}_{i(\ell)}^*, \mathbf{z}_i)}, \tag{11.20}$$

where $\widetilde{\mathbf{x}}_{i(\ell)}^*$ represents random draws from the parametric likelihood $f(\mathbf{x}|\mathbf{z}, \eta)$. Since it is much easier to simulate $\mathbf{x}$ from a parametric $f(\mathbf{x}|\mathbf{z},\eta)$ than to simulate it from $f(y|\mathbf{x},\mathbf{z})$, this new numerical approximation leads to substantial reductions in computing time.

Plugging (11.20) into (11.18), we find the estimating equations (11.18) are equivalent to

$$\sum_{i=1}^{n_1} S(\mathbf{x},y_i,\mathbf{z}_i,\boldsymbol{\beta}) + \sum_{i=n_1+1}^{n} \sum_{\ell=1}^{m} w_{i,k} S(\tilde{\mathbf{x}}^*_{i(\ell)},y_i,\mathbf{z}_i,\boldsymbol{\beta}) = 0, \tag{11.21}$$

where

$$w_{i,k} = \frac{f(y_i|\tilde{\mathbf{x}}^*_{i(\ell)},\mathbf{z}_i)}{\sum_{\ell=1}^{m} f(y_i|\tilde{\mathbf{x}}^*_{i(\ell)},\mathbf{z}_i)}. \tag{11.22}$$

Consequently, the quantile coefficients $\boldsymbol{\beta}$ can be easily obtained from a weighted quantile regression. In what follows, we outline a specific fractional imputation-like imputation algorithm for quantile regression.

Step 1. Apply quantile regression with the complete data on a fine grid of $0 < \tau_1 < \cdots < \tau_k < \cdots < \tau_{k_n} < 1$ and derive the conditional density $f(y|\mathbf{x},\mathbf{z})$ from the estimated quantile process. In this step, IPW will be used if missing $\mathbf{x}$ values are related to y.

Step 2. Model the conditional density $f(x|z)$ parametrically as $f(\mathbf{x} \mid \mathbf{z},\eta)$ and estimate η based on the complete data. If missing $\mathbf{x}$ values are related to y, IPW is used.

Step 3. Simulate m copies of $\mathbf{x}$ from the estimated $f(\mathbf{x}|\mathbf{z},\hat{\eta})$ for each missing $\mathbf{x}_i$ and assemble the weighted estimating function as in (11.21) to obtain the final estimator.

11.3.1.3   EM algorithm

Under the MAR assumption, the complete-case analysis could be seriously biased when the omission is related to y. The aforementioned MI-IPW algorithm is one way to correct the bias, but the IPW approach often inflates variances. More efficient estimation can be achieved by using an EM-like algorithm to solve the estimating equations (11.18), as do Wei and Yang (2014).

Let ν denote the indicator of iteration steps. The main steps of the EM algorithm are outlined as follows:

Step 1. Set the initial values of $(\boldsymbol{\beta}^{(0)}_{\tau_k},\boldsymbol{\gamma}^{(0)}_{\tau_k})_{k=1}^{k_n}$ based on the quantile regression with the complete data.

Step 2. Update the distribution $f^{(\nu)}(\mathbf{x}|y_i,\mathbf{z}_i)$ for missing $\mathbf{x}_i$ based on $(\boldsymbol{\beta}^{(\nu-1)}_{\tau_k},\boldsymbol{\gamma}^{(\nu-1)}_{\tau_k})_{k=1}^{k_n}$ from the previous iteration. In other words,

$$f^{(\nu)}(\mathbf{x}|y_i,\mathbf{z}_i) = \frac{f(y_i|\mathbf{x},\mathbf{z}_i,\boldsymbol{\beta}^{(\nu-1)}(\tau),\boldsymbol{\gamma}^{(\nu-1)}(\tau))f(\mathbf{x}|\mathbf{z}_i)}{\int_x f(y_i|\mathbf{x},\mathbf{z}_i,\boldsymbol{\beta}^{(\nu-1)}(\tau),\boldsymbol{\gamma}^{(\nu-1)}(\tau))f(\mathbf{x}|\mathbf{z}_i)dx},$$

where $\boldsymbol{\beta}^{(\nu)}(\tau)$ and $\boldsymbol{\gamma}^{(\nu)}(\tau)$ are natural linear splines expanded from the $\boldsymbol{\beta}^{(\nu)}$ and $\boldsymbol{\gamma}^{(\nu)}$ values and $f(y_i|\mathbf{x},\mathbf{z}_i,\boldsymbol{\beta}^{(\nu)}(\tau),\boldsymbol{\gamma}^{(\nu)}(\tau))$ is the conditional density of y_i given $(\mathbf{x},\mathbf{z}_i)$ at the νth iteration.

Step 3. Update $(\boldsymbol{\beta}^{(\nu)}_{\tau_k},\boldsymbol{\gamma}^{(\nu)}_{\tau_k})_{k=1}^{k_n}$ based on the new estimating functions with $f^{(\nu)}(\mathbf{x}|y_i,\mathbf{z}_i)$. Numerical integration can be used to perform this step. Let $\tilde{x}_i = (\tilde{x}_{i,1},\ldots,\tilde{x}_{i,m})$ be a fine grid of possible $\mathbf{x}_i$ values. The estimating equations are equivalent to

$$\sum_{i=1}^{n} \Big\{ \delta_i \Psi_{\tau_k}(\boldsymbol{\beta}_{\tau_k},\boldsymbol{\gamma}_{\tau_k},y_i,\mathbf{x}_i,\mathbf{z}_i) + (1-\delta_i) \sum_{j=1}^{m} \Big[\Psi_{\tau_k}(\boldsymbol{\beta}_{\tau_k},\boldsymbol{\gamma}_{\tau_k},y_i,\tilde{\mathbf{x}}_{i,j},\mathbf{z}_i)$$

$$\times f^{(\nu)}(\tilde{\mathbf{x}}_{i,j}|y_i,\mathbf{z}_i) \times (\tilde{\mathbf{x}}_{i,j+1}-\tilde{\mathbf{x}}_{i,j}) \Big] \Big\} = 0, \tag{11.23}$$

where $k = 1, \ldots, k_n$. Solving (11.23) can be achieved by weighted quantile regression with response y_i, covariates $(\tilde{\mathbf{x}}_{i,j}, \mathbf{z}_i)$, and weights δ_i or $(1 - \delta_i) f^{(\nu)}(\tilde{\mathbf{x}}_{i,j} | y_i, z_i) \times (\tilde{\mathbf{x}}_{i,j+1} - \tilde{\mathbf{x}}_{i,j})$.

Step 4. Repeat Steps 2 and 3 until the algorithm converges.

To further improve the robustness of the estimation, one could adapt the proposed estimation algorithm with Robins's estimating function (Robins et al., 1995)

$$\Omega(y, \mathbf{x}, \mathbf{z}, \delta, \boldsymbol{\beta}) = \frac{\delta}{\pi(y, z)} S(y, \mathbf{x}, \mathbf{z}, \boldsymbol{\beta}) + \frac{\delta - \pi(y, \mathbf{z})}{\pi(y, \mathbf{z})} E\{S(y, \mathbf{x}, \mathbf{z}, \boldsymbol{\beta}) \mid y, \mathbf{z}\},$$

where $\pi(y, \mathbf{z})$ is the probability of X being missing, which can be estimated using logistic regression, and $E\{S(y, \mathbf{x}, \mathbf{z}, \delta, \boldsymbol{\beta}) \mid y, \mathbf{z}\}$ can be evaluated using a similar approach. The resulting estimators inherit the double robustness from Robins's estimating equation, that is, if either $\pi(y, \mathbf{z})$ or $E\{S(y, \mathbf{x}, \mathbf{z}, \boldsymbol{\beta}) | y, \mathbf{z}\}$ is estimated consistently, the resulting estimators are consistent as well.

11.3.1.4   IPW algorithms

Another appealing approach to handle missing covariates is Inverse probability weighting. This method minimizes a weighted loss function

$$\frac{1}{n} \sum_{i=1}^{n} \frac{\delta_i}{p(\mathbf{x}_i)} \rho_\tau (y_i - \mathbf{x}_i^\top \boldsymbol{\beta} - \mathbf{z}_i^\gamma),$$

where $p(y_i, \mathbf{z}_i) = \text{Prob}(\delta_i = 1 | y_i, \mathbf{z}_i)$ is the probability of being observed, given the observed $y_i, \mathbf{z}_i$. The probability $p(y_i, \mathbf{z}_i)$ is often modeled parametrically through logistic or probit regressions. Chen et al. (2015) proposed estimating $p(y_i, \mathbf{z}_i)$ using nonparametric kernel smoothing,

$$\hat{p}(y_i, \mathbf{z}_i) = \frac{\sum_{j=1}^{n} K_h(y_i - y_j, \mathbf{z}_i - \mathbf{z}_j) \delta_j}{\sum_{j=1}^{n} K_h(y_i - y_j, \mathbf{z}_i - \mathbf{z}_j)},$$

where K_h is a multidimensional kernel function with bandwidth h.

Chen et al. (2015) also proposed an augmented inverse probability weighting (AIPW) algorithm that reweights the corrected estimating function (11.18) by

$$\sum_{i=1}^{n} \left\{ \frac{\delta_i}{p(y_i, \mathbf{z}_i)} S(\mathbf{x}_i, y_i, \mathbf{z}_i, \boldsymbol{\beta}) + \frac{1 - \delta_i}{1 - p(y_i, \mathbf{z}_i)} E_\mathbf{x}\{S(\mathbf{x}, y_i, \mathbf{z}_i \boldsymbol{\beta}) | y_i, \mathbf{z}_i\} \right\}, \quad (11.24)$$

and approximates the conditional expectation $E_\mathbf{x}\{S(\mathbf{x}, y_i, \mathbf{z}_i \boldsymbol{\beta}) | y_i, \mathbf{z}_i\}$ by kernel smoothing,

$$\frac{\sum_{j=1}^{n} \mathfrak{K}_h(y_i - y_j, \mathbf{z}_i - \mathbf{z}_j) S(\mathbf{x}, y_i, \mathbf{z}_i \boldsymbol{\beta}) \delta_j}{\sum_{j=1}^{n} \mathfrak{K}_h(y_i - y_j, \mathbf{z}_i - \mathbf{z}_j) \delta_j},$$

where $\mathfrak{K}_h(\mathbf{u})$ is some kernel smoother. The EM algorithm outlined in Section 11.3.1.3 can also be extended to solve the weighted and corrected estimating equations.

11.3.2   Statistical methods handling missing outcomes in quantile regression

11.3.2.1   Imputation approaches for missing outcomes

In this subsection, we assume that $(y_i, \mathbf{x}_i, \delta_i)$ is a random sample from a linear quantile regression model

$$Q_y(\tau | \mathbf{X}) = \mathbf{X}^\top \boldsymbol{\beta}_0(\tau). \quad (11.25)$$

The covariates $\mathbf{x}_i$ are completely observed, but some of the responses y_i are MAR. Using the same notation, we use δ_i to denote whether the ith response is observed. Under the MAR assumption, $\text{Prob}(\delta = 1|Y, \mathbf{X}) = \text{Prob}(\delta = 1|\mathbf{X})$, which implies a distributional restriction $F_y(\tau|\mathbf{X}, \delta = 1) = F_y(\tau|\mathbf{X}, \delta = 0)$. Let us assume that the quantile model (11.25) holds at all quantile levels. The distributional restriction further implies that the conditional quantile process can be estimated without bias using the complete data only and those missing responses can be simulated from the estimated conditional quantile process. An MI algorithm can easily be developed to achieve higher estimation efficiency, as Yoon (2010) does:

Step 1. Estimate quantile model (11.25),

$$\hat{\boldsymbol{\beta}}_{n_1}(\tau) = \arg\min_{\boldsymbol{\beta}} \sum_{i=1}^{n} \delta_i \rho_\tau (y_i - \mathbf{x}_i^\top \boldsymbol{\beta}).$$

Step 2. Simulate the missing y_i from the estimated conditional quantile process in Step 1,

$$\tilde{y}_{i(\ell)} = \mathbf{x}_i^\top \hat{\boldsymbol{\beta}}_{n_1}(u_\ell), \tag{11.26}$$

where u_ℓ is the ℓth random draw from Uniform (0,1) distribution.

Step 3. Re-estimate the quantile model (11.25) using assembled data,

$$\tilde{\boldsymbol{\beta}}_n(\tau) = \arg\min_{\boldsymbol{\beta}} \left\{ \sum_{i=1}^{n} \delta_i \rho_\tau (y_i - \mathbf{x}_i^\top \boldsymbol{\beta}) + \sum_i (1 - \delta_i) m^{-1} \sum_{\ell=1}^{m} \rho_\tau (\tilde{y}_{i(\ell)} - \mathbf{x}_i^\top \boldsymbol{\beta}) \right\}.$$

In some applications, researchers are interested in understanding the marginal distributions of Y in a population, so their goal is to estimate the marginal mean, median, and quantiles of Y, instead of the conditional associations. For example, pediatricians use growth charts, which consist of population percentiles of height and weight, to monitor childhood growth. Econometricians often evaluate the treatment effect of a policy in a population by comparing the marginal mean or quantiles of Y between the populations with and without that policy. Although the research interest in those applications is in estimating the marginal distributions of Y, imputing missing outcomes from the conditional quantile model (11.25) is one effective way to improve the estimation efficiency when those missing observations have available information in X. For example, the marginal mean of Y can be estimated by $m^{-1} \sum_{\ell=1}^{m} [n^{-1} \sum_{i=1}^{n} \{\delta_i y_i + (1 - \delta_i)\tilde{y}_{i(\ell)}\}]$, where $\tilde{y}_{i(\ell)}$ are simulated outcomes as defined in (11.26). Such a strategy is particularly useful in applications where the number of observed Y_i is limited, such as small-area estimations in survey analysis as in Chambers and Tzavidis (2006).

In a similar setting, Díaz (2015) proposed an efficient estimation of the marginal quantiles of Y that is constructed from efficient influence functions. Let θ_τ be the marginal τth quantile of Y. Díaz (2015) derived that the efficient influence function of θ_τ with respect to the joint distribution of $(\mathbf{X}, Y, \delta)$ is

$$\frac{\delta}{\pi(\mathbf{X})} \{I\{Y < \theta_\tau\} - F(\theta_\tau|\mathbf{X})\} + F(\theta_\tau|\mathbf{X}) - \tau,$$

where $\pi(\mathbf{X})$ is the probability of missing Y given $\mathbf{X}$, and F is the conditional distribution of Y given $\mathbf{X}$. Consequently, a doubly robust AIPW estimate (Bang and Robins, 2005) of θ_τ can be estimated by solving the estimating equations

$$\frac{1}{n} \sum_{i=1}^{n} \left[\frac{\delta_i}{\pi(\mathbf{x}_i)} \{I\{y_i < \theta_\tau\} - \hat{F}(\theta_\tau|\mathbf{x}_i)\} + \hat{F}(\theta_\tau|x_i) \right] = \tau,$$

where $\widehat{F}$ is the estimated conditional distribution of Y given $\mathbf{x}_i$ derived from $\mathbf{x}_i^\top \widetilde{\boldsymbol{\beta}}_n(\tau)$.

Using the efficient influence functions, Díaz (2015) also outlined a targeted minimum loss based estimator (TMLE) following van der Laan and Rubin (2006). The algorithm literately updates the estimate of θ_τ and the conditional density of Y given $\mathbf{X}$. The resulting TMLE estimate is asymptotically equivalent to the AIPW estimate, but has better performance with finite sample sizes.

11.3.2.2   Statistical methods for longitudinal dropout

The vast majority of clinical and public health research uses longitudinal designs, where patients or subjects are followed up multiple times over the course of study. In any longitudinal study, some subjects are likely to have missed a follow-up visit for various reasons and some subjects may drop out before the study ends. Handling those longitudinal missing outcomes appropriately is important to achieve accurate inferences in these studies.

Let $Y_{i,j}$ be the continuous response for the ith subject at time point j, where $\delta_{i,j}$ is the binary indicator of whether $Y_{i,j}$ is being observed and $\mathbf{x}_{i,j}$ is the $p \times 1$ covariate vector collected at time point j, $j = 1, \ldots, m$; $i = 1, \ldots, n..$ We assume that the quantiles of $Y_{i,j}$ are linear in $\mathbf{x}_{i,j}$,

$$Y_{i,j} = \mathbf{x}_{i,j}^\top \boldsymbol{\beta} + \epsilon_{i,j}, \quad \text{where } Q_\epsilon(\tau|\mathbf{x}_{i,j}) = 0.$$

Let D_i be the time when the ith subject drops out. Lipsitz et al. (1997) considered monotonic missing-data patterns, assuming that, if $D_i = d$, then $Y_{i,j}$ for $j \geqslant d$ are missing. The authors further assumed that the conditional probability of dropout on occasion d depends only on the previous $d-1$ responses and covariates, which is an MAR assumption. Denote $\pi_{i,d} = \text{Prob}(D_i = d)$ as the conditional probability of dropout on occasion d for the ith subject. One can then construct a set of weighted estimated equations

$$\sum_{i=1}^n \frac{1}{\pi_{i,D_i}} \sum_{j=1}^{D_i} S(y_i, \mathbf{x}_{i,j}, \boldsymbol{\beta}) \approx 0 \tag{11.27}$$

to solve for $\boldsymbol{\beta}$. The probability of dropout can be modeled recursively by

$$
\begin{aligned}
\pi_{i,d} &= \text{Prob}(\delta_{i,2} = 1, \delta_{i,3} = 1, \ldots, \delta_{i,d-1} = 1, \delta_d = 0 | y_{i,1}, .., y_{i,d-1}, \mathbf{x}_i) \\
&= \left(\prod_{t=2}^{D_i-1} \eta_{i,t} \right) (1 - \eta_{i,D_i})^{I\{D_i \leqslant m\}}
\end{aligned}
$$

where

$$\eta_{i,t} = \text{Prob}(\delta_{i,t} = 1 | \delta_{i,1} = 1, \ldots, \delta_{i,t-1} = 1, Y_{i,1}, \ldots, Y_{i,t-1}, \mathbf{x}_i).$$

Generalized logistic regression can be used to estimate $\eta_{i,t}$. Under the same MAR model for longitudinal dropout, Yi and He (2009) incorporated within-subject correlations into the estimating equation (11.27) to further improve the estimating efficiency.

Yuan and Yin (2010) allowed the response omissions to be nonrandom and non-monotonic. An L_2-penalized loss function was proposed to obtain regression quantiles while taking subject effects into account:

$$\sum_{i=1}^n \sum_{j=1}^m \delta_{i,j} \rho_\tau(y_i - \mathbf{x}_{i,j}^\top \boldsymbol{\beta} - \mathbf{z}_{i,j}^\top \mathbf{b}_i) + \frac{1}{2} \sum_{i=1}^n \mathbf{b}_i^\top \Lambda^{-1} \mathbf{b}_i, \tag{11.28}$$

where $\mathbf{x}_{i,j}$ and $\mathbf{z}_{i,j}$ are covariates and $\mathbf{b}_i$ represents subject-specific effects. The penalized estimating equation can be mapped to the likelihood function of the random effect model

$$y_{i,j} | \mathbf{b}_i \sim ALD(\tau, \mathbf{x}_{i,j}^\top \boldsymbol{\beta} - \mathbf{z}_{i,j}^\top \mathbf{b}_i), \quad \mathbf{b}_i \sim N(0, \Lambda). \tag{11.29}$$

Note that the Laplace likelihood is simply a working likelihood for the quantile model that utilizes the relation between the quantile regression loss function and the Laplace distribution function.

On the other hand, Yuan and Yin (2010) defined $s_{i,j}$ as the missing status on the jth follow-up, which takes on three possible values:

$$
s_{i,j} = \begin{cases} O, & y_{i,j} \text{ is observed,} \\ I, & y_{i,j} \text{ is missing but intermittent,} \\ D, & \text{subject } i \text{ drops out at the } j\text{th visit.} \end{cases}
$$

By the definition of dropout, the transition probability from D to I or O is zero, the transition probability from I to D is zero, and the transition probability from D to D is one. The rest of the transition probabilities are modeled by logistic regression with the shared individual random effect $\mathbf{b}_i$:

$$
\begin{aligned}
\pi_{i,j}^{(O)} &= \text{Prob}(s_{i,j} = O | s_{i,j-1} \neq D, \mathbf{b}_i) \\
&= \frac{1}{1 + \exp\{\mathbf{x}_{i,j}^\top \boldsymbol{\alpha}^{(I)} + \mathbf{b}_i^\top \boldsymbol{\gamma}^{(I)}\} + \exp\{\mathbf{x}_{i,j}^\top \boldsymbol{\alpha}^{(D)} + \mathbf{b}_i^\top \boldsymbol{\gamma}^{(D)}\}}
\end{aligned}
$$

$$
\begin{aligned}
\pi_{i,j}^{(I)} &= \text{Prob}(s_{i,j} = I | s_{i,j-1} \neq D, \mathbf{b}_i) \\
&= \frac{\exp\{\mathbf{x}_{i,j}^\top \boldsymbol{\alpha}^{(I)} + \mathbf{b}_i^\top \boldsymbol{\gamma}^{(I)}\}}{1 + \exp\{\mathbf{x}_{i,j}^\top \boldsymbol{\alpha}^{(I)} + \mathbf{b}_i^\top \boldsymbol{\gamma}^{(I)}\} + \exp\{\mathbf{x}_{i,j}^\top \boldsymbol{\alpha}^{(D)} + \mathbf{b}_i^\top \boldsymbol{\gamma}^{(D)}\}}
\end{aligned}
$$

$$
\begin{aligned}
\pi_{i,j}^{(D)} &= \text{Prob}(s_{i,j} = D | s_{i,j-1} = O, \mathbf{b}_i) \\
&= \frac{\exp\{\mathbf{x}_{i,j}^\top \boldsymbol{\alpha}^{(D)} + \mathbf{b}_i^\top \boldsymbol{\gamma}^{(D)}\}}{1 + \exp\{\mathbf{x}_{i,j}^\top \boldsymbol{\alpha}^{(I)} + \mathbf{b}_i^\top \boldsymbol{\gamma}^{(I)}\} + \exp\{\mathbf{x}_{i,j}^\top \boldsymbol{\alpha}^{(D)} + \mathbf{b}_i^\top \boldsymbol{\gamma}^{(D)}\}}
\end{aligned}
$$

Based on the transition models outlined, one could write out the conditional likelihood for the missing-data process of subject i. The missing-data model accounts for a non-ignorable missing-data mechanism by using the shared random effects $\mathbf{b}_i$, which is a common strategy for nonignorable dropout (Little, 1995).

The working likelihood of the observed data can be obtained by combining (11.29) with the transition models for $s_{i,j}$. With a noninformative prior being assigned to the related parameters, one can use the Gibbs sampler to obtain the posterior distributions of the parameters of interest.

Bibliography

A. A. Afifi and R. M. Elashoff. Missing observations in multivariate statistics III: Large sample analysis of simple linear regression. *Journal of the American Statistical Association*, 64(325):337–358, 1969a.

A. A. Afifi and R. M. Elashoff. Missing observations in multivariate statistics IV: A note on simple linear regression. *Journal of the American Statistical Association*, 64(325): 359–365, 1969b.

H. Bang and J. M Robins. Doubly robust estimation in missing data and causal inference models. *Biometrics*, 61(4):962–973, 2005.

R. J. Carroll and L. A. Stefanski. Approximate quasi-likelihood estimation in models with surrogate predictors. *Journal of the American Statistical Association*, 85:652–663, 1990.

R. J. Carroll, D. Ruppert, L. A. Stefanski, and C. M. Crainiceanu. *Measurement Error in Nonlinear Models: A Modern Perspective*. Chapman & Hall/CRC, Boca Raton, FL, 2nd edition, 2006.

R. Chambers and N. Tzavidis. M-quantile models for small area estimation. *Biometrika*, 93(2):255–268, 2006.

X. Chen, A. T. K. Wan, and Y. Zhou. Efficient quantile regression analysis with missing observations. *Journal of the American Statistical Association*, 110(510):723–741, 2015.

Y. H. Chen, N. Chatterjee, and R. J. Carroll. Retrospective analysis of haplotype-based case-control studies under a flexible model for gene-environment association. *Biostatistics*, 9(1):81–99, 2008.

L. M. Collins, J. L. Schafer, and C. M. Kam. A comparison of inclusive and restrictive strategies in modern missing data procedures. *Psychological Methods*, 6:330–351, 2001.

C. de Boor. *A Practical Guide to Splines*. Springer-Verlag, New York, 2001.

I. Díaz. Efficient estimation of quantiles in missing data models. arXiv:1512.08110, 2015.

J. Fan and R. Li. New estimation and model selection procedures for semiparametric modelling in longitudinal data analysis. *Journal of the American Statistical Association*, 99(467):710–723, 2004.

X. He and H. Liang. Quantile regression estimates for a class of linear and partially linear errors-in-variables models. *Statistica Sinica*, 10:129–140, 2000.

J. K. Kim. Parametric fractional imputation for missing data analysis. *Biometrika*, 98(1): 119–132, 2011.

S. Kotz, T. J. Kozubowkski, and K. Podgórski. *The Laplace distribution and generalizations*. Birkhäuser, Boston, 2001.

D. Y. Lin and Z. Ying. Semiparametric and nonparametric regression analysis of longitudinal data. *Journal of the American Statistical Association*, 96(453):103–113, 2001.

S. R. Lipsitz, G. M. Fitzmaurice, G. Molenberghs, and L. P. Zhao. Quantile regression methods for longitudinal data with drop-outs: Application to CD4 cell counts of patients infected with the human immunodeficiency virus. *Applied Statistics*, 46:463–476, 1997.

R. J. A. Little. Modeling the drop-out mechanism in repeated-measures studies. *Applied Statistics*, 90:1112–1121, 1995.

R. J. A. Little and D. B. Rubin. *Statistical analysis with missing data*. John Wiley & Sons, 2014.

J. M. Robins, A. Rotnitzky, and L. P. Zhao. Analysis of semiparametric regression models for repeated outcomes in the presence of missing data. *Journal of the American Statistical Association*, 90:106–121, 1995.

B. Rosner, W. C. Willett, and D Spiegelman. Correction of logistic regression relative risk estimates for systematic within-person measurement error. *Statistics in Medicine*, 8: 1051–1069, 1989.

D. B. Rubin. Multiple imputation after 18+ years. *Journal of the American Statistical Association*, 434:473–489, 1996.

D. B. Rubin. *Multiple Imputation for Nonresponse in Surveys.* John Wiley & Sons, Hoboken, NJ, 2004.

D. B. Rubin and J. L. Schafer. Efficiently creating multiple imputations for incomplete multivariate normal data. *Proceedings of the American Statistical Association Statistical Computing Section*, 1990.

S. M. Schennach. Quantile regression with mismeasured covariates. *Econometric Theory*, 24(04):1010–1043, 2008.

S. R. Seaman and I. R. White. Review of inverse probability weighting for dealing with missing data. *Statistical Methods in Medical Research*, 22(3):278–295, 2013.

J. Staudenmayer, D. Ruppert, and J. P. Buonaccorsi. Density estimation in the presence of heteroskedastic measurement error. *Journal of the American Statistical Association*, 103:726–736, 2008.

L. A. Stefanski and J. R. Cook. Simulation-extrapolation: The measurement error jackknife. *Journal of the American Statistical Association*, 90:1247–1256, 1995.

A. Tsiatis. *Semiparametric Theory and Missing Data.* Springer, New York, 2007.

M. J. van der Laan and D. Rubin. Targeted maximum likelihood learning. *International Journal of Biostatistics*, 2(1), 2006.

H. J. Wang, L. A. Stefanski, and Z. Zhu. Corrected-loss estimation for quantile regression with covariate measurement errors. *Biometrika*, 99(2):405–421, 2012.

Y. Wei and R. J. Carroll. Quantile regression with measurement error. *Journal of the American Statistical Association*, 104(497):1129–1143, 2009.

Y. Wei and Y. J. Yang. Quantile regression with missing covariates. *Statistica Sinica*, 24: 1277–1299, 2014.

Y. Wei, Y. Ma, and R. J. Carroll. Multiple imputation in quantile regression. *Biometrka*, 99:423–438, 2012.

Y. Wu, Y. Ma, and G. Yin. Smoothed and corrected score approach to censored quantile regression with measurement errors. *Journal of the American Statistical Association*, 110(512):1670–1683, 2015.

G. Y. Yi and W. He. Median regression models for longitudinal data with dropouts. *Biometrics*, 65(2):618–625, 2009.

J. Yoon. Quantile regression analysis with missing response with applications to inequality measures and data combination. Working paper, Robert Day School of Economics and Finance, Claremont McKenna College, Claremont, CA. Available at http://www.cmc.edu/pages/faculty/jyoon/, 2010.

Y. Yuan and G. Yin. Bayesian quantile regression for longitudinal studies with nonignorable missing data. *Biometrics*, 66(1):105–114, 2010.

12

Multiple-Output Quantile Regression

Marc Hallin

ECARES, Université Libre de Bruxelles, Belgium

Miroslav Šiman

Institute of Information Theory and Automation, Czech Academy of Sciences, Prague, Czech Republic

CONTENTS

12.1 Multivariate quantiles, and the ordering of $\mathbb{R}^d$, $d \geqslant 2$

Quantile regression is about estimating the quantiles of some d-dimensional response $\mathbf{Y}$ conditional on the values $\mathbf{x} \in \mathbb{R}^p$ of some covariates $\mathbf{X}$. The problem is well understood when $d = 1$ (*single-output* case, where Y is used instead of $\mathbf{Y}$): for a (conditional) probability distribution $P^Y = P^Y_{X=x}$ on $\mathbb{R}$, with distribution function $F = F_{X=x}$, the (conditional on $X = x$) quantile of order τ of Y is

$$q_\tau(x) := \inf\{y \; : \; F(y) \geqslant \tau\}, \quad \tau \in [0, 1).$$

This, under absolute continuity with nonvanishing density (which, for simplicity, we henceforth assume throughout), yields, for $\tau \in (0, 1)$, what we will call the "traditional definition:"

(a) *Regression quantiles: traditional definition.* The *regression quantile of order τ* of Y (relative to the vector of covariates X with values in $\mathbb{R}^p$) is the mapping

$$x \mapsto q_\tau(x) := F^{-1}(\tau), \quad x \in \mathbb{R}^p, \tag{12.1}$$

where $F(y) := P(Y \leqslant y | X = x)$.

The same concept, under finite moments of order 1, also admits the L_1 characterization:

(b) *Regression quantiles: L_1 definition.* The *regression quantile of order τ* of Y (relative to the vector of covariates X with values in $\mathbb{R}^p$) is the mapping $x \mapsto q_\tau(x)$, where $q_\tau(x)$ minimizes, for $x \in \mathbb{R}^p$,

$$E[\rho_\tau(Y - q) | X = x] \tag{12.2}$$

over $q \in \mathbb{R}$; the function $z \mapsto \rho_\tau(z) := (1-\tau)|z| \, I_{[z<0]} + \tau z \, I_{[z\geqslant 0]}$, as usual, stands for the so-called *check function.*

Neither this L_1 definition (b), nor the "traditional" one (a), leads to a straightforward empirical version (as there are no empirical versions of conditional distributions). The L_1 characterization, however, allows for a linear version of quantile regression:

(c) *Linear regression quantiles.* The *regression quantile hyperplane of order τ* of Y (relative to the vector of covariates X with values in $\mathbb{R}^p$) is the hyperplane with equation

$$y = q_\tau(x) = \alpha_\tau + \beta'_\tau x,$$

where α_τ and β_τ are the minimizers, over $(a, b') \in \mathbb{R}^{p+1}$, of

$$E[\rho_\tau(Y - a - b'X)]. \tag{12.3}$$

Contrary to the general concept defined in (a)–(b), regression quantile hyperplanes, thanks to their parametric form, also admit straightforward (substituting empirical distributions for the theoretical ones) empirical counterparts:

(d) *Empirical linear regression quantiles.* Denote by $(Y_1, X_1), \ldots, (Y_n, X_n)$ an n-tuple of points in $\mathbb{R}^{p+1}$. The corresponding empirical regression quantile hyperplane of order τ is the hyperplane with equation

$$y = q^{(n)}_\tau(x) = \alpha^{(n)}_\tau + \beta^{(n)'}_\tau x,$$

where $\alpha^{(n)}_\tau$ and $\beta^{(n)}_\tau$ are the minimizers, over $(a, b') \in \mathbb{R}^{p+1}$, of

$$\sum_{i=1}^n \rho_\tau(Y_i - a - b'X_i). \tag{12.4}$$

Note that, for $p = 0$, all definitions above yield "location quantiles," as opposed to "regression quantiles."

Now, a response seldom comes as an isolated quantity, and $\mathbf{Y}$, in most situations of practical interest, takes values in $\mathbb{R}^d$, with $d \geqslant 2$ (*multiple-output* case). An extension to $d \geqslant 2$ of the definitions above is thus extremely desirable. Unfortunately, those definitions all exploit the canonical ordering of $\mathbb{R}$. Such an ordering no longer exists in $\mathbb{R}^d$, $d \geqslant 2$. As a consequence, (location and regression) quantiles and check functions, but also equally basic univariate concepts such as distribution functions, signs, and ranks – all playing a fundamental role in statistical inference – do not straightforwardly extend to higher dimensions.

That problem of ordering $\mathbb{R}^d$ – hence that of defining pertinent concepts of multivariate quantiles, ranks and signs – has attracted much interest in the literature, and many solutions have been proposed. The whole theory of statistical depth and also, in a sense, the theory of copulas, aim at similar objectives. For obvious reasons of space constraints, we cannot provide here an extensive coverage of those theories. For insightful and extensive surveys of statistical depth, we refer to Zuo and Serfling (2000) or Serfling (2002a, 2012).

12.2   Directional approaches

Since the univariate concept of a quantile is well understood and enjoys all the properties one is expecting, a natural idea, in dimension $d \geqslant 2$, involves trying to reduce the multivariate problem to a collection of univariate ones by considering univariate distributions (such as marginal or projected distributions) associated with the d-dimensional ones. This is what we call a *directional approach*, as opposed to *global* approaches, where definitions are of a direct nature. We start with the pure location case ($p = 0$, no covariates) and projection ideas.

12.2.1   Projection methods

12.2.1.1   Marginal (coordinatewise) quantiles

If a coordinate system is adopted, $\mathbf{Y}$ is written as $(Y_1, \dots, Y_d)'$. The d marginal distribution functions characterize marginal quantiles $q_{\tau;j}$, $j = 1, \dots, d$, hence a coordinatewise multivariate quantile

$$\mathbf{q}_{\tau_1,\dots,\tau_d} := (q_{\tau_1;1}, \dots, q_{\tau_d;d})'.$$

The mapping $(\tau_1, \dots, \tau_d) \mapsto \mathbf{q}_{\tau_1,\dots,\tau_d}$ is actually the inverse of the copula transform; in particular, $\mathbf{q}_{1/2,\dots,1/2}$ yields the componentwise median. An empirical version of that mapping is readily obtained by considering the empirical marginal distributions of any observed n-tuple $\mathbf{Y}_1, \dots, \mathbf{Y}_n$.

This definition, however, does not provide an ordering of $\mathbb{R}^d$, but rather a d-tuple of (marginal) orderings. Moreover, $\mathbf{q}_{\tau_1,\dots,\tau_d}$ crucially depends on the coordinate system adopted, and is not even rotation-equivariant. Its empirical version, moreover, does not enjoy any of the properties that make univariate empirical quantiles a successful tool for statistical inference.

The concept being unsatisfactory in the location case, its regression extensions will not be examined.

12.2.1.2   Quantile biplots

Marginal quantiles actually are obtained by projecting P (equivalently, $\mathbf{Y}$) on d mutually orthogonal straight lines (characterized by the canonical orthonormal basis $(\mathbf{u}_1, \dots, \mathbf{u}_d)$)

through some origin. If the influence of the arbitrary choice of a basis is to be removed, one might also like to look at projections onto *all* unit vectors $\mathbf{u} \in \mathcal{S}^{d-1}$ through some given origin. This approach is investigated in Kong and Mizera (2008) and, in some detail, in Ahidar (2015, Section 2); see also Ahidar-Coutrix and Berthet (2016).

For each $\mathbf{u}$ and $\tau \in (0, 1)$, the univariate distribution of $\mathbf{u}'\mathbf{Y}$ yields a well-defined quantile of order τ ($q_{\tau\mathbf{u}}$, say). Define, for $\tau \in [1/2, 1)$, the *directional quantile* of order τ for direction $\mathbf{u}$ as the point $\mathbf{q}_{\tau\mathbf{u}} := q_{\tau\mathbf{u}}\mathbf{u}$. Substituting empirical quantiles $q_{\tau\mathbf{u}}^{(n)}$ for the theoretical ones yields empirical counterparts $\mathbf{q}_{\tau\mathbf{u}}^{(n)}$. The collection, for $\mathbf{u}$ ranging over the unit sphere $\mathcal{S}^{d-1}$ in $\mathbb{R}^d$, of all those directional quantiles yields what Kong and Mizera (2008) call a *quantile biplot*.

Intuitively appealing as it may be, however, this concept exhibits somewhat weird properties: quantile biplots are very sensitive to the (arbitrary) choice of an origin; they are neither translation- nor rotation-equivariant, and yield strange, often self-intersecting contours. Although the computation of each particular $\mathbf{q}_{\tau\mathbf{u}}^{(n)}$ is quite straightforward, the construction of empirical biplots, in principle, requires considering "infinitely many" directions $\mathbf{u}$. For all those reasons, the concept – which no longer appears in Kong and Mizera (2012) – will not be examined any further.

12.2.1.3   Directional quantile hyperplanes and contours

Instead of quantile biplots associated with the point-valued quantiles $q_{\tau\mathbf{u}}\mathbf{u}$, Kong and Mizera (2008, 2012) also suggest considering, for each direction $\mathbf{u}$ in $\mathcal{S}^{d-1}$ and each $\tau \in (0, 1/2)$, the *directional quantile hyperplane* $H_{\tau\mathbf{u}}$, with equation $\mathbf{u}'\mathbf{y} = q_{\tau\mathbf{u}}$.

Intuitively, that hyperplane is obtained by looking at the collection of all hyperplanes orthogonal to $\mathbf{u}$; the quantile hyperplane $H_{\tau\mathbf{u}}$ of order τ is the (uniquely defined, for an absolutely continuous distribution with nonvanishing density) hyperplane in that collection dividing $\mathbb{R}^d$ into halfspaces with $\mathbf{P}^{\mathbf{Y}}$-probabilities τ ("below" $H_{\tau\mathbf{u}}$) and $1 - \tau$ ("above" $H_{\tau\mathbf{u}}$), respectively.

Denote by $\mathcal{H}_{\tau\mathbf{u}}$ the halfspace lying above $H_{\tau\mathbf{u}}$. The intersection $\mathcal{H}(\tau) := \bigcap_{\mathbf{u} \in \mathcal{S}^{d-1}} \mathcal{H}_{\tau\mathbf{u}}$ of those halfspaces characterizes (for given $\tau \in (0, 1/2)$) an inner envelope. We propose the convenient terminology "quantile region" and "quantile contour" for those inner envelopes and their boundaries $H(\tau)$, which enjoy much nicer properties than the quantile biplots: quantile hyperplanes do not depend on any origin; quantile regions and contours are unique (population case) under Lebesgue-absolutely continuous distributions with connected support; they are convex and nested as τ increases, and affine-equivariant.

The index τ associated with a contour $H(\tau)$ or a region $\mathcal{H}(\tau)$ represents a "tangent probability mass;" indexation by "probability content" might be preferable, and, in view of nestedness, is quite possible – but there is no canonical relation between τ and the $\mathbf{P}^{\mathbf{Y}}$-probability of the quantile region $\mathcal{H}(\tau)$.

Empirical versions $H_{\tau\mathbf{u}}^{(n)}$, $\mathcal{H}^{(n)}(\tau)$, and $H^{(n)}(\tau)$, as usual, are obtained by replacing the distribution $\mathbf{P}^{\mathbf{Y}}$ of $\mathbf{Y}$ with the empirical measure associated with some observed n-tuple $\mathbf{Y}_1, \ldots, \mathbf{Y}_n$. However, the characterization of a given quantile contour $H^{(n)}(\tau)$ involves an infinite number of directions $\mathbf{u}$, which of course is not implementable. In order to overcome this, one can compute the N hyperplanes $H_{\tau\mathbf{u}_i}^{(n)}$ associated with a sample (random or systematic) of directions $\mathbf{u}_i$, $i = 1, \ldots, N$: for absolutely continuous $\mathbf{Y}_1, \ldots, \mathbf{Y}_n$, the resulting region $\mathcal{H}_N^{(n)}(\tau) := \bigcap_{i=1,\ldots,N} \mathcal{H}_{\tau\mathbf{u}_i}^{(n)}$ is an approximation of the actual quantile region $\mathcal{H}^{(n)}(\tau)$ (to which, under mild conditions, it converges as $N \to \infty$) – a "biased" one, though, since, with probability 1, $\mathcal{H}_N^{(n)}(\tau)$ strictly includes $\mathcal{H}^{(n)}(\tau)$ for all N.

12.2.1.4 Relation to halfspace depth

Kong and Mizera (2012) then establish a most interesting result that the quantile contours/regions thus defined (as envelopes) and the halfspace depth contours/regions, coincide (in the empirical case as well as in the population case).

Recall that the halfspace depth of a point $\mathbf{y}$ with respect to a probability distribution $P^{\mathbf{Y}}$ (Tukey, 1975) is the minimum, over all hyperplanes running through $\mathbf{y}$, of the $P^{\mathbf{Y}}$-probabilities of the halfspaces determined by those hyperplanes. The halfspace depth regions $\mathcal{D}(\delta)$ (halfspace depth contours $D(\delta)$) are the collections of points with halfspace depth larger than or equal to δ (for given δ); those regions are convex and nested as depth decreases. The empirical depth of $\mathbf{y}$ with respect to the n-tuple $\mathbf{Y}_1, \ldots, \mathbf{Y}_n$ is defined similarly, with the empirical distribution of the $\mathbf{Y}_i$ playing the role of $P^{\mathbf{Y}}$. The empirical halfspace depth contours $D^{(n)}(\delta)$ are polyhedrons, the facet hyperplanes of which typically run through d sample points.

An important byproduct of that result is the hint that only a finite number of directions do characterize a given empirical contour $H^{(n)}(\delta)$, namely, those directions that are orthogonal to the facets of $H^{(n)}(\delta)$. The definition adopted so far, which is related to the traditional univariate definition (12.1), does not readily provide a way to identify those directions, though. The directional Koenker–Bassett approach of Section 12.2.2, which extends the univariate L_1 definition (12.2), also leads to a numerical determination of the relevant directions.

In the presence of covariates ($p \geqslant 1$), the connection with halfspace depth does not help much, as most existing regression depth concepts, inspired by Rousseeuw and Hubert (1999), are limited to the single-output setting; a few exceptions can be found, however, in Mizera (2002) and Šiman (2011).

12.2.2 Directional Koenker–Bassett methods

12.2.2.1 Location case ($p = 0$)

Another directional approach is proposed in Hallin et al. (2010), based on a directional adaptation of the L_1 definition (12.2) rather than on Kong and Mizera's directional version of the traditional definition (12.1).

Instead of projecting $\mathbf{Y}$ on a direction $\mathbf{u} \in \mathcal{S}^{d-1}$, Hallin et al. (2010) propose minimizing the usual L_1 residual distance along a direction $\mathbf{u}$ ranging over $\mathcal{S}^{d-1}$: the usual Koenker–Bassett quantile hyperplane construction (Koenker and Bassett, 1978), with "vertical direction" $\mathbf{u}$. More precisely, denoting by $\boldsymbol{\Gamma}_{\mathbf{u}}$ an arbitrary $d \times (d-1)$ matrix of unit vectors such that $(\mathbf{u}, \boldsymbol{\Gamma}_{\mathbf{u}})$ constitutes an orthonormal basis of $\mathbb{R}^d$, decompose $\mathbf{Y}$ into $\mathbf{Y}_{\mathbf{u}} + \mathbf{Y}_{\mathbf{u}}^{\perp}$, where $\mathbf{Y}_{\mathbf{u}} := \mathbf{u}'\mathbf{Y}$ and $\mathbf{Y}_{\mathbf{u}}^{\perp} := \boldsymbol{\Gamma}_{\mathbf{u}}'\mathbf{Y}$. Hallin et al. (2010) define the directional τ-quantile hyperplane of $\mathbf{Y}$ (equivalently, of $P^{\mathbf{Y}}$) in the direction $\mathbf{u}$ as the hyperplane $\Pi_{\tau\mathbf{u}}$ with equation $\mathbf{u}'\mathbf{y} = \mathbf{b}_{\tau\mathbf{u}}'\boldsymbol{\Gamma}_{\mathbf{u}}'\mathbf{y} + a_{\tau\mathbf{u}}$, where ($\rho_\tau$ as usual stands for the τ-quantile *check function*)

$$(a_{\tau\mathbf{u}}, \mathbf{b}_{\tau\mathbf{u}}')' = \mathrm{argmin}_{(a,\mathbf{b}')' \in \mathbb{R}^d} \mathrm{E}[\rho_\tau(\mathbf{Y}_{\mathbf{u}} - \mathbf{b}'\mathbf{Y}_{\mathbf{u}}^{\perp} - a)]. \tag{12.5}$$

The empirical version $\Pi_{\tau\mathbf{u}}^{(n)}$ of $\Pi_{\tau\mathbf{u}}$, with equation $\mathbf{u}'\mathbf{y} = \mathbf{b}_{\tau\mathbf{u}}^{(n)'}\boldsymbol{\Gamma}_{\mathbf{u}}'\mathbf{y} + a_{\tau\mathbf{u}}^{(n)}$, is obtained by replacing the distribution $P^{\mathbf{Y}}$ of $\mathbf{Y}$ with the empirical measure associated with an observed n-tuple $\mathbf{Y}_1, \ldots, \mathbf{Y}_n$:

$$(a_{\tau\mathbf{u}}^{(n)}, \mathbf{b}_{\tau\mathbf{u}}^{(n)'})' = \mathrm{argmin}_{(a,\mathbf{b}')' \in \mathbb{R}^d} \sum_{i=1}^{n} \rho_\tau(\mathbf{Y}_{i,\mathbf{u}} - \mathbf{b}'\mathbf{Y}_{i,\mathbf{u}}^{\perp} - a). \tag{12.6}$$

For any fixed τ, the hyperplanes $\{\Pi_{\tau\mathbf{u}} : \mathbf{u} \in \mathcal{S}^{q-1}\}$ determine a quantile contour $R(\tau)$

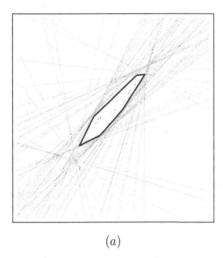 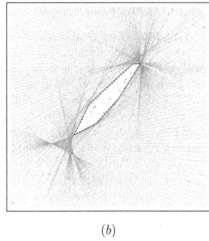

(a)   (b)

FIGURE 12.1
(a) All 0.2-quantile hyperplanes (thin gray lines) obtained, in a sample of $n = 49$ observations, from the directional Koenker–Bassett definition and the resulting directional 0.2-quantile contours (thick black lines). (b) Approximation (thin black lines) of the same directional 0.2-quantile contours (thick black dotted lines), based on Kong and Mizera's directional quantile hyperplanes (thin gray lines) corresponding to $N = 256$ equispaced $\mathbf{u}$ values on $\mathcal{S}^1$.

and a quantile region $\mathcal{R}(\tau)$ (for the empirical hyperplanes $\{\Pi_{\tau\mathbf{u}}^{(n)} : \mathbf{u} \in \mathcal{S}^{d-1}\}$, $R^{(n)}(\tau)$ and $\mathcal{R}^{(n)}(\tau)$). The hyperplanes constituting an empirical contour can be obtained as the solutions of a linear program parametrized by $\mathbf{u}$; see Hallin et al. (2010) for details, Paindaveine and Šiman (2012a,b) for further computational insights, and Boček and Šiman (2016) for software implementation. The linear programming structure of the problem implies that only a few critical values of $\mathbf{u}$ play a role. More precisely, the unit ball in $\mathbb{R}^d$ is partitioned into a finite number of cones with vertex at the origin, with all the $\mathbf{u}$ in a given cone determining the same quantile hyperplane $\Pi_{\tau\mathbf{u}}^{(n)}$. Those cones are obtained via standard parametric linear programming algorithms.

Hallin et al. (2010) show, moreover, that those quantile contours also coincide with the halfspace depth contours, hence with Kong and Mizera's directional quantile contours. The huge advantage with respect to the projection approach of Section 12.2.1.3 is that, thanks to the "analytical" nature of the L_1 definition, a given empirical contour here can be computed exactly in a finite number of steps. That advantage may disappear, however, as the size of the problem increases: when n and d become too large, linear programming algorithms eventually run into problems, and the approximate contours of Section 12.2.1.3 may be the only feasible solution. The two points of view, however, can be reconciled (see Paindaveine and Šiman, 2011).

Figure 12.1 shows (a) the empirical quantile contour of order $\tau = 0.2$, obtained from a data set of $n = 49$ observations and the directional Koenker–Bassett definition just described, and (b) the approximation of the same contour, based on the Kong and Mizera's directional quantile hyperplanes associated with $N = 256$ equispaced $\mathbf{u}$ values on $\mathcal{S}^1$.

The multivariate quantile contours resulting from this directional Koenker–Bassett approach inherit from their relation to halfspace depth the rich geometric features – convexity, connectedness, nestedness, affine equivariance – of the latter, while bringing to halfspace depth the nice analytical, computational, and probabilistic features of L_1 optimization –

tractable asymptotics (Bahadur representation, root-n consistency, asymptotic normality, etc.), L_1 characterization/optimality, implementable linear programming algorithms, and byproducts of optimization problems (duality and Lagrange multipliers); see Hallin et al. (2010) for explicit results and details.

12.2.2.2 (Nonparametric) regression case ($p \geqslant 1$)

A linear regression extension ($p \geqslant 1$), based on the L_1 approach, of the location concept described in the previous section is quite straightforward. Denote by $(\mathbf{X}_1', \mathbf{Y}_1')', \ldots, (\mathbf{X}_n', \mathbf{Y}_n')'$ an observed n-tuple of independent copies of $(\mathbf{X}', \mathbf{Y}')'$, where $\mathbf{Y} := (Y_1, \ldots, Y_d)'$ is a d-dimensional response and the random vector $\mathbf{X} := (X_1, \ldots, X_p)'$ is a p-tuple of covariates. Definitions (12.5) and (12.6) readily generalize into

$$(a_{\tau\mathbf{u}}, \mathbf{b}_{\tau\mathbf{u}}', \boldsymbol{\beta}_{\tau\mathbf{u}}')' = \operatorname{argmin}_{(a, \mathbf{b}', \boldsymbol{\beta}')' \in \mathbb{R}^{d+p}} \mathrm{E}[\rho_\tau(\mathbf{Y}_\mathbf{u} - \mathbf{b}'\mathbf{Y}_\mathbf{u}^\perp - \boldsymbol{\beta}'\mathbf{X} - a)] \tag{12.7}$$

and

$$(a_{\tau\mathbf{u}}^{(n)}, \mathbf{b}_{\tau\mathbf{u}}^{(n)\prime}, \boldsymbol{\beta}_{\tau\mathbf{u}}^{(n)\prime})' = \operatorname{argmin}_{(a, \mathbf{b}', \boldsymbol{\beta}')' \in \mathbb{R}^{d+p}} \sum_{i=1}^{n} \rho_\tau(\mathbf{Y}_{i,\mathbf{u}} - \mathbf{b}'\mathbf{Y}_{i,\mathbf{u}}^\perp - \boldsymbol{\beta}'\mathbf{X}_i - a), \tag{12.8}$$

characterizing hyperplanes (now in $\mathbb{R}^{d+p}$) $\Pi_{\tau\mathbf{u}}$, with equation

$$\mathbf{u}'\mathbf{y} = \mathbf{b}_{\tau\mathbf{u}}'\boldsymbol{\Gamma}_\mathbf{u}'\mathbf{y} + \boldsymbol{\beta}_{\tau\mathbf{u}}'\mathbf{x} + a_{\tau\mathbf{u}},$$

and empirical hyperplanes $\Pi_{\tau\mathbf{u}}^{(n)}$, with equation

$$\mathbf{u}'\mathbf{y} = \mathbf{b}_{\tau\mathbf{u}}^{(n)\prime}\boldsymbol{\Gamma}_\mathbf{u}'\mathbf{y} + \boldsymbol{\beta}_{\tau\mathbf{u}}^{(n)\prime}\mathbf{x} + a_{\tau\mathbf{u}}^{(n)},$$

respectively.

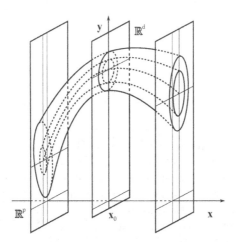

FIGURE 12.2
Directional Koenker–Bassett quantile regression: population quantile regression tubes ($d = 2$, $p = 1$).

To the best of our knowledge, the asymptotic properties of such extensions have not been worked out in the literature, although asymptotic results of the same type as those obtained for the location case certainly can be established under appropriate assumptions. The major problem is the interpretation of the contours characterized by the collection of hyperplanes $\Pi_{\tau\mathbf{u}}$ as $\mathbf{u}$ ranges over the unit sphere $\mathcal{S}^{d-1}$ in $\mathbb{R}^d$. The relevant quantile hyperplanes, quantile/depth regions and contours of interest are the location quantile hyperplanes, quantile/depth regions and contours associated with the d-dimensional distributions of $\mathbf{Y}$ conditional on $\mathbf{X}$ – namely, the collection, for $\mathbf{x}$ ranging over $\mathbb{R}^p$, of the hyperplanes, regions and contours associated with the distributions $\mathrm{P}^{\mathbf{Y}|\mathbf{X}=\mathbf{x}}$ of $\mathbf{Y}$ conditional on $\mathbf{X} = \mathbf{x}$. When plotted against $\mathbf{x}$ (which is possible for $d + p \leqslant 3$ only), those contours yield quantile regression "tubes" (see Figure 12.2). Unless very severe restrictions[1] are put on the data-generating process, the contours resulting from (12.7) and (12.8) are not the depth contours associated with the conditional distributions $\mathrm{P}^{\mathbf{Y}|\mathbf{X}=\mathbf{x}}$, but some averaged version of the latter; their interpretation is thus somewhat problematic. This is the reason why Hallin et al. (2015) consider a fully general nonparametric regression setup (rather than linear regression and definitions (12.7) and (12.8)) with the objective of reconstructing the collection of *conditional* (on the value $\mathbf{X} = \mathbf{x}$) quantile contours of $\mathbf{Y}$.

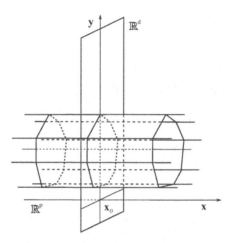

FIGURE 12.3
Directional Koenker–Bassett quantile regression: local constant empirical quantile regression tube at $\mathbf{x}_0$.

Two consistent estimation methods are provided in Hallin et al. (2015): a local constant method and a local bilinear one. In both cases, the estimators are based on a weighted (kernel-based) version of the location contour estimators, computationally still leading to a parametrized linear programming problem with directional parameter $\mathbf{u}$ ranging over the unit sphere $\mathcal{S}^{d-1}$. Bahadur representations of the resulting estimators are established under appropriate technical conditions on the joint distributions of $(\mathbf{X}', \mathbf{Y}')'$, the kernel and the bandwidth defining the weights; those representations entail consistency and asymptotic normality.

Local constant quantile contours yield, for given τ and a selected value $\mathbf{x}_0$ of $\mathbf{x}$, a "horizontal polygonal tube" (Figure 12.3) in $\mathbb{R}^{d+p}$, the interpretation of which is valid at $\mathbf{x}_0$ only, and provides no information on the way the conditional distribution of $\mathbf{Y}$ is varying in the neighborhood of $\mathbf{x}_0$.

---

[1] For instance, requiring that the distribution of $\mathbf{Y} - \mathbf{B}\mathbf{x}$ (for $\mathbf{B}$ some $d \times p$ regression matrix), conditional on $\mathbf{X} = \mathbf{x}$, does not depend on $\mathbf{x}$.

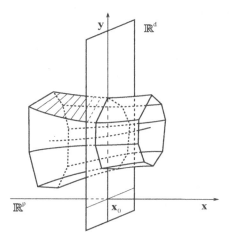

FIGURE 12.4
Directional Koenker–Bassett quantile regression: local bilinear empirical quantile regression tube at $\mathbf{x}_0$.

Local bilinear quantile contours are more informative, since they incorporate information on the derivatives with respect to $\mathbf{x}$ of the coefficients of the conditional quantile hyperplanes; they should also be more reliable at boundary points. The price to be paid is an increase of the number of free parameters involved. Note, however, that the smoothing features of the problem, namely the dimension p of kernels, remains unaffected, irrespective of d). The resulting empirical tubes, as shown in Figure 12.4, are no longer polygonal cylinders, but piecewise ruled quadrics, and a local (in the vicinity of $\mathbf{x}_0$) interpretation as "quantile regression tubes".

Figure 12.5 shows the empirical contours ($\tau = 0.2$ and 0.4) constructed, via (a) the local constant method and (b) the local bilinear method, for various values of x_0, for a set of $n = 4899$ observations ($d = 2$, $p = 1$) simulated from the bivariate heteroskedastic regression model

$$(Y_1, Y_2)' = (X, X^2)' + 0.5\big(1 + 3|\sin(\pi X/2)|\big)(e_1, e_2)', \qquad (12.9)$$

where X is uniform over $(-2, 2)$ and $(e_1, e_2)' \sim \mathcal{N}(\mathbf{0}, \mathbf{I})$ is bivariate normal. The axes are those of the response space, and (unlike in Figure 12.4) the contours associated with various values of X are shown side by side. The "parabolic regression median" and the periodic conditional scale are well picked up by both methods. The local bilinear contours are less sensitive, as expected, to boundary effects.

We refer the reader to Hallin et al. (2015) for details.

12.3 Direct approaches

All multiple-output quantile regression concepts presented so far were based on directional extensions of the usual single-output ones. Direct approaches are possible, though, along three main lines. The first is based on a spatial extension of the definition of the check function, leading to "spatial" quantiles (also called "geometric" quantiles). The second

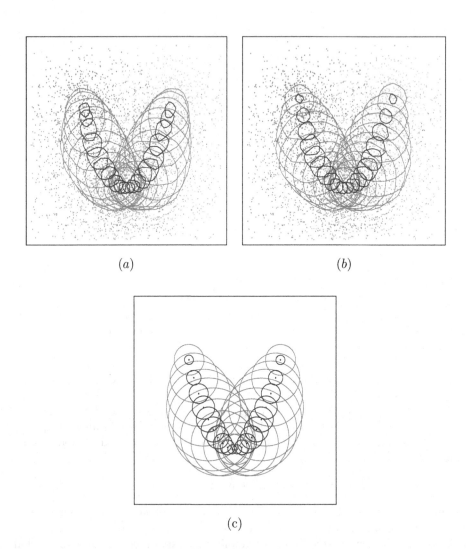

(a)   (b)

(c)

FIGURE 12.5
Directional Koenker–Bassett quantile regression: the empirical contours ($\tau = 0.2$ and 0.4)
obtained, for various values of x_0, via (a) the local constant method and (b) the local bilinear
method, for $n = 4899$ observations from model (12.9), along with (c) their (exact) population
counterparts; the dot at the center of the population contours is both the conditional mean
and the conditionally deepest point.

involves substituting, in the traditional Koenker and Bassett (1978) definition, ellipsoids for hyperplanes, and the "above/below" indicators with "outside/inside" ones. The third approach, inspired by the relation between directional quantiles and halfspace depth (Section 12.2.1.4), is based on recent measure transportation-related concepts (Carlier et al., 2016; Chernozhukov et al., 2017) of Monge–Kantorovich depth and quantiles.

12.3.1 Spatial (geometric) quantile methods

The concepts of *spatial* or *geometric* quantiles, generalizing the well-known spatial median (Haldane, 1948; Gower, 1974; Brown, 1983), are based on a multivariate extension of the L_1 approach outlined in Section 12.1; see Chaudhuri (1996), Koltchinskii (1997), and the references therein.

12.3.1.1 A spatial check function

The concepts of spatial median and quantiles are based on an alternative form for the check function that "naturally" extends to a multivariate context, possibly combined with transformation–retransformation ideas.

It is easy to see that the check function ρ_τ, in univariate or single-output regression settings, can be rewritten as

$$\begin{aligned}\rho_\tau(z) \quad &:= \quad (1-\tau)|z|I_{[z<0]} + \tau|z|I_{[z\geqslant 0]} \\ &= \quad \frac{1}{2}(|z| + vz) =: \frac{1}{2}\Phi_v(z)\end{aligned}$$

with $v := 2\tau - 1$. While τ ranges over $(0,1)$, v ranges over the open unit ball $(-1,1)$ in $\mathbb{R}$. Substituting Φ_v for ρ_τ in (12.2), (12.3), and (12.4) thus leads to the same concepts as in Section 12.1, with, however, a different index v. That index v has a centre-outward directional flavour, with $|v|$ measuring centrality and $v/|v|$ characterizing a direction (either $+1$ or -1).

It is tempting, therefore, to extend the univariate and single-output regression concepts (12.2)–(12.4) to the multivariate and multiple-output context by minimizing a criterion based on the "spatial check function"

$$\Phi_{\mathbf{v}}(\mathbf{z}) := \|\mathbf{z}\| + \mathbf{v}'\mathbf{z}, \quad \mathbf{z} \in \mathbb{R}^d, \tag{12.10}$$

where $\mathbf{v}$ ranges over the open unit ball in $\mathbb{R}^d$, hence takes the form $\mathbf{v} = \tau\mathbf{u}$, where $\tau = \|\mathbf{v}\|$ and $\mathbf{u} = \mathbf{v}/\|\mathbf{v}\| \in \mathcal{S}^{d-1}$.

This, in the location case ($p = 0$, no covariates), yields the population spatial quantile of order $\mathbf{v}$ and its empirical counterpart,

$$\mathbf{Q_v} := \operatorname{argmin}_{\mathbf{q}\in\mathbb{R}^d}\mathrm{E}[\Phi_{\mathbf{v}}(\mathbf{Y} - \mathbf{q})], \quad \mathbf{Q}_{\mathbf{v}}^{(n)} := \operatorname{argmin}_{\mathbf{q}\in\mathbb{R}^d}\sum_{i=1}^{n}[\Phi_{\mathbf{v}}(\mathbf{Y}_i - \mathbf{q})], \tag{12.11}$$

respectively. For $\tau = 0$ (hence $\mathbf{v} = \mathbf{0}$), one obtains the traditional *spatial median*. Although an intuitive justification of (12.10) is not straightforward, the solutions of (12.11) are such that

$$\mathrm{E}\left[(\mathbf{Y} - \mathbf{Q_v})/\|\mathbf{Y} - \mathbf{Q_v}\|\right] = -\mathbf{v} \quad \text{and} \quad \frac{1}{n}\sum_{i=1}^{n}(\mathbf{Y}_i - \mathbf{Q}_{\mathbf{v}}^{(n)})/\|(\mathbf{Y}_i - \mathbf{Q}_{\mathbf{v}}^{(n)})\| = -\mathbf{v},$$

which provides an interpretation in terms of the unit vectors originating in $\mathbf{Q_v}$ or $\mathbf{Q}_{\mathbf{v}}^{(n)}$ and pointing at $\mathbf{Y}$ or the observations $\mathbf{Y}_i$. An interesting discussion of this can be found in Serfling (2002b).

Finally, spatial quantiles are also quite robust. They characterize the underlying distribution, and their definition nicely extends (Chakraborty and Chaudhuri, 2014) to Banach spaces.

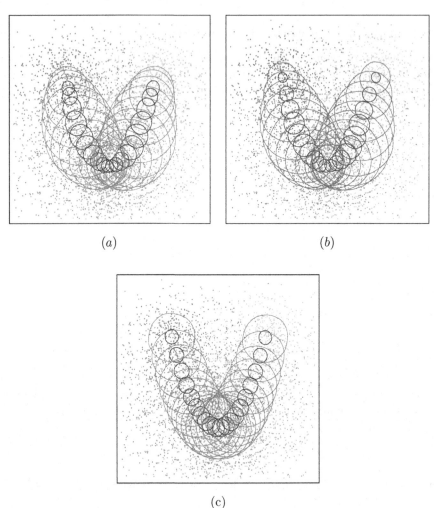

(a) (b)

(c)

FIGURE 12.6
Spatial quantile regression: empirical contours obtained, for various values of τ and x_0, via (a) the (nonparametric) local constant method, (b) the (nonparametric) local linear method, and (c) the (parametric) linear regression model with regressors X and X^2, for the same $n = 4899$ observations as in Figure 12.5.

12.3.1.2 Linear spatial quantile regression

In the (multiple-output) linear regression case, the observations $(\mathbf{Y}_i, \mathbf{X}_i)$ satisfy an equation of the form

$$\mathbf{Y}_i = \mathbf{a} + \mathbf{B}\mathbf{X}_i + \mathbf{E}_i, \quad i = 1, \ldots, n, \tag{12.12}$$

where $\mathbf{a}$ and $\mathbf{B}$ are $d \times 1$ and $d \times p$, respectively. The definitions in (12.11) straightforwardly generalize into

$$(\mathbf{a}_{\tau \mathbf{u}}, \mathbf{B}_{\tau \mathbf{u}}) := \operatorname{argmin}_{(\mathbf{a}, \mathbf{B}) \in \mathbb{R}^d \times \mathbb{R}^{d \times p}} \mathrm{E}[\Phi_{\tau \mathbf{u}}(\mathbf{Y} - \mathbf{a} - \mathbf{B} \mathbf{X})] \qquad (12.13)$$

and

$$(\mathbf{a}_{\tau \mathbf{u}}^{(n)}, \mathbf{B}_{\tau \mathbf{u}}^{(n)}) := \operatorname{argmin}_{(\mathbf{a}, \mathbf{B}) \in \mathbb{R}^d \times \mathbb{R}^{d \times p}} \sum_{i=1}^{n} \Phi_{\tau \mathbf{u}}(\mathbf{Y}_i - \mathbf{a} - \mathbf{B} \mathbf{X}_i), \qquad (12.14)$$

with $\mathbf{v} = \tau \mathbf{u}$ ranging over the open unit ball.

Explicit values of individual (fixed τ and $\mathbf{u}$) empirical spatial regression quantiles are easily obtained via convex optimization techniques, but the corresponding contours (viz., the envelopes of the $\tau \mathbf{u}$-quantile regression hyperplanes associated with fixed τ as $\mathbf{u}$ ranges over the unit sphere $\mathcal{S}^{d-1}$) can only be approximated by considering a large number of $\mathbf{u}$ values. A numerical illustration of the method, based on the same simulated data set as in Figure 12.5, is provided in Figure 12.6(c).

The spatial regression quantiles partially order the regression hyperplanes. On the other hand, they have no obvious probabilistic interpretation, and perform quite poorly for distributions with nonconvex level sets (see Figure 12.8(b)). They also fail to be fully affine-equivariant – a feature that can be taken care of via the transformation-retransformation technique advocated by Chakraborty and Chaudhuri (1996); see Chaudhuri (1996), Koltchinskii (1997), and Chakraborty (2003) for details, and Bai et al. (1990) for the special case of median regression.

12.3.1.3 Nonparametric spatial quantile regression

Taking expectations in (12.11) and (12.13) with respect to a Nadaraya–Watson estimate of the conditional distribution function of $\mathrm{P}^{\mathbf{Y}|\mathbf{X}=\mathbf{x}}$, one can obtain, similarly as in Section 12.2.2.2, locally constant and locally linear (nonparametric) estimators of the conditional spatial quantiles. Such methods and their properties are investigated in Cheng and de Gooijer (2007) and Chaouch et al. (2009), where we refer the reader to for details. An illustration is provided in Figure 12.6(a) and (b), for the same data as in Figure 12.5; the results look quite similar to those obtained from nonparametric directional quantile regression methods.

12.3.2 Elliptical quantiles

A concept of elliptical regression quantiles was proposed, very much in the same spirit as Koenker and Bassett's original definition, by Hlubinka and Šiman (2013, 2015) and Hallin and Šiman (2016). As in Section 12.2, we start with the pure location case ($p = 0$, no covariates) before turning to the general linear regression case.

12.3.2.1 Location case

The basic idea behind the concept is intuitively quite simple and appealing: instead of the inner envelope of a collection of directional Koenker–Bassett τ-quantile hyperplanes minimizing the expected value of a directional check function penalty (Section 12.2.2.1), consider an ellipsoid minimizing the same weighted L_1 objective function where, however, "above/below" the hyperplane is replaced with "outside/inside" the ellipsoid. This leads to the definition of a multivariate (location) elliptical τ-quantile as the ellipsoid

$$\mathcal{E}_{\tau}^{\mathrm{loc}} = \mathcal{E}_{\tau}^{\mathrm{loc}}(\mathbf{Y}) := \{\mathbf{y} \in \mathbb{R}^d : \mathbf{y}' \mathbf{A}_\tau \mathbf{y} + \mathbf{y}' \mathbf{b}_\tau - c_\tau = 0\},$$

where $\mathbf{A}_\tau \in \mathbb{R}^{d \times d}$, $\mathbf{b}_\tau \in \mathbb{R}^{d \times 1}$, and $c_\tau \geqslant 0$ minimize, subject to $\mathbf{A}$ being symmetric and positive semidefinite with determinant 1 – a *shape matrix* in the sense of Paindaveine (2008) – the objective function

$$\Psi_\tau^{loc}(\mathbf{A}, \mathbf{b}, c) := \mathrm{E}\rho_\tau(\mathbf{Y}'\mathbf{A}\mathbf{Y} + \mathbf{Y}'\mathbf{b} - c), \tag{12.15}$$

ρ_τ, as usual, stands for the check function

$$z \mapsto \rho_\tau(z) := z(\tau - I(z < 0)) = \max\{(\tau - 1)z, \tau z\}.$$

Note that the argument of ρ_τ in (12.15) is positive or negative according to whether $\mathbf{Y}$ takes value inside or outside the ellipsoid with equation $\mathbf{y}'\mathbf{A}\mathbf{y} + \mathbf{y}'\mathbf{b} = c$.

The positive semidefiniteness of $\mathbf{A}_\tau$ and the condition on its determinant ensure that $\mathcal{E}_\tau^{loc}$ is indeed an ellipsoid, centered at $\mathbf{s}_\tau := -\mathbf{A}_\tau^{-1}\mathbf{b}_\tau/2$, with equation $(\mathbf{y} - \mathbf{s}_\tau)'\mathbf{A}_\tau(\mathbf{y} - \mathbf{s}_\tau) = \kappa_\tau$, where $\kappa_\tau := c_\tau + \mathbf{b}_\tau'\mathbf{A}_\tau^{-1}\mathbf{b}_\tau/4$. The condition $\det(\mathbf{A}) = 1$ can be viewed as an identification constraint: for any $K > 0$, the triples $(\mathbf{A}, \mathbf{b}, c)$ and $(K\mathbf{A}, K\mathbf{b}, Kc)$ indeed define the same ellipsoid.

This definition certainly does not characterize an elliptical quantile as the solution of a linear programming problem; nor, as it stands, does it take the form of a convex optimization problem. The same concept, however, can be characterized as the unique solution of a convex optimization problem by relaxing the constraint $\det(\mathbf{A}) = 1$ into $(\det(\mathbf{A}))^{1/d} \geqslant 1$: unlike $\mathbf{A} \mapsto \det(\mathbf{A})$, the function $\mathbf{A} \mapsto (\det(\mathbf{A}))^{1/d}$ is concave on the cone of symmetric positive semidefinite matrices, and it can be shown that this convex optimization problem and the original nonconvex one share the same solution. That solution, moreover, is unique under absolutely continuous distributions with nonvanishing densities and finite moments of order 2.

12.3.2.2   Linear regression case

In the presence of covariates ($p \geqslant 1$), the traditional homoskedastic multiple-output linear regression model suggests, for an elliptical multiple-output regression quantile of order τ, a simple equation of the form

$$(\mathbf{y} - \boldsymbol{\beta}_\tau - \mathbf{B}_\tau\mathbf{x})'\mathbf{A}_\tau(\mathbf{y} - \boldsymbol{\beta}_\tau - \mathbf{B}_\tau\mathbf{x}) - \gamma_\tau = 0$$

with some $\mathbf{A}_\tau \in \mathbb{R}^{d \times d}$, $\boldsymbol{\beta}_\tau \in \mathbb{R}^{d \times 1}$, $\mathbf{B}_\tau \in \mathbb{R}^{d \times p}$, and $\gamma_\tau \geqslant 0$. The trouble is that the corresponding objective function

$$\mathrm{E}\rho_\tau((\mathbf{Y} - \boldsymbol{\beta} - \mathbf{B}\mathbf{X})'\mathbf{A}(\mathbf{Y} - \boldsymbol{\beta} - \mathbf{B}\mathbf{X}) - \gamma)$$

is not convex in $\boldsymbol{\beta}$ and $\mathbf{B}$, so that its minimization with respect to $\mathbf{A}$, $\boldsymbol{\beta}$, $\mathbf{B}$, and γ is not a *convex* optimization problem (see Hlubinka and Šiman, 2015).

In order to restore convexity, Hallin and Šiman (2016) consider instead the more general definition

$$\mathcal{E}_\tau^{reg} := \{(\mathbf{y}', \mathbf{x}')' \in \mathbb{R}^{d+p} : \tag{12.16}$$
$$(\mathbf{y} - \boldsymbol{\beta}_\tau - \mathbf{B}_\tau\mathbf{x})'\mathbf{A}_\tau(\mathbf{y} - \boldsymbol{\beta}_\tau - \mathbf{B}_\tau\mathbf{x}) - (\gamma_\tau + \mathbf{c}_\tau'\mathbf{x} + \mathbf{x}'\mathbf{C}_\tau\mathbf{x}) = 0\}$$

of an elliptical regression quantile $\mathcal{E}_\tau^{reg} = \mathcal{E}_\tau^{reg}(\mathbf{Y}, \mathbf{X})$, where a quadratic form of covariate-driven scale is allowed, and $\mathbf{A}_\tau, \boldsymbol{\beta}_\tau, \mathbf{B}_\tau, \gamma_\tau, \mathbf{c}_\tau$, and $\mathbf{C}_\tau$ jointly minimize

$$\Psi_\tau^{reg} := \mathrm{E}\rho_\tau((\mathbf{Y} - \boldsymbol{\beta} - \mathbf{B}\mathbf{X})'\mathbf{A}(\mathbf{Y} - \boldsymbol{\beta} - \mathbf{B}\mathbf{X}) - (\gamma + \mathbf{c}'\mathbf{X} + \mathbf{X}'\mathbf{C}\mathbf{X})) \tag{12.17}$$

under the constraint that $\mathbf{C} \in \mathbb{R}^{p \times p}$ is symmetric and $\mathbf{A} \in \mathbb{R}^{d \times d}$ is symmetric positive semidefinite with $\det(\mathbf{A}) = 1$.

This minimization, again, does not take the form of a convex optimization problem. Let, therefore, $\mathbf{M} := (\mathbf{M}^1, \ldots, \mathbf{M}^6)$, with

$$
\begin{aligned}
\mathbf{M}^1 &:= \mathbf{A} \in \mathbb{R}^{d \times d} \text{ symmetric positive semidefinite,} \\
\mathbf{M}^2 &:= \mathbf{B}'\mathbf{A}\mathbf{B} - \mathbf{C} \in \mathbb{R}^{p \times p} \text{ symmetric,} \\
\mathbf{M}^3 &:= -2\mathbf{B}'\mathbf{A} \in \mathbb{R}^{p \times d}, \\
\mathbf{M}^4 &:= -2\boldsymbol{\beta}'\mathbf{A} \in \mathbb{R}^{1 \times d}, \\
\mathbf{M}^5 &:= 2\boldsymbol{\beta}'\mathbf{A}\mathbf{B} - \mathbf{c}' \in \mathbb{R}^{1 \times p}, \\
\mathbf{M}^6 &:= \boldsymbol{\beta}'\mathbf{A}\boldsymbol{\beta} - \gamma \in \mathbb{R}.
\end{aligned}
$$

The correspondence between $\mathbf{M}$ and $(\mathbf{A}, \boldsymbol{\beta}, \mathbf{B}, \gamma, \mathbf{c}, \mathbf{C})$ is one-to-one, and $\mathbf{M}$ thus provides a reparametrization of the problem.

In this new parametrization, the elliptical regression quantile $\mathcal{E}_\tau^{\mathrm{reg}}$ can be expressed as

$$
\mathcal{E}_\tau^{\mathrm{reg}} = \{(\mathbf{y}', \mathbf{x}')' \in \mathbb{R}^{d+p} : r(\mathbf{y}, \mathbf{x}, \mathbf{M}_\tau) = 0\},
$$

where

$$
\begin{aligned}
r(\mathbf{y}, \mathbf{x}, \mathbf{M}) &:= \mathbf{y}'\mathbf{M}^1\mathbf{y} + \mathbf{x}'\mathbf{M}^2\mathbf{x} + \mathbf{x}'\mathbf{M}^3\mathbf{y} + \mathbf{M}^4\mathbf{y} + \mathbf{M}^5\mathbf{x} + \mathbf{M}^6 \\
&= (\mathbf{y} - \boldsymbol{\beta} - \mathbf{B}\mathbf{x})'\mathbf{A}(\mathbf{y} - \boldsymbol{\beta} - \mathbf{B}\mathbf{x}) - (\gamma + \mathbf{c}'\mathbf{x} + \mathbf{x}'\mathbf{C}\mathbf{x})
\end{aligned}
$$

(r is thus positive outside, and negative inside, the ellipsoid with equation $r = 0$), and $\mathbf{M}_\tau := (\mathbf{M}_\tau^1, \ldots, \mathbf{M}_\tau^6)$ jointly minimize

$$
\Psi_\tau^{\mathrm{reg}} = \Psi_\tau^{\mathrm{reg}}(\mathbf{M}) := \Psi_\tau^{\mathrm{reg}}(\mathbf{M}^1, \ldots, \mathbf{M}^6) = \mathrm{E}\rho_\tau\big(r(\mathbf{Y}, \mathbf{X}, \mathbf{M})\big),
$$

subject to $(\det(\mathbf{M}^1))^{1/d} \geqslant 1$; as in the location case, positive homogeneity of $\Psi_\tau^{\mathrm{reg}}(\mathbf{M}^1, \ldots, \mathbf{M}^6)$ implies $\det(\mathbf{M}_\tau) = 1$. The considerable advantage of this parametrization in terms of $\mathbf{M}$ is that it leads to a *convex* optimization problem, hence to a *unique* minimum under the assumptions made (which include the existence of finite second-order moments).

The (Karush–)Kuhn–Tucker necessary and sufficient conditions characterizing the solution imply, in particular, that the probability content of $\mathcal{E}_\tau^{\mathrm{reg}}$ is τ, and that

$$
\mathrm{E}[(\mathbf{Y}', \mathbf{X}')'|r \geqslant 0] = \mathrm{E}[(\mathbf{Y}', \mathbf{X}')'|r < 0],
$$

so that the probability mass centers of the interior and the exterior of $\mathcal{E}_\tau^{\mathrm{reg}}$ coincide. It is easy to see, moreover, that the elliptical regression quantiles $\mathcal{E}_\tau^{\mathrm{reg}}$ are both regression-equivariant and fully affine-equivariant.

In the sample case with n observations $(\mathbf{Y}_i', \mathbf{X}_i')'$, $i = 1, \ldots, n$, empirical versions $\mathcal{E}_{\tau;n}^{\mathrm{reg}}$ of the elliptical regression quantiles $\mathcal{E}_\tau^{\mathrm{reg}}$ are based on the empirical counterparts of (12.17). Classical results (such as van der Vaart and Wellner, 1996, Theorem 5.14) then guarantee basic convergence, as $n \to \infty$, of the vector

$$
\mathbf{m}_{\tau;n} := \big(\mathrm{vec}(\mathbf{M}_{\tau;n}^1)', \mathrm{vec}(\mathbf{M}_{\tau;n}^2)', \mathrm{vec}(\mathbf{M}_{\tau;n}^3)', \mathbf{M}_{\tau;n}^4, \mathbf{M}_{\tau;n}^5, \mathbf{M}_{\tau;n}^6\big)'
$$

of coefficients of the sample elliptical regression quantile to its (uniquely defined) population counterpart

$$
\mathbf{m}_\tau := \big(\mathrm{vec}(\mathbf{M}_\tau^1)', \mathrm{vec}(\mathbf{M}_\tau^2)', \mathrm{vec}(\mathbf{M}_\tau^3)', \mathbf{M}_\tau^4, \mathbf{M}_\tau^5, \mathbf{M}_\tau^6\big)'.
$$

Figure 12.7 illustrates the ability of elliptical regression quantiles to correctly estimate the trend and heteroskedasticity in linear regression models.

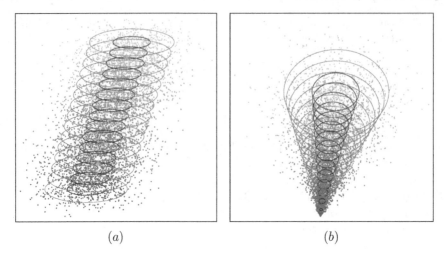

(a) (b)

FIGURE 12.7
Elliptical linear quantile regression: the empirical contours estimated, for equidistant regressor values and two quantile levels, from (a) a homoskedastic, and (b) a heteroskedastic regression model with linear trend.

12.3.3   Depth-based quantiles

As mentioned in the introduction, depth contours (preferably indexed by their probability content) naturally provide a plausible concept of quantile contours. Many depth concepts are available in the literature, and we will restrict ourselves to two important cases: halfspace depth, the relation of which to directional quantiles has been outlined in Section 12.2; and the more recent concept of Monge–Kantorovich depth.

12.3.3.1   Halfspace depth quantiles

Halfspace depth contours, as we have seen, coincide with the directional quantile contours of Section 12.2. This, from many points of view, is a very appealing property. However, it also has some less attractive consequences, which mainly originate in the linear foundations of both concepts – namely, the very special role of hyperplanes in their definition. Among those disturbing consequences are the affine invariance (equivariance) and convexity of the depth/quantile contours. For $d \geqslant 2$, those features indeed do not resist any nonlinear transformation, even the continuous monotone increasing marginal ones. This is incompatible with one of the core properties of univariate quantiles: equivariance under *order-preserving transformations* – namely, the fact that the quantile of order τ of a continuous monotone increasing transformation $T(Y)$ of Y is the value $T(q_\tau^Y)$ of the same transformation computed at q_τ^Y, the quantile of order τ of Y.

Convexity, moreover, leads to quite unnatural quantile contours, for example, for distributions with non-convex level sets. Figure 12.8(a) exhibits some empirical halfspace depth contours for a sample from a "banana-shaped" distribution. As quantile contours, they clearly cannot account for the banana shape of the distribution, and the deepest point (playing, in the quantile terminology, the role of a median) is not really central to the sample.

Those drawbacks of halfspace depth, hence of the directional quantiles described in Section 12.2, also extend to most other concepts of statistical depth; spatial quantiles similarly do a pretty poor job (see Figure 12.8(b)). These problems were the main motivation behind

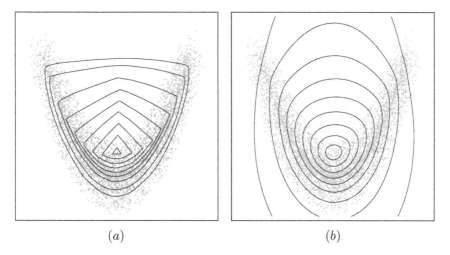

(a)   (b)

FIGURE 12.8
(a) Some empirical halfspace depth contours and (b) some spatial quantile contours obtained from 4899 simulated observations from a "banana-shaped" distribution.

the concept of Monge–Kantorovich depth, proposed by Chernozhukov et al. (2017) and based on measure transportation ideas, which we now briefly describe.

12.3.3.2   Monge–Kantorovich quantiles

The simplest and most intuitive formulation of the measure transportation problem is as follows. Let P_1 and P_2 be two probability measures over (for simplicity) $(\mathbb{R}^d, \mathcal{B}^d)$. Denote by $L : \mathbb{R}^{2d} \to [0, \infty]$ a Borel-measurable loss function such that $L(\mathbf{x}_1, \mathbf{x}_2)$ represents the cost of transporting $\mathbf{x}_1$ to $\mathbf{x}_2$. Monge's formulation of the optimal transportation problem is: find a measurable transport map $T_{P_1;P_2} : \mathbb{R}^d \to \mathbb{R}^d$ achieving the infimum

$$\inf_T \int_{\mathbb{R}^d} L(\mathbf{x}, T(\mathbf{x})) \mathrm{d}P_1, \quad T \text{ subject to } T * P_1 = P_2,$$

where $T * P_1$ denotes the "push-forward of P_1 by T" (more classical statistical notation for this constraint would be $P_1^{T\mathbf{X}} = P_2$). A map $T_{P_1;P_2}$ that attains this infimum is called an "optimal transport map," or an "optimal transport" for short, mapping P_1 to P_2.

In the sequel, we restrict ourselves to the L^2 loss function $L(\mathbf{x}_1, \mathbf{x}_2) = \|\mathbf{x}_1 - \mathbf{x}_2\|^2$. The results obtained by Kantorovich imply that, for that L^2 loss, if P_1 and P_2 are absolutely continuous with finite second-order moments, the solution exists, is (almost everywhere) unique, and the gradient of a convex (potential) function – a form of multivariate monotonicity. That type of result is further enhanced in a remarkable theorem by McCann (1995), itself extending a famous result by Brenier (1991), which implies that for any given (absolutely continuous) P_1 and P_2, there exists a P_1-essentially unique element in the class of gradients of convex functions mapping P_1 to P_2; under the existence of finite moments of order 2, moreover, that mapping coincides with the L^2-optimal transport of P_1 to P_2.

In dimension 1, halfspace depth contours are pairs of points of the form

$$\{F^{-1}(\tau), F^{-1}(1 - \tau)\}, \quad \tau \in (0, 1/2];$$

equivalently, letting $F_\pm := 2F - 1$, they are the $F_\pm$-inverse images of the one-dimensional spheres $\{2\tau - 1, 1 - 2\tau\}$, $\tau \in (0, 1/2]$, that is, the pairs

$$F_{\pm}^{-1}\left(\{-t, t\}\right), \quad t \in (0, 1).$$

The function $F_{\pm}$ maps $\mathbb{R}$ to the open one-dimensional unit ball $(-1, 1)$ and P to the uniform distribution over the unit ball; it is monotone increasing, hence the gradient (derivative) of a convex function. It thus follows from McCann's theorem that $F_{\pm}$ is the unique gradient of a convex function mapping P to the uniform distribution over the unit ball. Summing up, in dimension 1, halfspace depth contours are the images, by $F_{\pm}^{-1}$, where $F_{\pm}$ is the unique gradient of a convex function mapping P to the uniform distribution over the unit ball, of the spheres $\{\mathbf{t}|\ \|\mathbf{t}\| = \tau\}$ with probability content τ, $\tau \in (0, 1)$.

Turning to dimension d, define $F_{\pm}$ (from $\mathbb{R}^d$ to the open d-dimensional unit ball) as the unique gradient of a convex function mapping P to the uniform distribution over the unit ball;[2] that such an $F_{\pm}$ exists follows from McCann's theorem. The inverse $F_{\pm}^{-1}$ of $F_{\pm}$ qualifies as a quantile function (the Monge–Kantorovich quantile function), and the images, by $F_{\pm}^{-1}$, of the spheres $\{\mathbf{t}|\ \|\mathbf{t}\| = \tau\}$, with probability content $\tau \in (0, 1)$, as quantile contours (the Monge–Kantorovich quantile contours). Figure 12.9 (compare with Figure 12.8) shows that Monge–Kantorovich quantile contours, in contrast to the spatial and directional quantile ones, do pick up the nonconvex features of a distribution.

Each (absolutely continuous) distribution P on $\mathbb{R}^d$ is entirely characterized by the corresponding $F_{\pm}$, which induces a distribution-specific ordering of $\mathbb{R}^d$; that ordering is the combination of

(i) a center-outward ordering $\mathbf{y}_1 \preceq_{\mathrm{P}} \mathbf{y}_2$ if and only if $\|F_{\pm}(\mathbf{y}_1)\| \leqslant \|F_{\pm}(\mathbf{y}_2)\|$, and

(ii) an angular ordering, associated with the cosines

$$\cos_{\mathrm{P}}(\mathbf{y}_1, \mathbf{y}_2) := \left(F_{\pm}(\mathbf{y}_1)\right)'\left(F_{\pm}(\mathbf{y}_2)\right)/\|F_{\pm}(\mathbf{y}_1)\|\,\|F_{\pm}(\mathbf{y}_2)\|.$$

No moment conditions are required.

Unlike the directional quantile contours of Section 12.2, the Monge–Kantorovich ones are equivariant under order-preserving transformations – here, the class of transformations preserving, for some given P, (i) and (ii) above, that is, any T such that

$$\mathbf{y}_1 \preceq_{\mathrm{P}} \mathbf{y}_2 \quad \text{if and only if} \quad T(\mathbf{y}_1) \preceq_{T*\mathrm{P}} T(\mathbf{y}_2)$$

and

$$\cos_{\mathrm{P}}(\mathbf{y}_1, \mathbf{y}_2) = \cos_{T*\mathrm{P}}(T(\mathbf{y}_1), T(\mathbf{y}_2))$$

for any $\mathbf{y}_1, \mathbf{y}_2 \in \mathbb{R}^d$. Those transformations are of the form

$$T_{\mathrm{P},\mathrm{Q}} = \left(F_{\pm}^{\mathrm{Q}}\right)^{-1} \circ F_{\pm}^{\mathrm{P}},$$

where Q ranges over the family of absolutely continuous distributions over $\mathbb{R}^d$, $\left(F_{\pm}^{\mathrm{Q}}\right)^{-1}$ stands for the corresponding Monge–Kantorovich quantile function, and $\left(F_{\pm}^{\mathrm{P}}\right)^{-1}$ for the one associated with P. Equivariance trivially follows from the fact that

$$T_{\mathrm{P},\mathrm{Q}} * \mathrm{P} = \mathrm{Q}, \quad \text{hence} \quad F_{\pm}^{T_{\mathrm{P},\mathrm{Q}}*\mathrm{P}} \circ T_{\mathrm{P},\mathrm{Q}} = F_{\pm}^{\mathrm{P}}.$$

Note that affine equivariance in general does not hold, since affine transformations, for general P, are no longer order-preserving.

---

[2]Here and in the sequel, "uniform over the unit ball" means the product measure of a uniform over the unit sphere $\mathcal{S}^{d-1}$ with a uniform over the unit interval of radial distances.

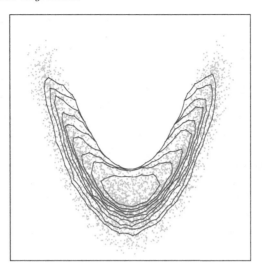

FIGURE 12.9
The same "banana-shaped" data as in Figure 12.8, with Monge–Kantorovich contours (for some selected quantile levels).

We refer the reader to Chernozhukov et al. (2017) for a description of the empirical versions of $F_{\pm}$ and their consistency properties.

All definitions above are about location ($p = 0$) only. Regression versions (and much more) are the subject of ongoing research, either along the lines of the local linear approach of Section 12.2.2.2, or along those developed by Carlier et al. (2016) who are considering another measure transportation-based approach where the reference uniform distribution is over the unit cube rather than the unit ball (see Section 12.4).

12.4 Some other concepts, and applications

Depth-based quantiles, as well as the elliptical ones, refer to center-outward orderings of $\mathbb{R}^d$: quantile regions are nested, and the reference structure is that of the unit ball. Other approaches are closer to the spirit of the coordinatewise definition of Section 12.2.1.1, where quantiles are indexed by d-tuples $(\tau_1, \ldots, \tau_d) \in (0,1)^d$; the reference structure there is that of the unit cube. The prototype of that approach is the so-called *Rosenblatt transformation* (Rosenblatt, 1952). Cai (2010), extending ideas by Gilchrist (2000), recently proposed, in a Bayesian context, a quantile concept based on general mappings from the unit cube $(0,1)^d$ to $\mathbb{R}^d$. Combining similar measure transportation ideas as in Chernozhukov et al. (2017) with the uniform distribution over the unit cube rather than the unit ball, Carlier et al. (2016) and Decurninge (2014) define multivariate quantile functions (associated with a distribution P over $\mathbb{R}^d$) and multiple-output quantile regression (in a linear regression setting) based on the inverse of the optimal transports mapping P to the uniform distribution over the unit cube $(0,1)^d$.[3] The location version of the latter can be seen as a nonlinear version of the very

---

[3]Note that the Rosenblatt transformation, in general, is not the gradient of any convex function, hence does not belong to the class of optimal transports considered in this context; see, however, Carlier et al. (2010).

popular independent component analysis models. All those approaches crucially depend on the choice of a coordinate system (hence a unit cube). Yet another approach, where quantiles are constructed on the basis of some preexisting partial ordering $\leq_0$ of $\mathbb{R}^d$, has been proposed by Belloni and Winkler (2011); not surprisingly, the result depends on the choice of $\leq_0$.

The applications of multiple-output quantile regression methods are without number, in a virtually unlimited number of domains. An immediate byproduct is the detection of multivariate outliers, for example in growth charts or medical diagnoses. Growth charts so far have essentially been based on marginal quantile plots. The multiple-output quantile concepts described here allow for spotting multivariate outliers that do not outlie in any marginal direction, thus providing a much more powerful diagnostic tool; see McKeague et al. (2011) and the references therein, as well as Chakraborty (2003), Cheng and de Gooijer (2007), or Wei (2008) for related applications and empirical studies. Another obvious and so far largely unexplored application is the problem of multivariate value-at-risk assessment in financial and actuarial statistics.

12.5 Conclusion

Quantile regression methods, by aiming to reconstruct the collection of distributions of a d-dimensional response $\mathbf{Y}$ conditional on the values of a set of covariates $\mathbf{X} = \mathbf{x}$, irrespective of the field of application, address one of the central problems of statistics. The major obstacle to extending traditional single-output quantile regression methods to the multiple-output setting has been the lack of a canonical concept of multivariate quantile, itself related to the lack of a canonical ordering of the Euclidean space with dimension $d \geqslant 2$. Ordering $\mathbb{R}^d$ has remained an open problem and an active domain of research for many years, but recent contributions, and the introduction of measure transportation concepts, are bringing appealing solutions to that problem, hence appealing concepts of quantile regression. Many developments can be expected along those promising lines, although it is clear that much work still remains to be done.

The picture provided in this contribution is inevitably incomplete, as the subject is rapidly developing. Desirable as it is, a clear discussion of the advantages and disadvantages of the various approaches described here would be premature, due to the lack of computational and empirical experience with most of them, and the lack of methodological results about the most recent ones. Inferential statistics, moreover, have their limits, and empirical versions of quantile regression methods, at best, provide an alternative and legible version of the data under study: while (by far) more readable, a collection of empirical quantile contours indeed yields roughly the same complexity as the data themselves. Quantile regression, in that respect, is nothing more – but certainly nothing less – than a sophisticated, powerful, and most meaningful tool for data analysis.

Acknowledgments

Marc Hallin acknowledges the support of the IAP research network grant P7/06 of the Belgian government (Belgian Science Policy). The research of Miroslav Šiman was supported by the Czech Science Foundation project GA17-07384S. We both thank Pavel Boček for technical support and Probal Chaudhuri for constructive remarks that helped improve this chapter.

Bibliography

A. Ahidar. *Surfaces quantiles: propriétés convergences et applications*. PhD thesis, Université de Toulouse, 2015.

A. Ahidar-Coutrix and P. Berthet. Convergence of multivariate quantile surfaces. Available from https://arxiv.org/abs/1607.02604, 2016.

Z. D. Bai, N. R. Chen, B. Q. Miao, and C. R. Rao. Asymptotic theory of least distances estimates in multivariate linear models. *Statistics*, 21:503–519, 1990.

A. Belloni and R. L. Winkler. On multivariate quantiles under partial orders. *Annals of Statistics*, 39:1125–1179, 2011.

P. Boček and M. Šiman. Directional quantile regression in Octave (and MATLAB). *Kybernetika*, 52:28–51, 2016.

Y. Brenier. Polar factorization and monotone rearrangement of vector-valued functions. *Communications on Pure and Applied Mathematics*, 44:375–417, 1991.

B. M. Brown. Statistical use of the spatial median. *Journal of the Royal Statistical Society, Series B*, 45:25–30, 1983.

Y. Cai. Multivariate quantile function models. *Statistica Sinica*, 20:481–496, 2010.

G. Carlier, A. Galichon, and F. Santambrogio. From Knothe's transport to Brenier's map. *SIAM Journal on Mathematical Analysis*, 41:2554–2576, 2010.

G. Carlier, V. Chernozhukov, and A. Galichon. Vector quantile regression. *Annals of Statistics*, 44:1165–1192, 2016.

A. Chakraborty and P. Chaudhuri. The spatial distribution in infinite-dimensional spaces and related quantiles and depths. *Annals of Statistics*, 42:1203–1231, 2014.

B. Chakraborty. On multivariate quantile regression. *Journal of Statistical Planning and Inference*, 110:109–132, 2003.

B. Chakraborty and P. Chaudhuri. On a transformation and re-transformation technique for constructing an affine equivariant multivariate median. *Proceedings of the American Mathematical Society*, 124:2539–2547, 1996.

M. Chaouch, A. Gannoun, and J. Saracco. Estimation de quantiles géométriques conditionnels et non conditionnels. *Journal de la Société Française de Statistique*, 150:1–27, 2009.

P. Chaudhuri. On a geometric notion of quantiles for multivariate data. *Journal of the American Statistical Association*, 91:862–872, 1996.

Y. Cheng and J. de Gooijer. On the uth geometric conditional quantile. *Journal of Statistical Planning and Inference*, 137:1914–1930, 2007.

V. Chernozhukov, A. Galichon, M. Hallin, and M. Henry. Monge-Kantorovich depth, quantiles, ranks and signs. *Annals of Statistics*, 45:223–256, 2017.

A. Decurninge. Multivariate quantiles and multivariate l-moments. arXiv:1409.6013, 2014.

W. G. Gilchrist. *Statistical Modelling with Quantile Functions*. Chapman & Hall/CRC, Boca Raton, FL, 2000.

J. C. Gower. Algorithm AS78: the mediancenter. *Applied Statistics*, 23:446–470, 1974.

J. B. S. Haldane. Note on the median of a multivariate distribution. *Biometrika*, 35:414–415, 1948.

M. Hallin and M. Šiman. Elliptical multiple-output quantile regression and convex optimization. *Statistics & Probability Letters*, 109:232–237, 2016.

M. Hallin, D. Paindaveine, and M. Šiman. Multivariate quantiles and multiple-output regression quantiles: From l_1 optimization to halfspace depth (with discussion). *Annals of Statistics*, 38:635–669, 2010.

M. Hallin, Z. Lu, D. Paindaveine, and M. Šiman. Local constant and local bilinear multiple-output quantile regression. *Bernoulli*, 21:1435–1466, 2015.

D. Hlubinka and M. Šiman. On elliptical quantiles in the quantile regression setup. *Journal of Multivariate Analysis*, 116:163–171, 2013.

D. Hlubinka and M. Šiman. On generalized elliptical quantiles in the nonlinear quantile regression setup. *TEST*, 24:249–264, 2015.

R. Koenker and G. Bassett. Regression quantiles. *Econometrica*, 46:33–50, 1978.

V. Koltchinskii. M-estimation, convexity and quantiles. *Annals of Statistics*, 25:435–477, 1997.

L. Kong and I. Mizera. Quantile tomography: using quantiles with multivariate data. arXiv:0805.0056v1, 2008.

L. Kong and I. Mizera. Quantile tomography: using quantiles with multivariate data. *Statistica Sinica*, 22:1589–1610, 2012.

R. J. McCann. Existence and uniqueness of monotone measure-preserving maps. *Duke Mathematical Journal*, 80:309–324, 1995.

I. W. McKeague, S. López-Pintado, M. Hallin, and M. Šiman. Analyzing growth trajectories. *Journal of Developmental Origins of Health and Disease*, 2:322–329, 2011.

I. Mizera. On depth and deep points: A calculus. *Annals of Statistics*, 30:1681–1736, 2002.

D. Paindaveine. A canonical definition of shape. *Statistics and Probability Letters*, 78: 2240–2247, 2008.

D. Paindaveine and M. Šiman. On directional multiple-output quantile regression. *Journal of Multivariate Analysis*, 102:193–212, 2011.

D. Paindaveine and M. Šiman. Computing multiple-output regression quantile regions. *Computational Statistics and Data Analysis*, 56:841–853, 2012a.

D. Paindaveine and M. Šiman. Computing multiple-output regression quantile regions from projection quantiles. *Computational Statistics*, 27:29–49, 2012b.

M. Rosenblatt. Remarks on a multivariate transformation. *Annals of Mathematical Statistics*, 23:470–472, 1952.

P. J. Rousseeuw and M. Hubert. Regression depth. *Journal of the American Statistical Association*, 94:388–402, 1999.

R. Serfling. Quantile functions for multivariate analysis: Approaches and applications. *Statistica Neerlandica*, 56:214–232, 2002a.

R. Serfling. A depth function and a scale curve based on spatial quantiles. In Y. Dodge, editor, *Statistical Data Analysis Based on the L_1-Norm and Related Methods*, pages 25–38. Birkhäuser, Basel, 2002b.

R. Serfling. Depth. In A. H. El-Shaarawi and W. Piegorsch, editors, *Encyclopedia of Environmetrics*, pages 636–641. Wiley, Chichester, 2nd edition, 2012.

M. Šiman. On exact computation of some statistics based on projection pursuit in a general regression context. *Communications in Statistics: Simulation and Computation*, 40:948–956, 2011.

J. W. Tukey. Mathematics and the picturing of data. In R. D. James, editor, *Proceedings of the International Congress of Mathematicians*, pages 523–531. Canadian Mathematical Congress, Vancouver, 1975.

A. W. van der Vaart and J. A. Wellner. *Weak Convergence*. Springer, New York, 1996.

Y. Wei. An approach to multivariate covariate-dependent quantile contours with application to bivariate conditional growth charts. *Journal of the American Statistical Association*, 103:397–409, 2008.

Y. Zuo and R. Serfling. General notions of statistical depth function. *Annals of Statistics*, 28:461–482, 2000.

13

Sample Selection in Quantile Regression: A Survey

Manuel Arellano

CEMFI, Madrid, Spain

Stéphane Bonhomme

University of Chicago, Chicago, USA

CONTENTS

13.1   Introduction

Nonrandom sample selection is a pervasive issue in applied work. Selection may arise because of data collection by the analyst. It may also be due to the fact that agents self-select into particular options, the latter form of selection being very common in economics.

A prototypical example of sample selection, due to Gronau (1974) and Heckman (1974), concerns selection into labor market participation. When interest centers on estimating determinants of wage offer functions, standard regression approaches will result in biased estimates if selection into work is not random. In this example, selection may arise due to individuals with low wage potential choosing not to participate to the labor market.

Selection biases show up in a variety of ways in the literature. A situation related to the previous example is the estimation of sector-specific wage offer functions, or more generally alternative-specific payoff functions, when alternatives are chosen in part based on some forecasts of payoffs. Other related settings are missing-data problems, and nonrandom attrition in longitudinal data sets. Although references are too many to be mentioned here, recent influential studies where sample selection plays a central role are Mulligan and Rubinstein (2008), Helpman et al. (2008), and Jiménez et al. (2014).

In linear models, Heckman (1976, 1979) proposed a method which has become very popular in empirical work. The assumptions of the Heckman model rely on Gaussianity, while allowing the errors in the outcome and participation equations to be correlated. The Heckman two-step estimator, which we briefly review below, provides a practical alternative to full maximum likelihood.

Building on the control function approach implicit in the Heckman method, a large econometric literature has since then extended the model by relaxing parametric assumptions and proposing semiparametric estimators. Influential examples include Heckman (1990), Ahn and Powell (1993), Donald (1995), Andrews and Schafgans (1998), Chen and Khan (2003), and Das et al. (2003). See also the references in Vella (1998). Additivity of the outcome in observed covariates and unobservables is key in all these approaches.

Much less is known regarding sample selection in nonlinear models. Manski (1994, 2003) derived worst-case bounds on quantiles of potential outcomes; see also Kitagawa (2010). Blundell et al. (2007) applied the bounds approach to document gender differences in wage inequality in the UK. However, the literature on parametric or semiparametric selection corrections in nonlinear models is scarce.

In this chapter we focus on the question of correcting quantile regression estimates for nonrandom sample selection. Quantile regression (Koenker and Bassett, 1978) is a versatile estimation approach which has been extensively studied. However, little work has been done at the intersection of quantile methods and sample selection methods. We review the approach of Arellano and Bonhomme (2017). A central observation is that, in quantile models, even linear ones, quantile curves on the selected sample are generally not linear. However, a correction is available which involves "rotating" the check function of quantile regression by an amount that is observation-specific and depends on the strength of selection.

Implementing this method requires estimating the degree of sample selection. Formally, the latter is defined as the dependence between the rank error in the equation of interest and the rank error in the selection equation. Working with a parametric copula, Arellano and Bonhomme derive moment restrictions on the copula parameter. As in linear models, "excluded" covariates affecting participation without entering the potential outcome equation are required for credible identification. The method then consists of three steps: estimation of the propensity score, estimation of the degree of selection (i.e. the copula parameter), and computation of quantile estimates through rotated quantile regression.

The sample selection problem we focus on here differs from censoring. Censored quantile regression is a well-studied problem (e.g. Powell, 1986; Chamberlain, 1993; Buchinsky, 1994; Buchinsky and Hahn, 1998; Chernozhukov and Hong, 2002; Portnoy, 2003; Chernozhukov et al., 2015); see also Chapters 6 and 7 in this handbook on censoring in survival analysis. It turns out that the Buchinsky and Hahn censoring correction may be interpreted as a selection correction based on a degenerate (Fréchet) copula.

We also review two other approaches to sample selection. Buchinsky (1998, 2001) proposed a control function approach to correct quantile regression estimates for sample selection. This method has been used by Albrecht et al. (2009) and Bollinger et al. (2011), among others. However, control function methods impose conditions on the data-generating process which may be inconsistent with quantile models unless the model is additive and quantile curves are parallel to each other, or selection is random (Huber and Melly, 2015). Lastly, the method in Arellano and Bonhomme (2017) is only one possibility for estimating selection-corrected quantile coefficients, and we briefly review an alternative approach based on maximum likelihood.

A nonquantile regression based approach to selection correction is to parametrically specify both outcome and selection equations, thus providing non-Gaussian extensions to the Heckman model. See, for example, Lee (1983), Smith (2003), or the recent application in Van Kerm (2013). Relative to fully parametric approaches, quantile regression provides added flexibility in the modeling of outcome variables.

In the final part of this chapter we revisit the empirical illustration in Huber and Melly (2015), and estimate uncorrected and selection-corrected wage returns to experience and education based on data on female wages and employment status from the 1991 Current Population Survey. Huber and Melly provided evidence of wages being nonadditive in covariates and unobservables in this setting. We complement their analysis by providing selection-corrected quantile regression estimates, which remain consistent under nonadditivity.

The outline of the chapter is as follows. In Sections 13.2 and 13.3 we review the approaches of Heckman (1979) and Arellano and Bonhomme (2017). In Sections 13.4 and 13.5 we discuss identification in the absence of parametric assumptions, and review several extensions. We present the empirical illustration in Section 13.6, and conclude in Section 13.7.

13.2   Heckman's parametric selection model

Consider an additive latent response model of the form

$$Y^* = X'\beta + \varepsilon, \tag{13.1}$$

where ε has mean zero and is identically distributed given X. If a random sample from (Y^*, X) were available, β could be consistently estimated under standard conditions using ordinary least squares (OLS). Sample selection arises as Y^* is only observed when the binary selection indicator D is equal to 1 (hence the star superscript, which refers to Y^* being a *latent* variable). In turn, D is given by

$$D = \mathbf{1}\{\eta \leqslant Z'\gamma\}, \tag{13.2}$$

where X is a subset of Z, and $\mathbf{1}\{\cdot\}$ is an indicator function. The scalar unobservable η is independent of Z, and possibly correlated with ε. Let $Y = DY^*$. A random sample from (Z, D, Y) is available, but Y^* is not observed when $D = 0$.

A textbook example is the following (Heckman, 1974). Y^* are wage offers, X are determinants of wages (such as education and experience), and D denotes labor force participation. Wage offers are not observed unless they have been accepted. In applications, in addition to X, Z typically contains *excluded* determinants of participation that do not appear in the wage offer equation, such as the number of children, marital status, or potential income when out of work, all of which aim to capture costs of working unrelated to potential wages. In this example, one expects dependence between η and ε if participation decisions are influenced by unobserved determinants of potential wage offers.

As latent outcomes Y^* are not observed for nonparticipants, it is not possible to directly estimate an empirical counterpart to $\mathbb{E}(Y^* \mid X)$. Instead, the conditional mean for participants, $\mathbb{E}(Y^* \mid D = 1, Z)$, which is identified from data on participants only, is instrumental in developing a selection correction method. Following Heckman (1979), we have

$$
\begin{aligned}
\mathbb{E}\left(Y^* \mid D = 1, Z\right) &= X'\beta + \mathbb{E}\left(\varepsilon \mid D = 1, Z\right) \\
&= X'\beta + \Lambda(Z),
\end{aligned}
\tag{13.3}
$$

where $\Lambda(Z) = \mathbb{E}\left(\varepsilon \mid \eta \leqslant Z'\gamma, Z\right)$ is a selection correction factor.

From (13.3) it follows that an OLS regression of Y on X on participants $D = 1$ will generally be inconsistent for β when ε and η are statistically dependent. This case precisely corresponds to nonrandom sample selection. Note that (13.3) suggests a strategy to consistently estimate β, by regressing Y on a linear function of X and an additive nonlinear function of Z. In the case where distributions are multivariate Gaussian, such a strategy simplifies to the Heckman (1979) method.

13.2.1   Two-step estimation in Gaussian models

Let us now assume that (ε, η) is bivariate Gaussian, independent of Z, with variances σ^2 and 1, respectively, and correlation ρ. In this case

$$\Lambda(Z) = -\rho\sigma\lambda(Z'\gamma), \quad \text{with } \lambda(u) = \frac{\phi(u)}{\Phi(u)},$$

where ϕ and Φ denote the standard Gaussian pdf and cdf, respectively. Note that the propensity score is $p(Z) = \Pr(D = 1 \mid Z) = \Phi(Z'\gamma)$, so we also have $\Lambda(Z) = -\rho\sigma\lambda\left[\Phi^{-1}\left(p(Z)\right)\right]$.

Heckman (1976, 1979) proposes a two-step estimator. In the first step, γ is estimated by a probit regression of D on Z. Letting $\hat{\gamma}$ denote the parameter estimate, the selection factor is estimated (up to scale) as $\hat{\lambda} = \lambda\left(Z'\hat{\gamma}\right)$. In the second step, β and $\rho\sigma$ are estimated by an OLS regression of Y on X and $\hat{\lambda}$ in the subsample of participants $D = 1$.

Formulas are available to correct the standard errors of the second-step estimator $\hat{\beta}$ for estimation error in the first step. In the Gaussian model, this two-step "control function" method provides an alternative to maximum likelihood, albeit at some efficiency cost (e.g. Nelson, 1984). An attractive feature of the method is that it can be extended to allow for semi- or nonparametric specifications, as reviewed in the introduction, provided additivity of (13.1) in X and ε is maintained. However, nonadditive models such as quantile models cannot be studied using those techniques.

13.3   A quantile generalization

In this section we describe the approach introduced in Arellano and Bonhomme (2017).

13.3.1   A quantile selection model

Consider now the following linear quantile specification of outcomes:

$$Y^* = X'\beta(U), \tag{13.4}$$

where $\beta(u)$ is increasing in u, and U is uniformly distributed on the unit interval, independent of X. Model (13.4) is a linear quantile model (Koenker and Bassett, 1978). In particular, $Q(\tau, X) = X'\beta(\tau)$ is the τth conditional quantile of Y^* given X. If a random sample from (Y^*, X) were available, one could thus consistently estimate $\beta(\tau)$ for all $\tau \in (0, 1)$ by quantile regression, under standard assumptions.

Maintaining the other assumptions of the Heckman Gaussian model, we assume that (13.2) holds with a Gaussian η independent of Z so that, equivalently,

$$D = \mathbf{1}\{V \leqslant p(Z)\}, \tag{13.5}$$

where $p(Z) = \Phi(Z'\gamma)$, and $V = \Phi(\eta)$ is the rank of η, which is uniformly distributed on $(0,1)$ and independent of Z.

Lastly, we assume that (U,V) follows a bivariate Gaussian copula with dependence parameter ρ, independent of Z. We denote by $G(\tau,p;\rho) = C(\tau,p;\rho)/p$ the *conditional copula* of U given V, defined on $(0,1) \times (0,1)$, where $C(\tau,p;\rho)$ denotes the unconditional copula of (U,V).[1] Note that model (13.4)–(13.5) simplifies to the Heckman Gaussian model when $X'\beta(U) = X'\beta + \sigma\Phi^{-1}(U)$ is a location-shift Gaussian model. Although we consider a Gaussian copula to fix ideas, any other parametric specification could be used, such as the Gumbel, Frank, or Bernstein copulas.

In the nonadditive model (13.4)-(13.5), quantile curves are generally nonadditive in the propensity score $p(Z)$ and covariates X. To see this, denote $Z = (X,W)$ (where W are the "excluded" covariates), and note that the conditional cdf of Y^* given $Z = z = (x,w)$ for participants $D = 1$, evaluated at $x'\beta(\tau)$ for some τ in the unit interval, is

$$\Pr\left(Y^* \leqslant x'\beta(\tau) \mid D = 1, Z = z\right) = \Pr\left(U \leqslant \tau \mid V \leqslant \Phi\left(z'\gamma\right), Z = z\right),$$
$$= G\left(\tau, \Phi\left(z'\gamma\right);\rho\right), \qquad (13.6)$$

where $G(\cdot,\cdot;\rho)$ is the conditional Gaussian copula with parameter ρ. It follows that the τth conditional quantile of Y^* given $D = 1$ and Z is

$$Q^s(\tau,Z) = X'\beta\left(\tau^*(Z)\right), \qquad (13.7)$$

where $\tau^*(Z) = G^{-1}\left(\tau, \Phi\left(Z'\gamma\right);\rho\right)$, and $G^{-1}(\tau,p;\rho)$ denotes the inverse of $G(\tau,p;\rho)$ with respect to its first argument (i.e. the conditional quantile function of U given $V \leqslant p$).[2] The s superscript refers to the fact that $Q^s(\tau,Z)$ is conditional on selection.

Nonadditivity of quantile curves implies that existing control function strategies cannot be used in the quantile selection model. We next review a method recently proposed by Arellano and Bonhomme (2017) to achieve consistent estimation in this model.

13.3.2 Estimation

Let us start with the case where γ and ρ are known. Later we will show how these parameters may be consistently estimated. From (13.6), for every $\tau \in (0,1)$ the parameter vector $\beta(\tau)$ is then characterized as the solution to the population moment restriction

$$\mathbb{E}\left[\mathbf{1}\left\{Y \leqslant X'\beta(\tau)\right\} - G\left(\tau, \Phi\left(Z'\gamma\right);\rho\right) \mid D = 1, Z\right] = 0. \qquad (13.8)$$

Hence, using DX as instruments and taking expectations,

$$\mathbb{E}\left[DX\left(\mathbf{1}\left\{Y \leqslant X'\beta(\tau)\right\} - G\left(\tau, \Phi\left(Z'\gamma\right);\rho\right)\right)\right] = 0. \qquad (13.9)$$

Arellano and Bonhomme noticed that (13.9) is the system of first-order conditions in the following optimization:

$$\beta(\tau) = \operatorname*{argmin}_{b(\tau)} \mathbb{E}\left[D\left(G_{\tau Z}\left(Y - X'b(\tau)\right)^+ + (1 - G_{\tau Z})\left(Y - X'b(\tau)\right)^-\right)\right], \qquad (13.10)$$

where $a^+ = \max(a,0)$, $a^- = \max(-a,0)$, and $G_{\tau z} = G\left(\tau, \Phi\left(z'\gamma\right);\rho\right)$ denotes the rank of $x'\beta(\tau)$ in the selected sample $D = 1$, conditional on $Z = z$. The function G plays a key role

---

[1]Letting $\Phi_2(\cdot,\cdot;\rho)$ denote the bivariate Gaussian cdf with parameter ρ, $C(\tau,p;\rho) = \Phi_2\left(\Phi^{-1}(\tau),\Phi^{-1}(p);\rho\right)$ and $G(\tau,p;\rho) = \Phi_2\left(\Phi^{-1}(\tau),\Phi^{-1}(p);\rho\right)/p$.

[2]The assumption that G is strictly monotone in its first argument is not without loss of generality. For example, it is not satisfied by Fréchet copulas; see Section 13.5.

here, as it maps ranks in the latent distribution (τs) into ranks in the selected distribution ($G_{\tau z}$s).

It is instructive to compare (13.10) with the optimization problem which would characterize $\beta(\tau)$ were a sample from (Y^*, X) available, that is,

$$\min_{b(\tau)} \mathbb{E}\left[\tau \left(Y^* - X'b(\tau)\right)^+ + (1-\tau)\left(Y^* - X'b(\tau)\right)^-\right]. \tag{13.11}$$

The function inside the expectation in (13.11) is the check function. In (13.10) we see that, in order to account for nonrandom sample selection, one needs to *rotate* the check function. The rotation angle depends on the amount of selection, and it is Z-specific. Such a rotation is needed unless U and V were independent, hence $G_{\tau Z} = \tau$, in which case standard quantile regression in the selected sample would be consistent for $\beta(\tau)$.

Interestingly, like (13.11), (13.10) is a linear program, hence in particular convex, and so is its sample counterpart. This implies that, given γ and ρ, one can estimate $\beta(\tau)$ for all τ in a τ-by-τ fashion by solving linear programs.

In practice γ and ρ need to be estimated. In the Gaussian specification for η, γ may be consistently estimated by a probit regression, as in the first step in the Heckman method.

In turn, the copula parameter ρ may be consistently estimated by taking advantage of the fact that (13.8) implies a number of moment restrictions (in fact, a continuum of such restrictions when covariates are continuously distributed), by using functions of Z as instruments. To describe the method to recover ρ, let us change the notation slightly and explicitly indicate the dependence on ρ and γ in $G_{\tau Z}^{\rho\gamma} = G\left(\tau, \Phi\left(Z'\gamma\right); \rho\right)$. For a given vector of instruments $\varphi(\tau, Z)$, ρ satisfies the moment restrictions

$$\mathbb{E}\left[D\varphi(\tau, Z)\left(\mathbf{1}\left\{Y \leqslant X'\overline{\beta}(\tau; \rho, \gamma)\right\} - G_{\tau Z}^{\rho\gamma}\right)\right] = 0, \tag{13.12}$$

where

$$\overline{\beta}(\tau; \rho, \gamma) = \operatorname*{argmin}_{b(\tau)} \mathbb{E}\left[D\left(G_{\tau Z}^{\rho\gamma}\left(Y - X'b(\tau)\right)^+ + (1 - G_{\tau Z}^{\rho\gamma})\left(Y - X'b(\tau)\right)^-\right)\right]. \tag{13.13}$$

The copula parameter ρ can thus be estimated in (13.12) for a finite set of τ values, based on the generalized method of moments (Hansen, 1982), by profiling out the $\overline{\beta}(\tau; \rho, \gamma)$ using (13.13).

In sum, given an independent and identically distributed sample (Y_i, Z_i, D_i), $i = 1, \ldots, N$ (where $Z_i = (X_i, W_i)$), Arellano and Bonhomme's three-step estimation algorithm is as follows. Code written in Matlab is available from the authors on request.

Algorithm 1

1. *Estimate γ by a probit regression,*

$$\widehat{\gamma} = \operatorname*{argmax}_{a} \sum_{i=1}^{N} D_i \ln \Phi(Z_i'a) + (1 - D_i) \ln \Phi(-Z_i'a).$$

2. *Estimate ρ by profiled GMM,*

$$\widehat{\rho} = \operatorname*{argmin}_{c} \left\| \sum_{i=1}^{N} \sum_{\ell=1}^{L} D_i \varphi\left(\tau_\ell, Z_i\right)\left[\mathbf{1}\left\{Y_i \leqslant X_i'\widehat{\beta}\left(\tau_\ell, c\right)\right\} - G\left(\tau_\ell, \Phi(Z_i'\widehat{\gamma}); c\right)\right]\right\|,$$

$$\tag{13.14}$$

where $\|\cdot\|$ is the Euclidean norm, $\tau_1 < \tau_2 < \ldots < \tau_L$ is a finite grid on $(0,1)$, $\varphi(\tau, Z_i)$ are instrument functions with $\dim \varphi \geqslant \dim \rho$, and

$$\hat{\beta}(\tau, c) = \underset{b(\tau)}{\operatorname{argmin}} \sum_{i=1}^{N} D_i \Big[G\left(\tau, \Phi(Z_i'\hat{\gamma}); c\right) \left(Y_i - X_i'b(\tau)\right)^{+}$$
$$+ \left(1 - G\left(\tau, \Phi(Z_i'\hat{\gamma}); c\right)\right) \left(Y_i - X_i'b(\tau)\right)^{-} \Big]. \quad (13.15)$$

3. *For any desired* $\tau \in (0,1)$, *compute* $\hat{G}_{\tau i} = G\left(\tau, \Phi(Z_i'\hat{\gamma}); \hat{\rho}\right)$ *for all* i, *and estimate* $\beta(\tau)$ *by rotated quantile regression,*

$$\hat{\beta}(\tau) = \underset{b(\tau)}{\operatorname{argmin}} \sum_{i=1}^{N} D_i \Big[\hat{G}_{\tau i} \left(Y_i - X_i'b(\tau)\right)^{+} + \left(1 - \hat{G}_{\tau i}\right) \left(Y_i - X_i'b(\tau)\right)^{-} \Big].$$
$$(13.16)$$

Note that Step 3 is not needed when the researcher is only interested in $\beta(\tau_1), \ldots, \beta(\tau_L)$, in which case Steps 1 and 2 suffice.

The main computational cost of this algorithm is in Step 2. The objective function in (13.14) is neither continuous nor convex, because it features indicator functions. When modeling selection through a Gaussian copula, one can rely on grid search for computation. Evaluating the objective function is usually fast and straightforward, because (13.15) is a linear program. In addition, using many percentile values τ_ℓ in (13.14) may smooth the objective function, hence aid computation.

The grid of τ values on the unit interval, and the instrument function $\varphi(\tau, Z)$, are to be chosen by the researcher. Although large grids slow down computation, it seems desirable to exploit a large number of restrictions to increase precision. Regarding the instruments, with a scalar ρ a possibility is to take φ to be the propensity score, or the propensity score multiplied by a function of τ. Optimal instruments may be constructed, given a finite grid of τs. However, characterizing efficiency properties in quantile selection models would require working under a continuum of moment restrictions.

Steps 1 and 2 in Algorithm 1 amount to profiled GMM estimation of a finite number of parameters: γ, ρ, and $\beta(\tau_1), \ldots, \beta(\tau_L)$. This is a well-understood estimation problem based on nonsmooth moment functions. For example, the methods described in Newey and McFadden (1994) can be used to show root-N consistency and asymptotic normality, and characterize and estimate asymptotic variances. The asymptotic distribution of $\hat{\beta}(\tau)$ in (13.16) may be derived using the same techniques. See Arellano and Bonhomme (2017) for a derivation of asymptotic variances. In fact, inference on the $\beta(\tau)$ process would also follow from standard arguments (Koenker and Xiao, 2002). Nonanalytical methods, such as subsampling (Chernozhukov and Fernández-Val, 2005), may also be used for inference. An important condition for estimator consistency is identification, which we discuss in the next section in a nonparametric setting.

Lastly, given estimates of conditional quantile functions, unconditional quantiles of latent outcomes and counterfactual distributions may be constructed using standard methods (Machado and Mata, 2005; Chernozhukov et al., 2013).

13.4   Identification

The methods described in the previous section rely on parametric assumptions on the propensity score and the copula, in addition to the assumed linear quantile specification for outcomes. It is possible to formulate and analyze a nonparametric quantile selection model where these assumptions are relaxed.

To proceed, let us replace (13.4) by a general quantile representation $Y^* = Q(\tau, X)$, where Q is increasing in its first argument, and let us allow for a nonparametric propensity score $p(Z)$ in (13.5). Lastly, let us assume that (U, V) is conditionally independent of Z given X, and denote the conditional copula of U given V and $X = x$ by $G_x(\tau, p)$. Here this function is also nonparametric.

Using arguments similar to those in Section 13.3, one may derive the following set of restrictions:

$$\Pr\left(Y \leqslant Q\left(\tau, x\right) \middle| D = 1, Z = z\right) = G_x\left(\tau, p(z)\right). \qquad (13.17)$$

The aim is to recover Q and G from (13.17). The propensity score $p(Z)$ is clearly identified based on data on participation and covariates. However, the quantile function Q and the conditional copula G consistent with (13.17) are not unique in general. This reflects the fact that the nonparametric quantile selection model is generally *set-identified*.

Arellano and Bonhomme (2017) emphasize two situations where the model is nonparametrically identified. A first case where $Q(\cdot, x)$ and G_x are identified is when $p(Z) = 1$ with positive probability conditional on $X = x$.[3] This case corresponds to "identification at infinity" (Chamberlain, 1986; Heckman, 1990). A second situation where identification holds is when the conditional copula G_x is real analytic. This case could be called "identification by extrapolation."

Given identification of a nonparametric Q for a parametric or analytic G, nonparametric rotated quantile regression methods based on kernel or series versions of (13.16) may then be used for estimation. Such methods may be combined with a flexible specification for G, based on Bernstein copulas, for example.

In other situations, the quantile function and conditional copula are generally partially identified. That is, a set of such functions is consistent with the population distribution. In a quantile model, failure of point identification affects the entire quantile curve. This contrasts with semiparametric linear models, where arguments such as "identification at infinity" are only needed to point-identify intercept parameters (Andrews and Schafgans, 1998; Das et al., 2003). Bounds on these functions may be constructed following Manski (1994, 2003). In practice, estimating such bounds may help assess the impact of parametric forms on the results, as described in Arellano and Bonhomme.

13.5   Other approaches

In this section we review several related approaches. We start with an alternative approach to Arellano and Bonhomme based on maximum likelihood. We then review control function approaches to selection correction in quantile models. Lastly, we clarify the link between selection correction and censoring correction in quantile regression.

---

[3]To see why this is the case, take $z = (x, w)$ such that $p(z) = 1$. As $G_x(\tau, 1) = \tau$, evaluating (13.17) at $Z = z$ shows that $Q(\tau, x)$ is identified as the τth conditional quantile of Y given $D = 1$ and $Z = z$. This is intuitive, as conditioning on the propensity score being 1 removes the sample selection problem.

13.5.1   A likelihood approach

The approach outlined in Section 13.3 is only one of several estimation possibilities in quantile selection models. As an example, a principled alternative would be to estimate the parameters of interest using a maximum likelihood approach. To see how this would work, note that evaluating (13.6) at $\tau = F(y \mid x)$ yields

$$\Pr\left(Y^* \leqslant y | D = 1, Z = z\right) = G\left(F(y \mid x; \beta(\cdot)), \Phi\left(z'\gamma\right); \rho\right),$$

where we have indicated the dependence of $F(y \mid x)$ on the quantile process $\beta(\tau)$. A joint (semiparametric) maximum likelihood estimator would thus maximize

$$\sum_{i=1}^{N} D_i \ln \Phi(Z_i'a) + (1 - D_i) \ln \Phi(-Z_i'a) + \sum_{i=1}^{N} D_i \ln f(Y_i \mid X_i; b(\cdot))$$

$$+ \sum_{i=1}^{N} D_i \ln \nabla G\left(F(Y_i \mid X_i; b(\cdot)), \Phi\left(Z_i'a\right); c\right)$$

with respect to a, c, and all $b(\tau)$ for $\tau \in (0,1)$, where f is the conditional pdf of Y^* given X, and ∇G denotes the derivative of G with respect to its first argument. An intermediate approach would be to profile out the $\beta(\tau)$ using (13.15), and estimate ρ based on the profiled likelihood.

In contrast to a likelihood-based approach, the estimator described in Section 13.3 exploits the τ-by-τ separability of the rotated quantile regression problems, as well as the convexity of the rotated quantile regression objective functions. On the other hand, such sequential estimators are not asymptotically efficient in general.

13.5.2   Control function approaches

Buchinsky (1998, 2001) introduced a control function method to correct for sample selection in quantile regression models. The method involves controlling for functions of the propensity score in the quantile regression. This approach delivers consistent estimates in additive models with independent errors, as in this case quantile functions of selected outcomes are indeed additive in covariates X and propensity score $p(Z)$. As an example, in the Gaussian Heckman model, (13.7) becomes

$$Q^s(\tau, Z) = X'\beta + \sigma\Phi^{-1}\left[G^{-1}\left(\tau, p(Z); \rho\right)\right], \tag{13.18}$$

where $p(Z) = \Phi(Z'\gamma)$, and $G^{-1}(\cdot, \cdot; \rho)$ is the inverse of the conditional Gaussian copula with respect to its first argument.

However, as (13.7) shows, in nonadditive models such as quantile selection models, quantile curves are generally not additive in X and $p(Z)$. As a result, as pointed out by Huber and Melly (2015), additive control function methods will not be consistent in general. Huber and Melly use this observation to develop tests of the additivity assumption (which they call "conditional independence") that make use of the Buchinsky control function estimator.

13.5.3   Link to censoring corrections

The selection correction problem reviewed in this chapter is different from other censoring corrections that have been extensively studied in the quantile regression literature. To see the link between these two problems, consider an outcome variable modeled as in (13.4), observed only when $Y^* \leqslant \mu$, where μ is a known constant. That is, Y^* is censored above

μ. In this case, denoting censoring as $D = 1$, the censoring rule takes the form in (13.5), with $Z = X$, $p(X) = F(\mu \mid X)$ (where F is the conditional cdf of Y^* given X), and $V = U$. The threshold μ need not be known. Moreover, it could be a (known or unknown) function $\mu(X)$ of covariates.

The conditional copula of (U, V) is thus known in this case, but it is degenerate. It coincides with the conditional upper Fréchet bound (Fréchet, 1951), whose expression is

$$G^+(\tau, p) = \min\left\{\frac{\tau}{p}, 1\right\}.$$

Note that G^+ is not strictly monotone in its first argument. Analogously as in (13.10) one may base estimation of $\beta(\tau)$, for any given τ, on the optimization problem

$$\min_{b(\tau)} \mathbb{E}\left[D\mathbf{1}\{p(X) > \tau\}\left(\frac{\tau}{p(X)}\left(Y - X'b(\tau)\right)^+ + \left(1 - \frac{\tau}{p(X)}\right)\left(Y - X'b(\tau)\right)^-\right)\right],$$

$$(13.19)$$

focusing on the subpopulation with $p(X) > \tau$ where the τth conditional quantile is identified

Equation (13.19) is the basis for the censored quantile regression estimator of Buchinsky and Hahn (1998). A slight difference from the analysis in Buchinsky and Hahn is that they consider a case where outcomes are censored from below, so the relevant conditional copula in their case is the lower Fréchet bound $G^-(\tau, p) = \max\left\{\frac{\tau + p - 1}{p}, 0\right\}$. Buchinsky and Hahn propose estimating the propensity score nonparametrically. Their estimator solves a convex problem, like the sample selection estimator reviewed in Section 13.3. This contrasts with the estimator of Powell (1986), which is based on a nonconvex objective function.

One difference between this model and the bivariate sample selection model is that, in the censoring model, the propensity score $p(X) = F(\mu \mid X)$ depends on the unknown distribution of latent outcomes. This explains the need for nonparametric estimation of $p(X)$, in contrast with the quantile selection model where the propensity score may be parametrically specified while preserving statistical coherency.

13.6   Empirical illustration

In this section we revisit the empirical illustration in Huber and Melly (2015), who study the returns to education and experience for women in the USA. We take their sample from the 2011 Merged Outgoing Rotation Groups of the Current Population Survey (CPS). The sample consists of 44,562 white women, 20,055 of whom are working outside of self-employment, the military, agriculture, and the public sector. The only difference from the sample in Huber and Melly is that we drop working women with missing wage values (1.6% of observations). Working is defined as having worked more than 35 hours in the week preceding the survey.

Huber and Melly test, and reject, the assumption that quantile functions are additive in X and τ on these data. Using the methods described in Section 13.3, here we estimate quantile regression specifications that account for the presence of sample selection. The dependent variable Y is the log-hourly wage, covariates X contain general labor market experience (i.e. age minus the number of years of schooling) and its square, five education indicators (more than 7, 8, 9, 11, and 13 years of education), interactions of experience and its square with years of schooling, and indicators for marital status and region of residence

FIGURE 13.1

Female wages (CPS, 1991), quantile regression curves

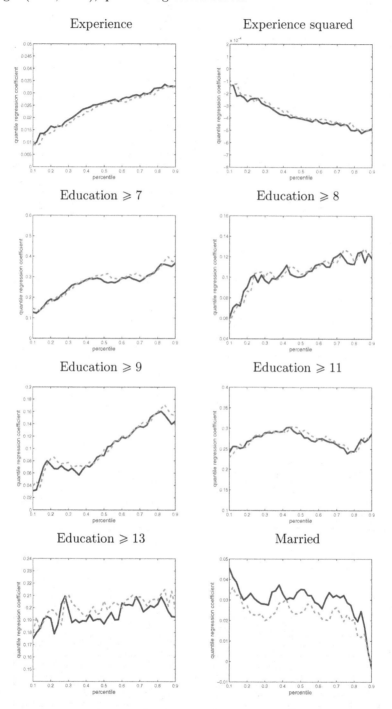

Notes. CPS, 1991. Sample comprising 44,562 women, 20,055 of whom are working. Quantile regression estimates. Dashed line: corrected for selection. Solid line: uncorrected.

TABLE 13.1

Female wages (CPS, 1991), parameter estimates

Uncorrected estimates

	Regression	Quantile regression $\tau = 0.25$	Quantile regression $\tau = 0.50$	Quantile regression $\tau = 0.75$
Experience	0.022 (0.0024)	0.016 (0.0026)	0.026 (0.0024)	0.029 (0.0026)
Experience squared	−0.00035 (0.000061)	−0.00024 (0.000057)	−0.00041 (0.000062)	−0.00043 (0.000057)
Education ≥ 7	0.30 (0.028)	0.20 (0.031)	0.28 (0.029)	0.31 (0.031)
Education ≥ 8	0.10 (0.014)	0.10 (0.015)	0.10 (0.014)	0.12 (0.015)
Education ≥ 9	0.10 (0.015)	0.067 (0.016)	0.089 (0.015)	0.14 (0.017)
Education ≥ 11	0.26 (0.017)	0.28 (0.018)	0.29 (0.017)	0.24 (0.019)
Education ≥ 13	0.19 (0.018)	0.20 (0.019)	0.19 (0.017)	0.20 (0.019)
Experience × education	0.0020 (0.00053)	0.0013 (0.00063)	0.0024 (0.00058)	0.0034 (0.00064)
Experience squared × education	−0.000044 (0.000013)	−0.000024 (0.000015)	−0.000045 (0.000014)	−0.000075 (0.000015)
Married	0.027 (0.0091)	0.032 (0.0092)	0.035 (0.0085)	0.030 (0.0093)

Selection-corrected estimates

	Regression	Quantile regression $\tau = 0.25$	Quantile regression $\tau = 0.50$	Quantile regression $\tau = 0.75$
Experience	0.022 (0.0026)	0.016 (0.0041)	0.026 (0.0030)	0.030 (0.0034)
Experience squared	−0.00034 (0.000061)	−0.00024 (0.00010)	−0.00040 (0.000071)	−0.00046 (0.000079)
Education ≥ 7	0.32 (0.037)	0.20 (0.15)	0.31 (0.089)	0.30 (0.037)
Education ≥ 8	0.11 (0.014)	0.10 (0.026)	0.10 (0.014)	0.13 (0.017)
Education ≥ 9	0.10 (0.017)	0.075 (0.026)	0.098 (0.019)	0.14 (0.021)
Education ≥ 11	0.27 (0.018)	0.28 (0.029)	0.30 (0.020)	0.25 (0.025)
Education ≥ 13	0.19 (0.021)	0.19 (0.029)	0.21 (0.022)	0.20 (0.024)
Experience × education	0.0020 (0.00063)	0.0011 (0.0012)	0.0020 (0.00076)	0.0033 (0.00089)
Experience squared × education	−0.000042 (0.000016)	−0.000021 (0.000033)	−0.000041 (0.000019)	−0.000070 (0.000022)
Married	0.022 (0.011)	0.023 (0.017)	0.024 (0.013)	0.021 (0.013)

Notes. CPS, 1991. Sample comprising 44,562 women. Regression and quantile regression estimates, corrected for selection and uncorrected. Covariates also include regional indicators and an intercept (not included). Robust standard errors in parentheses. Standard errors in bottom panel based on subsampling (subsample size 1000, 500 replications).

(four regions). As determinants of participation assumed not to enter the wage equation, we take the number of children in three age ranges and their interactions with marital status. We use CPS sample weights in all the computations. Finally, our specification is based on a probit propensity score and a Gaussian copula, we take $\varphi(\tau, Z) = \sqrt{\tau(1-\tau)}\hat{p}(Z)$, with $\hat{p}(Z)$ the estimated propensity score, and we take a grid τ_ℓ of deciles when estimating the copula parameter ρ. Computation of $\hat{\rho}$ in (13.14) is based on grid search.

Figure 13.1 presents the estimates for a selection of covariates. Quantiles corrected for sample selection are shown as dashed, while uncorrected ones are shown as solid lines. Similarly to Huber and Melly, excluded covariates (i.e. number of children and interactions with marital status) are strongly significant in the participation equation, estimates being omitted here for brevity. We see that quantile regression estimates vary substantially along the distribution. This is in line with Huber and Melly's finding that an additive model is not appropriate for this data. At the same time, correcting for sample selection tends to make small differences to the results. Although the coefficients for marital status differ by some margin, most of the uncorrected and corrected estimates are close to each other. This is so in spite of an estimated correlation $\hat{\rho} = -0.10$ (standard error 0.064), which reflects some positive selection of women into participation.[4] The variation along the distribution, and the similarity between uncorrected and selection-corrected estimates, are confirmed by the parameter estimates reported in Table 13.1.

We performed a number of robustness checks. We experimented with grids of τs of different sizes (equidistant grids with 2–50 knots), a different choice for the instrument function $(\varphi(\tau, Z) = \hat{p}(Z))$, and a different choice for G (based on the Frank copula). These choices all led to results very similar to Figure 13.1 and Table 13.1. Moreover, the minimum of the objective function in ρ was easy to identify in all experiments, although the objective function tended to be erratic for large values of $|\rho|$.

It is to be noted, however, that leaving the functional form of the copula fully unrestricted in this application would most likely lead to lack of point identification. Due to the restricted support of the propensity score (99% of the estimated propensity score being below 0.73 in the sample), nonparametric bounds based on a worst-case analysis would be wide. If desired, the method described in Arellano and Bonhomme (2017) may be used to compute estimates of worst-case bounds in a given application.

Overall, on this data set correcting for sample selection confirms the findings from standard quantile regression. It would be of interest to estimate similar specifications for other periods, especially since Mulligan and Rubinstein (2008) argue that the direction and intensity of females' selection into employment has changed since the 1970s in the USA.

13.7   Conclusion

Sample selection correction methods for linear models remain very popular in applied work. Since James Heckman's pioneering work, the use of these methods has led to uncovering important empirical regularities. In this chapter we have reviewed recently proposed selection correction approaches in nonlinear quantile models. The hope is that these methods could help document distributional effects in a variety of empirical settings where nonrandom sample selection arises.

Existing work on nonlinear selection models is scarce, however, and there remains a lot

---

[4]Note that ρ is the correlation between U in (13.4) and V in (13.5). Hence, a negative ρ means that high-U women have a higher propensity to participate to the labor market.

to be done. A particular issue is the reliance on parametric functional forms. In general, relaxing these assumptions results in lack of point identification. This is an area where recently developed methods allowing for uniform inference in the presence of partial identification (e.g. Tamer, 2010) might prove particularly useful.

Acknowledgments

We thank Roger Koenker, Blaise Melly, and participants at the December 2015 conference in Cambridge for comments. Arellano acknowledges research funding from the Ministerio de Economía y Competitividad, Grant ECO2016-79848-P.

Bibliography

H. Ahn and J. L. Powell. Semiparametric estimation of censored selection models with a nonparametric selection mechanism. *Journal of Econometrics*, 58:3–29, 1993.

J. Albrecht, A. van Vuuren, and S. Vroman. Counterfactual distributions with sample selection adjustments: Econometric theory and an application to the Netherlands. *Labour Economics*, 16(4):383–396, 2009.

D. W. Andrews and M. M. Schafgans. Semiparametric estimation of the intercept of a sample selection model. *Review of Economic Studies*, 65(3):497–517, 1998.

M. Arellano and S. Bonhomme. Quantile selection models with an application to understanding changes in wage inequality. *Econometrica*, 85(1):1–28, 2017.

R. Blundell, A. Gosling, H. Ichimura, and C. Meghir. Changes in the distribution of male and female wages accounting for employment composition using bounds. *Econometrica*, 75:323–364, 2007.

C. Bollinger, J. P. Ziliak, and K. R. Troske. Down from the mountain: Skill upgrading and wages in Appalachia. *Journal of Labor Economics*, 29(4):819–857, 2011.

M. Buchinsky. Changes in the U.S. wage structure 1963 to 1987: An application of quantile regressions. *Econometrica*, 62:405–458, 1994.

M. Buchinsky. The dynamics of changes in the female wage distribution in the USA: A quantile regression approach. *Journal of Applied Econometrics*, 13:1–30, 1998.

M. Buchinsky. Quantile regression with sample selection: Estimating women's return to education in the US. *Empirical Economics*, 26:87–113, 2001.

M. Buchinsky and J. Hahn. An alternative estimator for the censored regression model. *Econometrica*, 66:653–671, 1998.

G. Chamberlain. Asymptotic efficiency in semiparametric models with censoring. *Journal of Econometrics*, 32:189–218, 1986.

G. Chamberlain. Quantile regressions, censoring and the structure of wages. In C. Sims, editor, *Advances in Econometrics: Proceedings of the 6th World Congress in Barcelona*. Cambridge University Press, Cambridge, 1993.

S. Chen and S. Khan. Semiparametric estimation of a heteroskedastic sample selection model. *Econometric Theory*, 19(6):1040–1064, 2003.

V. Chernozhukov and I. Fernández-Val. Subsampling inference on quantile regression processes. *Sankhyā*, 67(2):253–276, 2005.

V. Chernozhukov and H. Hong. Three-step censored quantile regression and extramarital affairs. *Journal of the American Statistical Association*, 97(459):872–882, 2002.

V. Chernozhukov, I. Fernández-Val, and B. Melly. Inference on counterfactual distributions. *Econometrica*, 81(6):2205–68, 2013.

V. Chernozhukov, I. Fernández-Val, and A. E. Kowalski. Quantile regression with censoring and endogeneity. *Journal of Econometrics*, 186(1):201–221, 2015.

M. Das, W. K. Newey, and F. Vella. Nonparametric estimation of sample selection models. *Review of Economic Studies*, 70:33–58, 2003.

S. G. Donald. Two-step estimation of heteroskedastic sample selection models. *Journal of Econometrics*, 65(2):347–380, 1995.

M. Fréchet. Sur les tableaux de corrélation dont les marges sont données. *Annales de l'Université de Lyon, Série 3, Section A*, 14:53–77, 1951.

R. Gronau. Wage comparison – a selectivity bias. *Journal of Political Economy*, 82:1119–1143, 1974.

L. P. Hansen. Large sample properties of generalized method of moments estimators. *Econometrica*, 50:1029–1054, 1982.

J. J. Heckman. Shadow prices, market wages and labour supply. *Econometrica*, 42:679–694, 1974.

J. J. Heckman. The common structure of statistical models of truncation, sample selection and limited dependent variables and a simple estimator for such models. *Annals of Economic and Social Measurement*, 5:475–492, 1976.

J. J. Heckman. Sample selection bias as a specification error. *Econometrica*, 47:153–161, 1979.

J. J. Heckman. Varieties of selection bias. *American Economic Review*, 80:313–318, 1990.

E. Helpman, M. Melitz, and Y. Rubinstein. Estimating trade flows: Trading partners and trading volumes. *Quarterly Journal of Economics*, 123(2):441–487, 2008.

M. Huber and B. Melly. A test of the conditional independence assumption in sample selection models. *Journal of Applied Econometrics*, 30(7):1144–1168, 2015.

G. Jiménez, S. Ongena, J. L. Peydró, and J. Saurina. Hazardous times for monetary policy: What do twenty-three million bank loans say about the effects of monetary policy on credit risk-taking? *Econometrica*, 82(2):463–505, 2014.

T. Kitagawa. Testing for instrument independence in the selection model. Unpublished manuscript, 2010.

R. Koenker and G. Bassett. Regression quantiles. *Econometrica*, 46:33–50, 1978.

R. Koenker and Z. Xiao. Inference on the quantile regression process. *Econometrica*, 70: 1583–1612, 2002.

L.-f. Lee. Generalized econometric models with selectivity. *Econometrica*, 51(2):507–512, 1983.

J. A. F. Machado and J. Mata. Counterfactual decomposition of changes in wage distributions using quantile regression. *Journal of Applied Econometrics*, 20:445–465, 2005.

C. F. Manski. The selection problem. In C. Sims, editor, *Advances in Econometrics: Proceedings of the Sixth World Congress*, pages 143–170. Cambridge University Press, Cambridge, 1994.

C. F. Manski. *Partial Identification of Probability Distributions*. Springer-Verlag, Berlin, 2003.

C. Mulligan and Y. Rubinstein. Selection, investment, and women's relative wages over time. *Quarterly Journal of Economics*, 123(3):1061–1110, 2008.

F. D. Nelson. Efficiency of the two-step estimator for models with endogenous sample selection. *Journal of Econometrics*, 24:181–196, 1984.

W. K. Newey and D. McFadden. Large sample estimation and hypothesis testing. In R. F. Engle and D. McFadden, editors, *Handbook of Econometrics, Vol. IV*, pages 2111–2245. North-Holland, Amsterdam, 1994.

S. Portnoy. Censored regression quantiles. *Journal of the American Statistical Association*, 98(464):1001–1012, 2003.

J. L. Powell. Censored regression quantiles. *Journal of Econometrics*, 32:143–155, 1986.

M. D. Smith. Modelling sample selection using archimedean copulas. *Econometrics Journal*, 6:99–123, 2003.

E. Tamer. Partial identification in econometrics. *Annual Review of Economics*, 2:167–195, 2010.

P. Van Kerm. Generalized measures of wage differentials. *Empirical Economics*, 45(1): 465–482, 2013.

F. Vella. Estimating models with sample selection bias: A survey. *Journal of Human Resources*, 33(1):127–169, 1998.

14

Nonparametric Quantile Regression for Banach-Valued Response

Joydeep Chowdhury and Probal Chaudhuri

Indian Statistical Institute, Kolkata, India

CONTENTS

14.1   Introduction

Quantile regression for data involving covariates that are functions has been extensively considered in the recent literature. Linear quantile regression with real response and functional covariate is considered by Kato (2012) and Cardot et al. (2005). Nonparametric quantile regression with real response and functional covariate is investigated in Ferraty and Vieu (2006) and Gardes et al. (2010). Semiparametric quantile regression with real response and functional covariate is explored in Chen and Müller (2012). Nonparametric quantile regression with finite-dimensional response and functional covariate is studied in Chaouch and Laïb (2013, 2015). The examples below illustrate how the usual mean regression or median regression, which focuses on the center of the conditional distribution, sometimes fails to detect important features in the data, while quantile regression adequately captures those. In these examples, the responses are real-valued and the covariates are functions.

Example 1 (Simulated data) *We consider an example, where the response Y is real-valued and the covariate X is a random function, defined as $X(t) = U(1 + t^2)$ for $0 \leqslant t \leqslant 1$ and $U \sim U(0,1)$. Here the response $Y = \|X\|Z$, where $Z \sim N(0,1)$ and the norm is the L_2 norm on $[0,1]$. The sample size is 100. We construct local boxplots, where the radius of the neighborhood of each covariate curve is kept fixed at 0.5. We select six covariate curves for*

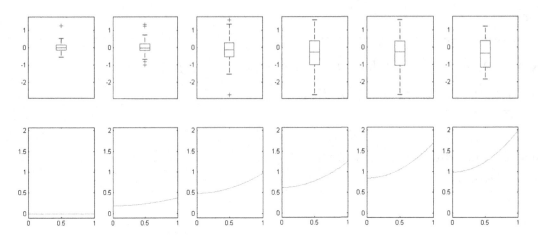

FIGURE 14.1
Local boxplots for the simulated data of Example 1: (top) local boxplots of the response;
(bottom) corresponding covariate curves.

demonstration. The numbers of observations in the neighborhoods of the six selected covari-
ate curves are 35, 50, 69, 70, 60, and 39. The local boxplots of the response corresponding to
the selected covariate curves are plotted in Figure 14.1. We note that the local conditional
median is almost same for different covariate curves. However, the local upper and lower
quartiles, which determine the upper and lower boundaries of the boxes, nicely capture the
variation of the conditional distribution of Y given X.

Example 2 (Tecator data) *This data set is available in the R packages caret and fda.usc.*
It contains the values of fat and protein contents and 100-channel spectrum of absorbances
for 215 observations on finely chopped pure meat. The fat and protein contents are mea-
sured by analytical chemistry. A Tecator Infratec Food and Feed Analyzer working in the
wavelength range 850–1050 nm was used to measure the transmittance of each sample by
the near infrared transmission (NIT) principle. The absorbance is the negative of the loga-
rithm to base 10 of the transmittance. Obtaining the spectrum of a sample is cost-efficient,
whereas getting the nutritional values is expensive. It is economically important to be able
to predict the fat and protein contents from the absorbance spectrum of a sample. Hence, we
consider the spectrum as a functional covariate and the fat and protein contents as response
variables. Here, the response may be viewed as real-valued (if we analyze the fat or protein
content individually), or considered as bivariate (if the fat and protein contents are analyzed
simultaneously).

As in the preceding example, we construct local boxplots for both protein content and
fat content, taking the curve of absorbance spectrum as the covariate. The radius of the
neighborhood of each covariate curve is fixed at 0.25. The local boxplots for some selected
covariate curves are plotted in Figure 14.2. The numbers of observations in the neighbor-
hoods of the selected covariate curves are 43, 74, 91, 74, 39, and 22. We notice that though
the local median regression detects the change in the center of the conditional distribution
with the change in the covariate, it misses some other important features of the conditional
distributions like the variation in the conditional spread of the response. The upper and
lower boundaries of the boxplots, which are the local first and the third quartiles respectively,
provide an idea of the changes in the conditional spread of the response with the change in
the covariate.

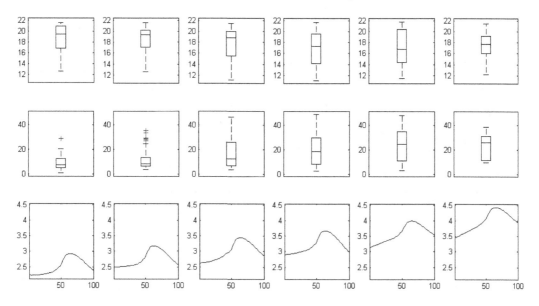

FIGURE 14.2
Local boxplots for the protein and the fat contents of the Tecator data of Example 2: (top) protein content; (middle) fat content; (bottom) corresponding covariate curves.

In both the preceding examples the responses are real-valued and the covariates are functions. The next example, however, involves a data set where the response is function-valued and the covariate is real-valued.

Example 3 (Pediatric airway data) *This data set was kindly provided to us by J. S. Marron and Yi Hong of the University of North Carolina at Chapel Hill. It was analyzed in Hong et al. (2013) (research supported by NIH 5P41EB002025-28, NIH 1R01HL105241-01, NSF EECS-1148870, NSF EECS-0925875, and NIH 2P41EB002025-26A1). It consists of cross-sectional areas of airways and the ages (in months) of 68 normal subjects and 17 patients with subglottic stenosis. The cross-sectional area is recorded at 168 points along the central line of the airway between the true vocal cord and the trachea carina. The ages are between 6 and 185 months. The objective here is to compare the curves of cross-sectional areas of subglottic stenosis patients to those of the normal individuals. The cross-sectional area of the airway depends on age. So we need to estimate the curve of cross-sectional areas of the airway of a subject with a given age, by taking the curve of cross-sectional areas as a functional response, and the age of the individual as a real-valued covariate.*

We take the ages of the subjects as the covariate and compute the local mean and the local pointwise median curves for the response, which is the curve of cross-sectional areas of the airway of a normal subject as well as the patients. We choose six values of the covariate for the normal subjects as well as the patients for demonstration. For the normal subjects, we choose the radius of the neighborhoods of each value of the covariate as 15 to compute the local mean and median curves. The numbers of observations in the neighborhoods of the six selected values of the covariate in the case of the normal subjects are 9, 8, 16, 15, 17, and 5. In the case of the patients, the radius of the neighborhood of each value of the covariate is taken as 30, and the corresponding numbers of observations in the six neighborhoods are 6, 8, 7, 5, 5, and 5. The data curves in a neighborhood along with the local mean and the local median curves for the selected values of the covariate for the normal subjects and the patients are plotted in Figures 14.3 and 14.4, respectively. For both the normal individuals and the patients, we observe that while the local mean and local median

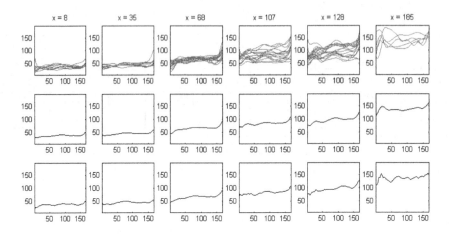

FIGURE 14.3
Plots for the pediatric airway data corresponding to the normal subjects: (top) local response curves; (middle) local mean regression curves; (bottom) local pointwise median regression curves. Corresponding covariate (X) values in months are given at the top of each column.

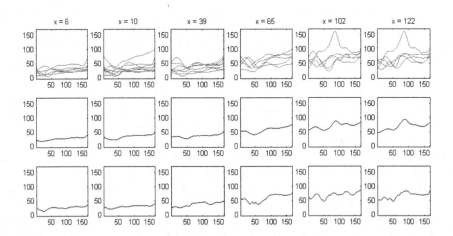

FIGURE 14.4
Plots for the pediatric airway data corresponding to the patients: (top) local response curves; (middle) local mean regression curves; (bottom) local pointwise median regression curves. Corresponding covariate (X) values in months are given at the top of each column.

curves detect some changes of the centers of the conditional distributions of the response for the different covariate values, these local curves fail to detect many other features (e.g., the variations in the conditional spread of the response given the covariate) that are visible in the data curves lying in the neighborhood of the covariate. One would expect to be able to capture those features by carrying out quantile regression analysis with some appropriate choice of quantile for such functional response.

We consider kernel-based nonparametric quantile regression, where the response lies in a normed linear space and the covariate lies in a semi-metric space. Existence of conditional quantiles in this setup and construction of their estimates are described in Section 14.2. Nonparametric estimation is discussed in Section 14.3. Sample conditional quantiles and their uses are demonstrated using simulation and real data examples in Section 14.4. Finally, consistency for the sample quantiles is established under some assumptions in Section 14.5.

14.2   Regression quantiles in Banach spaces

In this section, we define conditional quantiles in Banach spaces and discuss the requirements for their existence and uniqueness. The response variable, Y, is a random element in some Banach space $\mathcal{V}$, and the covariate, X, is a random element in a semi-metric space $(\mathcal{E}, d)$, where d is a semi-metric.

Let us first consider the case where the response Y is a real-valued random variable with conditional distribution function $F(\cdot \,|\, x)$ given $X = x$. Let $\alpha \in (0, 1)$. A conditional α-quantile of Y given $X = x$ is $Q_Y(\alpha \,|\, x)$ such that $F(Q_Y(\alpha \,|\, x) - \,|\, x) \leqslant \alpha$ and $F(Q_Y(\alpha \,|\, x) \,|\, x) \geqslant \alpha$. A conditional quantile $Q_Y(\alpha \,|\, x)$ also minimizes the function $\rho_\alpha(Q \,|\, x) = E\left[|Y - Q| + (2\alpha - 1)(Y - Q) \,|\, X = x \right]$ with respect to Q if $E\left[|Y| \,|\, X = x \right] < \infty$ (see, for example, Ferguson, 1967). Note that we need $E\left[|Y| \,|\, X = x \right] < \infty$ for the existence of $\rho_\alpha(Q \,|\, x)$. Modifying the expression for $\rho_\alpha(Q \,|\, x)$, we define the function $g_\tau(Q \,|\, x)$ as

$$g_\tau(Q \,|\, x) = E\left[|Y - Q| - |Y| \,\big|\, X = x \right] - \tau Q, \tag{14.1}$$

where $-1 < \tau = 2\alpha - 1 < 1$. The function $g_\tau(Q \,|\, x)$ is finite for each Q and x even when $E\left[|Y| \,|\, X = x \right] = \infty$ because $\big| |Y - Q| - |Y| \big| \leqslant |Q|$. Clearly, $Q_Y(\alpha \,|\, x)$ minimizes $g_\tau(Q \,|\, x)$ with respect to Q. The real-valued function $g_\tau(\cdot \,|\, x)$ can be defined on a Banach space $\mathcal{B}$ for τ lying in the open unit ball of $\mathcal{B}^*$, the dual space of $\mathcal{B}$ (see Chaudhuri, 1996; Chakraborty and Chaudhuri, 2014). We now define conditional quantiles of Y given $X = x$ as an element in $\mathcal{B}$, which minimizes an extension of $g_\tau(\cdot \,|\, x)$ in $\mathcal{B}$.

Definition 1 *Let the response variable Y be a random element in a Banach space $\mathcal{B}$, and the covariate X be a random element in a semi-metric space $(\mathcal{E}, d)$. For $\tau \in \mathcal{B}^*$ with $\|\tau\| < 1$, and $x \in \mathcal{E}$, we define the function $g_\tau(\cdot \,|\, x) : \mathcal{B} \to \mathbb{R}$ by*

$$g_\tau(Q \,|\, x) = E\left[\|Y - Q\| - \|Y\| \,\big|\, X = x \right] - \tau(Q). \tag{14.2}$$

The regression τ-quantile of Y given $X = x$ is defined as a minimizer of $g_\tau(\cdot \,|\, x)$. We shall denote this conditional τ-quantile by $Q_Y(\tau \,|\, x)$.

Note that, when the response Y is a real random variable and $-1 < \tau < 1$, $Q_Y(\tau \,|\, x)$ is same as the usual α-quantile of Y given $X = x$, where $\alpha = (\tau + 1)/2$. The value of the function $g_\tau(Q\,|\,x)$ is finite for every Q and x, even when the conditional expectation of Y given $X = x$ does not exist. Note that, if $E\left[\|Y\| \mid X = x\right] < \infty$, the conditional τ-quantile $Q_Y(\tau\,|\,x)$ can be defined as a minimizer of $g_0(Q\,|\,x) = E\left[\|Y - Q\| \mid X = x\right] - \tau(Q)$. The problem of existence and uniqueness of the spatial median in Banach spaces was considered earlier by Valadier (1984) and Kemperman (1987), and our proof of the following theorem closely follows the main arguments used by these authors.

Theorem 1 *Let the response space $\mathcal{B}$ be the dual of some Banach space. Then $Q_Y(\tau\,|\,x)$ exists in $\mathcal{B}$. Further, $Q_Y(\tau\,|\,x)$ is unique if $\mathcal{B}$ is strictly convex and the support of the conditional distribution of Y given $X = x$ is not contained in a straight line in $\mathcal{B}$.*

Proof *Since $\mathcal{B}$ is the dual of some Banach space, by the Banach–Alaoglu theorem, the closed ball $B(0,c)$ centered at $0 \in \mathcal{B}$ with radius c is weakly compact. As the function $g_\tau(Q\,|\,x)$ is continuous in Q, it is weakly continuous. Hence, $g_\tau(Q\,|\,x)$ has a minimizer Q in $B(0,c)$ for every $c > 0$, and for a minimizer Q we have $g_\tau(Q\,|\,x) \leqslant 0$, since $g_\tau(0\,|\,x) = 0$. Note that*

$$g_\tau(Q\,|\,x) = E\left[\|Y - Q\| - \|Y\| \mid X = x\right] - \tau(Q)$$

$$= \|Q\|\, E\left[\left\|\|Q\|^{-1}Y - \|Q\|^{-1}Q\right\| - \|Q\|^{-1}\|Y\| \,\middle|\, X = x\right] - \|Q\|\,\tau\left(\|Q\|^{-1}Q\right)$$

$$= \|Q\|\, E\left[\left\|\|Q\|^{-1}Y - V\right\| - \|Q\|^{-1}\|Y\| - \tau(V) \,\middle|\, X = x\right],$$

where $V \in \mathcal{B}$ and $\|V\| = 1$. Clearly, $g_\tau(Q\,|\,x) \to \infty$ as $\|Q\| \to \infty$ because $\|\tau\| < 1$. So, we can get a constant $c_1 > 0$ such that $g_\tau(Q\,|\,x) > 1$ whenever $\|Q\| > c_1$. So, the minimizer Q of $g_\tau(Q\,|\,x)$ in $B(0, c_1)$ must be the global minimizer of $g_\tau(Q\,|\,x)$. Hence, $Q_Y(\tau\,|\,x)$ exists in $\mathcal{B}$. Moreover, if $\mathcal{B}$ is strictly convex and the support of the conditional distribution of Y given $X = x$ is not contained in a straight line in $\mathcal{B}$, then the function $g_\tau(Q\,|\,x)$ becomes strictly convex in view of the arguments in the proof of Kemperman (1987, Theorem 2.17). This rules out the possibility of more than one minimizer, implying that $Q_Y(\tau\,|\,x)$ is unique.

The above theorem ensures the existence of $Q_Y(\tau\,|\,x)$ in many Banach spaces. For example, if $\mathcal{B} = L_p(\mathcal{X}, \mathcal{A}, \mu)$ with $1 < p < \infty$, where $(\mathcal{X}, \mathcal{A}, \mu)$ is some measure space, then $\mathcal{B}$ is the dual of $L_q(\mathcal{X}, \mathcal{A}, \mu)$, where $p^{-1} + q^{-1} = 1$. Moreover, if the measure μ is sigma-finite, then the space $L_\infty(\mathcal{X}, \mathcal{A}, \mu)$ is the dual of $L_1(\mathcal{X}, \mathcal{A}, \mu)$. Any Hilbert space is self-dual. So, the existence of $Q_Y(\tau\,|\,x)$ is guaranteed if the response space $\mathcal{B}$ is one of these spaces. Also, any Hilbert space or L_p-space ($0 < p < \infty$) is strictly convex. So, if the support of the conditional distribution of Y given $X = x$ is not contained in a straight line in these spaces, then $Q_Y(\tau\,|\,x)$ is unique. It is easy to show that the sequence space l_1 is the dual of the Banach space c_0 consisting of the bounded sequences converging to 0, equipped with the l_∞ norm (see Bhatia, 2009, p. 51). So, we can also apply the above theorem when the response is in the l_1 space.

It can be shown that $L_1[a, b]$ is not the dual of any Banach space (see Albiac and Kalton, 2006, p. 147). So, we cannot guarantee the existence of $Q_Y(\tau\,|\,x)$ in $L_1[a, b]$ using the above theorem. However, it is easy to show that a minimizer $Q_Y(\tau\,|\,x)$ of $g_\tau(Q\,|\,x)$ exists. Let $\tau \in \mathcal{B}^* = L_\infty[a, b]$ with $\tau(t) \in (-1, 1)$ for all t and $\|\tau\|_\infty = \sup_{a \leqslant t \leqslant b} |\tau(t)| < 1$. Then

$$g_\tau(Q\,|\,x) = E\left[\|Q - Y\|_1 \mid X = x\right] - \tau(Q)$$

$$= \int_a^b \left[E\left[|Y(t) - Q(t)| \mid X = x\right] - \tau(t)\,Q(t)\right]\,dt.$$

It is obvious that a minimizer of $g_\tau(Q\,|\,x)$ is $Q_Y(\tau\,|\,x)(t) = Q_{Y(t)}(\tau(t)\,|\,x)$, where

$Q_{Y(t)}\left(\tau\left(t\right)\mid x\right)$ is a $\tau\left(t\right)$-quantile of $Y\left(t\right)$ given $X=x$. In other words, $Q_Y\left(\tau\mid x\right)$ is nothing but a conditional pointwise (with respect to t) τ-quantile. Also, $Q_Y\left(\tau\mid x\right)$ is unique if $Q_{Y(t)}\left(\tau\left(t\right)\mid x\right)$ is unique for every t. Further, if the conditional distribution function $F_{Y(t)}(y\mid x)$ of $Y(t)$ given $X=x$ is continuous in both y and t, and the function $\tau\left(t\right)$ is continuous, then it is easy to see that $Q_Y\left(\tau\mid x\right)\left(t\right)$ is also continuous in t, implying that $g_\tau\left(Q\mid x\right)$ has a minimizer in $L_1\left[a,b\right]$.

As we have seen in the first two examples in Section 14.1 with real-valued responses, quantile regression is useful in detecting variations in the conditional spread of the response given the covariate. The quantile regression for Banach-valued response developed in this section can also be used to investigate the presence of heteroskedasticity in the data. A measure of conditional spread based on conditional quantiles can be defined as

$$D\left(\tau\mid x\right)=\left\|Q\left(\tau\mid x\right)-Q\left(-\tau\mid x\right)\right\|,$$

where $\left\|\tau\right\|<1$. Clearly, $D\left(\tau\mid x\right)$ generalizes the conditional inter-quantile range of a real-valued response.

14.3   Nonparametric estimation

We construct a nonparametric estimate of the conditional quantile, using a kernel function $K\left(\cdot\right):\left[0,\infty\right)\rightarrow\left[0,\infty\right)$ and bandwidth $h_n>0$. Let $\left(X_1,Y_1\right),\ldots,\left(X_n,Y_n\right)$ be independent random elements with the same joint distribution as $\left(X,Y\right)$ constituting the sample. The kernel estimate $Q_{n,Y}\left(\tau\mid x\right)$ of $Q_Y\left(\tau\mid x\right)$ can be defined as the minimizer of

$$g_{n,\tau}\left(Q\mid x\right)=\frac{\sum_{i=1}^n\left\|Y_i-Q\right\|K(h_n^{-1}d(x,X_i))}{\sum_{i=1}^n K(h_n^{-1}d(x,X_i))}-\tau\left(Q\right).$$

The computation of this minimizer, when the response space is infinite-dimensional, is difficult. We shall use finite-dimensional approximations (cf. Chakraborty and Chaudhuri, 2014).

Recall that a Banach space is said to have a Schauder basis if if contains a sequence $\left\{e_n\right\}$ such that for every element v in that Banach space, $v=\sum_{n=1}^\infty v_n e_n$ for some sequence of real numbers $\left\{v_n\right\}$. We now describe the computation of the estimate of $Q_Y\left(\tau\mid x\right)$, when the response space $\mathcal{B}$ is the dual of some Banach space, strictly convex and reflexive, and has a Schauder basis. Let $\left\{e_n\right\}$ be a Schauder basis of $\mathcal{B}$. For $v\in\mathcal{B}$, let $v=\sum_{n=1}^\infty v_n e_n$. Define $\psi_n\in\mathcal{B}^*$ by $\psi_n\left(v\right)=v_n$ for each $n=1,2,\ldots$. Assume that $\left\{\psi_n\right\}$ constitutes a Schauder basis of $\mathcal{B}^*$. This assumption holds in separable Hilbert spaces and L_p spaces with $p\in\left(1,\infty\right)$. Note that if $\mathcal{B}$ is a separable Hilbert space, it has a countable orthonormal basis, and $e_n=\psi_n$ for all n.

Let $\left\{d_n\right\}$ be a sequence of positive integers increasing to infinity. Let $\mathcal{Z}_n$ be the linear span of $\left\{e_1,e_2,\ldots,e_{d_n}\right\}$. For $v\in\mathcal{B}$, define $v^{(n)}=\sum_{i=1}^{d_n}v_i e_i$. So, if $Q_Y\left(\tau\mid x\right)=\sum_{n=1}^\infty q_n e_n$, we have $Q_Y^{(n)}\left(\tau\mid x\right)=\sum_{i=1}^{d_n}q_i e_i$. Similarly, for $\tau\in\mathcal{B}^*$, $\tau^{(n)}=\sum_{i=1}^{d_n}\tau_i\psi_i$. Finally, we define $g_{n,\tau}\left(\cdot\mid x\right)$ as

$$g_{n,\tau}\left(Q\mid x\right)=\frac{\sum_{i=1}^n\left(\left\|Y_i^{(n)}-Q\right\|\right)K(h_n^{-1}d(x,X_i))}{\sum_{i=1}^n K(h_n^{-1}d(x,X_i))}-\tau^{(n)}(Q). \tag{14.3}$$

The conditional sample τ-quantile $Q_{n,Y}\left(\tau\mid x\right)$ is computed as the minimizer with respect to $Q\in\mathcal{Z}_n$ of $g_{n,\tau}\left(Q\mid x\right)$.

Since the estimate $Q_{n,Y}(\tau \,|\, x)$ of $Q_Y(\tau \,|\, x)$ is defined on a finite-dimensional space, its computation is essentially the same as the computation of finite-dimensional multivariate quantiles, which has been described in Chaudhuri (1996). If $K(z) = I(0 \leqslant z \leqslant 1)$, then following Chaudhuri (1996), for each $i \in C_n = \{k \,|\, d(x, X_k) \leqslant h_n\}$, we first check whether the condition

$$\left\| \sum_{j:j \in C_n, j \neq i} \left\{ \|Y_j - Y_i\|^{-1}(Y_j - Y_i) \right\} + (|C_n| - 1)\tau \right\| \leqslant (1 + \|\tau\|) \tag{14.4}$$

is satisfied. If (14.4) is satisfied for some $i \in C_n$, then we set $Q_{n,Y}(\tau \,|\, x) = Y_i$. Otherwise, we try to solve $g_n^{(1)}(Q \,|\, x) = 0$ in $\mathcal{Z}_n$ using a Newton–Raphson type iterative method.

It is easier to compute the estimate of $Q_Y(\tau \,|\, x)$ when $\mathcal{B} \equiv L_1[a, b]$ and $\tau \in \mathcal{B}^* \equiv L_\infty[a, b]$ with $\|\tau\|_\infty < 1$. We first choose $a = t_0 < t_1 < \ldots < t_s = b$ and estimate $Q_{Y(t)}(\tau(t) \,|\, x)$ for each $t \in \{t_0, t_1, \ldots, t_s\}$. We use a Nadaraya–Watson type estimate for $Q_{Y(t)}(\tau(t) \,|\, x)$ by choosing a kernel function $K(\cdot) : [0, \infty) \to [0, \infty)$ and a positive number h_n as the bandwidth of the kernel. Define the conditional sample distribution function $F_{n,Y(t)}(y \,|\, x)$ of $Y(t)$ given $X = x$ by

$$F_{n,Y(t)}(y \,|\, x) = \frac{\sum_{i=1}^n I(Y(t) \leqslant y) \, K(h_n^{-1} d(x, X_i))}{\sum_{i=1}^n K(h_n^{-1} d(x, X_i))}$$

for $t \in \{t_0, t_1, \ldots, t_s\}$. We take $Q_{n,Y(t)}(\tau(t) \,|\, x) = \inf\{y \,|\, F_{n,Y(t)}(y \,|\, x) \geqslant (\tau(t) + 1)/2\}$, which is a $\tau(t)$-quantile of $F_{n,Y(t)}(y \,|\, x)$. Finally, for any $t_{i-1} < t < t_i$ with $i = 1, \ldots, s$, we set $Q_{n,Y(t)}(\tau(t) \,|\, x) = [(t - t_{i-1})/(t_i - t_{i-1})]Q_{n,Y(t_i)}(\tau(t_i) \,|\, x) + [(t_i - t)/(t_i - t_{i-1})]Q_{n,Y(t_{i-1})}(\tau(t_{i-1}) \,|\, x)$.

We estimate the conditional spread measure $D(\tau \,|\, x)$ by its natural sample analogue $D_n(\tau \,|\, x) = \|Q_{n,Y}(\tau \,|\, x) - Q_{n,Y}(-\tau \,|\, x)\|$.

The consistencies of the estimates $Q_{n,Y}(\tau \,|\, x)$ and $D_n(\tau \,|\, x)$ are established in Section 14.5.

14.4 Data analysis

In this section, we use the methods of quantile regression to analyze two simulated and three real data sets. In our analysis, we always model the response as a random element in some appropriate L_2 space, and quantile regression is done using spatial quantiles. All computations are done using the kernel $K(z) = I(0 \leqslant z \leqslant 1)$. The bandwidth h_n is chosen by leave-one-out cross-validation such that

$$h_n = \arg\min_h \sum_{i=1}^n \left\| Y_i - \tilde{m}_n^{(-i)}(X_i, h) \right\|,$$

where $\tilde{m}_n^{(-i)}(z, h)$ is the conditional sample spatial median at $X = z$, constructed with bandwidth h and leaving out the sample point (X_i, Y_i).

14.4.1 Simulation

We consider a heteroskedastic model and a homoskedastic model together to demonstrate how quantile regression can be employed to investigate heteroskedasticity in the sample.

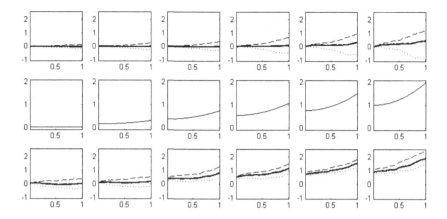

FIGURE 14.5

Plots of conditional quantiles for the heteroskedastic and homoskedastic models: (top) conditional quantile plots for the heteroskedastic model; (bottom) conditional quantile plots for the homoskedastic models; (middle) the corresponding covariate curves. The dashed, solid, and dotted curves in the plots in the top and bottom rows are $Q_{n,Y}(\tau \mid x)$, $Q_{n,Y}(0 \mid x)$ and $Q_{n,Y}(-\tau \mid x)$ respectively, for the heteroskedastic and the homoskedastic models.

In both models, the covariate $X(t) = (1 + t^2) U$ for $1 \leqslant t \leqslant 1$, where $U \sim U(0,1)$. In the heteroskedastic model, the response $Y(t) = X(t) B(t)$ for $0 \leqslant t \leqslant 1$, while in the homoskedastic model $Y(t) = X(t) + 0.5 B(t)$ for $0 \leqslant t \leqslant 1$. Here $B(t)$ is standard Brownian motion on $[0,1]$. We take the sample size to be 100 in both cases.

For a given x, let N_n be the number of X_i such that $d(x, X_i) \leqslant h_n$. We take d_n to be the largest integer less than or equal to $\sqrt{N_n}$ (cf. Chakraborty and Chaudhuri, 2014). The conditional covariance operator is estimated by the weighted sample dispersion operator, where the weights are based on the aforementioned kernel function and the bandwidth. The collection of eigenfunctions corresponding to the d_n largest eigenvalues of the estimated conditional covariance operator of Y given $X = x$ is chosen as the basis $\{e_1(x), e_2(x), \ldots, e_{d_n}(x)\}$. We take $\tau = 0.5 e_1(x)$, where $e_1(x)$ is nothing but the estimated first principal component of Y given $X = x$. The estimated conditional median $Q_{n,Y}(0 \mid x)$ and the quantiles $Q_{n,Y}(\tau \mid x)$ and $Q_{n,Y}(-\tau \mid x)$ are curves. To plot the conditional quantile curves, we first order the covariate curves by their L_2 norm, and then select six covariate curves whose ranks are equispaced in that ordering. The estimated conditional quantiles for the heteroskedastic and homoskedastic models along with the corresponding covariate curves are plotted in Figure 14.5.

The conditional spread measures $D_n(\tau \mid x)$ for the heteroskedastic model as well as for the homoskedastic model are plotted Figure 14.6. The plots for the homoskedastic and heteroskedastic data are very different in nature. The plot of $D_n(\tau \mid x)$ clearly reflects the variation of the conditional spread over the covariate space in the case of the heteroskedastic model.

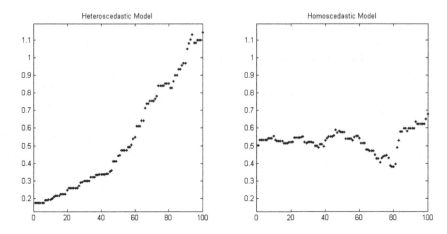

FIGURE 14.6
Conditional spread plots for the heteroskedastic model and the homoskedastic model: (left) heteroskedastic model; (right) homoskedastic model.

14.4.2 Tecator data

The Tecator data have already been described in Example 2. There, we used conditional boxplots to analyze the data, and the plots were presented in Figure 14.2. Here, the response Y is bivariate in nature, its two elements being the fat and protein contents of a meat sample, and the covariate X is the corresponding absorbance spectrum. For bivariate data, Marden (1998) introduced spider web plots using bivariate spatial quantiles. We construct the conditional spider web plots of the response for some selected covariate curves, for the angles $(2\pi k/10)$, $k = 1, \ldots, 10$, and the radius lengths $0.25, 0.5, 0.75$, respectively. The conditional spider web plots, along with the conditional boxplots for the fat and protein contents individually, are presented in Figure 14.7.

We notice from the conditional spider web plots that the conditional distribution of the response is somewhat skewed, but the degree of skewness varies with the covariate curves. The conditional spider web plots also reflect the heteroskedasticity present in the sample.

14.4.3 Pediatric airway data

This data set has been discussed in Example 3. Recall that here the curve of cross-sectional areas of the airway is a functional response (Y), and the age is a real-valued covariate (X).

We consider the response as a random function in an L_2 space and compute the sample conditional quantiles using the procedure described in Section 14.4.1. In Figure 14.8, the conditional quantile curves are plotted for the normal individuals and the patients for some selected ages. The conditional spread measures $D_n(\tau \mid x)$ for both groups of individuals are plotted in Figure 14.9.

The differences in the data corresponding to the two groups of individuals are clearly visible in the plots, which indicate the presence of heteroskedasticity in the normal individuals as well as the patients. Some of the extreme cases in the response corresponding to the patients appear to be outliers. This further justifies the use of median and quantile regression for this data set instead of mean regression, which is heavily influenced by outliers in the data.

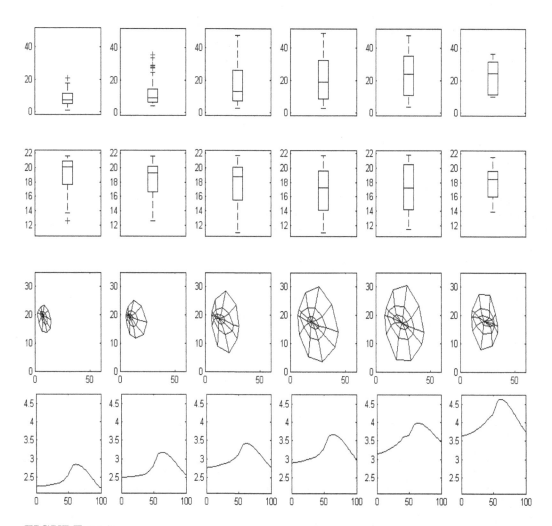

FIGURE 14.7

Conditional boxplots and spider web plots for the Tecator data. The first and second rows contain the conditional boxplots for the fat and protein contents, respectively. The third row contains the conditional spider web plots. The corresponding covariate curves are plotted in the fourth row.

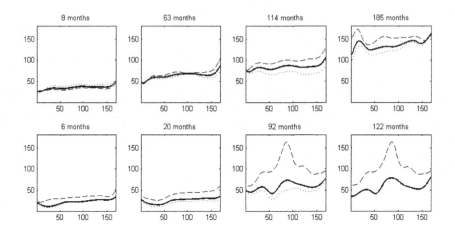

FIGURE 14.8
Plots of conditional quantiles for the pediatric airway data: (top) normal individuals; (bottom) patients. The dashed, solid, and dotted curves are $Q_{n,Y}(\tau \mid x)$, $Q_{n,Y}(0 \mid x)$ and $Q_{n,Y}(-\tau \mid x)$, respectively.

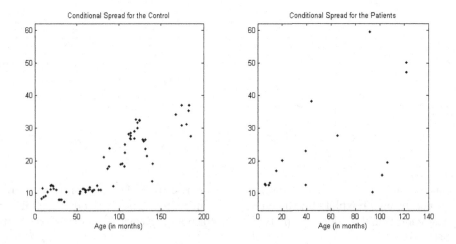

FIGURE 14.9
Conditional spread plots for (left) normal individuals and (right) patients in the pediatric airway data.

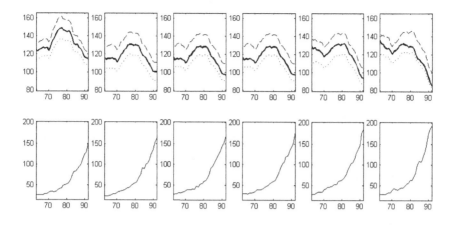

FIGURE 14.10
Plots of conditional quantiles of sales given prices for the cigarette data. The dashed, solid, and dotted curves in the plots in the top row are $Q_{n,Y}(\tau \mid x)$, $Q_{n,Y}(0 \mid x)$ and $Q_{n,Y}(-\tau \mid x)$, respectively, for the sales of cigarettes. The corresponding price curves are plotted in the bottom row.

14.4.4 Cigarette data

As an example where the response and covariate are both functions we consider the cigarette data set available in the Ecdat package in R. This contains the sales in packs per capita and the price per pack of cigarettes in 46 states of the USA for the years 1963–1992. We consider the curves of yearly prices and yearly sales over the 30-year time period as functions of time. So, we have 46 pairs of functional observations and we treat them as 46 independent and identically distributed observations on pairs of random functions in appropriate L_2 spaces. We first investigate the dependence of sales on prices by taking the curve of the sale of cigarettes as the response and the curve of the price per pack as the covariate. The method of estimating the conditional quantiles is the same as in Section 14.4.1.

In Figure 14.10 we plot the conditional quantiles of sales of cigarettes for six selected price curves that can be ranked by their L_2 norms, and the figures are arranged in increasing order of those norms. We note that over all the price levels, the sales of cigarettes decreased a little around 1970, then increased to reach a maximum level in the latter half of the 1970s. After 1980, the sales decreased monotonically. The sales appear to be higher in states with price curves having lower ranks in their L_2 norms. However, the sales do not appear to differ much among the states with mid-level or high prices. We also observe that during the 1970s in states with price curves with very low ranks in their L_2 norms, the increase in sales was much more pronounced than in the states with mid-level or higher price levels. The conditional spread measure $D_n(\tau \mid x)$ for yearly sales given the price per pack of cigarettes is plotted in Figure 14.11. The homoskedastic nature of the data is reflected in the plot.

14.4.4.1 Regression of price curve on sales curve

We next consider the curve of the price per pack of cigarettes as the response and the curve of the sales of cigarettes as the covariate to study the dependence of prices on the sales of cigarettes. We undertake this study to explore a possible causal relationship between prices and sales. For example, if the dependence of sales on prices is in some sense much stronger

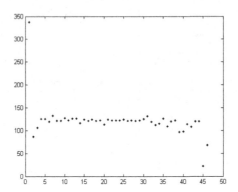

FIGURE 14.11
Conditional spread plots for sales given prices in the cigarette data.

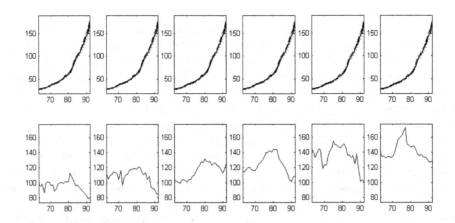

FIGURE 14.12
Plots of conditional quantiles of prices given sales for the cigarette data for $\tau = 0.5e_1(x)$. The dashed, solid, and dotted curves in the plots in the top row are $Q_{n,Y}(\tau \mid x)$, $Q_{n,Y}(0 \mid x)$, and $Q_{n,Y}(-\tau \mid x)$, respectively, for the prices of cigarettes. The corresponding sales curves are plotted in the bottom row.

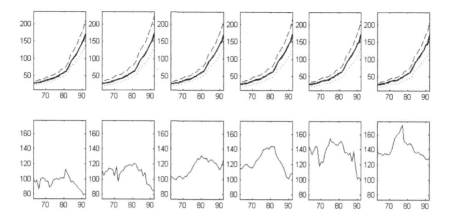

FIGURE 14.13
Plots of conditional quantiles of prices given sales for the cigarette data for $\tau = 0.99e_1(x)$. The dashed, solid, and dotted curves in the plots in the top row are $Q_{n,Y}(\tau \mid x)$, $Q_{n,Y}(0 \mid x)$, and $Q_{n,Y}(-\tau \mid x)$, respectively, for the prices of cigarettes. The corresponding sales curves are plotted in the bottom row.

than the dependence of prices on sales, then we can infer that prices drive sales and not vice versa. We plot the conditional quantiles for prices per pack of cigarettes for six selected sales curves in Figure 14.12. As in the earlier case, we choose six sales curves that are ranked by their L_2 norms, and the figures are arranged from lower ranks to higher ones.

We observe that the plots of regression quantiles of the prices given the sales corresponding to τ, 0 and $-\tau$ do not show significant variation over the selected covariate curves. Note that here $\tau = 0.5e_1(x)$ as in all the preceding quantile plots, where $e_1(x)$ is the estimated first principal component of Y given $X = x$. The dependence between the two variables that was reflected in the previous analysis does not seem to be reflected here. On further investigation, it is found that the conditional spatial median, as well as a substantial part of the central region of the conditional distribution, does not vary significantly over the covariate curves. The cross-validated choice of h becomes too big and the resulting neighborhoods include almost all the observations. As a result, we cannot see any variation of conditional quantiles across the covariate curves. The cross-validated choice of h turns out to be 596.24, and the numbers of observations in the neighborhoods for the six selected covariate curves are 45, 45, 45, 45, 46, and 46 out of a sample size of 46.

In this case, to investigate the variation of the conditional quantiles over the covariate space, one has to select a smaller bandwidth h and also look into more extreme conditional quantiles than those considered earlier. For that, we decided to choose $h = 100$ and take $\tau = 0.99e_1(x)$. We plot the conditional quantiles, with these choices for the same six selected covariate curves as before, in Figure 14.13. Now we can detect some variation in the conditional extreme quantiles over the covariate space. This demonstrates the usefulness of conditional extreme quantiles in detecting variation in the conditional distribution of the response, which may be hard to detect otherwise. The conditional spread measure $D_n(\tau \mid x)$, where $\tau = 0.99e_1(x)$, is plotted in Figure 14.14. It is observable from this plot that there is variation in the conditional spread across the covariate space. However, for $\tau = 0.5e_1(x)$ and $h = 100$, if one computes the conditional quantiles for τ and $-\tau$ and the associated

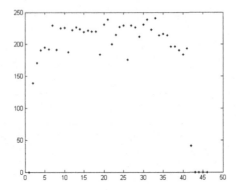

FIGURE 14.14
Conditional spread plots for prices given sales in the cigarette data for $\tau = 0.99 e_1(x)$.

conditional spread measure $D_n(\tau \mid x)$, there will be almost no observable variation in the quantiles as well as the spread measure over the covariate curves.

From the plots, we note that the dependence of prices on sales is considerably weaker than the dependence of sales on prices. This is an indication that it is price that is the cause in the variation in sales, and not vice versa.

14.5   Consistency

In this section we prove the consistency of the conditional sample quantile $Q_{n,Y}(\tau \mid x)$. The norm in a Banach space is called locally uniformly rotund if, for any sequence $\{v_n\}$ and any $v \in \mathcal{B}$ with $\|v_n\| = \|v\| = 1$ for all $n \geq 1$, $\lim_{n \to \infty} \|v_n + v\| = 2$ implies $\lim_{n \to \infty} \|v_n - v\| = 0$. The norm in any Hilbert space or any L_p-space satisfies this property. This assumption ensures that the Banach space is strictly convex. We proceed to show that the conditional sample spatial τ-quantile $Q_{n,Y}(\tau \mid x)$ converges to the conditional spatial τ-quantile $Q_Y(\tau \mid x)$ *almost surely*. We need the following additional assumptions.

1. *Assumptions on response space.* The response space $\mathcal{B}$ is a reflexive Banach space with a Schauder basis, and its norm is locally uniformly rotund.

2. *Assumption on kernel.* The kernel function $K(\cdot)$ has support $[0,1]$ and is bounded and bounded away from zero, that is, $0 < m \leqslant K(\cdot) \leqslant M < \infty$ for some constants m and M.

3. *Assumption on bandwidth.* Define $\phi(h \mid x) = P[d(x,X) \leqslant h]$. For sample size n, we denote the bandwidth used in the kernel estimate by h_n. Then $h_n \to 0$ and $\frac{\log n}{n\phi(h_n \mid x)} \to 0$ as $n \to \infty$.

4. *Assumption on conditional distributions.* The conditional distribution $\mu_Y(\cdot \mid x)$ of Y given $X = x$ is nonatomic, and its support is not contained in a straight line in $\mathcal{B}$. Further, the conditional distribution $\mu_Y(\cdot \mid z)$ of Y given $X = z$ converges *weakly* to $\mu_Y(\cdot \mid x)$ as $d(x,z) \to 0$.

We prove the consistency of the sample conditional spatial quantile in the following theorem. Consistency of spatial quantiles in a Banach space was stated in Chakraborty and Chaudhuri (2014, Theorem 3.2) in an unconditional setup. The consistency of conditional spatial quantiles presented in the following theorem can be viewed as an extension of that result.

Theorem 2 *Let $d_n \to \infty$ as $n \to \infty$ and assumptions 1–4 hold. Then*

$$\|Q_{n,Y}(\tau \mid x) - Q_Y(\tau \mid x)\| \to 0$$

almost surely as $n \to \infty$.

    Proof *Recall the definition of $g_\tau(Q \mid x)$ in (14.2). In this proof, we redefine $g_{n,\tau}(Q \mid x)$ as*

$$g_{n,\tau}(Q \mid x) = \frac{\sum_{i=1}^n [\|Y_i^{(n)} - Q\| - \|Y_i^{(n)}\|] K(h_n^{-1} d(x, X_i))}{\sum_{i=1}^n K(h_n^{-1} d(x, X_i))} - \tau^{(n)}(Q).$$

Note that this modification of $g_{n,\tau}(Q \mid x)$ does not alter its minimizer $Q_{n,Y}(\tau \mid x)$. In view of Asplund (1968, Theorems 1 and 3), it is enough to prove that

$$g_\tau(Q_{n,Y}(\tau \mid x) \mid x) \to g_\tau(Q_Y(\tau \mid x) \mid x) \tag{14.5}$$

almost surely as $n \to \infty$.

In the first step, we show that (14.5) holds when the response Y is bounded almost surely. We define the following function $\tilde{g}_{n,\tau}(\cdot \mid x)$, which will be required in subsequent arguments:

$$\tilde{g}_{n,\tau}(Q \mid x) = E[\|Y^{(n)} - Q\| - \|Y^{(n)}\| \mid X = x] - \tau^{(n)}(Q).$$

Since $g_\tau(Q \mid x)$ is minimized at $Q_Y(\tau \mid x)$, we get

$$
\begin{aligned}
0 &\leqslant g_\tau(Q_{n,Y}(\tau \mid x) \mid x) - g_\tau(Q_Y(\tau \mid x) \mid x) \\
&= [g_\tau(Q_{n,Y}(\tau \mid x) \mid x) - \tilde{g}_{n,\tau}(Q_{n,Y}(\tau \mid x) \mid x)] \\
&\quad + [\tilde{g}_{n,\tau}(Q_{n,Y}(\tau \mid x) \mid x) - \tilde{g}_{n,\tau}(Q_Y(\tau \mid x) \mid x)] \\
&\quad + [\tilde{g}_{n,\tau}(Q_Y(\tau \mid x) \mid x) - g_\tau(Q_Y(\tau \mid x) \mid x)].
\end{aligned}
\tag{14.6}
$$

The first and the third terms on the right-hand side of (14.6) are bounded by $\sup_{\|Q\| \leqslant C'} |g_\tau(Q \mid x) - \tilde{g}_{n,\tau}(Q \mid x)|$, where $C' = C_0 + \|Q_Y(\tau \mid x)\|$, C_0 being the constant in Lemma 2 stated and proved in the next subsection. We have $g_\tau(Q \mid x) - \tilde{g}_{n,\tau}(Q \mid x) \leqslant 2E[\|Y - Y^{(n)}\| \mid X = x] + \|Q\| \|\tau - \tau^{(n)}\|$ for any Q. Hence,

$$\sup_{\|Q\| \leqslant C} |g_\tau(Q \mid x) - \tilde{g}_{n,\tau}(Q \mid x)| \to 0$$

as $n \to \infty$, because $E[\|Y - Y^{(n)}\| \mid X = x] \to 0$ as $n \to \infty$ in view of the dominated convergence theorem and the facts that Y is almost surely bounded and $Y^{(n)} \to Y$ almost surely as $n \to \infty$. So, the first and the third terms on the right-hand side of (14.6) converge to 0 almost surely as $n \to \infty$.

We next proceed to analyze the behavior of the second term

$$[\tilde{g}_{n,\tau}(Q_{n,Y}(\tau \mid x) \mid x) - \tilde{g}_{n,\tau}(Q_Y(\tau \mid x) \mid x)]$$

on the right-hand side of (14.6) for large enough n. We have

$$
\begin{aligned}
&\tilde{g}_{n,\tau}\left(Q_{n,Y}\left(\tau\mid x\right)\mid x\right) - \tilde{g}_{n,\tau}\left(Q_Y\left(\tau\mid x\right)\mid x\right) \\
&= \left[\tilde{g}_{n,\tau}\left(Q_{n,Y}\left(\tau\mid x\right)\mid x\right) - g_{n,\tau}\left(Q_{n,Y}\left(\tau\mid x\right)\mid x\right)\right] \\
&\quad + \left[g_{n,\tau}\left(Q_{n,Y}\left(\tau\mid x\right)\mid x\right) - g_{n,\tau}(Q_Y^{(n)}\left(\tau\mid x\right)\mid x)\right] \\
&\quad + \left[g_{n,\tau}(Q_Y^{(n)}\left(\tau\mid x\right)\mid x) - \tilde{g}_{n,\tau}(Q_Y^{(n)}\left(\tau\mid x\right)\mid x)\right] \\
&\quad + \left[\tilde{g}_{n,\tau}(Q_Y^{(n)}\left(\tau\mid x\right)\mid x) - \tilde{g}_{n,\tau}\left(Q_Y\left(\tau\mid x\right)\mid x\right)\right].
\end{aligned}
\tag{14.7}
$$

Since $Q_{n,Y}\left(\tau\mid x\right)$ minimizes $g_{n,\tau}\left(Q\mid x\right)$, for the second term on the right-hand side of (14.7) we get

$$
g_{n,\tau}\left(Q_{n,Y}\left(\tau\mid x\right)\mid x\right) - g_{n,\tau}(Q_Y^{(n)}\left(\tau\mid x\right)\mid x) \leqslant 0
$$

for every n. For the fourth term, we get

$$
\left|\tilde{g}_{n,\tau}(Q_Y^{(n)}\left(\tau\mid x\right)\mid x) - \tilde{g}_{n,\tau}\left(Q_Y\left(\tau\mid x\right)\mid x\right)\right| \leqslant 2\|Q_Y^{(n)}\left(\tau\mid x\right) - Q_Y\left(\tau\mid x\right)\| \to 0
$$

as $n \to \infty$.

    We now examine the first and the third terms. Let $\|Q_Y\left(\tau\mid x\right)\| = K < \infty$. So $\|Q_Y^{(n)}\left(\tau\mid x\right)\| \leqslant 2K$ for all sufficiently large n. Using Lemma 2, we get that $\|Q_{n,Y}\left(\tau\mid x\right)\| \leqslant C_0 + \|Q_Y\left(\tau\mid x\right)\| = C_0 + K$. Hence, the first and third terms on the right-hand side of (14.7) are bounded above by $\sup_{\|Q\|\leqslant K_1}|g_{n,\tau}\left(Q\mid x\right) - \tilde{g}_{n,\tau}\left(Q\mid x\right)|$ for all sufficiently large n, where $K_1 = C_0 + 2K$. Let $\mu_n\left(\cdot\mid x\right)$ be the weighted empirical measure on $\mathcal{Z}_n$ defined as

$$
\mu_n\left(B\mid x\right) = \frac{\sum_{i=1}^n I(Y_i^{(n)} \in B)K(h_n^{-1}d(x,X_i))}{\sum_{i=1}^n K(h_n^{-1}d(x,X_i))}
$$

for any measurable subset B of $\mathcal{Z}_n$. Let $\tilde{\mu}_n\left(\cdot\mid x\right)$ be the conditional probability measure on $\mathcal{Z}_n$ defined by $\tilde{\mu}_n\left(B\mid x\right) = P[Y_i^{(n)} \in B\mid X = x]$, where B is a measurable subset of $\mathcal{Z}_n$. By assumptions 2 and 3, we get $\sum_{i=1}^n K(h_n^{-1}d(x,X_i)) > 0$ almost surely for all sufficiently large n, and so, $\mu_n\left(\cdot\mid x\right)$ is almost surely well defined for all sufficiently large n. For $\|Q\| \leqslant K_1$, we have

$$
\begin{aligned}
|g_{n,\tau}\left(Q\mid x\right) - \tilde{g}_{n,\tau}\left(Q\mid x\right)| &= \left|\int_{\mathcal{Z}_n} \left[\|y - Q\| - \|y\|\right]\left(\mu_n\left(\cdot\mid x\right) - \tilde{\mu}_n\left(\cdot\mid x\right)\right)(dy)\right| \\
&= K_1\left|\int_{\mathcal{Z}_n} \left[\frac{\|y - Q\| - \|y\|}{K_1}\right]\left(\mu_n\left(\cdot\mid x\right) - \tilde{\mu}_n\left(\cdot\mid x\right)\right)(dy)\right| \\
&\leqslant K_1 d_{\mathrm{BL}}\left(\mu_n\left(\cdot\mid x\right), \tilde{\mu}_n\left(\cdot\mid x\right)\right),
\end{aligned}
$$

where $d_{\mathrm{BL}}\left(\cdot,\cdot\right)$ denotes the bounded Lipschitz metric (see, for example, Pollard, 1984, p. 74). Hence, $\sup_{\|Q\|\leqslant K_1}|g_{n,\tau}\left(Q\mid x\right) - \tilde{g}_{n,\tau}\left(Q\mid x\right)| \leqslant K_1 d_{\mathrm{BL}}\left(\mu_n\left(\cdot\mid x\right), \tilde{\mu}_n\left(\cdot\mid x\right)\right)$ for all n. Using assumptions 3 and 4, we get the weak convergence of $\mu_n\left(\cdot\mid x\right)$ to $\tilde{\mu}_n\left(\cdot\mid x\right)$ almost surely as $n \to \infty$. Since $\mathcal{B}$ is a separable Banach space, we have $d_{\mathrm{BL}}\left(\mu_n\left(\cdot\mid x\right), \tilde{\mu}_n\left(\cdot\mid x\right)\right) \to 0$ and $\sup_{\|Q\|\leqslant K_1}|g_{n,\tau}\left(Q\mid x\right) - \tilde{g}_{n,\tau}\left(Q\mid x\right)| \to 0$ almost surely as $n \to \infty$. Consequently, combining the behavior of the four terms on the right-hand side of (14.7), we get that for any $\epsilon > 0$, $\tilde{g}_{n,\tau}\left(Q_{n,Y}\left(\tau\mid x\right)\mid x\right) - \tilde{g}_{n,\tau}\left(Q_Y\left(\tau\mid x\right)\mid x\right) < \epsilon$ for all sufficiently large n almost surely. Using this in (14.6), we finally get that for any $\epsilon > 0$,

$$
0 \leqslant g_\tau\left(Q_{n,Y}\left(\tau\mid x\right)\mid x\right) - g_\tau\left(Q_Y\left(\tau\mid x\right)\mid x\right) < \epsilon
$$

for all sufficiently large n almost surely. Hence, $g_\tau (Q_{n,Y} (\tau \,|\, x) \,|\, x) - g_\tau (Q_Y (\tau \,|\, x) \,|\, x) \to 0$ almost surely as $n \to \infty$, when the response Y is bounded almost surely.

In the second step, we prove that (14.5) holds when the response Y is not necessarily bounded. Given any $\epsilon > 0$, we can get $M_\epsilon > 0$ such that $P\,[\|Y\| > M_\epsilon] < C_0^{-1}\epsilon$. So, we have

$$|g_\tau (Q_{n,Y} (\tau \,|\, x) \,|\, x) - g_\tau (Q_Y (\tau \,|\, x) \,|\, x)| \leqslant 2\epsilon + |\bar{g} (Q_{n,Y} (\tau \,|\, x) \,|\, x) - \bar{g} (Q_Y (\tau \,|\, x) \,|\, x)|$$

for all sufficiently large n almost surely, where

$$\bar{g} (Q \,|\, x) = E\,[(\|Y - Q\| - \|Y\|)\, I\,(\|Y\| \leqslant M_\epsilon) \,|\, X = x] - \tau (Q).$$

Using the first step, we have $|\bar{g} (Q_{n,Y} (\tau \,|\, x) \,|\, x) - \bar{g} (Q_Y (\tau \,|\, x) \,|\, x)| \to 0$ almost surely as $n \to \infty$. Therefore, $|g_\tau (Q_{n,Y} (\tau \,|\, x) \,|\, x) - g_\tau (Q_Y (\tau \,|\, x) \,|\, x)| \to 0$ almost surely as $n \to \infty$.

Next we consider the case when the response space $\mathcal{B}$ is $L_1 [a, b]$. Note that the response space here is neither strictly convex nor a reflexive Banach space. But, as we shall show, here also the conditional sample τ-quantile $Q_{n,Y} (\tau \,|\, x)$ converges almost surely to the conditional pointwise τ-quantile $Q_Y (\tau \,|\, x)$ for any fixed x. For that, we need a new assumption which is stated below.

5. *Assumption on conditional distributions.* For all $t \in [a, b]$, the conditional distribution function $F_{Y(t)} (y \,|\, z)$ of $Y(t)$ given $X = z$ has a density $f_{Y(t)} (y \,|\, z)$, which is bounded and bounded away from 0 for z lying in a neighborhood of x, and y lying in a neighborhood of $Q_{Y(t)} (\tau (t) \,|\, z)$. Further, the conditional quantile function $Q_{Y(t)} (\tau (t) \,|\, x)$ is a continuous function of t, and we also have

$$\sup_{a \leqslant t \leqslant b} |Q_{Y(t)} (\tau (t) \,|\, z) - Q_{Y(t)} (\tau (t) \,|\, x)| \leqslant C \,(d\,(z, x))^p,$$

where $0 < p \leqslant 1$ and $C > 0$ are constants.

Theorem 3 *Suppose that the positive integer s, which appears in the construction of the conditional sample pointwise τ-quantile $Q_{n,Y} (\tau \,|\, x)$, is such that $n^{-k}s$ is bounded and bounded away from 0 as $n \to \infty$ for some positive integer k. Also assume that $\delta_n = \sup\{|t_{i+1} - t_i| \,|\, i = 1, \ldots, s\} \to 0$ as $n \to \infty$. Then, under assumptions 2, 3, and 5, $\sup_{t \in [a,b]} |Q_{n,Y} (\tau \,|\, x) (t) - Q_Y (\tau \,|\, x) (t)| \to 0$ as $n \to \infty$ almost surely.*

Proof *Denote $T_n = \{t_0, t_1, \ldots, t_s\} \subset [a, b]$. We first show that, under assumptions 2, 3, and 5, $\max_{t \in T_n} |Q_{n,Y(t)} (\tau (t) \,|\, x) - Q_{Y(t)} (\tau (t) \,|\, x)| \leqslant \epsilon_n$ except for only finitely many n almost surely, where $\epsilon_n = c_2 \sqrt{\frac{\log n}{n\phi(h_n \,|\, x)}} + c_3 h_n^p$ and $c_2, c_3 > 0$ are constants.*

It is easy to verify using assumptions 2 and 3 and the Bernstein inequality (see, for example, Pollard, 1984, p. 193) that $n^{-1} \sum_{i=1}^{n} (E[K(h_n^{-1}d(x, X))])^{-1} K(h_n^{-1}d(x, X_i)) \to 1$ almost surely as $n \to \infty$. Also, from assumptions 2 and 3, we get $nE\,[K(h_n^{-1}d(x, X))] \geqslant mn\phi (h_n \,|\, x) \to \infty$ as $n \to \infty$. Hence, $\sum_{i=1}^{n} K(h_n^{-1}d(x, X_i)) \to \infty$ almost surely as $n \to \infty$, and consequently, $\sum_{i=1}^{n} K(h_n^{-1}d(x, X_i)) > 0$ for all but finitely many n almost surely. This implies that the conditional sample distribution function $F_{n,Y(t)} (y \,|\, x)$ and the conditional pointwise sample quantiles are well defined for all sufficiently large n almost surely.

By assumption 3, we have $\epsilon_n \to 0$ as $n \to \infty$. We also have

$$P[\max_{t \in T_n} |Q_{n,Y(t)}(\tau(t) \mid x) - Q_{Y(t)}(\tau(t) \mid x)| > \epsilon_n \mid X_1, X_2, \ldots]$$

$$\leqslant \sum_{t \in T_n} P[|Q_{n,Y(t)}(\tau(t) \mid x) - Q_{Y(t)}(\tau(t) \mid x)| > \epsilon_n \mid X_1, X_2, \ldots]$$

$$\leqslant \sum_{t \in T_n} P[Q_{n,Y(t)}(\tau(t) \mid x) > Q_{Y(t)}(\tau(t) \mid x) + \epsilon_n \mid X_1, X_2, \ldots]$$

$$+ \sum_{t \in T_n} P[Q_{n,Y(t)}(\tau(t) \mid x) < Q_{Y(t)}(\tau(t) \mid x) - \epsilon_n \mid X_1, X_2, \ldots]$$

$$\leqslant \sum_{t \in T_n} P\left[\sum_{i=1}^{n} I\left(Y(t) > Q_{Y(t)}(\tau(t) \mid x) + \epsilon_n\right) W_{i,n} > \frac{1 - \tau(t)}{2} \sum_{i=1}^{n} W_{i,n} \,\middle|\, X_1, X_2, \ldots\right]$$

$$+ \sum_{t \in T_n} P\left[\sum_{i=1}^{n} I\left(Y(t) \leqslant Q_{Y(t)}(\tau(t) \mid x) - \epsilon_n\right) W_{i,n} \geqslant \frac{\tau(t) + 1}{2} \sum_{i=1}^{n} W_{i,n} \,\middle|\, X_1, X_2, \ldots\right],$$

where $W_{i,n} = K(h_n^{-1} d(x, X_i))$ for $i = 1, \ldots, n$. Denote $U_n = \sum_{i=1}^{n} W_{i,n}$. Then, using assumption 5 and the Bernstein inequality, we get

$$P\left[\sum_{i=1}^{n} I\left(Y(t) > Q_{Y(t)}(\tau(t) \mid x) + \epsilon_n\right) W_{i,n} > \frac{1 - \tau(t)}{2} U_n \,\middle|\, X_1, X_2, \ldots\right] \leqslant a_1 \exp\left[-c_4 \epsilon_n^2 U_n\right]$$

and

$$P\left[\sum_{i=1}^{n} I\left(Y(t) \leqslant Q_{Y(t)}(\tau(t) \mid x) - \epsilon_n\right) W_{i,n} \geqslant \frac{\tau(t) + 1}{2} U_n \,\middle|\, X_1, X_2, \ldots\right] \leqslant a_1 \exp\left[-c_4 \epsilon_n^2 U_n\right]$$

for all sufficiently large n, where a_1 and c_4 are positive constants. Again, using assumptions 2 and 3, we get

$$P\left[\max_{t \in T_n} |Q_{n,Y(t)}(\tau(t) \mid x) - Q_{Y(t)}(\tau(t) \mid x)| > \epsilon_n \,\middle|\, X_1, X_2, \ldots\right] \leqslant c_5 n^{-q}$$

for all sufficiently large n almost surely, where $q > 1$ and $c_5 > 0$. Hence,

$$\sum_{n=1}^{\infty} P\left[\max_{t \in T_n} |Q_{n,Y(t)}(\tau \mid x) - Q_{Y(t)}(\tau \mid x)| > \epsilon_n \,\middle|\, X_1, X_2, \ldots\right] < \infty$$

almost surely. Using the Borel–Cantelli lemma,

$$P[\limsup_{n \to \infty} \{\max_{t \in T_n} |Q_{n,Y(t)}(\tau \mid x) - Q_{Y(t)}(\tau \mid x)| > \epsilon_n\} \mid X_1, X_2, \ldots] = 0,$$

which implies

$$P[\limsup_{n \to \infty} \{\max_{t \in T_n} |Q_{n,Y(t)}(\tau \mid x) - Q_{Y(t)}(\tau \mid x)| > \epsilon_n\}] = 0.$$

Therefore, we get $\max_{t \in T_n} |Q_{n,Y(t)}(\tau \mid x) - Q_{Y(t)}(\tau \mid x)| \leqslant \epsilon_n$ except for finitely many n almost surely.

We now define a function

$$R_n(\tau \mid x)(t) = \left[\frac{t - t_{i-1}}{t_i - t_{i-1}}\right] Q_{Y(t_i)}(\tau(t_i) \mid x) + \left[\frac{t_i - t}{t_i - t_{i-1}}\right] Q_{Y(t_{i-1})}(\tau(t_{i-1}) \mid x)$$

if $t_{i-1} \leqslant t \leqslant t_i$, $i = 1, \ldots, s$. Then we get $\sup_{t \in [a,b]} |Q_{n,Y}(\tau \,|\, x)(t) - R_n(\tau \,|\, x)(t)| = O(\epsilon_n)$ as $n \to \infty$ almost surely from the above arguments, which implies $\sup_{t \in [a,b]} |Q_{n,Y}(\tau \,|\, x)(t) - R_n(\tau \,|\, x)(t)| \to 0$ as $n \to \infty$. Note that $Q_Y(\tau \,|\, x)(\cdot)$ is uniformly continuous on $[a,b]$ in view of assumption 5. Denote the modulus of continuity of a function $f(\cdot)$ on $[a,b]$ at δ as $w(f, \delta) = \sup \{|f(s) - f(t)| \,|\, |s - t| \leqslant \delta\}$. We get $w(Q_Y(\tau \,|\, x), \delta) \to 0$ as $\delta \to 0$. Also, note that

$$\sup_{t \in [a,b]} |R_n(\tau \,|\, x)(t) - Q_Y(\tau \,|\, x)(t)|$$

$$\leqslant \max \{|Q_{Y(t_{i+1})}(\tau(t_{i+1}) \,|\, x) - Q_{Y(t_i)}(\tau(t_i) \,|\, x)| \,|\, i = 1, \ldots, s\}$$

$$\leqslant \sup_{|s-t| \leqslant \delta_n} |Q_{Y(s)}(\tau(s) \,|\, x) - Q_{Y(t)}(\tau(t) \,|\, x)|$$

$$= w(Q_Y(\tau \,|\, x), \delta_n).$$

Then $w(Q_Y(\tau \,|\, x), \delta_n) \to 0$ as $n \to \infty$, and as a result $\sup_{t \in [a,b]} |R_n(\tau \,|\, x)(t) - Q_Y(\tau \,|\, x)(t)| \to 0$ as $n \to \infty$. Hence the proof is complete.

In view of the consistency of $Q_{n,Y}(\tau \,|\, x)$ established in Theorems 2 and 3, we get that $D_n(\tau \,|\, x)$ converges to $D(\tau \,|\, x)$ almost surely as $n \to \infty$. Further, in the case of sample conditional pointwise quantiles, it follows from Theorem 3 that the convergence rate of the estimate is determined by $\sqrt{\frac{\log n}{n \phi(h_n \,|\, x)}}$, $h_n{}^p$, and $w(Q_Y(\tau \,|\, x), \delta_n)$.

14.5.1 Additional mathematical details

We prove some additional mathematical results in this subsection, which were required in the development of the proof of Theorem 2.

Lemma 1 *Let v satisfy $P[\|Y\| = v \,|\, X = x] = 0$. Then, under assumption 4, $|P[\|Y^{(n)}\| \leqslant v \,|\, X = x] - P[\|Y^{(n)}\| \leqslant v \,|\, X = z]| \to 0$ as $n \to \infty$ and $d(x, z) \to 0$.*

Proof *Note that*

$$1 - \left(1 - \left(\delta^{-1}\left(v - \|y^{(n)}\|\right)\right)^+\right)^+ \leqslant I\left(\|y^{(n)}\| \leqslant v\right) \leqslant \left(1 - \left(\delta^{-1}\left(\|y^{(n)}\| - v\right)\right)^+\right)^+,$$

for any $\delta > 0$, where for a real number r, $(r)^+ = rI(r \geqslant 0)$. So,

$$P\left[\|Y^{(n)}\| \leqslant v \,\Big|\, X = x\right] - P\left[\|Y^{(n)}\| \leqslant v \,\Big|\, X = z\right]$$

$$= \int I\left(\|y^{(n)}\| \leqslant v\right) \mu_Y(dy \,|\, x) - \int I\left(\|y^{(n)}\| \leqslant v\right) \mu_Y(dy \,|\, z)$$

$$\leqslant \int \left(1 - \left(\delta^{-1}\left(\|y^{(n)}\| - v\right)\right)^+\right)^+ \mu_Y(dy \,|\, x) - \int \left[1 - \left(1 - \left(\delta^{-1}\left(v - \|y^{(n)}\|\right)\right)^+\right)^+\right]$$

$$\times \mu_Y(dy \,|\, z)$$

$$\leqslant \int \left[\left(1 - \left(\delta^{-1}\left(\|y^{(n)}\| - v\right)\right)^+\right)^+ - \left[1 - \left(1 - \left(\delta^{-1}\left(v - \|y^{(n)}\|\right)\right)^+\right)^+\right]\right] \mu_Y(dy \,|\, x)$$

$$+ \int \left[1 - \left(1 - \left(\delta^{-1} \left(v - \|y^{(n)}\| \right) \right)^{+} \right)^{+} \right] \left(\mu_Y \left(dy \,|\, x \right) - \mu_Y \left(dy \,|\, z \right) \right)$$

$$\leqslant \int \left[\left(1 - \left(\delta^{-1} \left(\|y^{(n)}\| - v \right) \right)^{+} \right)^{+} - \left[1 - \left(1 - \left(\delta^{-1} \left(v - \|y^{(n)}\| \right) \right)^{+} \right)^{+} \right] \right] \mu_Y \left(dy \,|\, x \right)$$

$$+ \int \left(1 - \left(\delta^{-1} \left(\|y^{(n)}\| - v \right) \right)^{+} \right)^{+} \left(\mu_Y \left(dy \,|\, x \right) - \mu_Y \left(dy \,|\, z \right) \right).$$

Note that

$$\int \left[\left(1 - \left(\delta^{-1} \left(\|y^{(n)}\| - v \right) \right)^{+} \right)^{+} - \left[1 - \left(1 - \left(\delta^{-1} \left(v - \|y^{(n)}\| \right) \right)^{+} \right)^{+} \right] \right] \mu_Y \left(dy \,|\, x \right)$$

$$\leqslant \int I \left(v - \delta < \|y^{(n)}\| \leqslant v + \delta \right) \mu_Y \left(dy \,|\, x \right)$$

$$= P \left[\|Y^{(n)}\| \leqslant v + \delta \,\Big|\, X = x \right] - P \left[\|Y^{(n)}\| \leqslant v - \delta \,\Big|\, X = x \right].$$

$$(14.8)$$

For the term

$$\int \left(1 - \left(\delta^{-1} \left(\|y^{(n)}\| - v \right) \right)^{+} \right)^{+} \left(\mu_Y \left(dy \,|\, x \right) - \mu_Y \left(dy \,|\, z \right) \right),$$

we first show that that function $f(y)$ on the response space $\mathcal{B}$ defined by

$$f(y) = \left(1 - \left(\delta^{-1} \left(\|y^{(n)}\| - v \right) \right)^{+} \right)^{+}$$

is a bounded Lipschitz function. Note that $0 \leqslant f(y) \leqslant 1$ and

$$|f(y_1) - f(y_2)| \leqslant \frac{\left| \|y_1^{(n)}\| - \|y_2^{(n)}\| \right|}{\delta} \leqslant \frac{\|y_1^{(n)} - y_2^{(n)}\|}{\delta}.$$

Define the basis projection map $P_n(\cdot)$ on $\mathcal{B}$ as $P_n(y) = y^{(n)}$, for every $n \geqslant 1$. It is clear that P_n is a bounded linear functional on $\mathcal{B}$ for every $n \geqslant 1$. The collection of bounded linear functionals $\{P_n\}$ satisfy $\sup_n \|P_n(y)\| < \infty$ for each $y \in \mathcal{B}$, since $P_n(y) \to y$ as $n \to \infty$. Therefore, by the uniform boundedness principle, $\sup_n \|P_n\| < \infty$. Denote $C_1 = \sup_n \|P_n\|$. Note that $C_1 \geqslant 1$ as $\|P_1\| \geqslant 1$. So we get $\|y^{(n)}\| = \|P_n(y)\| \leqslant C_1 \|y\|$ for every $n \geqslant 1$. Therefore,

$$|f(y_1) - f(y_2)| \leqslant \frac{\|y_1^{(n)} - y_2^{(n)}\|}{\delta} = \frac{\|P_n(y_1 - y_2)\|}{\delta} \leqslant \frac{C_1}{\delta} \|y_1 - y_2\|.$$

Hence, $f(y)$ is a bounded Lipschitz function. So we have

$$\int \left(1 - \left(\delta^{-1} \left(\|y^{(n)}\| - v \right) \right)^{+} \right)^{+} \left(\mu_Y \left(dy \,|\, x \right) - \mu_Y \left(dy \,|\, z \right) \right)$$

$$\leqslant \frac{C_1}{\delta} \left| \int \frac{\delta}{C_1} f(y) \left(\mu_Y \left(dy \,|\, x \right) - \mu_Y \left(dy \,|\, z \right) \right) \right|$$

$$\leqslant \frac{C_1}{\delta} d_{\mathrm{BL}} \left(\mu_Y \left(\cdot \,|\, x \right), \mu_Y \left(\cdot \,|\, z \right) \right)$$

$$(14.9)$$

if $\delta \leqslant C_1$, where $d_{\mathrm{BL}}\left(\cdot,\cdot\right)$ denotes the Bounded Lipschitz metric. Therefore, if $\delta \leqslant C_1$, combining (14.8) and (14.9), we get

$$P\left[\|Y^{(n)}\| \leqslant v \,\Big|\, X = x\right] - P\left[\|Y^{(n)}\| \leqslant v \,\Big|\, X = z\right]$$

$$\leqslant \left[P\left[\|Y^{(n)}\| \leqslant v + \delta \,\Big|\, X = x\right] - P\left[\|Y^{(n)}\| \leqslant v - \delta \,\Big|\, X = x\right]\right] \qquad (14.10)$$

$$+ \frac{C_1}{\delta} d_{\mathrm{BL}}\left(\mu_Y\left(\cdot \,|\, x\right), \mu_Y\left(\cdot \,|\, z\right)\right).$$

Now, given $\epsilon > 0$, choose $\delta_1 > 0$ such that $\delta_1 \leqslant C_1$, $P\left[v - \delta_1 < \|Y\| \leqslant v + \delta_1 \,|\, X = x\right] \leqslant \epsilon/6$ and $P\left[\|Y\| = v - \delta_1 \,|\, X = x\right] = P\left[\|Y\| = v + \delta_1 \,|\, X = x\right] = 0$. We can find such a δ_1, since $P\left[\|Y\| = v \,|\, X = x\right] = 0$, and there are at most countably many u for which $P\left[\|Y\| = u \,|\, X = x\right] > 0$. Since $\|Y^{(n)}\| \to \|Y\|$ almost surely as $n \to \infty$, $\|Y^{(n)}\| \to \|Y\|$ in distribution, and we can find N such that for all $n \geqslant N$, $\left|P\left[\|Y^{(n)}\| \leqslant v - \delta_1 \,|\, X = x\right] - P\left[\|Y\| \leqslant v - \delta_1 \,|\, X = x\right]\right| < \epsilon/6$ and $\left|P\left[\|Y^{(n)}\| \leqslant v + \delta_1 \,|\, X = x\right] - P\left[\|Y\| \leqslant v + \delta_1 \,|\, X = x\right]\right| < \epsilon/6$. Therefore, for all $n \geqslant N$, $\left|P\left[\|Y^{(n)}\| \leqslant v + \delta_1 \,|\, X = x\right] - P\left[\|Y^{(n)}\| \leqslant v - \delta_1 \,|\, X = x\right]\right| < \epsilon/2$. Since $\mathcal{B}$ is a separable Banach space, under assumption 4 we have $d_{\mathrm{BL}}\left(\mu_Y\left(\cdot \,|\, x\right), \mu_Y\left(\cdot \,|\, z\right)\right) \to 0$ as $d\left(x,z\right) \to 0$. So, we can choose $\delta_2 > 0$ such that $d_{\mathrm{BL}}\left(\mu_Y\left(\cdot \,|\, x\right), \mu_Y\left(\cdot \,|\, z\right)\right) < (\epsilon/2)C_1^{-1}\delta_1$ if $d\left(x,z\right) < \delta_2$. Therefore, from (14.10), we get that if $n \geqslant N$ and $d\left(x,z\right) < \delta_2$, then $P\left[\|Y^{(n)}\| \leqslant v \,|\, X = x\right] - P\left[\|Y^{(n)}\| \leqslant v \,|\, X = z\right] < \epsilon$.

Similarly we can show that $P\left[\|Y^{(n)}\| \leqslant v \,\Big|\, X = z\right] - P\left[\|Y^{(n)}\| \leqslant v \,\Big|\, X = x\right] < \epsilon$ for all sufficiently large n and sufficiently small $d\left(x,z\right)$. Therefore,

$$\left|P\left[\|Y^{(n)}\| \leqslant v \,\Big|\, X = x\right] - P\left[\|Y^{(n)}\| \leqslant v \,\Big|\, X = z\right]\right| \to 0$$

as $n \to \infty$ and $d\left(x,z\right) \to 0$.

Lemma 2 *Under assumptions 2–4, there exists a constant $C_0 > 0$ such that $\|Q_{n,Y}\left(\tau \,|\, x\right) - Q_Y\left(\tau \,|\, x\right)\| \leqslant C_0$ for all sufficiently large n almost surely.*

Proof *We shall show that if $\|Q\| > C_2$, where C_2 is some positive constant, then $g_{n,\tau}\left(Q \,|\, x\right) > g_{n,\tau}\left(0 \,|\, x\right)$ almost surely for all sufficiently large n. Therefore, we must have $\|Q_{n,Y}\left(\tau \,|\, x\right)\| \leqslant C_2$ for all sufficiently large n almost surely, as $Q_{n,Y}\left(\tau \,|\, x\right)$ minimizes $g_{n,\tau}\left(Q \,|\, x\right)$. This will complete the proof. Note that*

$$g_{n,\tau}\left(Q \,|\, x\right) - g_{n,\tau}\left(0 \,|\, x\right) = \frac{n^{-1}\sum_{i=1}^{n}[\|Y_i^{(n)} - Q\| - \|Y_i^{(n)}\| - \tau^{(n)}(Q)]E_n^{-1}K(h_n^{-1}d(x,X_i))}{n^{-1}\sum_{i=1}^{n} E_n^{-1}K(h_n^{-1}d(x,X_i))},$$

where $E_n = E[K(h_n^{-1}d(x,X))]$. Using assumptions 2 and 3 and the Bernstein inequality, we get that $n^{-1}\sum_{i=1}^{n} E_n^{-1}K(h_n^{-1}d(x,X_i)) \to 1$ almost surely as $n \to \infty$. So, we proceed to show that if $\|Q\| > C_2$ for some positive constant C_2, the numerator $n^{-1}\sum_{i=1}^{n}[\|Y_i^{(n)} - Q\| - \|Y_i^{(n)}\| - \tau^{(n)}(Q)]E_n^{-1}K(h_n^{-1}d(x,X_i)) > 0$ for all sufficiently large n almost surely.

Choose a positive integer m_1 such that $m_1 > \sqrt{2}[\sqrt{2} - \sqrt{1 + \|\tau\|}]^{-1}$. Choose $C_2 > 0$ such that $P\left[\|Y\| > m_1^{-1}C_2 \,|\, X = x\right] < (1/3)m_1^{-1}$, $P\left[\|Y\| = m_1^{-1}C_2 \,|\, X = x\right] = 0$ and $P\left[\|Y^{(n)}\| = m_1^{-1}C_2 \,|\, X = x\right] = 0$ for all $n \geqslant 1$. Such a constant C_2 exists since there are at most countably many y for which either $P\left[\|Y\| = y \,|\, X = x\right] > 0$ or

$P\left[\|Y^{(n)}\| = y \,|\, X = x\right] > 0$ *for some n. Now,*

$$n^{-1} \sum_{i=1}^{n} [\|Y_i^{(n)} - Q\| - \|Y_i^{(n)}\| - \tau^{(n)}(Q)] E_n^{-1} K(h_n^{-1} d(x, X_i))$$

$$= n^{-1} \sum_{i=1}^{n} [\|Y_i^{(n)} - Q\| - \|Y_i^{(n)}\| - \tau^{(n)}(Q)] E_n^{-1} K(h_n^{-1} d(x, X_i)) I[\|Y_i^{(n)}\| \leqslant m^{-1} C_2]$$

$$+ n^{-1} \sum_{i=1}^{n} [\|Y_i^{(n)} - Q\| - \|Y_i^{(n)}\| - \tau^{(n)}(Q)] E_n^{-1} K(h_n^{-1} d(x, X_i)) I[\|Y_i^{(n)}\| > m^{-1} C_2].$$

$$(14.11)$$

Consider $A_n = n^{-1} \sum_{i=1}^{n} E_n^{-1} K(h_n^{-1} d(x, X_i)) I[\|Y_i^{(n)}\| > m_1^{-1} C_2]$. *Now* $A_n - E[A_n] = \sum_{i=1}^{n} G_{i,n}$, *where*

$$G_{i,n} = \frac{1}{nE_n} \left[K(h_n^{-1} d(x, X_i)) I[\|Y_i^{(n)}\| > m_1^{-1} C_2] - E[K(h_n^{-1} d(x, X)) I[\|Y^{(n)}\| > m_1^{-1} C_2]] \right].$$

Since $E\left[K\left(h_n^{-1} d(x, X)\right) I\left[\|Y^{(n)}\| > m_1^{-1} C_2\right] \right] \geqslant 0$, *we get* $|G_{i,n}| \leqslant (nE_n)^{-1} M$ *for all i. Therefore, using the Bernstein inequality and assumption 2, we get* $P\left[|\sum_{i=1}^{n} G_{i,n}| > (1/3) m_1^{-1}\right] < 2 \exp\left[-c_6 m^{-2} n\phi\left(h_n \,|\, x\right)\right]$, *where* $c_6 > 0$ *is a constant. Using assumptions 2 and 3, we get that* $c_6 m^{-2} n\phi\left(h_n \,|\, x\right) > 2 \log n$ *for all sufficiently large n. Hence, for all sufficiently large n,* $P\left[|\sum_{i=1}^{n} G_{i,n}| > (1/3) m_1^{-1}\right] < 2 \exp\left[-c_6 m^{-2} n\phi\left(h_n \,|\, x\right)\right] < 2 \exp\left[-2 \log n\right] = 2n^{-2}$. *Therefore, using the Borel–Cantelli lemma, we get that* $|\sum_{i=1}^{n} G_{i,n}| > (1/3) m_1^{-1}$ *for only finitely many n almost surely. So,* $0 \leqslant A_n \leqslant E[A_n] + (1/3) m_1^{-1}$ *almost surely for all sufficiently large n. Note that* $E[A_n] = E_n^{-1} E\left[K\left(h_n^{-1} d(x, X)\right) P\left[\|Y^{(n)}\| > m_1^{-1} C_2 \,|\, X\right]\right]$. *Denote* $p_n = P\left[\|Y^{(n)}\| > m_1^{-1} C_2 \,|\, X = x\right]$. *Using assumption 3 and Lemma 1, we get* $E[A_n] - p_n \to 0$ *as* $n \to \infty$. *Also, since* $\|Y^{(n)}\| \to \|Y\|$ *almost surely as* $n \to \infty$, *we have* $p_n \to P\left[\|Y\| > m_1^{-1} C_2 \,|\, X = x\right] < (1/3) m_1^{-1}$. *Therefore, for all sufficiently large n,* $0 \leqslant A_n < (2/3) m_1^{-1} < m_1^{-1}$ *almost surely. Note that* $n^{-1} \sum_{i=1}^{n} E_n^{-1} K(h_n^{-1} d(x, X_i)) > 1 - (1/3) m_1^{-1}$ *almost surely for all sufficiently large n. So if* $\|Q\| > C_2$, *for the first term on the right-hand side of (14.11), we get*

$$n^{-1} \sum_{i=1}^{n} \left[\|Y_i^{(n)} - Q\| - \|Y_i^{(n)}\| - \tau^{(n)}(Q)\right] E_n^{-1} K(h_n^{-1} d(x, X_i)) I\left[\|Y_i^{(n)}\| \leqslant m_1^{-1} C_2\right]$$

$$\geqslant n^{-1} \sum_{i=1}^{n} \left[\|Q\| - \|Y_i^{(n)}\|\| - \|Y_i^{(n)}\| - \tau^{(n)}(Q)\right] E_n^{-1} K(h_n^{-1} d(x, X_i)) I\left[\|Y_i^{(n)}\| \leqslant m_1^{-1} C_2\right]$$

$$\geqslant n^{-1} \sum_{i=1}^{n} \left[\|Q\| \left(1 - 2m_1^{-1} - \|\tau\|\right)\right] E_n^{-1} K(h_n^{-1} d(x, X_i)) I\left[\|Y_i^{(n)}\| \leqslant m_1^{-1} C_2\right]$$

$$> \|Q\| \left(1 - 2m_1^{-1} - \|\tau\|\right) \left(1 - m_1^{-1}\right)$$

for all sufficiently large n almost surely. For the second term on the right-hand side of

(14.11), *we get*

$$\left| n^{-1} \sum_{i=1}^{n} \left[\|Y_i^{(n)} - Q\| - \|Y_i^{(n)}\| - \tau^{(n)}(Q) \right] E_n^{-1} K(h_n^{-1} d(x, X_i)) I \left[\|Y_i^{(n)}\| > m_1^{-1} C_2 \right] \right|$$

$$\leqslant n^{-1} \sum_{i=1}^{n} \left| \|Y_i^{(n)} - Q\| - \|Y_i^{(n)}\| - \tau^{(n)}(Q) \right| E_n^{-1} K(h_n^{-1} d(x, X_i)) I \left[\|Y_i^{(n)}\| > m_1^{-1} C_2 \right]$$

$$\leqslant n^{-1} \sum_{i=1}^{n} \|Q\| \left[1 + \|\tau\| \right] E_n^{-1} K(h_n^{-1} d(x, X_i)) I \left[\|Y_i^{(n)}\| > m_1^{-1} C_2 \right]$$

$$< \|Q\| \left[1 + \|\tau\| \right] m_1^{-1}$$

for all sufficiently large n almost surely. By the choice of m_1, it follows that $[(1 - 2m_1^{-1} - \|\tau\|)(1 - m_1^{-1}) - (1 + \|\tau\|)m_1^{-1}] = [2(1 - m_1^{-1})^2 - (1 + \|\tau\|)] > 0$. Therefore, if $\|Q\| > C_2$, comparing the two terms on the right-hand side of (14.11), we get that $n^{-1} \sum_{i=1}^{n} [\|Y_i^{(n)} - Q\| - \|Y_i^{(n)}\| - \tau^{(n)}(Q)] E_n^{-1} K(h_n^{-1} d(x, X_i)) > 0$ for all sufficiently large n almost surely. Hence, $g_{n,\tau}(Q \mid x) > g_{n,\tau}(0 \mid x)$ almost surely for all sufficiently large n if $\|Q\| > C_2$.

14.6 Concluding remarks

In this paper kernel-based nonparametric quantile regression is developed when we have a response in an infinite-dimensional space and a covariate lying in a semi-metric space. We have established the consistency of the sample conditional quantiles. An estimate of the conditional spread based on regression quantiles has been developed. The consistency of this conditional spread measure follows from the consistency of the conditional quantile estimates.

In the special case of l_1 and L_1 spaces, we have seen that our regression quantiles become pointwise regression quantiles, and they are very easy to compute. However, pointwise regression quantiles convey information about only marginals of the conditional distribution of the response. On the other hand, for finite-dimensional Euclidean spaces equipped with the l_2 norm, it follows from Koltchinskii (1997, Theorem 2.5) that the conditional spatial quantiles completely determine the conditional distribution of the response.

We have defined a measure of conditional spread $D(\tau \mid x)$ based on the conditional quantiles $Q_Y(\tau \mid x)$ and $Q_Y(-\tau \mid x)$. This measure depends on τ, and consequently, it can be viewed as a directional measure of conditional spread. A more global measure of conditional spread, which does not depend on the quantile direction τ, can be defined as $D_g(p \mid x) = \sup \{\|Q_Y(\tau_1 \mid x) - Q_Y(\tau_2 \mid x)\| \mid \|\tau_1\|, \|\tau_2\| \leqslant p\}$. One can estimate $D_g(p \mid x)$ by the maximum of $\|Q_{n,Y}(\tau_1 \mid x) - Q_{n,Y}(\tau_2 \mid x)\|$, where the maximum is computed over a finite grid of τ's inside the open ball in $\mathcal{B}^*$ with radius p. However, the construction of this estimate of $D_g(p \mid x)$ is highly computation intensive, and we have not pursued it here.

The order of convergence of an estimate provides an idea of the accuracy of that estimate for growing sample size. In the case of conditional pointwise quantiles, we have seen that the convergence rate of $Q_{n,Y}(\tau \mid x)$ depends on the small ball probability function $\phi(h_n \mid x)$. This function in turn depends on the dimension of the covariate space. If the covariate space is $\mathbb{R}^d$ and the covariate has a positive density in a neighborhood of x, then $\phi(h_n \mid x) = O(h_n^d)$. But if the covariate space is infinite-dimensional, for example, if the covariate is a continuous Gaussian Markov process on $[0, 1]$ with the L_p norm, then, following Li and

Shao (2001, Theorem 3.1) and Li (2001, Theorem 1.1), we get $\phi\left(h_n \mid x\right) = O\left(\exp\left(-ah_n^{-2}\right)\right)$ as $n \to \infty$, for some positive constant a, which makes the attainable rate of convergence of the regression quantile estimate very slow. However, the dimension of the response does not have much effect on this rate of convergence. To improve the asymptotic convergence rate of the nonparametric conditional quantile estimates, we may try to reduce the dimension of the covariate before carrying out the nonparametric regression analysis.

We intend to study the asymptotic distributions of sample conditional quantile estimates in future. Derivation of these distributions will help us to carry out tests of hypotheses and to construct confidence sets for functional parameters. When the response space is finite-dimensional, the asymptotic distribution of conditional spatial quantiles has been derived in Chaouch and Laïb (2013, 2015) for infinite-dimensional covariates. For observations in an infinite-dimensional separable Hilbert space, a Bahadur type asymptotic linear representation for the unconditional sample quantile is derived in Chakraborty and Chaudhuri (2014). If such a representation can be established in the regression setting with the response being a random element in an infinite-dimensional separable Hilbert space, the asymptotic distribution of the sample regression quantile will follow.

Bibliography

F. Albiac and N. J. Kalton. *Topics in Banach Space Theory*, volume 233 of *Graduate Texts in Mathematics*. Springer, New York, 2006.

E. Asplund. Fréchet differentiability of convex functions. *Acta Mathematica*, 121(1):31–47, 1968.

R. Bhatia. *Notes on Functional Analysis*. Hindustan Book Agency, New Delhi, 2009.

H. Cardot, C. Crambes, and P. Sarda. Quantile regression when the covariates are functions. *Nonparametric Statistics*, 17(7):841–856, 2005.

A. Chakraborty and P. Chaudhuri. The spatial distribution in infinite dimensional spaces and related quantiles and depths. *Annals of Statistics*, 42(3):1203–1231, 2014.

M. Chaouch and N. Laïb. Nonparametric multivariate l_1-median regression estimation with functional covariates. *Electronic Journal of Statistics*, 7:1553–1586, 2013.

M. Chaouch and N. Laïb. Vector-on-function quantile regression for stationary ergodic processes. *Journal of the Korean Statistical Society*, 44(2):161–178, 2015.

P. Chaudhuri. On a geometric notion of quantiles for multivariate data. *Journal of the American Statistical Association*, 91(434):862–872, 1996.

K. Chen and H.-G. Müller. Conditional quantile analysis when covariates are functions, with application to growth data. *Journal of the Royal Statistical Society, Series B*, 74: 67–89, 2012.

T. S. Ferguson. *Mathematical Statistics: A Decision Theoretic Approach*. Academic Press, New York, 1967.

F. Ferraty and P. Vieu. *Nonparametric functional data analysis: theory and practice*. Springer, New York, 2006.

L. Gardes, S. Girard, and A. Lekina. Functional nonparametric estimation of conditional extreme quantiles. *Journal of Multivariate Analysis*, 101(2):419–433, 2010.

Y. Hong, B. Davis, J. S. Marron, R. Kwitt, and M. Niethammer. Weighted functional boxplot with application to statistical atlas construction. In *International Conference on Medical Image Computing and Computer-Assisted Intervention*, pages 584–591. Springer, New York, 2013.

K. Kato. Estimation in functional linear quantile regression. *Annals of Statistics*, 40(6): 3108–3136, 2012.

J. H. B. Kemperman. The median of a finite measure on a Banach space. In Y. Dodge, editor, *Statistical Data Analysis Based on the L_1-Norm and Related Methods (Neuchâtel, 1987)*, pages 217–230. North-Holland, Amsterdam, 1987.

V. I. Koltchinskii. M-estimation, convexity and quantiles. *Annals of Statistics*, 25(2):435–477, 1997.

W. V. Li. Small ball probabilities for Gaussian Markov processes under the L_p-norm. *Stochastic Processes and Their Applications*, 92(1):87–102, 2001.

W. V. Li and Q.-M. Shao. Gaussian processes: Inequalities, small ball probabilities and applications. *Stochastic Processes: Theory and Methods*, 19:533–597, 2001.

J. I. Marden. Bivariate QQ-plots and spider web plots. *Statistica Sinica*, 8(3):813–826, 1998.

D. Pollard. *Convergence of Stochastic Processes*. Springer, New York, 1984.

M. Valadier. La multi-application médianes conditionnelles. *Probability Theory and Related Fields*, 67(3):279–282, 1984.

15

High-Dimensional Quantile Regression

Alexandre Belloni

Duke University, Durham, North Carolina, USA

Victor Chernozhukov

MIT, Cambridge, Massachusetts, USA

Kengo Kato

University of Tokyo, Tokyo, Japan

CONTENTS

15.1 Introduction

High-dimensional models arise from the need for practitioners to improve the accuracy and validity of current models and to handle the increasing availability of data. Large models can arise from using a very flexible specification with many parameters when the functional form is unknown or from having a data-rich environment with many variables that need to be incorporated into the model.

The use of such large models poses a challenge for the traditional asymptotic analysis where it is assumed that the number of parameters p is fixed as the sample size n increases. Quantile regression models in this fixed-p asymptotics have been extensively studied since the seminal work of Koenker and Bassett (1978), and Koenker (2005) provides an excellent review of the available results. However, the asymptotic results of the traditional analysis might not provide an accurate approximation for the behavior of the estimators in a high-dimensional setting where the dimension is not negligible relative to the sample size. To better capture finite-sample behavior of high-dimensional models, it has been proposed to conduct an asymptotic analysis where the model can change with the sample size. There are roughly two main asymptotic regimes: (i) p increasing but slowly relative to n; and (ii) p increasing fast with n. Regime (i) relies on the same estimators as in the fixed-p setting but fully accounts for the impact of the model size on the estimation characterizing sufficient conditions (e.g. restrictions on the growth of p relative to n) for its validity based on finite-sample analysis. Several authors have contributed work within this setting, including Welsh (1989), He and Shao (2000), and Belloni et al. (2011). Conditions are provided under which the quantile regression estimator achieves the (uniform) ℓ_2-rate of convergence of $\sqrt{p/n}$ and has a Bahadur representation that allows for the construction of uniform confidence bands. In regime (ii), new estimators are needed since the dimensionality is so large that quantile regression estimators are no longer consistent as they overfit the available data. In particular, regime (ii) allows for models with the number of parameters p being larger than the available sample size n (e.g., $p/n \to \infty$). The focus of this chapter is precisely on quantile regression models within this latter regime.

Throughout the chapter we consider an outcome of interest Y and conditioning variables Z where the conditional quantile function is given by $Q_Y(\tau \mid Z)$ for $\tau \in \mathcal{T} \subset (0,1)$. In order to estimate the conditional quantile function define $X = G(Z)$, where $G(Z)$ is a p-vector of known transformations of Z, and for a conformable vector β_τ we write

$$Q_Y(\tau \mid Z) = X'\beta_\tau + r_\tau(Z), \tag{15.1}$$

where $r_\tau(Z) := Q_Y(\tau \mid Z) - X'\beta_\tau$ denotes an approximation error function, and potentially $p \gg n$. In the parametric case we have $Z = X$ and $r_\tau(Z) = 0$, while in the nonparametric case the coefficient process $\{\beta_\tau : \tau \in \mathcal{T}\}$ is chosen to yield a good approximation of the conditional quantile function.

Although generically this problem is ill-posed for $p \gg n$, some data-generating processes have coefficient processes with specific patterns that make the model estimable. Regularization methods that exploit the knowledge of such patterns have emerged as a key approach to estimating high-dimensional models. The main pattern used in the literature is that, within the high-dimensional model, there is a small submodel $T_\tau \subset \{1, \dots, p\}$ with a manageable number of parameters ($|T_\tau| \leqslant s$) which yields a good enough approximation. These models are called approximate sparse models as only $s \ll n$ nonzero coefficients suffice to achieve a vanishing approximation error for each τ, namely,

$$\sup_{\tau \in \mathcal{T}} \|\beta_\tau\|_0 \leqslant s \quad \text{and} \quad \sup_{\tau \in \mathcal{T}} \mathbb{E}_n[r_\tau(Z)^2] \leqslant Cs \log(p \vee n)/n, \tag{15.2}$$

with high probability, where C is a fixed constant and $\|\cdot\|_0$ denotes the number of nonzero components of a vector. We note that the approximately sparse framework generalizes traditional series/sieve expansions by allowing the identity of the submodel $T_\tau := \mathrm{support}(\beta_\tau)$ to be unknown and to potentially vary across $\tau \in \mathcal{T}$. For example, if the conditional quantile function is given by an infinite series $Q_Y(\tau \mid Z) = \sum_{j=1}^{\infty} X_j \theta_{\tau j}$, we have $r_\tau(Z) = \sum_{j \notin T_\tau, j \leqslant p} X_j \theta_{\tau j} + \sum_{j > p} X_j \theta_{\tau j}$ which satisfies the requirements (15.2) provided coefficients $\{\theta_{\tau j} : j \geqslant 1\}$ decay to zero appropriately fast.

Our aim is to discuss estimators with good rates of convergence and to construct simultaneous confidence bands for the quantile process $\{\beta_\tau, \tau \in \mathcal{T}\}$ that are valid uniformly over a large class of data-generating processes. However, because of the high dimensionality of the model and the use of regularization, estimation of the conditional quantile function (prediction) and the construction of confidence regions for parameters (inference) require different estimation procedures. This is in sharp contrast to the low-dimensional case. Indeed, prediction relates to the ℓ_2 rate of convergence of the estimator of the coefficient process while inference pertains to the ℓ_∞ rate of convergence, which are fundamentally different when the dimension of the model is large relative to the sample size. There has been extensive work on these questions, mostly on linear regression models (Candes and Tao, 2007; Meinshausen and Bühlmann, 2006; Meinshausen and Yu, 2009; Bickel et al., 2009; Belloni and Chernozhukov, 2013; Belloni et al., 2012, 2014a,b; van de Geer et al., 2014; Zhang and Zhang, 2014) and more recently on nonlinear models (Belloni and Chernozhukov, 2011; Belloni et al., 2012, 2014a, 2013a,c, 2015b; van de Geer et al., 2014; Javanmard and Montanari, 2014a,b; Ning and Liu, 2014; Zhao et al., 2014). Together with the nondifferentiability of the check function and the desired uniform validity over a continuum of quantile indices, high-dimensional quantile regression models are among the most interesting and challenging cases to analyze.

Regularization can be achieved in various ways, but given the approximate sparse setting, shrinkage and model selection have been the predominant approaches in the literature. Throughout this chapter we focus on ℓ_1-norm regularization which generalizes the lasso estimator introduced by Frank and Friedman (1993), Tibshirani (1996), and Knight and Fu (2000) for linear regression models. In these linear regression models, ℓ_1-norm regularization can provide simultaneous shrinkage and model selection; we will discuss below in detail analogous properties for quantile regression models. In addition to the estimation property guarantees, we will discuss how these estimators can be characterized as the solution of convex problems which makes them suitable for reliable computational methods despite the high dimensionality.

All estimators discussed here achieve an important uniformity guarantee with respect to the (unknown) values of the coefficient process. Importantly, these uniform properties translate into more reliable finite-sample performance of inference procedures because they are robust with respect to (unavoidable) model selection mistakes. In contrast, there is now substantial theoretical and empirical evidence on the potentially poor finite-sample performance of estimators that rely on perfect model selection to build confidence regions when applied to models without separation from zero of the coefficients (i.e. small coefficients). We refer to Leeb and Pötscher (2005, 2008) and the references therein for a series of negative results.

The chapter is organized as follows. Section 15.2 is devoted to estimation of the conditional quantile function. We first derive ℓ_1 and ℓ_2 rates of convergence for the coefficient process based on ℓ_1-regularized quantile regression and refitted quantile regression post model selection. We discuss data-driven approaches to set the penalization parameters that are theoretically valid. Additionally, we discuss how to exploit group structure in quantile regression models. Section 15.3 focuses on the construction of confidence bands for the coefficient process that builds upon results of Section 15.2.

Regarding notation, the symbols P and E denote probability and expectation operators with respect to a generic probability measure. If we need to signify the dependence on a probability measure P, we use P as a subscript in P_P and E_P. For a function F we let $E_n[F(Y, X)] = \frac{1}{n}\sum_{i=1}^{n} F(Y_i, X_i)$. For a vector $v = (v_1, \ldots, v_p)' \in \mathbb{R}^p$, $\|v\|_0$ denotes the ℓ_0-"norm" of v, that is, the number of nonzero components of v, $\|v\|_1$ denotes the ℓ_1-norm of v, that is, $\|v\|_1 = |v_1| + \ldots + |v_p|$, $\|v\|_2$ denotes the Euclidean norm of v, that is, $\|v\|_2 = \sqrt{v'v}$, and $\|v\|_\infty$ denotes the ℓ_∞-norm, that is, $\|v\|_\infty = \max_{j=1,\ldots,p} |v_j|$. Moreover, for $T \subseteq \{1, \ldots, p\}$ we denote by $v_T = (v_j)_{j \in T}$ a subvector of v, and $v_{-j} = v_{\{1,\ldots,p\}\backslash\{j\}}$ as the vector v without its jth component. We use ∂_t to denote the partial derivative with respect to t. Finally, for sequences a_n, b_n, we use $a_n \lesssim b_n$ to denote that there is a constant C such $a_n \leqslant Cb_n$.

15.2   Estimation of the conditional quantile function

In this section we discuss estimators for the coefficient process $\{\beta_\tau : \tau \in \mathcal{T}\}$ that lead to reliable estimates of the conditional quantile function in (15.1). With that in mind we focus on estimators with near optimal rates of convergence in the ℓ_1- and ℓ_2-norms uniformly over the quantile indices $\tau \in \mathcal{T}$.

15.2.1   Regularity conditions

Throughout this chapter we will make the following assumptions on the (sequence of) data-generating process(es). Let δ_n and Δ_n be fixed sequences going to zero and let $c_1 > 0$ and $C_1 > 0$ be fixed constants.

Condition R. *We observe independent and identically distributed (i.i.d.) random vectors $\{Y_i, Z_i, X_i : i = 1, \ldots, n\}$ that obey the model described in (15.1).*

(i)   *There exist $s = s_n$ such that for all $\tau \in \mathcal{T} \subset [c_1, 1 - c_1]$, $\|\beta_\tau\|_0 \leqslant s$.*

(ii) *The conditional probability distribution function of Y given Z is absolutely continuous with continuously differentiable density $f_Y(\cdot \mid Z)$ such that $c_1 \leqslant \underline{f} \leqslant f_{\tau i} := f_Y(Q_Y(\tau \mid Z_i) \mid Z_i)$, $\sup_{t \in \mathbb{R}} f_Y(t \mid Z) \leqslant \bar{f} \leqslant C_1$, and $\sup_{t \in \mathbb{R}} |f_Y'(t \mid Z)| \leqslant \bar{f}' \leqslant C_1$ almost surely.*

(iii) *The covariates satisfy $c_1 \leqslant \mathrm{E}_P[\{X'\theta\}^2] \leqslant C_1$, for all $\|\theta\|_2 = 1$, $M_n \geqslant \{\mathrm{E}_P[\|X\|_\infty^{2q}]\}^{1/(2q)}$, and $M_n^2 s \log^{1/2}(p \vee n) \leqslant \delta_n n^{1/2 - 1/q}$ for some $q > 2$.*

(iv) *The approximation error satisfies $\sup_{\tau \in \mathcal{T}, i=1,\ldots,n} |r_\tau(Z_i)| \leqslant \delta_n$ and $\sup_{\tau \in \mathcal{T}} \mathrm{E}_n[r_\tau^2(Z)] \leqslant Cs \log(p \vee n)/n$ with probability $1 - \Delta_n$.*

Condition R is just a set of sufficient conditions that allows us to cover all the estimators described below. Some of the requirements can be relaxed, and we will briefly comment when appropriate. A central challenge in the high-dimensional setting is to handle design matrices that do not have full rank since potentially $p > n$, which creates a lack of identification. Despite that, Condition R ensures some good properties of the design matrix over restricted sets that will prove useful for our results. For example, the so-called sparse eigenvalues of order $k = Cs$,

$$\phi_{\min}(k) := \min_{1 \leqslant \|\delta\|_0 \leqslant k} \frac{\mathrm{E}_n[(X'\delta)^2]}{\|\delta\|_2^2} \quad \text{and} \quad \phi_{\max}(k) := \max_{1 \leqslant \|\delta\|_0 \leqslant k} \frac{\mathrm{E}_n[(X'\delta)^2]}{\|\delta\|_2^2}, \tag{15.3}$$

are bounded away from zero and from above with high probability under Condition R. A second challenge is that the neighborhood in which we obtain a quadratic minorant for the criterion function is only a small subset of $\mathbb{R}^p$. The analysis relies on the restricted nonlinear impact coefficient associated with a set $A \subseteq \mathbb{R}^p$,

$$q_A^* := \min_{\delta \in A} \frac{\{\mathrm{E}_n[(X'\delta)^2]\}^{3/2}}{\mathrm{E}_n[(X'\delta)^3]} \geqslant \min_{\delta \in A} \frac{\{\mathrm{E}_n[(X'\delta)^2]\}^{1/2}}{\|\delta\|_1 \max_{i \leqslant n} \|X_i\|_\infty} \vee \min_{\delta \in A} \left(\frac{\{\mathrm{E}_n[(X'\delta)^2]\}^{1/2}}{\|\delta\|_1 \{\mathrm{E}_n[\|X\|_\infty^3]\}^{1/3}} \right)^3,$$

which controls the quality of minoration of the quantile regression objective function by a quadratic function over the restricted set A. Technical results require that q_A^* does not go to zero too fast, which restricts the choice of A. For example, suppose we search over s-sparse vectors only to find an estimator $\hat{\beta}_\tau$ (which is computationally demanding even if s is known). Then $\delta = \hat{\beta}_\tau - \beta_\tau \in A = \{\delta \in \mathbb{R}^p : \|\delta\|_0 \leqslant 2s\}$, and $q_A^* \gtrsim_P \sqrt{\phi_{\min}(2s)}/\{s^{1/2} n^{1/(2q)} M_n\} \vee \phi_{\min}(2s)^{3/2}/\{s^{3/2} M_n^3\}$. The choice of set A varies with different estimators, as we discuss below.

15.2.2 ℓ_1-penalized quantile regression

The first estimator we consider is the ℓ_1-penalized quantile regression estimator. It has been considered in Li and Zhu (2008), Wu and Liu (2009), Belloni and Chernozhukov (2011), Kato (2011), and Wang (2013). For each $\tau \in \mathcal{T}$ define the estimator for β_τ as

$$\widehat{\beta}_\tau \in \arg\min_{\beta \in \mathbb{R}^p} \mathrm{E}_n[\rho_\tau(Y - X'\beta)] + \lambda_\tau \|\Gamma\beta\|_1, \tag{15.4}$$

where ρ_τ is the check function, Γ is a diagonal matrix with $\Gamma_{jj}^2 = \mathrm{E}_n[X_j^2]$, $j = 1, \ldots, p$, and λ_τ is the penalty parameter. In the high-dimensional case, the theoretical properties of this estimator have been studied in Belloni and Chernozhukov (2011), Kato (2011), and Wang (2013), where it was shown to have near optimal rate of convergence.

The main issue in the high-dimensional setting is the lack of full rank of the design matrix, which creates a lack of identification. A key condition for setting the parameters in the literature of ℓ_1-penalized M-estimators is that the penalty parameter needs to be larger than the ℓ_∞-norm of the gradient of the criterion at the true value. This ensures that the regularization is large enough to guarantee that the estimation error (e.g., $\widehat{\beta}_\tau - \beta_\tau$) belongs to a restricted set for which identification is possible. In the case of (15.4), the choice of penalty parameter is intended to satisfy (with high probability) that

$$\lambda_\tau \geqslant c \left\| \Gamma^{-1} \mathrm{E}_n[(\tau - 1\{Y \leqslant Q_Y(\tau \mid Z)\})X] \right\|_\infty \tag{15.5}$$

for any fixed $c > 1$. Importantly, the right-hand side of (15.5) has a pivotal distribution as $1\{Y_i \leqslant Q_Y(\tau \mid Z_i)\}$ equals 1 with probability τ and equals 0 with probability $1 - \tau$ conditionally on Z_i. This observation allows us to provide a data-driven way to set λ_τ so that, uniformly over $\tau \in \mathcal{T}$, (15.5) holds with probability $1 - \xi$. We recommend setting $c = 1.1$ and

$$\lambda_\tau := c\lambda\sqrt{\tau(1-\tau)}, \quad \text{with } \lambda := (1-\xi)\text{-quantile of } \sup_{\tau \in \mathcal{T}} \frac{\|\Gamma^{-1}\mathrm{E}_n[(\tau - 1\{U \leqslant \tau\})X]\|_\infty}{\sqrt{\tau(1-\tau)}}, \tag{15.6}$$

where $U_1, \ldots, U_n$ are i.i.d. uniform(0,1) random variables conditional on $(X_i)_{i=1}^n$. Such choice is adaptive to the correlation structure of the covariates, which typically helps to avoid overpenalization. We suggest using simulations to approximate λ. Under our conditions we will have $\lambda \lesssim \sqrt{\log(p/\xi)/n}$.

To see how the regularization restores identification consider the parametric case where $r_\tau(Z) = 0$ almost surely. By the definition of the estimator $\widehat{\beta}_\tau$ as the minimizer in (15.4), we have

$$\mathrm{E}_n[\rho_\tau(Y - X'\widehat{\beta}_\tau)] - \mathrm{E}_n[\rho_\tau(Y - X'\beta_\tau)] \leqslant \lambda_\tau \|\Gamma\beta_\tau\|_1 - \lambda_\tau \|\Gamma\widehat{\beta}_\tau\|_1, \tag{15.7}$$

and by convexity of the check function (via the subgradient), we have

$$\begin{aligned}
\mathrm{E}_n[\rho_\tau(Y - X'\widehat{\beta}_\tau)] - \mathrm{E}_n[\rho_\tau(Y - X'\beta_\tau)] &\geqslant \mathrm{E}_n[(\tau - 1\{Y \leqslant X'\beta_\tau\})X]'(\widehat{\beta}_\tau - \beta_\tau) \\
&= \{\Gamma^{-1}\mathrm{E}_n[(\tau - 1\{Y \leqslant X'\beta_\tau\})X]\}'\Gamma(\widehat{\beta}_\tau - \beta_\tau) \\
&\geqslant -(\lambda_\tau/c)\|\Gamma(\widehat{\beta}_\tau - \beta_\tau)\|_1,
\end{aligned} \tag{15.8}$$

where we used Hölder's inequality and (15.5). Since $c > 1$, (15.7) and (15.8) lead to the restriction that

$$\widehat{\beta}_\tau - \beta_\tau \in \Delta_{\tau,\mathbf{c}} := \{\delta \in \mathbb{R}^p : \|\Gamma_{T_\tau^c}\delta_{T_\tau^c}\|_1 \leqslant \mathbf{c}\|\Gamma_{T_\tau}\delta_{T_\tau}\|_1\},$$

where $\mathbf{c} = (c+1)/(c-1)$. Intuitively, the set $\Delta_{\tau,\mathbf{c}}$ contains vectors for which the ℓ_1-norm in T_τ^c (i.e. outside the support T_τ) is bounded by a constant times the ℓ_1-norm in the support

T_τ whose dimension is $s \ll p$. It follows that rates of convergence can be derived from the so-called restricted eigenvalue

$$\kappa_{\mathbf{c}} := \min_{\tau \in \mathcal{T}} \min_{\delta \in \Delta_{\tau,\mathbf{c}}} \frac{\{\mathrm{E}_n[(X'\delta)^2]\}^{1/2}}{\|\Gamma_{T_\tau} \delta_{T_\tau}\|_2}, \tag{15.9}$$

which can be bounded away from zero for many designs of interest even if $p \gg n$. For example, if the sparse eigenvalues (15.3) of order $k = Cs$ are bounded away from zero and from above for some large C, then the restricted eigenvalue $\kappa_{\mathbf{c}}$ is bounded away from zero for n sufficiently large. Note that this is implied by Condition R. We refer to Bickel et al. (2009) for other conditions that imply that $\kappa_{\mathbf{c}}$ is bounded away from zero. To handle the approximation errors, it is established that either $\widehat{\beta}_\tau - \beta_\tau \in \Delta_{2\mathbf{c}}$ or $\|\Gamma(\widehat{\beta}_\tau - \beta_\tau)\|_1 \leqslant C\bar{f}\mathrm{E}_n[r_\tau^2(Z)]/\lambda_\tau$. Under Condition R this suffices to obtain good behavior of the nonlinear impact coefficient, which leads the results we will state below.

Theorem 1 *Under Condition R, and the choice of penalty level $\{\lambda_\tau, \tau \in \mathcal{T}\}$ as in (15.6) with $\xi = 1/n$, with probability $1 - o(1)$ we have*

$$\sup_{\tau \in \mathcal{T}} \|\widehat{\beta}_\tau - \beta_\tau\|_2 \lesssim \sqrt{\frac{s \log(p \vee n)}{n}} \quad \text{and} \quad \sup_{\tau \in \mathcal{T}} \|\widehat{\beta}_\tau - \beta_\tau\|_1 \lesssim s\sqrt{\frac{\log(p \vee n)}{n}}.$$

Theorem 1 establishes ℓ_1 and ℓ_2 rates of convergence for the ℓ_1-penalized quantile regression estimator uniformly over $\tau \in \mathcal{T}$ in spite of possible model selection mistakes. The rate is remarkably close to the optimal rate of $\sqrt{s/n}$ if the support T_τ was known for every $\tau \in \mathcal{T}$ and there were no approximation errors. The impact of the high dimensionality p is only logarithmic, which allows for consistent estimates even if $p \gg n$. This parallels lasso results for linear regression models, although additional difficulties arise due to the nonlinearity. These results were first derived in Belloni and Chernozhukov (2011) for the parametric case requiring only $s \log(p \vee n) \leqslant \delta_n n$ for well-behaved designs. Extensions to the nonparametric case have recently been developed in Kato (2011) and Belloni et al. (2013c) under various conditions.

Remark 1 (Sparsity of ℓ_1-penalized quantile regression) *Next we discuss sparsity properties of the ℓ_1-penalized estimator. Sparse estimators are desirable as they have low entropy, which makes them suitable for refitting post selection and other transformations. For the case of the lasso for smooth criterion functions the penalty choice that bounds the ℓ_∞-norm of the gradient of the criterion at the true value automatically yields a sparsity bound of the form*

$$\widehat{s} \lesssim s \, \phi_{\max}(s)/\kappa_{\mathbf{c}}^2;$$

see Belloni and Chernozhukov (2013) for the case of linear regression, and Belloni et al. (2013d) for the case of logistic regression. In contrast, the penalty choice described above does not automatically yield sparsity guarantees in the quantile regression case. As shown in Belloni and Chernozhukov (2011), larger penalty choices can yield sparsity guarantees but the associated larger bias can lead to poor finite-sample performance. Here we discuss a simple trimming argument used in Belloni et al. (2013c, 2016) that yields sparse estimators and still achieves good estimation performance when good rates of convergence in the ℓ_1-norm are available.

Define the trimmed estimator as $\bar{\beta}_\tau$ as

$$\bar{\beta}_{\tau j} = \widehat{\beta}_{\tau j} \mathbf{1}\{\Gamma_{jj}|\widehat{\beta}_{\tau j}| > \lambda_\tau\}, \quad j = 1, \ldots, p, \ \tau \in \mathcal{T}. \tag{15.10}$$

It follows that if a component outside T_τ is trimmed, estimation error in ℓ_1-norm decreases.

Otherwise, there are at most s components in T_τ that could be trimmed so only an additional term of $\lambda_\tau s$ could be added to the estimation error in the ℓ_1-norm (similarly for the ℓ_2-norm). Therefore, since $|T_\tau| \leqslant s$ and if a component satisfies $|\bar{\beta}_{\tau j}| > 0$ it is at least $\lambda_\tau \Gamma_{jj}$, we have

$$\{\|\bar{\beta}_\tau\|_0 - s\}\lambda_\tau \leqslant \|\Gamma(\bar{\beta}_\tau - \beta_\tau)\|_1 \leqslant \|\Gamma(\hat{\beta}_\tau - \beta_\tau)\|_1 + \|\Gamma(\hat{\beta}_\tau - \bar{\beta}_\tau)\|_1 \leqslant \|\Gamma(\hat{\beta}_\tau - \beta_\tau)\|_1 + \lambda_\tau s. \tag{15.11}$$

Under the conditions of Theorem 1, the number of nonzero components after trimming is bounded by

$$\|\bar{\beta}_\tau\|_0 \leqslant 2s + \|\Gamma(\hat{\beta}_\tau - \beta_\tau)\|_1/\lambda_\tau \lesssim s,$$

where the last inequality holds with high probability uniformly over $\tau \in \mathcal{T}$ by Theorem 1. Therefore, provided that the restricted eigenvalue is bounded away from zero, the size of the selected model is (at most) of the same order as the model T_τ. ∎

15.2.3 Refitted quantile regression after selection

Next we discuss the common practice of refitting the quantile regression model after selecting a set of variables $\hat{T}_\tau$. The selection of the set of covariates $\hat{T}_\tau$ can combine prior knowledge of the practitioner and data-driven methods (sequence of statistical tests, ℓ_1-penalization, smoothly clipped absolute deviation (SCAD), trimming, etc.). Therefore, given a set of variables corresponding to the support $\hat{T}_\tau$, we define the associated post-selection refitted estimator as

$$\tilde{\beta}_\tau \in \arg\min_{\beta \in \mathbb{R}^p} E_n[\rho_\tau(Y - X'\beta)] : \text{support}(\beta) \subseteq \hat{T}_\tau. \tag{15.12}$$

For perfect model selection, $\hat{T}_\tau = T_\tau$, the estimator is the so-called oracle estimator. However, in many applications, it is very likely that model selection mistakes will occur regardless of the chosen procedure. (In fact, there is no method that performs consistent model selection uniformly over data-generating processes induced by $\beta_\tau \in \{\beta \in \mathbb{R}^p : \|\beta\|_2 \leqslant C/\sqrt{n}\}$ even if the dimension $p \geqslant 2$ is fixed and $\mathcal{T}$ is a singleton.) We are interested on the properties of the estimator obtained by refitting a quantile regression model with a possible misspecification due to model selection mistakes.

The performance of the refitted estimator defined in (15.12) can be captured by two aspects: the amount of misspecification, characterized by the best model within $\hat{T}_\tau$; and the amount of potential overfitting, characterized by the total number of components $|\hat{T}_\tau|$ and the number of total variables p from which $\hat{T}_\tau$ was selected. It follows that the dependence of p will be logarithmic, which allows us to select $\hat{T}_\tau$ from a very large pool of variables.

Theorem 2 *Under Condition R, for $\hat{Q} \geqslant \sup_{\tau \in \mathcal{T}} E_n[\rho_\tau(Y - X'\tilde{\beta}_\tau)] - E_n[\rho_\tau(Y - X'\beta_\tau)]$ and $\hat{s} \geqslant \sup_{\tau \in \mathcal{T}} |\hat{T}_\tau|$, provided that $M_n\hat{s}\log^{1/2}(p \vee n) \leqslant \delta_n n^{1/2-1/q}$, with probability $1 - o(1)$ we have*

$$\sup_{\tau \in \mathcal{T}} \|\tilde{\beta}_\tau - \beta_\tau\|_2 \lesssim \sqrt{\frac{(\hat{s} + s)\log(p \vee n)}{n}} + \hat{Q}^{1/2}.$$

Theorem 2 establishes the rate of convergence which depends on the number of variables used and on the quality of the model $\hat{T}_\tau$ (measured via $\hat{Q}$). The bound exploits the fact that under Condition R and $M_n\hat{s}\log^{1/2}(p \vee n) \leqslant \delta_n n^{1/2-1/q}$, the sparse eigenvalue $\phi_{\min}(\hat{s} + s)$ is bounded away from zero with high probability for n sufficiently large. Note that because of the intrinsic sparsity of $\tilde{\beta}_\tau$, the ℓ_1 rate of convergence follows from $\|\delta\|_1 \leqslant \{\|\delta\|_0\}^{1/2}\|\delta\|_2$.

It follows that using the model selected by ℓ_1-penalized quantile regression after trimming small components at the threshold λ_τ yields a suitable model that balances well the size of the model $\hat{s}$ and the misspecification $\hat{Q}$ relative to the model β_τ. The next corollary summarizes the performance of such approach.

Corollary 3 *Under the conditions of Theorem 1, refitting a quantile regression model based on the trimmed ℓ_1-penalized quantile regression estimator $\bar{\beta}_\tau$, with probability $1 - o(1)$ we have*

$$\sup_{\tau \in \mathcal{T}} \|\tilde{\beta}_\tau - \beta_\tau\|_2 \lesssim \sqrt{\frac{s \log(p \vee n)}{n}} \quad and \quad \sup_{\tau \in \mathcal{T}} \|\tilde{\beta}_\tau - \beta_\tau\|_1 \lesssim s\sqrt{\frac{\log(p \vee n)}{n}}.$$

Corollary 3 shows that the refitted quantile regression estimator based on support($\bar{\beta}_\tau$) (discussed in Remark 1) achieves the same rates of convergence as the ℓ_1-penalized quantile regression estimator. However, it achieves such rates having a smaller bias and better sparsity (but potentially larger variance).

15.2.4   Group lasso for quantile regression models

In many applications the variables that will be selected have some additional structures that one can potentially use to further improve the adaptivity of the estimator. A common structure is the "group" structure in which we select groups of variables instead of individual variables. This was first proposed in Yuan and Lin (2006) for linear regression models (named group lasso) which motivated an active literature (Bach, 2008; Nardi and Rinaldo, 2008; Meier et al., 2008; Huang and Zhang, 2010; Wei and Huang, 2010; Obozinski et al., 2011; Lounici et al., 2011; Bunea et al., 2014). Kato (2011) originally investigated how to exploit the group structure in high-dimensional quantile models. Let $\{G_1, \ldots, G_{\bar{g}}\}$ be a partition of $\{1, \ldots, p\}$ such that $G_1 = \{1\}$. Each G_k is a "group" of the components of x and $p_k = |G_k|$ denotes the cardinality of the group. The group lasso estimator is defined as

$$\widehat{\beta}_\tau^G \in \arg\min_{\beta \in \mathbb{R}^p} \mathrm{E}_n[\rho_\tau(Y - X'\beta)] + \lambda_\tau^G \sum_{k=2}^{\bar{g}} \sqrt{p_k} \|\widehat{\Sigma}_k^{1/2} \beta_{G_k}\|_2, \qquad (15.13)$$

where $\widehat{\Sigma}_k = \mathrm{E}_n[X_{G_k} X'_{G_k}]$ is an estimate of the covariance matrix of the covariates in the kth group. The use of ℓ_2-norms as the penalization for the groups selects either all of the variables in the group or none of the variables in the group. Importantly, we note that the group penalty can be formulated as a convex programming problem as it can be cast as a second-order conic programming problem for which efficient computational algorithms are available.

The penalty choice for this estimator can also be computed through a pivotal quantity by setting

$$\lambda_\tau^G := c \times \left\{ (1 - \xi)\text{-quantile of } \max_{k=2,\ldots,\bar{g}} \|\widehat{\Sigma}_k^{-1/2} \mathrm{E}_n[(\tau - 1\{U \leq \tau\}) X_{G_k}/\sqrt{p_k}]\|_2 \right\},$$

where $c > 1$ is a constant and $U_1, \ldots, U_n$ are i.i.d. uniform$(0, 1)$ random variables conditional on $(X_i)_{i=1}^n$. Under regularity conditions similar to Condition R, Kato (2011) proved in the parametric case (i.e. $r_\tau = 0$) that with probability $1 - \xi - o(1)$,

$$\|\widehat{\beta}_\tau^G - \beta_\tau\|_2 \lesssim \sqrt{\frac{\sum_{k:\|\beta_{\tau,G_k}\|_2 \neq 0} p_k}{n} \left(1 + \frac{\log(\bar{g} \vee n)}{\underline{p}} \right)},$$

where $\underline{p} = \min_{k=2,\ldots,\bar{g}} p_k$.[1] In applications for which $\sum_{k:\|\beta_{\tau,G_k}\|_2 \neq 0} p_k$ is close to s, the group penalty can improve upon the standard ℓ_1-penalty function provided groups are large so that $\log(\bar{g} \vee n)/\underline{p} = o(\log(p \vee n))$. The above result parallels those from Huang and Zhang (2010) and Lounici et al. (2011) in the mean regression case.

---

[1] Note that Kato (2011) derived a similar bound for the nonparametric case; see Kato (2011) for details.

15.2.5   Estimation of the conditional density

A byproduct of estimating the conditional quantile function is an estimate for the conditional probability density function in high dimensions. Following Koenker (2005), we use the identity

$$f_{\tau i} := f_Y(Q_Y(\tau \mid Z_i) \mid Z_i) = \frac{1}{\partial Q_Y(\tau \mid Z_i)/\partial \tau}.$$

Letting $\widehat{Q}_Y(\tau \mid Z_i)$ denote an estimate of the τ-conditional quantile function, based on ℓ_1-penalized quantile regression or post-selection refitted quantile regression, and for $h = h_n \to 0$ a bandwidth parameter, we let

$$\widehat{f}_{\tau i} = \frac{2h}{|\widehat{Q}_Y(\tau + h \mid Z_i) - \widehat{Q}_Y(\tau - h \mid Z_i)|} \tag{15.14}$$

be an estimator of $f_{\tau i}$. Provided that the conditional quantile function is three times continuously differentiable, the estimator (15.14) has a bias of order h^2 since it is based on the first-order partial difference of the estimated conditional quantile function.

We also note that higher-order estimators have been proposed under additional smoothness assumptions. For example, a smaller bias of order h^4 is achieved by setting

$$\widehat{f}_{\tau i} = h \left| \frac{3}{4} \{\widehat{Q}_Y(\tau + h \mid Z_i) - \widehat{Q}_Y(\tau - h \mid Z_i)\} - \frac{1}{12} \{\widehat{Q}_Y(\tau + 2h \mid Z_i) - \widehat{Q}_Y(\tau - 2h \mid Z_i)\} \right|^{-1}. \tag{15.15}$$

Under mild regularity conditions, the estimators (15.14) and (15.15) achieve

$$\widehat{f}_{\tau i} - f_{\tau i} = O\left(h^{\bar{k}} + \max_{k \in \{\pm 1, \pm \bar{k}/2\}} \frac{|\widehat{Q}_Y(\tau + kh \mid Z_i) - Q_Y(\tau + kh \mid Z_i)|}{h} \right), \tag{15.16}$$

where $\bar{k} = 2$ for (15.14) and $\bar{k} = 4$ for (15.15). Therefore, the uniform rates of convergence obtained in Theorems 1 and 2 can be used to derive uniform rates of convergence for $\widehat{f}_\tau$ over $\tau \in \mathcal{T}$.

15.3   Confidence bands for the coefficient process

In this section we discuss methods to construct confidence bands for parameters of the coefficient process in the high-dimensional setting. Our aim is to construct simultaneous confidence bands for these parameters that are valid uniformly over a large class of probability measures P, say $\mathcal{P}$. Namely, for a subset of variables $\mathcal{J} \subseteq \{1, \dots, p\}$, and for each $\tau \in \mathcal{T}$, we construct an appropriate estimator $\check{\beta}_{\tau j}$ of $\beta_{\tau j}$, along with an estimator $\widehat{\sigma}_{\tau j}$ of the standard deviation of $\sqrt{n}(\check{\beta}_{\tau j} - \beta_{\tau j})$, such that

$$\sup_{P \in \mathcal{P}} \left| \mathrm{P}_P \left(\check{\beta}_{\tau j} - \frac{c_{(1-\alpha)}\widehat{\sigma}_{\tau j}}{\sqrt{n}} \leqslant \beta_{\tau j} \leqslant \check{\beta}_{\tau j} + \frac{c_{(1-\alpha)}\widehat{\sigma}_{\tau j}}{\sqrt{n}}, \ \forall \tau \in \mathcal{T}, \ j \in \mathcal{J} \right) - (1-\alpha) \right| = o(1), \tag{15.17}$$

where $\alpha \in (0, 1)$ and $c_{(1-\alpha)}$ is an appropriate critical value. Several cases are of interest:

(i) $\mathcal{T}$ is a singleton and $|\mathcal{J}| = k$ is fixed,

(ii) $\mathcal{T} = [\underline{\tau}, \bar{\tau}]$ and $|\mathcal{J}| = k$ is fixed,

(iii) $\mathcal{T}$ is a singleton and $|\mathcal{J}| \to \infty$,

(iv) $\mathcal{T} = [\underline{\tau}, \bar{\tau}]$ and $|\mathcal{J}| \to \infty$,

where potentially $|\mathcal{J}| = p \gg n$ in (iii) and (iv). The underlying inferential theory of these cases is potentially very different. Case (i) corresponds to a standard normal limit $\sqrt{n}\Sigma_{\tau,\mathcal{J}}^{-1/2}(\check{\beta}_{\tau,\mathcal{J}} - \beta_{\tau,\mathcal{J}}) \rightsquigarrow N(0, I_{\mathcal{J}})$ as the sample size grows. Case (ii) corresponds to the quantile process case. In cases (i) and (ii) we can hope for the limit distribution to be a tight random element in $\ell^\infty(\mathcal{T} \times \mathcal{J})$. In contrast, in cases (iii) and (iv), because $|\mathcal{J}| \to \infty$, there will be no tight limit. We will discuss how the pivotality of the score will be crucial in the construction of critical values for all of these cases.

Because of the high dimensionality all these cases are delicate, even case (i) with $k = 1$. We stress that none of the estimators discussed in Section 15.2 are suitable for achieving (15.17). The use of regularization leads to a failure of asymptotic linearization of the estimators. In turn, this lack of asymptotic linearization typically translates into severe distortions in the coverage probability of the confidence bands constructed by ignoring (unavoidable) finite-sample model selection mistakes. Nonetheless, the estimators discussed in Section 15.2 can be used in the construction of new estimators that will achieve (15.17).

The key to achieving (15.17) is to consider a score function $\varphi_{\tau j}$ to estimate each $\beta_{\tau j}$ satisfying

$$\mathrm{E}_P[\varphi_{\tau j}(Y, X, \beta_{\tau j}, \eta_{\tau j})] = 0 \quad \text{and} \quad \partial_t \{\mathrm{E}_P[\varphi_{\tau j}(Y, X, \beta_{\tau j} + t, \eta_{\tau j})]\}|_{t=0} \neq 0, \quad (15.18)$$

where $\eta_{\tau j}$ is a nuisance parameter containing $\beta_{\tau,-j}$, and a near-orthogonality property, namely that for all $\tilde{\eta}$ in a pre-specified set we have

$$\partial_t \{\mathrm{E}_P[\varphi_{\tau j}(Y, X, \beta_{\tau j}, \eta_{\tau j} + t\tilde{\eta})]\}|_{t=0} = o_P(n^{-1/2}). \quad (15.19)$$

Such orthogonality conditions have played an important role in statistics and econometrics. In low-dimensional settings, a similar condition was used by Neyman (1979). In semiparametric models orthogonality conditions were used in Newey (1990, 1994), Andrews (1994), Robins and Rotnitzky (1995), Linton (1996), Ackerberg et al. (2014), and others. In high-dimensional settings, Belloni et al. (2010, 2012) were the first to use an orthogonality condition in an instrumental variables model with many instruments. More recently similar ideas have also been used in the literature to construct confidence bands in high-dimensional linear regression models, generalized linear models, and other nonlinear models; see Belloni et al. (2013a,b,c, 2014a,b, 2015b), Zhang and Zhang (2014), van de Geer et al. (2014), Javanmard and Montanari (2014a,b), Zhao et al. (2014), and Ning and Liu (2014). Next we discuss estimators that explicitly (or implicitly) construct a score satisfying (15.18) and (15.19). Although these estimators will have the same first-order asymptotic properties, different implementations lead to estimators with different finite-sample performance.

Remark 2 (Standard score function for quantile regression models) *It is instructive to revisit the standard score function associated with quantile regression models to estimate $\beta_{\tau j}$. For simplicity consider the parametric case where $r_\tau = 0$. In this case the score is given by the product of X_j and the "derivative" of the check function*

$$\tilde{\varphi}_{\tau j}(Y, X, \beta_{\tau j}, \eta_{\tau j}) = (\tau - 1\{Y \leqslant X_j\beta_{\tau j} + X'_{-j}\beta_{\tau,-j}\})X_j,$$

where the nuisance parameters are $\eta_{\tau j} := \beta_{\tau,-j}$. Since $Q_Y(\tau \mid Z) = X_j\beta_{\tau j} + X'_{-j}\beta_{\tau,-j}$, it satisfies the moment condition (15.18) but not the near-orthogonality condition (15.19). Indeed, note that for a direction $\tilde{\eta}$ we have

$$\partial_t \mathrm{E}_P[\tilde{\varphi}_{\tau j}(Y, X, \beta_{\tau j}, \eta_{\tau,j} + t\tilde{\eta})] = \mathrm{E}_P[f_\tau X_j X'_{-j}\tilde{\eta}]$$

which does not equal zero except in very special settings (e.g. consider a parametric location model, so that $f_\tau = f_\tau(Z)$ is constant, where X_j is uncorrelated to all the other variables). Of particular relevance to us is the consequence of this lack of orthogonality on the bias of using a plug-in approach to estimate $\beta_{\tau j}$ via

$$\widehat{\beta}_{\tau j}^b \in \operatorname*{argmin}_{\theta \in \mathbb{R}} | \, \mathrm{E}_n[\widetilde{\varphi}_{\tau j}(Y, X, \theta, \widehat{\beta}_{\tau, -j})] \, |$$

where $\widehat{\beta}_{\tau,-j}$ was obtained through regularization methods due to the high dimensionality. Under mild conditions, setting $\tilde{\eta} = \widehat{\beta}_{\tau,-j} - \beta_{\tau,-j}$, we have that the bias is potentially

$$
\begin{aligned}
\mathrm{E}_P[\widetilde{\varphi}_{\tau j}(Y, X, \beta_{\tau j}, \eta)]\big|_{\eta = \widehat{\beta}_{\tau,-j}} \; &= \mathrm{E}_P[\widetilde{\varphi}_{\tau j}(Y, X, \beta_{\tau j}, \eta_{\tau j})] + \partial_t \mathrm{E}_P[\widetilde{\varphi}_{\tau j}(Y, X, \beta_{\tau j}, \eta_{\tau,j} + t\tilde{\eta})] \\
&\quad + O(\|\tilde{\eta}\|_2^2) \\
&= 0 + \mathrm{E}_P[f_\tau X_j X'_{-j}]\tilde{\eta} + O(\|\tilde{\eta}\|_2^2) \\
&= O_P(\sqrt{(s \log p)/n}) + O_P((s \log p)/n),
\end{aligned}
$$

where we used the typical rate of convergence $\|\tilde{\eta}\|_2 = O_P(\sqrt{(s \log p)/n})$; see Section 15.2. Such distortions are likely to destroy $\sqrt{n}$-consistency for $\widehat{\beta}_{\tau j}^b$. The role of the orthogonality condition (15.19) is to make the first term of the last line zero, which allows one to control the bias, making it negligible for $\sqrt{n}$-asymptotics. ∎

Remark 3 (Estimated conditional density $\widehat{f}_\tau$) *For the sake of exposition we assume that the values of the conditional density function $f_{\tau i}$, $i = 1, \ldots, n$, are known in this section. We note that, since it suffices to know them up to a multiplicative constant, this formally covers the homoskedastic case, where $f_{\tau i} = f_\tau$, $i = 1, \ldots, n$. Nonetheless, in heteroskedastic settings one needs to estimate $f_{\tau i}$ possibly via the method discussed in Section 15.2.5 that handles the high dimensionality. We suggest constructing $\widehat{f}_{\tau i}$ as in (15.14) with bandwidth $h := n^{-1/6}\tau(1 - \tau)/2$. See Belloni et al. (2013c) for a detailed analysis.* ∎

15.3.1 Construction of an orthogonal score function

In this section we provide a construction to achieve a score with the near-orthogonality condition (15.19). It is critical to adapt the score function for each $\tau \in \mathcal{T}$ and $j \in \mathcal{J}$. We will consider the score function

$$\varphi_{\tau j}(Y, X, \beta_{\tau j}, \eta_{\tau j}) = (\tau - 1\{y \leqslant X_j \beta_j + X'_{-j}\beta_{\tau,-j} + r_\tau(Z)\})f_\tau\{X_j - X'_{-j}\gamma_\tau^j\}, \quad (15.20)$$

where the nuisance parameter is $\eta_{\tau j} := (\beta_{\tau,-j}, r_\tau, \gamma_\tau^j)$. The additional $(p-1)$-dimensional nuisance parameter γ_τ^j solves

$$\gamma_\tau^j \in \arg\min_{\gamma \in \mathbb{R}^{p-1}} \mathrm{E}_P[f_\tau^2\{X_j - X'_{-j}\gamma\}^2] \quad (15.21)$$

whose first-order condition implies that $\mathrm{E}_P[f_\tau\{X_j - X'_{-j}\gamma\}f_\tau X_{-j}] = 0$. Therefore we have that

$$
\begin{aligned}
\partial_t \mathrm{E}_P[\varphi_{\tau j}(Y, X, \beta_{\tau j}, \eta_{\tau,j} + t(\tilde{\eta}, 0, 0))] &= -\mathrm{E}_P[f_\tau\{X_j - X_{-j}\gamma_\tau^j\}f_\tau X'_{-j}\tilde{\eta}] = 0, \\
\partial_t \mathrm{E}_P[\varphi_{\tau j}(Y, X, \beta_{\tau j}, \eta_{\tau,j} + t(0, 0, \tilde{\eta}))] &= -\mathrm{E}_P[\{\tau - F_Y(Q_Y(\tau \mid Z) \mid Z)\}X'_{-j}\tilde{\eta}] = 0, \\
\partial_t \mathrm{E}_P[\varphi_{\tau j}(Y, X, \beta_{\tau j}, \eta_{\tau,j} + t(0, -r_\tau, 0))] &= -\mathrm{E}_P[f_\tau\{X_j - X'_{-j}\gamma_\tau^j\}f_\tau r_\tau(Z)] = o_P(n^{-1/2}),
\end{aligned}
$$

where the first condition follows from the definition of γ_τ^j, the second from the definition of the conditional quantile function, and the last condition is assumed (see Remark 5 for a discussion).

In order to estimate γ_τ^j, $\tau \in \mathcal{T}$, $j \in \mathcal{J}$, defined in (15.21), given its dimensionality, we will rely on regularization methods under approximate sparsity assumptions. Indeed, the parameter γ_τ^j is associated with a linear regression model based on the weighted covariates

$$f_\tau X_j = f_\tau X'_{-j} \gamma_\tau^j + v_\tau^j, \quad \mathrm{E}[f_\tau v_\tau^j X_{-j}] = 0, \tag{15.22}$$

where $v_\tau^j := f_\tau X_j - f_\tau X'_{-j} \gamma_\tau^j$. Consequently, a suitable estimator is the (weighted) lasso estimator,

$$\hat{\gamma}_\tau^j \in \arg \min_{\gamma \in \mathbb{R}^{p-1}} \tfrac{1}{2} \mathrm{E}_n [f_\tau^2 \{X_j - X'_{-j} \gamma\}^2] + \lambda_L \|\Gamma_{(\tau j)} \gamma\|_1, \tag{15.23}$$

or the associated post-lasso estimator,

$$\tilde{\gamma}_\tau^j \in \arg \min_{\gamma \in \mathbb{R}^{p-1}} \tfrac{1}{2} \mathrm{E}_n [f_\tau^2 \{X_j - X'_{-j} \gamma\}^2] : \mathrm{support}(\gamma) \subseteq \mathrm{support}(\hat{\gamma}_\tau^j). \tag{15.24}$$

Importantly, we need the rates of convergence to be uniform over $(\tau, j) \in \mathcal{T} \times \mathcal{J}$. As such, the theoretical analysis will require that the penalty level λ_L satisfies

$$\lambda_L \geqslant c \sup_{\tau \in \mathcal{T}, j \in \mathcal{J}} \left\| \Gamma_{(\tau j)}^{-1} \mathrm{E}_n [v_\tau^j f_\tau X_{-j}] \right\|_\infty \tag{15.25}$$

with high probability for some fixed constant $c > 1$. Although the right-hand side in (15.25) is not pivotal, by using penalty loadings $\Gamma_{(\tau j)}$ that induce a self-normalized sum, Belloni et al. (2013a) obtained a theoretically valid choice using self-normalized moderate deviation theory. That motivates our recommendation to set the penalty levels as

$$\lambda_L := 1.1 \sqrt{n} 2 \Phi^{-1} (1 - \xi / \{np|\mathcal{J}|\}). \tag{15.26}$$

The penalty loading $\Gamma_{(\tau j)}$ is a diagonal matrix defined by the following procedure:

1. Compute the post-lasso estimator $\tilde{\gamma}_\tau^j$ based on λ_L and initial values $\Gamma_{(\tau j)kk} = \max_{i \leqslant n} \|f_{\tau i} X_i\|_\infty \{\mathrm{E}_n [f_\tau^2 X_j^2]\}^{1/2}$, $k = 1, \ldots, p-1$.

2. Set $\hat{v}_{\tau i}^j = f_{\tau i}(X_{ij} - X'_{i,-j} \tilde{\gamma}_\tau^j)$ and update $\Gamma_{(\tau j)kk} = \{\mathrm{E}_n [f_\tau^2 X_k^2 (\hat{v}_\tau^j)^2]\}^{1/2}$, $k \in \{1, \ldots, p\} \backslash \{j\}$.

We note that the proposed choice of penalty parameters will lead to sparse estimates and there is no additional need for trimming (unlike in the ℓ_1-penalized quantile regression case, see Remark 1).

Remark 4 (Alternative score functions with orthogonality property) *We point out that alternative score functions satisfying the orthogonality property can be constructed. One avenue is to use different weights $\tilde{f}_\tau$ instead of the conditional probability density f_τ in equation (15.22) and set*

$$\varphi_{\tau j}(Y, X, \beta_{\tau j}, \eta_{\tau j}) = (\tau - 1\{Y \leqslant X_j \beta_j + X'_{-j} \beta_{\tau,-j} + r_\tau(Z)\}) \tilde{v}_{\tau j} (\tilde{f}_\tau / f_\tau),$$

where $\tilde{v}_{\tau i}^j$ is the new residual in (15.22) corresponding to $\tilde{f}_\tau$. It turns out that the choice $\tilde{f}_\tau = f_\tau$ minimizes the asymptotic variance of the estimators we will discuss based upon the empirical analog of (15.18), among all the score functions satisfying (15.18) and (15.19). ∎

15.3.2 Regularity conditions

Next we state sufficient conditions for the validity of the confidence regions proposed below. These requirements are used in conjunction with Condition R stated earlier.

Condition I

1. *The parameters satisfy the following condition. For all $\tau \in \mathcal{T}$, we have $\|\beta_\tau\|_2 + \max_{j \in \mathcal{J}} \|\gamma_\tau^j\|_2 \leqslant C_1$. There exist $s = s_n$ and $\bar{\gamma}_\tau^j$, with $\tau \in \mathcal{T}$ and $j \in \mathcal{J}$, such that $\sup_{\tau \in \mathcal{T}, j \in \mathcal{J}} \|\bar{\gamma}_\tau^j\|_0 \leqslant s$ and $\sup_{\tau \in \mathcal{T}, j \in \mathcal{J}} (\|\bar{\gamma}_u^j - \gamma_u^j\|_2 + s^{-1/2}\|\bar{\gamma}_u^j - \gamma_u^j\|_1) \leqslant C_1 (s \log(p \vee n)/n)^{1/2}$.*

2. *The covariates satisfy the following moment conditions for all $\tau \in \mathcal{T}$:*

   (i) $\min_{j \neq k}(\mathrm{E}_P[|f_\tau^2 v_\tau^j X_k|^2] \wedge \mathrm{E}_P[|f_\tau^2 X_j X_k|^2]) \geqslant c_1$;

   (ii) $\max_{j \neq k} \mathrm{E}_P[|v_\tau^j X_k|^3]^{1/3} \log^{1/2}(p \vee n) \leqslant \delta_n n^{1/6}$;

   (iii) $\sup_{\|\theta\|_2 = 1} \mathrm{E}_P[\{X'\theta\}^4] \leqslant C_1$.

3. *The approximation error r_τ in (15.1) satisfies for all $\tau \in \mathcal{T}$:*

   (i) $\sup_{\|\theta\|_2 = 1} \mathrm{E}_P[r_\tau^2(Z)\{X'\theta\}^2] \leqslant C_1 \mathrm{E}_P[r_\tau^2(Z)]$;

   (ii) $\mathrm{E}_P[r_\tau^2(Z)] \leqslant C_1 s \log(p \vee n)/n$;

   (iii) $\sup_{j \in \mathcal{J}} |\mathrm{E}_P[f_\tau r_\tau(Z) v_\tau^j]| \leqslant \delta_n n^{-1/2}$.

   Furthermore, the following conditions hold:

   (iv) $M_n^2 s \log(p \vee n) \leqslant \delta_n n^{1/2 - 1/q}$;

   (v) $M_n^6 s \leqslant \delta_n n^{1 - 3/q}$;

   (vi) $(\sqrt{s \log p/n} + \sqrt{\log |\mathcal{J}|} \log n/\sqrt{n})^{1/2} (s \log(p \vee n))^{1/2} \leqslant C_0 \delta_n$.

Condition I.1 assumes that the high-dimensional nuisance parameter γ_τ^j is also approximately sparse. Therefore it allows for regularization methods like (weighted) lasso to effectively estimate it. Condition I.2 gives sufficient moments conditions. In particular, these are used to invoke results for self-normalized sums which are important to set up the penalty parameters for the weighted lasso estimator. Finally, Condition I.3 imposes conditions on the approximation error r_τ and restrictions on how fast s and p can grow.

Remark 5 (Negligible approximation error bias) *The condition*

$$\mathrm{E}[f_\tau\{X_j - X'_{-j}\gamma_\tau^j\}f_\tau r_\tau(Z)] = o(n^{-1/2})$$

is needed to avoid an identification problem that can distort coverage of $\sqrt{n}$-consistent confidence regions. Indeed, without additional restrictions we can always define $r_\tau(Z) = n^{-1/2}X_j$ which would alter $\beta_{\tau j}$ by $n^{-1/2}$. Although this modification is not consequential for the rates of convergence, it would create severe distortions on the coverage. Importantly, the high-dimensionality can help in the plausibility as discussed in Belloni et al. (2013c). Let $p \gg n$ and suppose $Q_Y(\tau \mid Z) = \sum_{j=1}^\infty \theta_{\tau j} X_j$, where coefficients satisfy $|\theta_{\tau j}| \leqslant Cj^{-2}$. In that case $r_\tau(Z) = r_\tau^{(1)}(Z) + r_\tau^{(2)}(Z) := \sum_{k \leqslant p, k \notin T_\tau} \theta_{\tau k} X_k + \sum_{k \geqslant p} \theta_{\tau k} X_k$. By the orthogonality condition we automatically have $\mathrm{E}_P[f_\tau\{X_j - X'_{-j}\gamma_\tau^j\}f_\tau r_\tau^{(1)}(Z)] = 0$, and

$$
\begin{aligned}
\mathrm{E}_P[f_\tau\{X_j - X'_{-j}\gamma_\tau^j\}f_\tau r_\tau^{(2)}(Z)] &\leqslant \{\mathrm{E}_P[f_\tau^2\{X_j - X'_{-j}\gamma_\tau^j\}^2]\}^{1/2}\{\mathrm{E}_P[f_\tau^2(\textstyle\sum_{k>p}\theta_{\tau k}X_k)^2]\}^{1/2} \\
&\lesssim \{\textstyle\sum_{k>p}\theta_{\tau k}^2\}^{1/2} \lesssim 1/p = o(n^{-1/2})
\end{aligned}
$$

under assumptions similar to Condition R. Moreover, a definition of $\beta_\tau \in \mathrm{argmin}_{\beta \in \mathbb{R}^p} \mathrm{E}_P[f_\tau \rho_\tau(Y - X'\beta)]$ would imply $\mathrm{E}[\psi_\tau(Y - X'\beta_\tau)f_\tau X] = 0$. Therefore

$$
\begin{aligned}
0 &= \mathrm{E}_P[\psi_\tau(Y - X'\beta_\tau)f_\tau\{X_j - X'_{-j}\gamma_\tau^j\}] \\
&= \mathrm{E}_P[\{F_Y(X'\beta_\tau + r_\tau(Z) \mid Z) - F_Y(X'\beta_\tau \mid Z)\}f_\tau\{X_j - X'_{-j}\gamma_\tau^j\}] \\
&= \mathrm{E}_P[f_\tau r_\tau(Z)f_\tau\{X_j - X'_{-j}\gamma_\tau^j\}] + O(\bar{f}'\bar{f}\mathrm{E}_P[r_\tau^2(Z)|X_j - X'_{-j}\gamma_\tau^j|]).
\end{aligned}
$$

Under Condition R, Condition I and $|r_\tau(Z)| \leqslant \delta_n$ *almost surely, we have*

$$\mathrm{E}_P[r_\tau^{8/3}(Z)]^{3/4}\mathrm{E}_P[|X_j - X'_{-j}\gamma_\tau^j|^4]^{1/4} \lesssim \delta_n^{1/2}\{n^{-1}s\log(p \vee n)\}^{3/4}$$

and the condition $\mathrm{E}_P[f_\tau\{X_j - X'_{-j}\gamma_\tau^j\}f_\tau r_\tau(Z)] = o(n^{-1/2})$ *is satisfied provided that* $s^3\log^3(p \vee n) \leqslant \delta_n n.$ ∎

15.3.3 Score function estimator

Next we formalize an estimator based on the estimated score. Algorithm 4 describes the method in detail. It aims to explicitly construct the score $\varphi_{\tau j}$ defined in (15.20) using post-regularization procedures to estimate the high-dimensional nuisance parameters.

Algorithm 4 (Estimates based on score function) *For each* $\tau \in \mathcal{T}$ *and* $j \in \mathcal{J}$:
Step 1. *Run* ℓ_1-*penalized quantile regression* (15.4) *of* Y *on* X, *to compute* $\widehat{\beta}_\tau$.
 Threshold the estimator $\widehat{\beta}_\tau$ *as in* (15.10) *to compute* $\bar{\beta}_\tau$.
 Run post-ℓ_1-*penalized quantile regression* (15.12) *of* Y *on* X *to compute* $\widetilde{\beta}_\tau$ *with* $\widehat{T}_\tau = \mathrm{support}(\bar{\beta}_\tau)$.
Step 2. *Run the (weighted) lasso estimator* (15.23) *of* $f_\tau X_j$ *on* $f_\tau X_{-j}$ *to compute* $\widehat{\gamma}_\tau^j$.
 Run the post-lasso estimator (15.24) *of* $f_\tau X_j$ *on* $f_\tau X_{-j}$ *to compute* $\widetilde{\gamma}_\tau^j$ *with* $\widehat{T} = \mathrm{support}(\widehat{\gamma}_\tau^j)$.
Step 3. *Define* $\widehat{\varphi}_{\tau j}(Y, X) := (1\{Y \leqslant X_j\theta + X'_{-j}\widetilde{\beta}_{\tau,-j}\} - \tau)f_\tau\{X_j - X'_{-j}\widetilde{\gamma}_\tau^j\}$ *and compute* $\breve{\beta}_{\tau j}$ *as*

$$\breve{\beta}_{\tau j} \in \arg\min_{\theta \in \mathcal{A}_{\tau j}} L_{\tau j}(\theta) := \frac{\{\mathrm{E}_n[\widehat{\varphi}_{\tau j}(Y, X)]\}^2}{\mathrm{E}_n[\widehat{\varphi}_{\tau j}(Y, X)^2]},$$

where $\mathcal{A}_{\tau j} := \{\theta \in \mathbb{R} : |\theta - \widetilde{\beta}_{\tau j}| \leqslant \Gamma_{jj}^{-1}\log^{-1} n\}$.

Step 1 estimates the nuisance parameters from the quantile model which are needed for identification. Step 2 estimates the nuisance parameters that are needed for the orthogonality condition which reduces the bias due to model selection mistakes. Step 3 solves (approximately) the moment condition. Each step requires a different method but all steps are computationally feasible despite the high dimensionality. (The search in the Step 3 only needs to verify at most n points.) We note that Step 1 only depends on $\tau \in \mathcal{T}$ and it is the same for all $j \in \mathcal{J}$. However, Step 2 adapts to each quantile index and each coefficient of interest $(\tau, j) \in \mathcal{T} \times \mathcal{J}$. Next we state the main representation result regarding the estimators constructed via Algorithm 4.

Theorem 5 (Uniform Bahadur representation, estimates via score function) *Under Conditions R and I, uniformly over* $P \in \mathcal{P}_n$, *the estimator for the quantile process* $(\breve{\beta}_{\tau j})_{\tau \in \mathcal{T}, j \in \mathcal{J}}$ *based on Algorithm 4 satisfies*

$$\sqrt{n}\sigma_{\tau j}^{-1}(\breve{\beta}_{\tau j} - \beta_{\tau j}) = \frac{1}{\sqrt{n}}\sum_{i=1}^n \bar{\psi}_{\tau j}(Y_i, X_i) + O_P(\delta_n) \text{ in } \ell^\infty(\mathcal{T} \times \mathcal{J}),$$

where we define $\bar{\psi}_{\tau j}(Y, X) := -\sigma_{\tau j}^{-1}J_{\tau j}^{-1}\varphi_{\tau j}(Y, X, \beta_{\tau j}, \eta_{\tau j})$, $J_{\tau j} = \partial_t\mathrm{E}_P[\varphi_{\tau j}(Y, X, t, \eta_{\tau j})]|_{t=\beta_{\tau j}}$, *and* $\sigma_{\tau j}^2 := J_{\tau j}^{-2}\mathrm{E}_P[\varphi_{\tau j}^2(Y, X, \beta_{\tau j}, \eta_{\tau j})]$.

Theorem 5 obtains a linear representation for the estimates. Under our conditions the estimates $\breve{\beta}_{\tau j}$ are suitable for the construction of confidence intervals. Indeed, for a fixed $\tau \in \mathcal{T}$ and $j \in \mathcal{J}$, we have $\sqrt{n}\sigma_{\tau j}^{-1}(\breve{\beta}_{\tau j} - \beta_{\tau j}) \rightsquigarrow N(0, 1)$. Section 15.3.5 below discusses the construction of confidence bands more generally.

15.3.4 Double selection estimator

Next we describe an alternative estimator based on the double selection idea introduced in Belloni et al. (2014a). Again it relies on regularization methods to estimate the nuisance parameters. The estimator is described in Algorithm 6.

Algorithm 6 (Estimates based on double selection) *For each $\tau \in \mathcal{T}$ and $j \in \mathcal{J}$:*
Step 1. Run ℓ_1-penalized quantile regression (15.4) of Y on X, to compute $\widehat{\beta}_\tau$.
   Threshold the estimator $\widehat{\beta}_\tau$ as in (15.10) to compute $\bar{\beta}_\tau$.
Step 2. Run the (weighted) lasso estimator (15.23) of $f_\tau X_j$ on $f_\tau X_{-j}$ to compute $\widehat{\gamma}_\tau^j$.
Step 3. Compute $\breve{\beta}_{\tau j}$ as

$$(\breve{\beta}_{\tau j}, \breve{\theta}_{\tau, -j}) \in \arg\min_{\beta \in \mathbb{R}^p} \mathrm{E}_n[f_\tau \rho_\tau(Y - X'\beta)] : \operatorname{support}(\beta) \subseteq \{j\} \cup \operatorname{support}(\bar{\beta}_\tau) \cup \operatorname{support}(\widehat{\gamma}_\tau^j).$$

The method selects variables that are good to estimate the conditional quantile function in Step 1. Step 2 selects variables that can potentially create omitted variable bias (the density as weights accounts for the nonlinearity of the quantile regression model). This additional adaptivity creates the robustness due to potential model selection mistakes.

As will be shown below, the double selection estimator has the same first-order asymptotics as the estimator based on the score. However, finite-sample performance can differ. The connection between the estimators can be seen from the first-order optimality condition of Step 3. Letting $\widehat{T}_{\tau j} = \operatorname{support}(\bar{\beta}_\tau) \cup \operatorname{support}(\widehat{\gamma}_\tau^j)$, the optimality of $(\breve{\beta}_{\tau j}, \breve{\theta}_{\tau, -j})$ in Step 3 implies

$$\mathrm{E}_n[\psi_\tau(Y - X_j\breve{\beta}_{\tau j} - X'_{-j}\breve{\theta}_{\tau, -j})f_\tau(X_j, X'_{\widehat{T}_{\tau j}})] \approx 0.$$

Thus, multiplying the vector by $(1, -\widehat{\gamma}_\tau^j)$, we obtain

$$\mathrm{E}_n[\psi_\tau(Y - X_j\breve{\beta}_{\tau j} - X'_{-j}\breve{\theta}_{\tau, -j})f_\tau\{X_j - X'_{\widehat{T}_{\tau j}}\widehat{\gamma}_\tau^j\}] \approx 0.$$

This is precisely the estimated score used in Algorithm 4 but with $\breve{\theta}_{\tau, -j}$ instead of $\widetilde{\beta}_{\tau, -j}$. Therefore the double selection creates a score with the desired orthogonality condition implicitly and updates the estimate for the nuisance parameter using the support $\widehat{T}_{\tau j}$. The following theorem formally states its Bahadur representation.

Theorem 7 (Uniform Bahadur representation, estimates via double selection) *Under Conditions R and I, uniformly over $P \in \mathcal{P}_n$, the estimator for the quantile process $(\breve{\beta}_{\tau j})_{\tau \in \mathcal{T}, j \in \mathcal{J}}$ based on Algorithm 6 satisfies*

$$\sqrt{n}\sigma_{\tau j}^{-1}(\breve{\beta}_{\tau j} - \beta_{\tau j}) = \frac{1}{\sqrt{n}} \sum_{i=1}^n \bar{\psi}_{\tau j}(Y_i, X_i) + O_P(\delta_n) \text{ in } \ell^\infty(\mathcal{T} \times \mathcal{J}),$$

where we define $\bar{\psi}_{\tau j}(Y, X) := -\sigma_{\tau j}^{-1} J_{\tau j}^{-1} \varphi_{\tau j}(Y, X, \beta_{\tau j}, \eta_{\tau j})$, $J_{\tau j} = \partial_t \mathrm{E}_P[\varphi_{\tau j}(Y, X, t, \eta_{\tau j})]|_{t=\beta_{\tau j}}$ and $\sigma_{\tau j}^2 := J_{\tau j}^{-2} \mathrm{E}_P[\varphi_{\tau j}^2(Y, X, \beta_{\tau j}, \eta_{\tau j})]$.

15.3.5 Confidence bands

In this section we build confidence bands around the estimates discussed in Sections 15.3.3 and 15.3.4. We begin with the estimates of the associated standard errors $\sigma_{\tau j}$. Here we discuss three possible estimators which are based on estimates computed in Algorithm 4 or 6. They are as follows:

$$
\begin{aligned}
\hat{\sigma}^2_{\tau j(1)} &:= \tau(1-\tau)\mathrm{E}_n[(\tilde{v}^j_\tau)^2]^{-1}, \\
\hat{\sigma}^2_{\tau j(2)} &:= \tau(1-\tau)\left[\{\mathrm{E}_n[\hat{f}^2_\tau(X_j, X'_{\check{T}})'(X_j, X'_{\check{T}})]\}^{-1}\right]_{11}, \\
\hat{\sigma}^2_{\tau j(3)} &:= \mathrm{E}_n[\hat{f}_\tau X_j \tilde{v}^j_\tau]^{-2}\mathrm{E}_n[\psi^2_\tau(Y - X_j\check{\beta}_{\tau j} - X'_{-j}\check{\theta}_{\tau,-j})(\tilde{v}^j_\tau)^2],
\end{aligned}
\tag{15.27}
$$

where $\tilde{v}^j_\tau := \hat{f}_\tau\{X_j - X'_{-j}\tilde{\gamma}^j_\tau\}$, and $\check{T} = \mathrm{support}(\bar{\beta}_\tau) \cup \mathrm{support}(\check{\gamma}^j_\tau)$ is the set of controls used in the double selection quantile regression. Option (3) corresponds to a plug-in estimate of $\sigma^2_{\tau j}$ given in Theorems 5 and 7. Option (2) corresponds to the first entry of the inverse of the weighted design matrix $\mathrm{E}_n[\hat{f}^2_\tau(X_j, X'_{\check{T}})'(X_j, X'_{\check{T}})]$. Option (1) relies on the orthogonality condition that implies $\mathrm{E}_P[f_\tau X_j v^j_\tau] = \mathrm{E}_P[f_\tau(v^j_\tau)^2]$. Although all three estimates are consistent under similar regularity conditions, their finite-sample behaviour might differ. Based on the small-sample performance in computational experiments, we recommend the use of $\hat{\sigma}_{\tau j}$ equal to $\hat{\sigma}_{\tau j(3)}$ for the score estimator and equal to $\hat{\sigma}_{\tau j(2)}$ for the double selection estimator. (Alternatively, $\hat{\sigma}_{\tau j} := \max\{\hat{\sigma}_{\tau j(2)}, \hat{\sigma}_{\tau j(3)}\}$ seems to perform reliably for both estimators.)

It follows from Theorems 5 and 7 that marginal confidence intervals are valid for each $\beta_{\tau j}, \tau \in \mathcal{T}, j \in \mathcal{J}$, namely

$$
\lim_{n\to\infty}\sup_{P\in\mathcal{P}_n}\sup_{t\in\mathbb{R}}\sup_{\tau\in\mathcal{T}, j\in\mathcal{J}}|\mathrm{P}(\sigma^{-1}_{\tau j}\sqrt{n}(\check{\beta}_{\tau j} - \beta_{\tau j}) \leq t) - \mathrm{P}(N(0,1) \leq t)| = 0.
$$

Of particular relevance is the uniform validity over data-generating processes $\mathrm{P} \in \mathcal{P}_n$ satisfying Conditions R and I. These data-generating processes include cases in which coefficients are small so that model selection mistakes are unavoidable in finite samples and are capable of creating severe coverage distortions (e.g. of order $1/\sqrt{n}$).

Next we turn to the construction of simultaneous confidence bands. We will focus on studentized bands of the form stated in (15.17), namely for some confidence level $1 - \alpha \in (0,1)$,

$$
\check{\beta}_{\tau j} - \frac{c^*_{(1-\alpha)}\hat{\sigma}_{\tau j}}{\sqrt{n}} \leq \beta_{\tau j} \leq \check{\beta}_{\tau j} + \frac{c^*_{(1-\alpha)}\hat{\sigma}_{\tau j}}{\sqrt{n}}, \quad \text{for all } \tau \in \mathcal{T} \text{ and } j \in \mathcal{J},
$$

where $c^*_{(1-\alpha)}$ is a suitable critical value defined below. We build upon the linear representations derived in Theorems 5 and 7, and the pivotality of $\bar{\psi}_{\tau j}(Y, X)$. Pivotality is crucial in the high-dimensional setting as it allows us to bypass approximations via Gaussian processes which are attractive in other settings.[2] The definition of the critical value is based on the process $\mathbb{U}_n : \mathcal{T} \times \mathcal{J} \to \mathbb{R}$ given by

$$
\mathbb{U}_n(\tau, j) := \frac{\sigma^{-1}_{\tau j}}{\sqrt{n}}\sum_{i=1}^n \psi_\tau(U_i)v^j_{\tau i},
$$

where U_i are i.i.d. uniform$(0,1)$ random variables, independent of $\{X_i\}^n_{i=1}$. Clearly this process is pivotal conditional on the data. Then we define the critical value as

$$
c^*_{(1-\alpha)} = \inf\left\{t : \mathrm{P}\left(\sup_{\tau\in\mathcal{T}, j\in\mathcal{J}}|\mathbb{U}_n(\tau, j)| \leq t \mid \{X_i\}^n_{i=1}\right) \geq 1 - \alpha\right\}.
$$

It follows that $c^*_{(1-\alpha)}$ can be estimated via simulations using plug-in estimates for v^j_τ and $\sigma_{\tau j}$

---

[2]Indeed, in other settings it is possible to apply recent high-dimensional central limit theorems (Chernozhukov et al., 2013, 2014a,b, 2015) to the linear representation to approximate its finite-sample distribution via a sequence of Gaussian process (see, for example, Belloni et al., 2015a,b).

for $\tau \in \mathcal{T}$ and $j \in \mathcal{J}$ provided the plug-in estimates have sufficiently fast rate of convergence. Analogous critical values can be constructed for one-sided confidence bands.

Once again, because the results in Theorems 5 and 7 hold uniformly over $P \in \mathcal{P}_n$, so do these confidence bands, in spite of possible model selection mistakes. The pivotality allows us to cover all four cases discussed earlier, even if $|\mathcal{J}| \gg n$.

15.3.6   Confidence bands via inverse statistics

An alternative approach to constructing confidence bands is based on the criterion function $L_{\tau j}$ used in Step 3 of Algorithm 4,

$$L_{\tau j}(\theta) := \frac{\{\mathrm{E}_n[\widehat{\varphi}_{\tau j}(Y,X)]\}^2}{\mathrm{E}_n[\widehat{\varphi}_{\tau j}(Y,X)^2]},$$

where $\widehat{\varphi}_{\tau j}(Y,X) := (1\{Y \leqslant X_j\theta + X'_{-j}\widetilde{\beta}_{\tau,-j}\} - \tau)f_\tau\{X_j - X'_{-j}\widetilde{\gamma}^j_\tau\}$ is an estimate of the score with the orthogonality property. For each $\tau \in \mathcal{T}$ and $j \in \mathcal{J}$, we have that the criterion $L_{\tau j}$ evaluated at $\theta = \beta_{\tau j}$ satisfies

$$nL_{\tau j}(\beta_{\tau j}) \rightsquigarrow \chi^2(1). \tag{15.28}$$

The following formal results holds.

Theorem 8 (Inverse statistics) *Under Conditions R and I, we have that*

$$nL_{\tau j}(\beta_{\tau j}) = \mathbb{U}^2_n(\tau,j) + o_P(1) \quad and \quad \mathbb{U}^2_n(\tau,j) \rightsquigarrow \chi^2(1)$$

uniformly over $P \in \mathcal{P}_n$.

This property allows the construction of a confidence region by inverse statistics:

$$\mathcal{I}_{\tau j}(\iota) := \{\theta \in \mathcal{A}_{\tau j} : nL_{\tau j}(\theta) \leqslant \iota\}. \tag{15.29}$$

This leads to marginal confidence intervals that are valid for all $\{\beta_{\tau j}, \tau \in \mathcal{T}, j \in \mathcal{J}\}$ with asymptotically correct coverage, namely, letting $\iota_{(1-\alpha)} = (1-\alpha)$ − quantile of $\chi^2(1)$, we have

$$\lim_{n\to\infty}\sup_{P\in\mathcal{P}_n}\sup_{\alpha\in(0,1)}\sup_{\tau\in\mathcal{T},j\in\mathcal{J}}|\mathrm{P}_P(\beta_{\tau j}\in\mathcal{I}_{\tau j}(\iota_{(1-\alpha)}))-(1-\alpha)|=0.$$

In order to construct simultaneous confidence regions based on the inverse statistics we need a suitable critical value. Letting

$$\iota^*_{(1-\alpha)} = \inf\left\{t : P\left(\sup_{\tau\in\mathcal{T},j\in\mathcal{J}}\mathbb{U}^2_n(\tau,j)\leqslant t \mid \{X_i\}^n_{i=1}\right)\geqslant 1-\alpha\right\},$$

we have that

$$\lim_{n\to\infty}\sup_{P\in\mathcal{P}_n}\left|\mathrm{P}_P\left(\beta_{\tau j}\in\mathcal{I}_{\tau j}(\iota^*_{(1-\alpha)})\text{ for all }\tau\in\mathcal{T}\text{ and }j\in\mathcal{J}\right)-(1-\alpha)\right|=o(1). \tag{15.30}$$

Bibliography

D. Ackerberg, X. Chen, J. Hahn, and Z. Liao. Asymptotic efficiency of semiparametric two-step GMM. *Review of Economic Studies*, 81:919–943, 2014.

D. W. K. Andrews. Asymptotics for semiparametric econometric models via stochastic equicontinuity. *Econometrica*, 62:43–72, 1994.

F. R. Bach. Consistency of the group lasso and multiple kernel learning. *Journal of Machine Learning Research*, 9:1179–1225, 2008.

A. Belloni and V. Chernozhukov. ℓ_1-penalized quantile regression in high-dimensional sparse models. *Annals of Statistics*, 39(1):82–130, 2011.

A. Belloni and V. Chernozhukov. Least squares after model selection in high-dimensional sparse models. *Bernoulli*, 19(2):521–547, 2013.

A. Belloni, V. Chernozhukov, and C. Hansen. LASSO methods for Gaussian instrumental variables models. arXiv:1012.1297, 2010.

A. Belloni, V. Chernozhukov, and I. Fernández-Val. Conditional quantile processes based on series or many regressors. arXiv:1105.6154, 2011.

A. Belloni, D. Chen, V. Chernozhukov, and C. Hansen. Sparse models and methods for optimal instruments with an application to eminent domain. *Econometrica*, 80(6):2369–2430, 2012.

A. Belloni, V. Chernozhukov, I. Fernández-Val, and C. Hansen. "Program Evaluation and Causal Inference With High-Dimensional Data." Econometrica 85.1 (2017): 233–298.

A. Belloni, V. Chernozhukov, and C. Hansen. Inference for high-dimensional sparse econometric models. In D. Acemoglu, M. Arellano, and E. Dekel, editors, *Advances in Economics and Econometrics: The 2010 World Congress of the Econometric Society, Volume 3: Econometrics*, pages 245–295. Cambridge University Press, New York, 2013b.

A. Belloni, V. Chernozhukov, and K. Kato. Robust inference in high-dimensional approximately sparse quantile regression models. arXiv:1312.7186, 2013c.

A. Belloni, V. Chernozhukov, and Y. Wei. "Post-selection infer- ence for generalized linear models with many controls." Journal of Business & Economic Statistics 34.4 (2016): 606–619.

A. Belloni, V. Chernozhukov, and C. Hansen. Inference on treatment effects after selection among high-dimensional controls. *Review of Economic Studies*, 81:608–650, 2014a.

A. Belloni, V. Chernozhukov, and L. Wang. Pivotal estimation via square-root lasso in nonparametric regression. *Annals of Statistics*, 42(2):757–788, 2014b.

A. Belloni, V. Chernozhukov, D. Chetverikov, and Y. Wei. Uniformly valid post-regularization confidence regions for many functional parameters in Z-estimation framework. arXiv:1512.07619, 2015a.

A. Belloni, V. Chernozhukov, and K. Kato. Uniform post-selection inference for least absolute deviation regression and other Z-estimation problems. *Biometrika*, 102(1):77–94, 2015b.

A. Belloni, M. Chen, and V. Chernozhukov. Quantile graphical models: Prediction and conditional independence with applications to financial risk management. arXiv:1607.00286, 2016.

P. J. Bickel, Y. Ritov, and A. B. Tsybakov. Simultaneous analysis of lasso and Dantzig selector. *Annals of Statistics*, 37(4):1705–1732, 2009.

F. Bunea, J. Lederer, and Y. She. The group square-root lasso: Theoretical properties and fast algorithms. *IEEE Transactions on Information Theory*, 60(2):1313–1325, 2014.

E. Candes and T. Tao. The Dantzig selector: Statistical estimation when p is much larger than n. *Annals of Statistics*, 35(6):2313–2351, 2007.

V. Chernozhukov, D. Chetverikov, and K. Kato. Gaussian approximations and multiplier bootstrap for maxima of sums of high-dimensional random vectors. *Annals of Statistics*, 41(6):2786–2819, 2013.

V. Chernozhukov, D. Chetverikov, and K. Kato. Gaussian approximation of suprema of empirical processes. *Annals of Statistics*, 42(4):1564–1597, 2014a.

V. Chernozhukov, D. Chetverikov, and K. Kato. Anti-concentration and honest, adaptive confidence bands. *Annals of Statistics*, 42(5):1787–1818, 2014b.

V. Chernozhukov, D. Chetverikov, and K. Kato. Comparison and anti-concentration bounds for maxima of Gaussian random vectors. *Probability Theory and Related Fields*, 162:47–70, 2015.

I. E. Frank and J. H. Friedman. A statistical view of some chemometrics regression tools. *Technometrics*, 35(2):109–135, 1993.

X. He and Q.-M. Shao. On parameters of increasing dimensions. *Journal of Multivariate Analysis*, 73(1):120–135, 2000.

J. Huang and T. Zhang. The benefit of group sparsity. *Annals of Statistics*, 38(4):1978–2004, 2010.

A. Javanmard and A. Montanari. Confidence intervals and hypothesis testing for high-dimensional regression. *Journal of Machine Learning Research*, 15(1):2869–2909, 2014a.

A. Javanmard and A. Montanari. Hypothesis testing in high-dimensional regression under the Gaussian random design model: Asymptotic theory. *IEEE Transactions on Information Theory*, 60(10):6522–6554, 2014b.

K. Kato. Group Lasso for high dimensional sparse quantile regression models. arXiv:1103.1458, 2011.

K. Knight and W. J. Fu. Asymptotics for lasso-type estimators. *Annals of Statistics*, 28:1356–1378, 2000.

R. Koenker. *Quantile Regression*. Cambridge University Press, Cambridge, 2005.

R. Koenker and G. Bassett. Regression quantiles. *Econometrica*, 46(1):33–50, 1978.

H. Leeb and B. M. Pötscher. Model selection and inference: Facts and fiction. *Econometric Theory*, 21:21–59, 2005.

H. Leeb and B. M. Pötscher. Sparse estimators and the oracle property, or the return of Hodges' estimator. *Journal of Econometrics*, 142(1):201–211, 2008.

Y. Li and J. Zhu. l_1-norm quantile regression. *Journal of Computational and Graphical Statistics*, 17(1):1–23, 2008.

O. Linton. Edgeworth approximation for MINPIN estimators in semiparametric regression models. *Econometric Theory*, 12(01):30–60, 1996.

K. Lounici, M. Pontil, S. van de Geer, and A. B. Tsybakov. Oracle inequalities and optimal inference under group sparsity. *Annals of Statistics*, 39:2164–2204, 2011.

L. Meier, S. van de Geer, and P. Bühlmann. The group lasso for logistic regression. *Journal of the Royal Statistical Society, Series B*, 70(1):53–71, 2008.

N. Meinshausen and P. Bühlmann. High dimensional graphs and variable selection with the Lasso. *Annals of Statistics*, 34:1436–1462, 2006.

N. Meinshausen and B. Yu. Lasso-type recovery of sparse representations for high-dimensional data. *Annals of Statistics*, 37(1):2246–2270, 2009.

Y. Nardi and A. Rinaldo. On the asymptotic properties of the group lasso estimator for linear models. *Electronic Journal of Statistics*, 2:605–633, 2008.

W. K. Newey. Semiparametric efficiency bounds. *Journal of Applied Econometrics*, 5(2): 99–135, 1990.

W. K. Newey. The asymptotic variance of semiparametric estimators. *Econometrica*, 62: 1349–1382, 1994.

J. Neyman. $C(\alpha)$ tests and their use. *Sankhyā*, 41:1–21, 1979.

Y. Ning and H. Liu. A general theory of hypothesis tests and confidence regions for sparse high dimensional models. arXiv:1412.8765, 2014.

G. Obozinski, L. Jacob, and J.-P. Vert. Group lasso with overlaps: The latent group lasso approach. arXiv:1110.0413, 2011.

J. M. Robins and A. Rotnitzky. Semiparametric efficiency in multivariate regression models with missing data. *Journal of the American Statistical Association*, 90(429):122–129, 1995.

R. J. Tibshirani. Regression shrinkage and selection via the Lasso. *Journal of the Royal Statistical Society, Series B*, 58:267–288, 1996.

S. A. van de Geer, P. Bühlmann, and Y. Ritov. On asymptotically optimal confidence regions and tests for high-dimensional models. *Annals of Statistics*, 42:1166–1202, 2014.

L. Wang. L_1 penalized LAD estimator for high dimensional linear regression. *Journal of Multivariate Analysis*, 120:135–151, 2013.

F. Wei and J. Huang. Consistent group selection in high-dimensional linear regression. *Bernoulli*, 16(4):1369–1384, 2010.

A. H. Welsh. On M-processes and M-estimation. *Annals of Statistics*, 17:337–361, 1989.

Y. Wu and Y. Liu. Variable selection in quantile regression. *Statistica Sinica*, 19(2):801, 2009.

M. Yuan and Y. Lin. Model selection and estimation in regression with grouped variables. *Journal of the Royal Statistical Society, Series B*, 68(1):49–67, 2006.

C.-H. Zhang and S. S. Zhang. Confidence intervals for low-dimensional parameters with high-dimensional data. *Journal of the Royal Statistical Society, Series B*, 76:217–242, 2014.

T. Zhao, M. Kolar, and H. Liu. A general framework for robust testing and confidence regions in high-dimensional quantile regression. arXiv:1412.8724, 2014.

16

Nonconvex Penalized Quantile Regression: A Review of Methods, Theory and Algorithms

Lan Wang

University of Minnesota, Minneapolis, Minnesota, USA

CONTENTS

16.1 Introduction

Quantile regression is now a widely recognized useful alternative to the classical least-squares regression. It was introduced in the seminal paper of Koenker and Bassett (1978b). Given a response variable Y and a vector of covariates $\mathbf{x}$, quantile regression estimates the effects of $\mathbf{x}$ on the conditional quantile of Y. Formally, the τth $(0 < \tau < 1)$ conditional quantile of Y given $\mathbf{x}$ is defined as $Q_Y(\tau|\mathbf{x}) = \inf\{t : F_{Y|\mathbf{x}}(t) \geq \tau\}$, where $F_{Y|\mathbf{x}}$ is the conditional cumulative distribution function of Y given $\mathbf{x}$. An important special case of quantile regression is the least absolute deviation (LAD) regression (Koenker and Bassett, 1978a), which estimates the conditional median $Q_Y(0.5|\mathbf{x})$.

The most prominent feature of quantile regression is its ability to incorporate heterogeneity, which can arise from heteroskedastic variances or other sources beyond the commonly used location–scale models. Quantile regression allows the covariates to influence the location, dispersion, and other aspects of the conditional distribution. As an example, in the analysis of the birthweight data set – see Abrevaya (2001) and Koenker (2005, Chapter 1) – researchers found evidence that the effects of mother's gender, race, and education level on the lower tail of baby's birthweight distribution are quite different from their effects on the central part of the conditional distribution. Such heterogeneity is likely to be overlooked by least-squares regression which focuses on modeling the conditional mean. In contrast, by examining the results of quantile regression at different choices of τ, we can obtain a more accurate understanding of the effects of $\mathbf{x}$ on Y. Furthermore, in many applications, the conditional quantile is of direct scientific interest, such as the lower quantile of baby's birthweight.

Computationally, quantile regression can be formulated as a convex optimization problem where the objective function has the form of asymmetrically weighted absolute values of residuals. It can be efficiently computed via linear programming for moderately large problems; see, for example, the quantreg package (Koenker, 2008) in R (R Core Team, 2008). Quantile regression enjoys several other appealing properties. It is naturally robust to outliers in the response space. The median regression is more efficient than least-squares regression in estimating the conditional mean when we have symmetric heavy-tailed random errors. For any monotone function $h(\cdot)$, we have $Q_{h(Y)}(\tau|\mathbf{x}) = h(Q_Y(\tau|\mathbf{x}))$. This equivariance property usually does not hold for the conditional mean function.

The methodology and theory of quantile regression have been thoroughly studied in the classical asymptotic framework where the number of covariates p is fixed while the sample size n goes to infinity. We refer to Koenker (2005) for a comprehensive introduction; see also the review articles Yu et al. (2003) and He (2009) for related applications. In the classical setting (fixed p), Wang et al. (2007a), Li and Zhu (2008), Zou and Yuan (2008), Wu and Liu (2009), Shows et al. (2010), Kai et al. (2011), Wagener et al. (2012), and Wang et al. (2013a), among others, investigated regularized quantile regression for variable selection. For unpenalized quantile regression, when p grows with n but $p = o(n)$, several authors (Welsh, 1989; Bai and Wu, 1994; He and Shao, 2000) established useful theory for M-estimators with nonsmooth objective function which applies to quantile regression; Belloni et al. (2011) established useful asymptotic theory uniformly over quantile level τ. However, these results no longer apply when $p > n$.

Recent advances in technology have led to greater accessibility of massive data in diverse fields such as genomics, economics, finance, and image analysis. In these contemporary data sets, the number of variables is often substantially larger than the sample size ($p \gg n$). For example, genomic studies often involve a small group of patients (several dozen or fewer) but the microarray expression levels are measured on thousands of genes. For insightful discussions on the statistical challenges of high dimensionality, we refer to the review articles of Donoho (2000), Fan and Li (2006), Fan et al. (2014b), and Horowitz (2015), among others. Quantile regression enjoys two distinctive advantages for analyzing high-dimensional data.

- *Quantile-adaptive sparsity.* Most of the existing work on high-dimensional regression relies on the notion of sparsity. In the quantile regression framework, we allow both the sparsity level (the number of nonzero coefficients) and sparsity locations (which covariates are relevant) to depend on τ, the quantile level of interest. For example, the effect of gender may be relevant for modeling the lower tail of the conditional distribution of Y given $\mathbf{x}$, but not so important if we consider the conditional median. This general notion of quantile-adaptive sparsity provides a more flexible and realistic framework for modeling high-dimensional heterogeneous data.

- *Weaker regularity conditions for asymptotic theory.* For high-dimensional regression, the conditions imposed to derive the asymptotic theory are as important as the theoretical results themselves. For the theory of sparse quantile regression, we do not need to impose restrictive distributional or moment conditions on the random errors and allow their distributions to depend on the covariates. In contrast, existing theory in the literature for high-dimensional penalized least squares usually requires Gaussian or sub-Gaussian tail behavior and the vast majority requires at least second moments for the random error. We consider this as a significant advantage of sparse quantile regression as model checking is a daunting task in high dimension.

The literature on penalized quantile regression has been growing very rapidly in recent years. In this chapter, we provide a selective review of recent developments on nonconvex penalized quantile regression in the $p \gg n$ setting, which requires the development of new asymptotic theory and algorithms. It is worth emphasizing that important progress has also been made on L_1 penalized quantile regression in high dimension. Belloni and Chernozhukov (2011) derived nice asymptotic results uniformly in τ; Wang (2013) also showed that with large probability, the L_2 estimation error of L_1 penalized median regression has a near-oracle rate. Weighted L_1 penalized quantile regression was recently considered in Bradic et al. (2011) and Fan et al. (2014a) for efficiency or robustness consideration when $p \gg n$. We refer to Chapter 15 in this handbook for an in-depth review.

The rest of the chapter is organized as follows. Section 16.2 reviews nonconvex penalized linear quantile regression in ultra-high dimension. Section 16.3 discusses nonconvex penalized semiparametric quantile regression. Section 16.4 considers computational aspects of nonconvex penalized quantile regression. Section 16.5 discusses simultaneous estimation and variable selection at multiple quantiles, and two-stage analysis with quantile-adaptive marginal screening. Section 16.6 concludes the chapter.

16.2 High-dimensional sparse linear quantile regression

16.2.1 Background on penalized high-dimensional regression and the choice of penalty function

We first consider the more familiar mean regression setting and briefly discuss the general intuition behind penalization/regularization for high-dimensional regression and the main motivation for the use of nonconvex penalty functions. The same intuition and motivation apply to high-dimensional quantile regression. Let $\{Y_i, \mathbf{x}_i\}$, $i = 1, \ldots, n$, be a random sample from the regression model

$$Y_i = \mathbf{x}_i' \boldsymbol{\beta}_0 + \epsilon_i, \tag{16.1}$$

where $\mathbf{x}_i = (x_{i0}, x_{i1}, \ldots, x_{ip})'$ is a vector of covariates with $x_{i0} = 1$, $\boldsymbol{\beta}_0 = (\beta_{00}, \beta_{01}, \ldots, \beta_{0p})'$ is a vector of unknown parameters, and the random error ϵ_i satisfies $E(\epsilon_i | \mathbf{x}_i) = 0$. In practice, it is common to standardize the design matrix such that each column (corresponding to the n observations on one covariate) has L_2 norm $\sqrt{n}$.

The primary challenge for high-dimensional regression ($p \gg n$) is that the estimation problem is ill-posed. Fortunately, in many applications it is reasonable to assume the true parameter β_0 is sparse, that is, most of its components are zero. Hence, the regression function resides in a low-dimensional manifold. Under the sparsity assumption, the use of penalization or regularization can help achieve both estimation accuracy and interpretability. There has been a large amount of literature on penalized least-squares procedures for conditional mean regression (i.e., $E(\epsilon_i | \mathbf{x}_i) = 0$); see Fan and Lv (2008) for many references. Penalized least-squares regression minimizes $n^{-1} \sum_{i=1}^{n} (Y_i - \mathbf{x}_i'\beta)^2 + \sum_{i=1}^{p} p_{\lambda_n}(|\beta_j|)$, where $p_{\lambda_n}(\cdot)$ denotes a penalty function with a positive tuning parameter λ_n. The choice of penalty function $p_{\lambda_n}(\cdot)$ is directly related to the goals of high-dimensional regression, which are often twofold (Bickel (2008)):

- *prediction:* to provide an accurate prediction of a future observation;

- *sparsity recovery:* to identify zero and nonzero coefficients, hence to accurately illustrate the relationship between $\mathbf{x}$ and Y.

The two objectives have subtle but important differences. In microarray studies, the first goal corresponds to constructing an effective prediction model for predicting the response of a future patient; the second goal aims to identify the set of relevant genes as therapeutic targets.

A popular choice of the penalty function is the L_1 or lasso penalty (Tibshirani, 1996) for which $p_{\lambda_n}(|\beta_j|) = \lambda_n |\beta_j|$. The use of the lasso penalty achieves accurate prediction under weak conditions (Greenshtein and Ritov, 2004) and is computationally convenient due to the convex structure. However, it often requires stringent conditions on the design matrix to consistently identify the underlying model (Zou, 2006; Zhao and Yu, 2006). This motivates the use of a nonconvex penalty, which alleviates the problem of over-penalization by the L_1 penalty and can consistency identify the underlying model under much more relaxed conditions on the design matrix (Fan and Li, 2001, fixed p). The work we review in this chapter emphasizes the second goal of sparsity recovery.

16.2.2 Nonconvex penalized high-dimensional linear quantile regression

16.2.2.1 Overview

Given a $0 < \tau < 1$, linear quantile regression imposes the model $Q_{Y_i}(\tau | \mathbf{x}_i) = \mathbf{x}_i'\beta_0(\tau)$. Equivalently, we can write the model in the form of (16.1) by taking $\epsilon_i = Y_i - Q_{Y_i}(\tau | \mathbf{x}_i)$, which implies that ϵ_i satisfies the quantile constraint $P(\epsilon_i \leqslant 0 | \mathbf{x}_i) = \tau$. Note that the quantile regression coefficients are allowed to depend on τ. In the special case where the ϵ_i are independent and identically distributed, the slopes of the conditional quantiles are constant across different values of τ.

We consider estimating the τth ($0 < \tau < 1$) conditional quantile of Y given a vector of high-dimensional covariates $\mathbf{x}$. For simplicity of notation, we write $\beta_0(\tau)$ as β_0 when no confusion will be caused. The penalized quantile regression minimizes

$$Q(\beta) = n^{-1} \sum_{i=1}^{n} \rho_\tau(Y_i - \mathbf{x}_i'\beta) + \sum_{j=1}^{p} p_{\lambda_n}(|\beta_j|), \tag{16.2}$$

where $\rho_\tau(u) = u\{\tau - I(u < 0)\}$ is the quantile loss function (or check function), and $p_{\lambda_n}(\cdot)$ is a penalty function with a tuning parameter λ_n. The tuning parameter controls the model

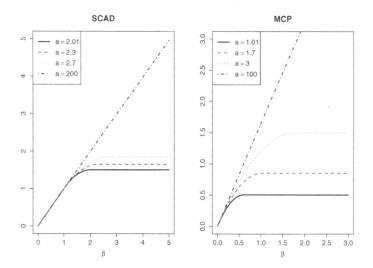

FIGURE 16.1
SCAD and MCP penalty functions ($\lambda = 1$)

complexity; in what follows, we denote it by λ for simplicity of notation. The role of the penalty function is to shrink the estimates of small coefficients toward zero. If the underlying model is sparse, when the penalty function and λ are appropriately chosen, many estimated coefficients will be shrunk to exactly zero, which results in simultaneous estimation and variable selection.

We focus on nonconvex penalty functions here for the purpose of sparsity recovery (see Section 16.2.1). For theoretical development, the penalty function only needs to satisfy some general conditions (see Section 16.2.2.2). Two popular choices of nonconvex penalty functions are the smoothly clipped absolute deviation (SCAD) penalty function (Fan and Li, 2001) and the minimax concave penalty (MCP) function (Zhang, 2010). The SCAD penalty function is given by

$$p_\lambda(|\beta|) = \lambda|\beta|I(0 \leq |\beta| < \lambda) + \frac{a\lambda|\beta| - (\beta^2 + \lambda^2)/2}{a-1}I(\lambda \leq |\beta| \leq a\lambda) + \frac{(a+1)\lambda^2}{2}I(|\beta| > a\lambda),$$

for some $a > 2$. The MCP function has the form

$$p_\lambda(|\beta|) = \lambda\left(|\beta| - \frac{\beta^2}{2a\lambda}\right)I(0 \leq |\beta| < a\lambda) + \frac{a\lambda^2}{2}I(|\beta| \geq a\lambda),$$

for some $a > 1$. Figure 16.2.2.1 depicts the two penalty functions. The use of nonconvex penalty functions requires the development of new asymptotic theory (see Section 16.3.3). Due to the nonsmoothness of the loss function and the nonconvexity of the penalty function, new algorithms are required to compute the penalized quantile regression estimator (see Section 16.4).

The problem of selecting the tuning parameter λ is crucial in practice. One popular approach is to use cross-validation to select λ. However, cross-validation is known to often lead to overfitting (Wang et al., 2007b; Zhang et al., 2010). When the goal is to identify the underlying sparsity pattern, Chen and Chen (2008), Wang et al. (2009a), Kim et al. (2012) and Wang et al. (2013b), among others, showed that a modified Bayesian information (BIC) type criterion achieves model selection consistency for the conditional mean regression model when $p > n$. Motivated by the recent work of Lee et al. (2013) on quantile regression, we choose the λ that minimizes

$$\text{HBIC}(\lambda) = \log\left(\sum_{i=1}^{n} \rho_\tau(Y_i - \mathbf{x}_i'\boldsymbol{\beta}_\lambda)\right) + |\mathcal{S}_\lambda|\frac{\log(\log n)}{n}C_n, \qquad (16.3)$$

where $\widehat{\boldsymbol{\beta}}_\lambda = (\widehat{\beta}_{\lambda,1},\ldots,\widehat{\beta}_{\lambda,p})'$ is the penalized quantile regression estimator obtained by minimizing (16.2) with the tuning parameter λ, $\mathcal{S}_\lambda \equiv \{j : \widehat{\beta}_{\lambda,j} \neq 0, 1 \leqslant j \leqslant p\}$ is the estimated support of the model, $|\mathcal{S}_\lambda|$ is its cardinality, and C_n is a sequence of positive constants diverging to infinity as n increases such that $C_n = O(\log(p))$.

16.2.2.2   Oracle property of the nonconvex penalized quantile regression estimator

Let $S_0 = \{1 \leqslant j \leqslant p_n : \beta_{0j} \neq 0\}$ be the index set of the nonzero coefficients in the unknown true quantile regression parameter $\boldsymbol{\beta}_0$, and $|S_0| = q_n$ be its cardinality, which may increase with n and satisfies $q_n = O(n^{c_1})$ for some $0 \leqslant c_1 < 1/2$. Note that both S_0 and q_n also depend on τ, but we omit the dependence in the notation for simplicity. Without loss of generality, we assume $\boldsymbol{\beta}_0 = (\boldsymbol{\beta}_{01}',\mathbf{0}')'$, where $\mathbf{0}$ denotes a $(p_n - q_n)$-dimensional vector of zeros. The oracle estimator is defined as $\widehat{\boldsymbol{\beta}} = (\widehat{\boldsymbol{\beta}}_1',\mathbf{0}')'$, where $\widehat{\boldsymbol{\beta}}_1$ is the quantile regression estimator obtained when the model is fitted using only the relevant covariates in S_0.

Fan and Li (2001) first studied nonconvex penalized likelihood and established the oracle property in the classical fixed p setting. An estimator of $\boldsymbol{\beta}_0$ is said to possess the *oracle property* if with, probability approaching 1, it estimates the zero coefficients to be exactly zero; and it asymptotically estimates $\boldsymbol{\beta}_{01}$ as efficiently as if S_0 were known in advance. Kim et al. (2008), Zhang (2010), and Fan and Lv (2011) further developed the oracle theory for nonconvex penalized least-squares regression when $p \gg n$.

The techniques developed in the aforementioned literature for penalized least-squares regression do not apply to quantile regression as we need to handle both the nonsmooth loss function and the nonconvex penalty function in ultra-high dimension. In particular, the penalized least-squares regression problem can be written as a constrained smooth optimization problem, for which the Karush–Kuhn–Tucker (KKT) condition is sufficient (Bertsekas, 2008) and plays a key role in establishing the oracle theory. For quantile regression with nonconvex penalty, the KKT local optimality condition is necessary but generally not sufficient. Wang et al. (2012) recently established the oracle theory for nonconvex penalized sparse linear quantile regression when p is allowed to grow at an exponential rate of n. They showed that the nonconvex penalized quantile regression objective function can be represented as the difference between two convex functions. They then made use of a novel sufficient optimality condition for the convex differencing algorithm (Tao and An, 1997) and employed empirical process techniques to derive various error bounds to establish the asymptotic theory.

The penalty function $p_\lambda(t)$ is assumed to be nondecreasing and concave for $t \in [0,+\infty)$, with a continuous derivative $\dot{p}_\lambda(t)$ on $(0,+\infty)$. Assume that there exist positive constants c_2 and M such that $2c_1 < c_2 \leqslant 1$ and $n^{(1-c_2)/2}\min_{1\leqslant j\leqslant q_n}|\beta_j| \geqslant M$. Then under some regularity conditions, for $\lambda = o(n^{-(1-c_2)/2})$, $n^{-1/2}q_n = o(\lambda)$, and $\log(p) = o(n\lambda^2)$, Wang et al. (2012) showed that

$$P(\widehat{\boldsymbol{\beta}} \in \mathcal{B}_n(\lambda)) \to 1$$

as $n \to \infty$, where $\mathcal{B}_n(\lambda)$ is the set of local minima of the nonconvex penalized quantile objective function (16.2). Note that if $\lambda = n^{-\frac{1}{2}+\delta}$ for some $c_1 < \delta < c_2/2$, then we have

$p = o(\exp(n^\delta))$, which is referred to as nonpolynomial order or NP-dimensionality in the statistical literature. Also, the above oracle property is derived without imposing restrictive distributional or moment conditions on the random errors which are often required for high-dimensional penalized mean regression. More specifically, we only assume that the conditional probability density function of the random error ϵ_i, denoted by $f_i(\cdot|\mathbf{x}_i)$, is uniformly bounded away from 0 and ∞ in a neighborhood around 0 for all i. This includes, for example, the Cauchy distribution, and is much milder than the Gaussian or sub-Gaussian assumption in the mean regression literature.

16.3 High-dimensional sparse semiparametric quantile regression

16.3.1 Overview

Semiparametric quantile regression is important for high-dimensional data analysis for several reasons. First of all, it possesses all the advantages we discussed earlier for the sparse quantile regression framework. Second, it incorporates nonlinear covariate effects, which often arise in real data analysis, and circumvents the curse of dimensionality associated with fully nonparametric models. Third, it alleviates the difficulty of model checking in high dimension by using a more flexible regression model.

Several authors have made important contributions to semiparametric quantile regression in the classical fixed p condition, including He and Shi (1996), He et al. (2002), Kim (2007), Wang et al. (2009b), Qian and Peng (2010), Koenker (2011), Noh et al. (2012), Zhu et al. (2010), Bücher et al. (2014), Yin et al. (2014), and Ma and He (2016). Recently, Kato (2011) studied grouped penalized quantile regression and derived a nonasymptotic bound on the estimation error; Tang et al. (2013) considered a two-step procedure for a nonparametric varying-coefficients quantile regression model with a diverging number of nonparametric functional coefficients; and Sherwood and Wang (2016) studied partially linear additive quantile regression in ultra-high dimension.

In the following, we focus on high-dimensional partially linear additive quantile regression, one of the most popular semiparametric regression models. Let Y be the response variable, and let $\mathbf{x} = (x_1, \ldots, x_{p_n})'$ and $\mathbf{z} = (z_1, \ldots, z_d)'$ be p_n- and d-dimensional vectors of covariates, respectively. We assume that the τth $(0 < \tau < 1)$ conditional quantile of Y given $(\mathbf{x}', \mathbf{z}')$ is

$$Q_Y(\tau|\mathbf{x}, \mathbf{z}) = \mathbf{x}'\boldsymbol{\beta}_0 + g(\mathbf{z}), \tag{16.4}$$

where $g(\mathbf{z}) = g_0 + \sum_{j=1}^d g_j(z_j)$, with $g_0 \in \mathbb{R}$. For identification purposes, g_j is assumed to satisfy $E[g_j(z_j)] = 0$, $j = 1, \ldots, d$. As an example of application in analyzing microarray data, the vector $\mathbf{x}$ may contain the gene expression levels of thousands of genes, while the vector $\mathbf{z}$ may contain clinical or environment variables of interest.

16.3.2 Nonconvex penalized partially linear additive quantile regression

Let $\{Y_i, \mathbf{x}_i, \mathbf{z}_i\}$, $i = 1, \ldots, n$, be a random sample generated from the partially linear additive quantile regression model in (16.4). We use a linear combination of B-spline basis functions to approximate $g(\cdot)$. Specifically, let $\boldsymbol{\pi}(t) = (b_1(t), \ldots, b_{k_n+l+1}(t))'$ be a vector of normalized B-spline basis functions of order $l+1$ with k_n quasi-uniform internal knots on $[0, 1]$. Then $g(\cdot)$ can be approximated using a linear combination of B-spline basis functions in $\boldsymbol{\Pi}(\mathbf{z}_i) = (1, \boldsymbol{\pi}(z_{i1})', \ldots, \boldsymbol{\pi}(z_{id})')'$. We refer to Schumaker (1981) for details of the B-spline construction, and the fact that there exists $\boldsymbol{\xi}_0 \in \mathbb{R}^{L_n}$, where $L_n = d(k_n + l + 1) + 1$, such that $\sup_{t \in \mathbb{R}^d} |\boldsymbol{\Pi}(t)'\boldsymbol{\xi}_0 - g(t)| = O(k_n^{-r})$. For ease of notation, in what follows we use the same number of basis functions for $g_j(\cdot)$, $j = 1, \ldots, d$. In practice, such restrictions are not necessary.

We are interested in the case where p_n is much larger than the sample size n. In what follows, we focus on the case the number of nonlinear components d is fixed. The penalized partially linear additive quantile regression estimator minimizes

$$Q^P(\boldsymbol{\beta}, \boldsymbol{\xi}) = n^{-1} \sum_{i=1}^{n} \rho_\tau (Y_i - \mathbf{x}_i'\boldsymbol{\beta} - \boldsymbol{\Pi}(\mathbf{z}_i)'\boldsymbol{\xi}) + \sum_{j=1}^{p_n} p_\lambda(|\beta_j|), \tag{16.5}$$

where $p_\lambda(\cdot)$ is a penalty function with tuning parameter λ. Denote the penalized quantile regression estimator by $(\widehat{\boldsymbol{\beta}}', \widehat{\boldsymbol{\xi}}')$ and write $\widehat{\boldsymbol{\xi}} = (\widehat{\xi}_0, \widehat{\boldsymbol{\xi}}_1', \ldots, \widehat{\boldsymbol{\xi}}_d')'$ with $\widehat{\xi}_0 \in \mathbb{R}$ and $\widehat{\boldsymbol{\xi}}_j \in \mathbb{R}^{k_n+l+1}$, $j = 1, \ldots, d$. We then estimate g_j by $\widehat{g}_j(z_{ij}) = \boldsymbol{\pi}(z_{ij})'\widehat{\boldsymbol{\xi}}_j - n^{-1} \sum_{i=1}^{n} \boldsymbol{\pi}(z_{ij})'\widehat{\boldsymbol{\xi}}_j$, for $j = 1, \ldots, d$; and estimate g_0 by $\widehat{g}_0 = \widehat{\xi}_0 + n^{-1} \sum_{i=1}^{n} \sum_{j=1}^{d} \boldsymbol{\pi}(z_{ij})'\widehat{\boldsymbol{\xi}}_j$. The centering of $\widehat{g}_j$ is the sample analog of the identifiability condition $E[g_j(\mathbf{z}_i)] = 0$. The estimator of $g(\mathbf{z}_i)$ is $\widehat{g}(\mathbf{z}_i) = \widehat{g}_0 + \sum_{j=1}^{d} \widehat{g}_j(z_{ij})$.

The practical performance of the B-spline approximation to nonlinear functions depends on the number of knots k_n. In practice, we found that a small number of knots, between three and five, usually works well in a variety of settings. A high-dimensional BIC criterion, similar to that discussed in Section 16.2.2.1, can be used to select the tuning parameter λ. We select the λ that minimizes

$$\text{QBIC}(\lambda) = \log \left(\sum_{i=1}^{n} \rho_\tau \left(Y_i - \mathbf{x}_i'\widehat{\boldsymbol{\beta}}_\lambda - \boldsymbol{\Pi}(\mathbf{z}_i)'\widehat{\boldsymbol{\xi}}_\lambda \right) \right) + \nu_\lambda \frac{\log(p_n) \log(\log(n))}{2n},$$

where ν_λ is the degrees of freedom of the fitted model, which is the number of interpolated fits for quantile regression.

16.3.3   Oracle properties

We briefly summarize the large-sample properties of the oracle estimator and the penalized quantile regression estimator. The former is of independent interest because it allows the dimension of the linear parameter of the true underlying model to diverge with the sample size.

In model (16.4), it is assumed that the vector of coefficients $\boldsymbol{\beta}_0 = (\beta_{01}, \beta_{02} \ldots, \beta_{0p_n})'$ is sparse. Let $A = \{1 \leqslant j \leqslant p_n : \beta_{0j} \neq 0\}$ be the index set of nonzero coefficients and $q_n = |A|$. Without loss of generality, we assume that the first q_n components of $\boldsymbol{\beta}_0$ are nonzero and the rest are zero. As before, both A and q_n depend on τ (A may also depend on n), but we omit the dependence in the notation. We write $\boldsymbol{\beta}_0 = (\boldsymbol{\beta}_{01}', \mathbf{0}_{p_n - q_n}')'$.

The oracle estimator for $\boldsymbol{\beta}_0$ assumes the set A is known in advance, and has the form $(\widehat{\boldsymbol{\beta}}_1', \mathbf{0}_{p_n - q_n}')'$, where

$$(\widehat{\boldsymbol{\beta}}_1, \widehat{\boldsymbol{\xi}}) = \underset{(\boldsymbol{\beta}_1, \boldsymbol{\xi})}{\text{argmin}} \frac{1}{n} \sum_{i=1}^{n} \rho_\tau (Y_i - \mathbf{x}_{A_i}'\boldsymbol{\beta}_1 - \boldsymbol{\Pi}(\mathbf{z}_i)'\boldsymbol{\xi}), \tag{16.6}$$

with $\mathbf{x}_{A_1}', \ldots, \mathbf{x}_{A_n}'$ denoting the row vectors of X_A, the submatrix consisting of the first q_n columns of design matrix X. Allowing q_n to diverge with n such that $q_n = O(n^C)$ for some

$C < \frac{1}{3}$, Sherwood and Wang (2016) showed that

$$\|\hat{\boldsymbol{\beta}}_1 - \boldsymbol{\beta}_{01}\| = O_p\left(\sqrt{n^{-1}q_n}\right),$$

$$n^{-1}\sum_{i=1}^{n}\left(\hat{g}(\mathbf{z}_i) - g_0(\mathbf{z}_i)\right)^2 = O_p\left(n^{-1}(q_n + k_n)\right).$$

They also showed that for $l \times q_n$ matrix A_n with l fixed and $A_n A_n' \to G$, a positive definite matrix, $\sqrt{n}A_n \Sigma_n^{-1/2}\left(\hat{\boldsymbol{\beta}}_1 - \boldsymbol{\beta}_{01}\right)$ converges to a multivariate normal distribution for some matrix Σ_n.

For the asymptotic theory of the nonconvex penalized partially linear additive quantile regression defined by (16.5), Sherwood and Wang (2016) explored the local optimality condition of the convex differencing program (Tao and An, 1997; Wang et al., 2012) and extended it to incorporating nonparametric components. Let $\hat{\boldsymbol{\eta}} \equiv \left(\hat{\boldsymbol{\beta}}', \hat{\boldsymbol{\xi}}'\right)$ be the oracle estimator. Sherwood and Wang (2016) showed that under some regularity condition, for either the SCAD penalty or MCP function with tuning parameter λ, if $\lambda = o\left(n^{-(1-C_4)/2}\right)$ for some positive constant C_4, $n^{-1/2}q_n = o(\lambda)$, $n^{-1/2}k_n = o(\lambda)$. and $\log(p_n) = o(n\lambda^2)$,

$$P\left(\hat{\boldsymbol{\eta}} \in \mathcal{E}_n(\lambda)\right) \to 1 \quad \text{as } n \to \infty,$$

where $\mathcal{E}_n(\lambda)$ is the set of local minima of the nonconvex penalized quantile objective function in (16.6).

16.4 Computational aspects of nonconvex penalized quantile regression

Due to the nonsmoothness of the quantile loss function and the nonconvexity of the penalty function, computation of the nonconvex penalized quantile regression estimator is significantly more challenging compared with penalized least-squares regression or unpenalized quantile regression.

Depending on the specific problem, some of the existing linear programming based algorithms may be adapted to nonconvex penalized quantile regression. We will review two such algorithms in Section 16.4.1. However, when p gets larger these algorithms slow down quickly. Section 16.4.2 reviews a new coordinate descent algorithm recently developed by Peng and Wang (2015), which can improve the computation speed substantially in high dimensions. We illustrate the algorithms using nonconvex penalized linear quantile regression, but they can be extended to semiparametric quantile regression. These algorithms can also be applied to lasso penalized quantile regression.

16.4.1 Linear programming based algorithms (moderately large p)

The first algorithm (Sherwood and Wang, 2016) can be implemented easily using the quantreg package (Koenker, 2008). Hence, it has the advantage that it does not require practitioners to do much programming on their own. For sparse regression, we often initiate the algorithm with $\hat{\boldsymbol{\beta}}^1 = 0$. For $t > 1$, let $\hat{\boldsymbol{\beta}}^{t-1} = (\hat{\beta}_1^{t-1}, \dots, \hat{\beta}_{p_n}^{t-1})'$ denote the estimator of

$\boldsymbol{\beta}$ at step $t-1$. We update the estimator at step t by

$$\widehat{\boldsymbol{\beta}}^t = \underset{\boldsymbol{\beta}}{\operatorname{argmin}} \left\{ n^{-1} \sum_{i=1}^{n} \rho_\tau(Y_i - \mathbf{x}_i'\boldsymbol{\beta}) + \sum_{j=1}^{p_n} p'_\lambda\left(|\widehat{\beta}_j^{t-1}|\right)|\beta_j| \right\}, \qquad (16.7)$$

where if $\tilde{\beta}_j^{(t-1)} = 0$ we take the derivative $p'_\lambda(0)$ as $p'_\lambda(0+) = \lambda$. This approx-imating the nonconvex penalty function locally using a linear function (this step will be omitted if the penalty function itself is L_1), which is the core idea of the MM algorithm (e.g. Lange, 2004; Hunter and Lange, 2000; Hunter and Li, 2005) or LLA algorithm (Zou and Li, 2008).

By observing that $|\beta_j| = \rho_\tau(\beta_j) + \rho_\tau(-\beta_j)$, we can formulate (16.7) as a weighted quantile regression problem on a set of augmented observations. More specifically, let $(Y_i^*, \mathbf{x}_i^*)$, $i = 1, \ldots, n + 2p_n$, where $(Y_i^*, \mathbf{x}_i^*) = (Y_i, \mathbf{x}_i,)$, $i = 1, \ldots, n$; $(Y_i^*, \mathbf{x}_i^*) = (0, 1)$, $i = n + 1, \ldots, n + p_n$; and $(Y_i^*, \mathbf{x}_i^*) = (0, -1)$, $i = n + p_n + 1, \ldots, n + 2p_n$. We solve (16.7) by fitting a weighted linear quantile regression model using these $n + 2p_n$ augmented observations with weights w_i^t, where $w_i^t = 1$, $i = 1, \ldots, n$; $w_{n+j}^t = p'_\lambda\left(|\widehat{\beta}_j^{t-1}|\right)$, $j = 1, \ldots, p_n$; and $w_{n+p_n+j}^t = -p'_\lambda\left(|\widehat{\beta}_j^{t-1}|\right)$, $j = 1, \ldots, p_n$.

Alternatively, we can solve (16.7) by introducing slack variables and directly use linear programming, as in the algorithm in Wang et al. (2012). That is, we write the optimization problem in (16.7) as

$$\underset{\boldsymbol{\xi},\boldsymbol{\zeta}}{\min} \left\{ \frac{1}{n} \sum_{i=1}^{n} (\tau \xi_i^+ + (1-\tau)\xi_i^-) + \sum_{j=1}^{p} w_j^{(t-1)} \zeta_j \right\}, \qquad (16.8)$$

subject to $\xi_i^+ - \xi_i^- = Y_i - \mathbf{x}_i'\boldsymbol{\beta}, \quad i = 1, 2, \ldots, n,$

$\xi_i^+ \geqslant 0, \xi_i^- \geqslant 0, \quad i = 1, 2, \ldots, n,$

$\zeta_j \geqslant \beta_j, \zeta_j \geqslant -\beta_j, \quad j = 1, 2, \ldots, p,$

where ξ_i^+, ξ_i^-, and ζ_j are slack variables. Note that (16.8) is a linear programming problem and can be solved using many existing software packages.

The above linear programming based algorithms are convenient to implement for mod-erately large p, but can slow down quickly when p gets larger (our own numerical experience indicates that this can happen when p is more than a few hundred). Also, the convergence theory of these algorithms has not been investigated.

16.4.2   New iterative coordinate descent algorithm (larger p)

Motivated by the recent development of coordinate descent algorithms for penalized least-squares regression (e.g. Friedman et al., 2007; Wu and Lange, 2008; Breheny and Huang, 2011; Mazumder et al., 2011; Jiang and Huang, 2014), Peng and Wang (2015) recently a considered iterative coordinate descent algorithm (QICD algorithm) for nonconvex penal-ized quantile regression, which combines the idea of the MM algorithm with that of the coordinate descent algorithm.

The QICD algorithm iterates between two steps:

- *A majorization minimization step.* The nonconvex objective function is replaced by its majorization function to create a surrogate objective function, which is updated at each iteration.

- *A coordinate descent step.* Within each iteration, the surrogate function is minimized by solving a sequence of univariate minimization subproblems, each of which minimizes along a selected coordinate with all other coordinates fixed.

The algorithm is remarkably fast as, for each univariate minimization problem, we only need to compute a weighted median, which can be efficiently computed using quicksort or partition-exchange sort. Extending the theory of Tseng (2001), Peng and Wang (2015) established that the QICD algorithm converges to a stationary point of the nonconvex penalized quantile regression objective function in (16.2) under some regularity conditions. The QICD algorithm is now implemented in the QICD package (Peng, 2016) and the rqPen package (Sherwood and Maidman, 2016) in R (R Core Team, 2008).

It is interesting to note that Li and Arce (2004) provided an example of the use of the coordinate descent method for unpenalized median regression and claimed that it converges to an "inferior" solution. This example was sometimes cited as evidence against the coordinate descent algorithm. On the other hand, good empirical performance was also reported in Wu and Lange (2008) for a fast greedy coordinate descent algorithm for median regression. However, they have not studied the convergence theory. Our study revealed that the situation is quite favorable when the sample size is moderately large; when n is small, coordinate descent algorithms are likely to get stuck at kinks. It is helpful to examine Li and Arce's example more carefully. It was based on a mere five observations ($n = 5$). We did some calculation using their five data points. The global minimum is $(-1.25, 0.83)$, which yields the objective function value 6.26; the coordinate descent algorithm gives the solution $(-0.7, 1.1)$. Although this solution appears to be some distance away from the global minimum, it yields the objective function value 6.49. This is an example where the objective function is quite flat around the global minimum. The coordinate descent algorithm still yields a reasonable solution and may not be declared a complete failure. In fact, in their more recent paper, Paredes and Arce (2011) applied the coordinate descent algorithm to l_0-regularized median regression and reported positive empirical performance.

In addition to the technical arguments in Peng and Wang (2015), we can provide some more intuitive rationale for the good performance of the iterative coordinate algorithm with reasonably large sample size. It is motivated by the arguments in Tseng (2001), or more specifically his Lemma 3.1. Tseng's setup allows the penalty function part to be nonsmooth as long as it is separable; but he assumes the loss function to be smooth. When n is large, the loss function is expected to become closer and closer to a smooth function with high probability. A consequence of this is that the directional derivative can be approximated using coordinate-wise directional derivatives with high probability (recall that in the smooth case, the vector of the derivative can be computed by the derivative with respect to each coordinate separately). Hence, his basic argument can still carry through. A longer note is available from me upon request for anyone who is interested.

16.5 Other related problems

16.5.1 Simultaneous estimation and variable selection at multiple quantiles

In some applications, researchers may be interested in simultaneous variable selection and estimation at multiple quantiles. In particular, if most of the linear covariates have zero coefficients across all the quantiles of interest, group selection is likely to help combine information across quantiles.

When $p \gg n$, Sherwood and Wang (2016) investigated this problem using nonconvex penalized partially linear additive quantile regression. Let $0 < \tau_1 < \tau_2 < \ldots < \tau_M < 1$ be a set of quantiles of interest, where M is a positive integer. Denote $Q_{Y_i}(\tau_m|\mathbf{x}_i, \mathbf{z}_i) = \mathbf{x}_i'\boldsymbol{\beta}_0^{(m)} + g_0^{(m)}(\mathbf{z}_i)$, where $g_0^{(m)}(\mathbf{z}_i) = g_{00}^{(m)} + \sum_{j=1}^{d} g_{0j}^{(m)}(z_{ij})$, with $g_{00}^{(m)} \in \mathbb{R}$ and $E(g_{0j}^{(m)}(z_{ij})) = 0$, $m = 1, \ldots, M$. Write $\boldsymbol{\beta}_0^{(m)} = (\beta_{01}^{(m)}, \beta_{02}^{(m)} \ldots, \beta_{0p_n}^{(m)})'$, $m = 1, \ldots, M$. Let $\bar{\boldsymbol{\beta}}_0^j$ be the M-vector $(\beta_{0j}^{(1)}, \ldots, \beta_{0j}^{(M)})'$, $1 \leq j \leq p_n$. The set $A = \{j : ||\bar{\boldsymbol{\beta}}_0^j||_1 \neq 0, 1 \leq j \leq p_n\}$ denotes the index set of active variables, where $||\cdot||_1$ denotes the L_1 norm.

Let $\boldsymbol{\beta} = (\boldsymbol{\beta}^{(1)'}, \ldots, \boldsymbol{\beta}^{(M)'})'$ and $\boldsymbol{\xi} = (\boldsymbol{\xi}^{(1)'}, \ldots, \boldsymbol{\xi}^{(M)'})$. For simultaneous variable selection and estimation at multiple quantiles, we estimate $(\boldsymbol{\beta}_0^{(m)}, \boldsymbol{\xi}_0^{(m)})$, $m = 1, \ldots, M$, by minimizing

$$n^{-1} \sum_{m=1}^{M} \sum_{i=1}^{n} \rho_{\tau_m}\left(Y_i - \mathbf{x}_i'\boldsymbol{\beta}^{(m)} - \boldsymbol{\Pi}(\mathbf{z}_i)'\boldsymbol{\xi}^{(m)}\right) + \sum_{j=1}^{p_n} p_\lambda(||\bar{\boldsymbol{\beta}}^j||_1). \quad (16.9)$$

In the above, we use a group penalty which encourages group-wise sparsity and forces the covariates that have no effect on any of the M quantiles to be excluded together (see also Yuan and Lin, 2007; Zou and Yuan, 2008; Liu and Wu, 2011). Sherwood and Wang (2016) showed that the above estimation procedure enjoys the oracle property under some regularity conditions.

Belloni and Chernozhukov (2011) derived rates of convergence that are uniform over a continuous set of quantile indices for L_1 penalized quantile regression. Zheng et al. (2015) studied a related but somewhat different problem. They employed adaptive L_1 penalties and proposed a uniform selector of the tuning parameter for a continuous range of quantile levels. They derived the oracle rate of uniform convergence and weak convergence of the parameter estimators.

16.5.2   Two-stage analysis with quantile-adaptive screening

16.5.2.1   Background

In big data applications, the use of penalized regression is often preceded by a screening procedure, the aim of which is to use a computationally expedient procedure to quickly reduce the dimensionality to a moderate size, which can still be larger than the sample size n but is more manageable for computation. There has been active work on variable screening for the mean regression model. Fan and Lv (2008) proposed the sure independence screening methodology for linear regression and established the sure screening property. See also Fan and Song (2009), Hall and Miller (2009), Fan et al. (2011), and Bühlmann et al. (2010).

For ultra-high-dimensional data, He et al. (2013) introduced a quantile-adaptive model-free variable screening procedure, which we will briefly review below. The model-free feature is in the same spirit as that of Zhu et al. (2011) which proposed carrying out variable screening without specifying a particular model structure. This is appealing in practice due to the difficulty of validating a statistical model in high dimension. The quantile-adaptive feature is the same as what we discussed earlier for penalized quantile regression, which allows the sets of active variables to be different when modeling different conditional quantiles. It is effective for analyzing high-dimensional heterogeneous data.

16.5.2.2   Quantile-adaptive model-free nonlinear screening

At a given quantile level τ ($0 < \tau < 1$), we define the set of active variables

$$M_\tau = \{j : Q_Y(\tau|\mathbf{x}) \text{ functionally depends on } X_j\},$$

where $\mathbf{x} = (X_1, \ldots, X_p)'$. The variable screening procedure proposed in He et al. (2013) ranks the importance of variables by a marginal quantile utility based on the observation

$$Y \text{ and } X_j \text{ are independent} \Leftrightarrow Q_Y(\tau|\mathbf{x}) - Q_Y(\tau) = 0, \quad \forall \tau \in (0,1),$$

where $Q_Y(\tau)$ is the τth unconditional quantile of Y.

We estimate the marginal condition quantile using B-spline approximation. We observe that the τth conditional quantile of Y given X_j is expressed as $f_j(X_j) = \operatorname{argmin}_f E[\rho_\tau(Y - f(X_j))]$. We then approximate $f_j(t)$ by $\boldsymbol{\pi}(t)'\boldsymbol{\beta}$, where $\boldsymbol{\pi}(t) = (B_1(t), \ldots, B_N(t))'$ is a vector of basis functions. Let $\widehat{\boldsymbol{\beta}}_j = \operatorname{argmin}_{\boldsymbol{\beta} \in \mathbb{R}^N} \sum_{i=1}^n \rho_\tau(Y_i - \boldsymbol{\pi}(X_{ij})'\boldsymbol{\beta})$, and define $\widehat{f}_{nj}(t) = \boldsymbol{\pi}(t)'\widehat{\boldsymbol{\beta}}_j - F_{Y,n}^{-1}(\tau)$ where $F_{Y,n}^{-1}(\tau)$ is the τth sample quantile function based on $Y_1, \ldots, Y_n$. Thus $\widehat{f}_{nj}(t)$ is a nonparametric estimator of $Q_Y(\tau|\mathbf{x}) - Q_Y(\tau)$, which is expected to be close to zero if X_j is independent of Y.

The screening procedure retains all variables in the set

$$\widehat{M}_\tau = \{1 \leqslant j \leqslant p : \|\widehat{f}_{nj}\|_n^2 \geqslant \nu_n\}$$

where $\|\widehat{f}_{nj}\|_n^2 = n^{-1} \sum_{i=1}^n \widehat{f}_{nj}(X_{ij})^2$ and ν_n is a threshold value. A rule of thumb in practice is to rank all the features by the magnitude of $\|\widehat{f}_{nj}\|_n^2$ and keep the top $[n/\log(n)]$ features. He et al. (2013) established the sure screening property of the proposed procedure, that is,

$$P\left(M_\tau \subset \widehat{M}_\tau\right) \to 1, \quad \text{as } n \to \infty.$$

Hence, with probability approaching 1, all important variables are retained. This is the most important property of marginal screening. Furthermore, they showed that

$$P\left(|\widehat{M}_\tau| \leqslant 2N^2 n^\tau \lambda_{\max}(\boldsymbol{\Sigma})/\delta\right) \to 1, \quad \text{as } n \to \infty.$$

where τ and δ are some positive constants, $\boldsymbol{\Sigma} = E(\boldsymbol{\Pi\Pi}')$ with $\boldsymbol{\Pi} = (\boldsymbol{\pi}(X_1), \ldots, \boldsymbol{\pi}(X_p))'$. If $\lambda_{\max}(\boldsymbol{\Sigma}) = O(n^\gamma)$ for some $\gamma > 0$, then the model obtained after screening is of polynomial size with high probability. He et al. (2013) also investigated quantile-adaptive nonlinear screening for the random censoring case.

16.6 Discussion

This chapter provides a selective overview of nonconvex penalized quantile regression in high dimension. Penalized quantile regression provides a valuable and powerful tool for analyzing high-dimensional heterogeneous data. It relies on the more flexible quantile-adaptive sparsity framework, and generally requires weaker conditions for the asymptotic theory than penalized least-squares regression.

Although high-dimensional data analysis has become the most active research area in statistics, there are still many challenging unsolved problems which call for the development of new methods, algorithms, and theory. For example, in today's big data era, one often has to deal with millions of observations and thousands of variables. Existing algorithms are still quite powerless in this situation. Developing scalable algorithms for quantile regression that can handle larger quantities of data is an urgent issue. Recently, Yu et al. (2016) explored a parallel algorithm using the alternating direction method of multiplier (ADMM: Boyd et al., 2011) approach for large-scale nonconvex penalized quantile regression and observe favorable performance when both n and p are large. Developing the methods and theory for high-dimensional quantile regression in areas such as survival analysis and longitudinal data analysis is also important. These exciting research areas pose both great challenges and opportunities.

As pointed out by a referee, an alternative procedure to reduce the bias of the lasso is to threshold the lasso estimator and then reestimate the model using only covariates with nonzero coefficients in the previous step. If the threshold parameter is selected appropriately, the resulting estimator also enjoys the oracle property asymptotically under regularity conditions. To achieve this, it usually requires an appropriate choice of two regularization parameters: one for the lasso and the other for thresholding the lasso. Thus, the tuning parameter selection is critical in such a procedure. If a covariate is mistakenly deleted in an earlier step, it will be excluded from the final fitted model. The numerical results for the mean regression model in Wang et al. (2013b) suggested that the refitted least-squares estimator based on thresholding the lasso performs similarly to the nonconvex penalized estimator in the large-sample setting; although Fan and Li (2001, Figure 5(c)) demonstrated that the hard-thresholding rule typically inflates the L_2 risk due to its discontinuity when the sample size is small.

As another referee pointed out, it was observed in Leeb and Pötscher (2006, 2008) that for a general class of sparse estimators, it is impossible to consistently estimate the distribution of these estimators uniformly with respect to the unknown true regression parameter β_0 in a small neighborhood of zero. In particular, the minimax risk behavior of this class of estimators may be undesirable if the true regression parameter β_0 has components not exactly zero but very close to zero. This class of estimators includes SCAD, lasso, other popular shrinkage estimators such as hard-thresholded estimators, and post-model-selection estimators such as refitted regression estimators after BIC or lasso model selection. This observation appears to have two immediate consequences for the theory and practice of penalized regression. The first is that a β-min condition is often imposed as part of the regularity conditions for both nonconvex-penalized regression and refitted least-squares regression after model selection to effectively distinguish between zero and nonzero coefficients. This condition requires the smallest signal to decay to zero at a rate with a certain lower bound (see, for example, Belloni and Chernozhukov, 2009; Wang et al., 2013b; Fan et al., 2015). The second implication is that it makes statistical inference challenging. Fortunately, the situation is not as dismal is it first looks. Andrews and Guggenberger (2009) noted that the existence of a uniformly consistent estimator of the sampling distribution is not necessary to achieve the goal of producing a uniformly valid confidence interval. In a moving-parameter framework in which the underlying distribution is allowed to depend on n, Hall et al. (2009) investigated using an m-out-of-n bootstrap for the lasso. Chatterjee and Lahiri (2011, 2013) showed that bootstrap and its variants can produce valid confidence intervals for a coefficient in an underlying sparse model no matter whether it is zero or nonzero. Alternatively, Zhang and Zhang (2014) and van de Geer et al. (2014) showed that a desparsifying approach can be used to construct asymptotically valid confidence intervals. Belloni et al. (2015) and Zhao et al. (2014) further derived uniformly valid inference for sparse high-dimensional quantile regression.

Bibliography

J. Abrevaya. The effects of demographics and maternal behavior on the distribution of birth outcomes. *Empirical Economics*, 25:247–257, 2001.

D. W. K. Andrews and P. Guggenberger. Incorrect asymptotic size of subsampling procedures based on post-consistent model selection estimators. *Journal of Econometrics*, 152 (1):19–27, 2009.

Z. Bai and Y. Wu. Limiting behavior of M-estimators of regression coefficients in high dimensional linear models I. Scale dependent case. *Journal of Multivariate Analysis*, 51: 211–239, 1994.

A. Belloni and V. Chernozhukov. Least squares after model selection in high-dimensional sparse models. arXiv:1001.0188, 2009.

A. Belloni and V. Chernozhukov. ℓ_1-penalized quantile regression in high-dimensional sparse models. *Annals of Statistics*, 39:82–130, 2011.

A. Belloni, V. Chernozhukov, and I. Fernández-Val. Conditional quantile processes based on series or many regressors. arXiv:1105.6154, 2011.

A. Belloni, V. Chernozhukov, and K. Kato. Uniform post-selection inference for least absolute deviation regression and other Z-estimation problems. *Biometrika*, 102:77–94, 2015.

D. P. Bertsekas. *Nonlinear Programming*. Athena Scientific, Belmont, MA, 3rd edition, 2008.

P. Bickel. Discussion on the paper "sure independence screening for ultrahigh dimensional feature space" by fan and lv. *Journal of the Royal Statistical Society, Series B*, 70: 883–884, 2008.

S. Boyd, N. Parikh, E. Chu, B. Peleato, and J. Eckstein. Distributed optimization and statistical learning via the alternating direction method of multipliers. *Foundations and Trends in Machine Learning*, 3(1):1–122, 2011.

J. Bradic, J. Fan, and W. Wang. Penalized composite quasi-likelihood for ultrahigh dimensional variable selection. *Journal of the Royal Statistical Society, Series B*, 73:325–349, 2011.

P. Breheny and J. Huang. Coordinate descent algorithms for nonconvex penalized regression, with applications to biological feature selection. *Annals of Applied Statistics*, 5:232–253, 2011.

A. Bücher, A. El Ghouch, M. Kalisch, and I. Van Keilegom. Single-index quantile regression models for censored data. ISBA Discussion Paper, 2014.

P. Bühlmann, M. Kalisch, and M. H. Maathuis. Variable selection in high-dimensional linear models: Partially faithful distributions and the PC-simple algorithm. *Biometrika*, 97:261–278, 2010.

A. Chatterjee and S. N. Lahiri. Bootstrapping lasso estimators. *Journal of the American Statistical Association*, 106(494):608–625, 2011.

A. Chatterjee and S. N. Lahiri. Rates of convergence of the adaptive LASSO estimators to the oracle distribution and higher order refinements by the bootstrap. *Annals of Statistics*, 41:1232–1259, 2013.

J. Chen and Z. Chen. Extended Bayesian information criterion for model selection with large model space. *Biometrika*, 95:759–771, 2008.

D. L. Donoho. High-dimensional data: The curses and blessings of dimensionality. Paper presented to the American Mathematical Society Conference on Mathematical Challenges of 21st Century, 2000.

J. Fan and R. Li. Variable selection via nonconcave penalized likelihood and its oracle property. *Journal of the American Statistical Association*, 96:1348–1360, 2001.

J. Fan and R. Li. Statistical challenges with high-dimensionality: Feature selection in knowledge discovery. In M. Sanz-Solé, J. Soria, J. L. Varona, and J. Verdera, editors, *Proceedings of International Congress of Mathematicians (ICM)*, volume II, pages 595–622. European Mathematical Society, Zürich, 2006.

J. Fan and J. Lv. Sure independence screening for ultrahigh dimensional feature space (with discussion). *Journal of the Royal Statistical Society, Series B*, 70:849–911, 2008.

J. Fan and J. Lv. Non-concave penalized likelihood with NP-dimensionality. *IEEE Transactions on Information Theory*, 57:5467–5484, 2011.

J. Fan and R. Song. Sure independence screening in generalized linear models with NP-dimensionality. *Annals of Statistics*, 38:3567–3604, 2009.

J. Fan, Y. Feng, and R. Song. Nonparametric independence screening in sparse ultra-high dimensional additive models. *Journal of the American Statistical Association*, 106:544–557, 2011.

J. Fan, Y. Fan, and E. Barut. Adaptive robust variable selection. *Annals of Statistics*, 42:324–351, 2014a.

J. Fan, F. Han, and H. Liu. Challenges of big data analysis. *National Science Review*, 1:293–314, 2014b.

J. Fan, H. Liu, Q. Sun, and T. Zhang. TAC for sparse learning: Simultaneous control of algorithmic complexity and statistical error. arXiv:1507.01037, 2015.

J.H. Friedman, T. Hastie, H. Höfling, and R. Tibshirani. Pathwise coordinate optimization. *Annals of Applied Statistics*, 1:302–332, 2007.

E. Greenshtein and Y. A. Ritov. Persistence in high-dimensional linear predictor selection and the virtue of overparametrization. *Bernoulli*, 10:971–988, 2004.

P. Hall and H. Miller. Using generalized correlation to effect variable selection in very high dimensional problems. *Journal of Computational and Graphical Statistics*, 18:533–550, 2009.

P. Hall, E. R. Lee, and B. U. Park. Bootstrap-based penalty choice for the lasso, achieving oracle performance. *Statistica Sinica*, pages 449–471, 2009.

X. He. Modeling and inference by quantile regression. Technical report, Department of Statistics, University of Illinois at Urbana-Champaign, 2009.

X. He and Q. M. Shao. On parameters of increasing dimensions. *Journal of Multivariate Analysis*, 73:120–135, 2000.

X. He and P. Shi. Bivariate tensor-product B-splines in a partly linear model. *Journal of Multivariate Analysis*, 58:162–181, 1996.

X. He, Z. Zhu, and W. Fung. Estimation in a semiparametric model for longitudinal data with unspecified dependence structure. *Biometrika*, 89:579–590, 2002.

X. He, L. Wang, and H. Hong. Quantile-adaptive model-free nonlinear feature screening for high-dimensional heterogeneous data. *Annals of Statistics*, 41:342–369, 2013.

J. L. Horowitz. Variable selection and estimation in high-dimensional models. *Canadian Journal of Economics*, 48:389–407, 2015.

D. R. Hunter and K. Lange. Quantile regression via an MM algorithm. *Journal of Computational and Graphical Statistics*, 9:60–77, 2000.

D. R. Hunter and R. Li. Variable selection using MM algorithms. *Annals of Statistics*, 33:1617–1642, 2005.

D. Jiang and J. Huang. Majorization minimization by coordinate descent for concave penalized generalized linear models. *Statistics and Computing*, 24:871–883, 2014.

B. Kai, R. Li, and H. Zou. New efficient estimation and variable selection methods for semiparametric varying-coefficient partially linear models. *Annals of Statistics*, 39:305–332, 2011.

K. Kato. Group Lasso for high dimensional sparse quantile regression models. arXiv:1103.1458, 2011.

M.-O. Kim. Quantile regression with varying coefficients. *Annals of Statistics*, 35:92–108, 2007.

Y. Kim, H. Choi, and H.-S. Oh. Smoothly clipped absolute deviation on high dimensions. *Journal of the American Statistical Association*, 103:1665–1673, 2008.

Y. Kim, S. Kwon, and H. Choi. Consistent model selection criteria on high dimensions. *Journal of Machine Learning Research*, 13:1037–1057, 2012.

R. Koenker. *Quantile Regression*. Cambridge University Press, Cambridge, 2005.

R. Koenker. quantreg: Quantile regression. R package version 5.11. http://CRAN.R-project.org/package=quantre, 2008.

R. Koenker. Additive models for quantile regression: Model selection and confidence bandaids. *Brazilian Journal of Statistics*, 25:239–262, 2011.

R. Koenker and G. W. Bassett. The asymptotic distribution of the least absolute error estimator. *Journal of the American Statistical Association*, 7:618–622, 1978a.

R. Koenker and G. W. Bassett. Regression quantiles. *Econometrica*, 46:33–50, 1978b.

K. Lange. *Optimization*. Springer, New York, 2004.

E. R. Lee, H. Noh, and B. U. Park. Model selection via Bayesian information criterion for quantile regression models. *Journal of the American Statistical Association*, 109:216–229, 2013.

H. Leeb and B. M. Pötscher. Performance limits for estimators of the risk or distribution of shrinkage-type estimators, and some general lower risk-bound results. *Econometric Theory*, 22(01):69–97, 2006.

H. Leeb and B. M. Pötscher. Sparse estimators and the oracle property, or the return of hodges? estimator. *Journal of Econometrics*, 142(1):201–211, 2008.

Y. Li and G. R. Arce. A maximum likelihood approach to least absolute deviation regression. *EURASIP Journal on Applied Signal Processing*, 12:1762–1769, 2004.

Y. J. Li and J. Zhu. L_1-norm quantile regression. *Journal of Computational and Graphical Statistics*, 17:163–185, 2008.

Y. Liu and Y. Wu. Simultaneous multiple non-crossing quantile regression estimation using kernel constraints. *Journal of Nonparametric Statistics*, 23:415–437, 2011.

S. Ma and X. He. Inference for single-index quantile regression models with profile optimization. *Annals of Statistics*, 44:1234–1268, 2016.

R. Mazumder, J. H. Friedman, and T. Hastie. Sparsenet: Coordinate descent with non-convex penalties. *Journal of American Statistical Association*, 106:1125–1138, 2011.

H. Noh, K. Chung, and I. Van Keilegom. Variable selection of varying coefficient models in quantile regression. *Electronic Journal of Statistics*, 6:1220–1238, 2012.

J. L. Paredes and G. R. Arce. Compressive sensing signal reconstruction by weighted median regression estimates. *IEEE Transactions on Signal Processing*, 59:2585–2601, 2011.

B. Peng and L. Wang. An iterative coordinate-descent algorithm for high-dimensional non-convex penalized quantile regression. *Journal of Computational and Graphical Statistics*, 24:676–694, 2015.

Bo Peng. *QICD: Estimate the Coefficients for Non-Convex Penalized Quantile Regression Model by using QICD Algorithm*, 2016. URL http://CRAN.R-project.org/package=QICD. R package version 1.0.1.

J. Qian and L. Peng. Censored quantile regression with partially functional effects. *Biometrika*, 97:839–850, 2010.

R Core Team. *R: A Language and Environment for Statistical Computing*. R Foundation for Statistical Computing, Vienna, Austria, 2008. URL http://www.R-project.org/.

L. Schumaker. *Spline Functions: Basic Theory*. Wiley: New York, 1981.

B. Sherwood and A. Maidman. *rqPen: Penalized Quantile Regression*, 2016. URL http://CRAN.R-project.org/package=rqPen. R package version 1.4.

B. Sherwood and L. Wang. Partially linear additive quantile regression in ultra-high dimension. *Annals of Statistics*, 44:288–317, 2016.

J. H. Shows, W. Lu, and H. H. Zhang. Sparse estimation and inference for censored median regression. *Journal of Statistical Planning and Inference*, 140:1903–1917, 2010.

Y. L. Tang, X. Y. Song, H. X. Wang, and Z. Y. Zhu. Variable selection in high-dimensional quantile varying coefficient models. *Journal of Multivariate Analysis*, 122:115–132, 2013.

P. D. Tao and L.T.H. An. Convex analysis approach to D.C. programming: Theory, algorithms and applications. *Acta Mathematica Vietnamica*, 22:289–355, 1997.

R. Tibshirani. Regression shrinkage and selection via lasso. *Journal of the Royal Statistical Society, Series B*, 58:267–288, 1996.

P. Tseng. Convergence of a block coordinate descent method for nondifferentiable minimization. *Journal of Optimization Theory and Applications*, 109:475–494, 2001.

S. van de Geer, P. Bühlmann, Y. Ritov, and R. Dezeure. On asymptotically optimal confidence regions and tests for high-dimensional models. *Annals of Statistics*, 42(3):1166–1202, 2014.

J. Wagener, S. Volgushev, and H. Dette. The quantile process under random censoring. *Mathematical Methods of Statistics*, 21:127–141, 2012.

H. Wang, G. Li, and G. Jiang. Robust regression shrinkage and consistent variable selection through the LAD-Lasso. *Journal of Business & Economic Statistics*, 25:347–355, 2007a.

H. Wang, R. Li, and C. L Tsai. Tuning parameter selectors for the smoothly clipped absolute deviation method. *Biometrika*, 94:553–568, 2007b.

H. Wang, B. Li, and C. Leng. Shrinkage tuning parameter selection with a diverging number of parameters. *Journal of the Royal Statistical Society: Series B*, 71:671–683, 2009a.

H. J. Wang, Z. Zhu, and J. Zhou. Quantile regression in partially linear varying coefficient models. *Annals of Statistics*, 37:3841–3866, 2009b.

H. J. Wang, J. Zhou, and Y. Li. Variable selection for censored quantile regression. *Statistica Sinica*, 23:145–167, 2013a.

L. Wang. The L_1 penalized LAD estimator for high dimensional linear regression. *Journal of Multivariate Analysis*, 120:135–151, 2013.

L. Wang, Y. Wu, and R. Li. Quantile regression for analyzing heterogeneity in ultra-high dimension. *Journal of American Statistical Association*, 107:214–222, 2012.

L. Wang, Y. Kim, and R Li. Calibrating non-convex penalized regression in ultra-high dimension. *Annals of Statistics*, 41:2505–2536, 2013b.

A. Welsh. On *M*-processes and *M*-estimation. *Annals of Statistics*, 17:337–361, 1989.

T. T. Wu and K. Lange. Coordinate descent algorithms for lasso penalized regression. *Annals of Applied Statistics*, 2:224–244, 2008.

Y. C. Wu and Y. F. Liu. Variable selection in quantile regression. *Statistica Sinica*, 19:801–817, 2009.

G. S. Yin, D. L. Zeng, and H. Li. Censored quantile regression with varying coefficients. *Statistica Sinica*, 24:855–870, 2014.

K. Yu, Z. Lu, and J. Stander. Quantile regression: Applications and current research areas. *The Statistician*, 52:331–350, 2003.

L. Yu, N. Lin, and L. Wang. A parallel algorithm for large-scale nonconvex penalized quantile regression. Technical report, Washington University in St. Louis and University of Minnesota, 2016.

M. Yuan and Y. Lin. Model selection and estimation in regression with grouped variables. *Journal of the Royal Statistical Society, Series B*, 68:49–67, 2007.

C. H. Zhang. Nearly unbiased variable selection under minimax concave penalty. *Annals of Statistics*, 38:894–942, 2010.

C.-H. Zhang and S. S. Zhang. Confidence intervals for low dimensional parameters in high dimensional linear models. *Journal of the Royal Statistical Society, Series B*, 76(1):217–242, 2014.

Y. Zhang, R. Li, and C.-L. Tsai. Regularization parameter selections via generalized information criterion. *Journal of American Statistical Association*, 105:312–323, 2010.

P. Zhao and B. Yu. On model selection consistency of lasso. *Journal of Machine Learning Research*, 7:2541–2563, 2006.

T. Zhao, M. Kolar, and H. Liu. A general framework for robust testing and confidence regions in high-dimensional quantile regression. arXiv:1412.8724, 2014.

Q. Zheng, L. Peng, and X. He. Globally adaptive quantile regression with ultra-high dimensional data. *Annals of Statistics*, 42:2225–2258, 2015.

L. Zhu, M. Huang, and R. Li. Semiparametric quantile regression with high dimensional covariates. *Statistica Sinica*, 22:1379–1401, 2010.

L. P. Zhu, L. X. Li, R. Li, and L. X. Zhu. Model-free feature screening for ultrahigh dimensional data. *Journal of American Statistical Association*, 106:1464–1475, 2011.

H. Zou. The adaptive lasso and its oracle properties. *Journal of the American Statistical Association*, 101:1418–1429, 2006.

H. Zou and R. Li. One-step sparse estimates in nonconcave penalized likelihood models. *Annals of Statistics*, 36:1509–1566, 2008.

H. Zou and M. Yuan. Composite quantile regression and the oracle model selection theory. *Annals of Statistics*, 36:1108–1126, 2008.

17

QAR and Quantile Time Series Analysis

Zhijie Xiao

Boston College, Chestnut Hill, Massachusetts, USA

CONTENTS

17.1   Introduction

Quantile regression and quantile domain analysis are important tools in the study of time series. Quantile regression provides a method of estimating the conditional quantile function, and thus the whole conditional distribution, of a time series, and also substantially expands the modeling options for time series analysis. Quantile domain analysis provides a convenient way to study time series dynamics beyond the first few moments. This chapter provides a selective review of quantile time series analysis, with a focus on parametric models, in particular, quantile autoregression (QAR) and related processes.

While the ordinary least-squares method estimates the conditional mean in dynamic models, quantile regression can be applied to traditional time series models to estimate the conditional quantiles. We start with quantile regression on traditional time series models. Popular models, such as the autoregressive and autoregressive moving average models, nonlinear regression, and time series regressions with serially correlated errors, are considered in Section 17.2. In Section 17.3 we talk about quantile regressions on time series models with conditional heteroskedasticity. Quantile regressions with heavy-tailed errors are discussed in Section 17.4. Quantile time series regressions with deterministic and/or stochastic trend, including unit root and cointegrating regressions, are discussed in Section 17.5.

Traditional time series models and methods focus on capturing the dynamics in the central part of a conditional distribution. This is effective for understanding the central tendency within a dataset, but will often be less useful for assessing the behavior of dependence close to the upper or lower extremes within a population. However, from the perspective of investors or risk managers, the dynamics at the tails may be the ones of greatest interest. For such applications, quantile domain analysis can play an important role as an extension of the conventional methods. In particular, quantile domain models provide a local approach to directly model the dynamics of a time series at a specified quantile or the whole distribution. A leading model that has attracted a lot of research attention is the QAR model that we consider in Section 17.6. Other important dynamic quantile models, in particular the CAViaR models, are briefly discussed in Section 17.7.

An interesting direction of recent research in quantile time series analysis is in the frequency domain. Spectral density captures the correlation structure of a time series at different frequencies. However, as a Fourier transformation of the covariances, the conventional spectral density only looks at the second moments and may miss important dependence information beyond correlation. The quantile spectrum analysis can help address these issues. We discuss quantile spectrum methods in Section 17.8. Sections 17.9 briefly talks about quantile regression based forecasting. Section 17.10 concludes.

17.2 Quantile regression estimation of traditional time series models

Quantile regression provides a method of estimating the conditional quantiles (or conditional distribution) in traditional time series models. If we consider a time series model that captures the dynamics between a random variable Y_t and past information, say, contained in X_t, ordinary least-squares (OLS) regression based on a quadratic loss function delivers an estimate of the conditional mean $E(Y_t|X_t)$; least absolute deviation (LAD) regression based on the absolute deviation loss function delivers an estimate of the conditional median of Y given X; and quantile regression based on an asymmetric loss function, $\rho_\tau(u) = u(\tau - I(u < 0))$, gives an estimate of the τth conditional quantile of Y given X. In this section, we discuss quantile regressions on the autoregressive (AR) models, the general autoregressive moving average (ARMA) models, and general linear and nonlinear dynamic regressions.

17.2.1 Quantile regression estimation of the traditional AR model

Consider the following classical AR model of order p:

$$Y_t = \theta_0 + \theta_1 Y_{t-1} + \cdots + \theta_p Y_{t-p} + u_t, \tag{17.1}$$

where u_t is an independent identically distributed (i.i.d.) mean-zero sequence with distribution function $F(\cdot)$. Then the conditional distribution of Y_t (given past information) is simply a location shift of $F(\cdot)$, with conditional mean $\theta_0 + \theta_1 Y_{t-1} + \cdots + \theta_p Y_{t-p}$. Thus, the conditional quantile function of Y_t is given by

$$Q_{Y_t}(\tau|\mathcal{F}_{t-1}) = \theta_0 + \theta_1 Y_{t-1} + \cdots + \theta_p Y_{t-p} + F^{-1}(\tau),$$

where $\mathcal{F}_{t-1}$ denotes the σ-field containing information up to time $t-1$.

Let $\theta_0(\tau) = \theta_0 + F_u^{-1}(\tau)$, $\theta(\tau) = (\theta_0(\tau), \theta_1, \ldots, \theta_p)^\top$, and $X_t = (1, Y_{t-1}, \ldots, Y_{t-p})^\top$. We may write

$$Q_{Y_t}(\tau|\mathcal{F}_{t-1}) = \theta(\tau)^\top X_t.$$

Given time series observations $\{Y_t\}_{t=1}^n$, the vector $\theta(\tau)$ can be estimated by the quantile regression):

$$\hat{\theta}(\tau) = \min_\theta \sum_t \rho_\tau(Y_t - \theta^\top X_t), \tag{17.2}$$

where $\rho_\tau(u) = u(\tau - I(u < 0))$, $\tau \in (0,1)$ and $I(\cdot)$ is the indicator function. The criterion function $\rho_\tau(\cdot)$ is called the "check function" in Koenker and Bassett (1978), and the solutions, $\hat{\theta}(\tau)$, are called the autoregression quantiles. Given $\hat{\theta}(\tau)$, the τth conditional quantile function of Y_t given X_t can be estimated by

$$\hat{Q}_{Y_t}(\tau|X_t) = X_t^\top \hat{\theta}(\tau),$$

and the conditional density of Y_t at $y = Q_{Y_t}(\tau|X_t)$ can be estimated by the difference quotients,

$$\hat{f}_{Y_t}(y|X_t) = \frac{2h}{\hat{Q}_{Y_t}(\tau+h|X_t) - \hat{Q}_{Y_t}(\tau-h|X_t)},$$

for some appropriately chosen sequence of $h = h(n) \to 0$.

If $\{Y_t\}_{t=1}^n$ is an AR(p) process determined by (17.1) and $\{u_t\}$ are i.i.d. random variables with mean 0 and variance $\sigma^2 < \infty$, and the distribution function of u_t, F, has a continuous density f with $f(u) > 0$ on $\mathcal{U} = \{u : 0 < F(u) < 1\}$, then the autoregression quantiles $\hat{\theta}(\tau)$, defined as the solution of (17.2), have limit

$$f[F^{-1}(\tau)]\Omega_0^{1/2}\sqrt{n}(\hat{\theta}(\tau) - \theta(\tau)) \Rightarrow B_k(\tau), \tag{17.3}$$

where

$$\Omega_0 = E(X_t X_t^\top) = \begin{bmatrix} 1 & \boldsymbol{\mu}_y^\top \\ \boldsymbol{\mu}_y & \Omega_y \end{bmatrix}, \quad \Omega_y = \begin{bmatrix} E(Y_t^2) & \cdots & E(Y_t Y_{t-p+1}) \\ \vdots & \ddots & \vdots \\ E(Y_t Y_{t-p+1}) & \cdots & E(Y_t^2) \end{bmatrix}. \tag{17.4}$$

Here $\boldsymbol{\mu}_y = E(Y_t) \cdot 1_{p \times 1}$, and $B_k(\tau)$ represents a k-dimensional standard Brownian bridge, $k = p + 1$.

By definition, for any fixed τ, $B_k(\tau)$ is $N(0, \tau(1-\tau)I_k)$, thus

$$\sqrt{n}(\hat{\theta}(\tau) - \theta(\tau)) \rightsquigarrow N\left(0, \frac{\tau(1-\tau)}{f[F^{-1}(\tau)]^2}\Omega_0^{-1}\right).$$

17.2.2 Quantile regressions of other time series models with i.i.d. errors

Quantile regression analysis can be easily extended to many other traditional time series models. We may consider a general time series regression model

$$Y_t = \theta_0 + \sum_{j=1}^{k} \theta_i X_{jt} + u_t = \theta^\top X_t + u_t,$$

where $X_t = (1, X_{1t}, \ldots, X_{kt})^\top$ is a vector of regressors that may contain lag values of Y_t and lag values of other covariates, and u_t is an i.i.d. mean zero sequence with distribution function $F(\cdot)$. Then the conditional distribution of Y_t is given by

$$Q_{Y_t}(\tau|\mathcal{F}_{t-1}) = \theta_0 + F^{-1}(\tau) + \theta_1 X_{1t} + \cdots + \theta_k X_{kt} = \theta(\tau)^\top X_t,$$

and the model can again be estimated via (17.2). The asymptotic behavior of quantile regression on such time series models is given by (17.3) with a different Ω_0:

$$\Omega_0 = \lim_{n\to\infty} \left[\frac{1}{n} \sum_{t=1}^{n} X_t X_t^\top\right].$$

A leading case of such regressions is the autoregressive distributed lag (ADL) models that are widely used in empirical applications:

$$Y_t = \alpha_0 + \sum_{i=1}^{p} \alpha_i Y_{t-i} + \sum_{j=1}^{q} \beta_j Z_{t-j} + u_t.$$

The above analysis can also be extended to various nonlinear parametric regression models. For example, consider a nonlinear autoregressive process

$$Y_t = \alpha + h(Y_{t-1}, \ldots, Y_{t-p}, \beta) + u_t,$$

where again u_t is an i.i.d. mean-zero sequence with pdf $f(\cdot)$ and cdf $F(\cdot)$. Then the conditional quantile function of Y_t is given by

$$Q_{Y_t}(\tau|\mathcal{F}_{t-1}) = h(Y_{t-1}, \ldots, Y_{t-p}, \beta) + \alpha + Q_u(\tau) = H(X_t; \theta(\tau)),$$

where $X_t = (Y_{t-1}, \ldots, Y_{t-p})^\top$. We may estimate the vector of parameters $\theta(\tau) = (\alpha + Q_u(\tau), \beta^\top)^\top$ (and thus the conditional quantile of Y_t) by the following nonlinear quantile regression:

$$\min_{\theta} \sum_t \rho_\tau(Y_t - H(X_t, \theta)).$$

Let $\varepsilon_{t\tau} = y_t - Q_u(\tau) - \alpha - h(X_t, \beta)$, $\dot{H}_\theta(x_t, \theta) = \partial H(x_t; \theta)/\partial\theta$, $\psi_\tau(u) = \tau - I(u < 0)$, and assume that

$$\Omega_n(\tau) = \frac{1}{n} \sum_t \dot{H}_\theta(X_t, \theta(\tau))\dot{H}_\theta(X_t, \theta(\tau))^\top \xrightarrow{P} \Omega(\tau)$$

and

$$\frac{1}{\sqrt{n}} \sum_t \dot{H}_\theta(x_t, \theta(\tau))\psi_\tau(\varepsilon_{t\tau}) \Rightarrow N(0, \tau(1-\tau)\Omega(\tau)),$$

where $\Omega(\tau)$ is nonsingular. Then under appropriate assumptions, the nonlinear QAR estimator $\hat{\theta}(\tau)$ is root-n consistent and

$$\sqrt{n}\left(\hat{\theta}(\tau) - \theta(\tau)\right) \Rightarrow N\left(0, \frac{\tau(1-\tau)}{f_u(Q_u(\tau))^2}\Omega(\tau)^{-1}\right).$$

17.2.3 Quantile regression estimation of ARMA models

We consider the ARMA(p, q) process

$$y_t = \mu_0 + \alpha_{10}y_{t-1} + \cdots + \alpha_{p0}y_{t-p} + u_t + \beta_{10}u_{t-1} + \cdots + \beta_{q0}u_{t-q}, \qquad (17.5)$$

where $\mu_0, \alpha_{10}, \ldots, \alpha_{p0}, \beta_{10}, \ldots, \beta_{q0}$ are the true values of the ARMA parameters, and the $\{u_t\}$ are i.i.d. random variables with mean zero, variance $\sigma^2 < \infty$, and distribution function F. Denoting the τth quantile function y_t by $Q_{y_t}(\tau|x)$, we have

$$
\begin{aligned}
Q_{y_t}(\tau|\mathcal{F}_{t-1}) &= \mu_0 + Q_u(\tau) + \alpha_{10}y_{t-1} + \cdots + \alpha_{p0}y_{t-p} + \beta_{10}u_{t-1} + \cdots + \beta_{q0}u_{t-q} \\
&= (\mu_0 + Q_u(\tau), \alpha_{10}, \ldots, \alpha_{p0}, \beta_{10}, \ldots, \beta_{q0})
\begin{bmatrix}
1 \\
y_{t-1} \\
\vdots \\
y_{t-p} \\
u_{t-1} \\
\vdots \\
u_{t-q}
\end{bmatrix} \\
&= \theta_0(\tau)^\top X_t(\theta_0),
\end{aligned}
$$

where $\theta_0(\tau)^\top = (\lambda_0(\tau), \alpha_{10}, \ldots, \alpha_{p0}, \beta_{10}, \ldots, \beta_{q0})$, $\lambda_0(\tau) = \mu_0 + Q_u(\tau)$, and $\theta_0^\top = (\mu_0, \alpha_{10}, \ldots, \alpha_{p0}, \beta_{10}, \ldots, \beta_{q0})$. Notice that the u_{t-j} are not observable and thus depend on θ. Consequently, X_t is dependent on θ. Thus, we write the vector of regressors X_t as $X_t(\theta_0)$. Denote $\gamma_0(\tau) = (\lambda_0(\tau), \mu_0, \alpha_{10}, \ldots, \alpha_{p0}, \beta_{10}, \ldots, \beta_{q0})^\top$. Then $Q_{y_t}(\tau|\mathcal{F}_{t-1}) = \theta_0(\tau)^\top X_t(\theta_0) = H(X_t; \gamma_0(\tau))$.

If we use $\gamma = (\lambda, \mu, \alpha_1, \ldots, \alpha_p, \beta_1, \ldots, \beta_q)^\top$ to signify a vector of general parameters, and denote

$$
\begin{aligned}
u_t(\gamma) &= y_t - H(X_t; \gamma) \\
&= y_t - H(X_t; \gamma_0(\tau)) + H(X_t; \gamma_0(\tau)) - H(X_t; \gamma) \\
&= u_t - Q_u(\tau) + (\lambda_0(\tau) - \lambda) + (\alpha_{10} - \alpha_1)y_{t-1} + \cdots + (\alpha_{p0} - \alpha_p)y_{t-p} \\
&\quad + (\beta_{10} - \beta_1)u_{t-1} + \cdots + (\beta_{q0} - \beta_q)u_{t-q},
\end{aligned}
$$

let

$$D_t(\gamma)^\top = \left[\ldots, -\frac{\partial u_t(\gamma)}{\partial \gamma_i}, \ldots \right].$$

Define the quantile regression estimators as

$$\hat{\gamma}(\tau) = \arg\min_{\gamma} \sum_{t=1}^{n} \rho_\tau(y_t - H(X_t; \gamma)).$$

Then it can be shown that

$$\sqrt{n}\,[\hat{\gamma}(\tau) - \gamma(\tau)] \Rightarrow N(0, \tau(1-\tau)\Sigma_f^{-1}\Sigma_D\Sigma_f^{-1}) = \frac{1}{f(F^{-1}(\tau))}N(0, \tau(1-\tau)\Sigma_D^{-1}),$$

where

$$
\begin{aligned}
\Sigma_D &= \operatorname{Cov}(D_t(\gamma_0(\tau))) = \lim \frac{1}{n}\sum_{t=1}^{n} D_t(\gamma_0(\tau))D_t(\gamma_0(\tau))^\top, \\
\Sigma_f &= \lim \frac{1}{n}\sum_{t=1}^{n} f(F^{-1}(\tau))D_t(\gamma_0(\tau))D_t(\gamma_0(\tau))^\top = f(F^{-1}(\tau))\Sigma_D.
\end{aligned}
$$

17.2.4   Quantile regressions with serially correlated errors

Quantile regression can also be applied to regression models with serially correlated errors. Consider the linear model

$$Y_t = \alpha + \beta^\top X_t + u_t = \theta^\top Z_t + u_t, \tag{17.6}$$

where X_t and u_t are k- and one-dimensional weakly dependent random variables, $\{X_t\}$ and $\{u_t\}$ are independent of each other, and $E(u_t) = 0$. If we denote the distribution function of u_t by $F_u(\cdot)$, then, conditional on X_t, the τth quantile of Y_t, given X_t, is

$$Q_{Y_t}(\tau|X_t) = \alpha + \beta^\top X_t + F_u^{-1}(\tau) = \theta(\tau)^\top Z_t,$$

where $\theta(\tau) = (\alpha + F_u^{-1}(\tau), \beta^\top)^\top$. The vector of parameters $\theta(\tau)$ can be estimated by solving the problem

$$\hat{\theta}(\tau) = \arg\min_{\theta \in \mathbb{R}^p} \sum_{t=1}^n \rho_\tau(Y_t - Z_t^\top \theta). \tag{17.7}$$

Let $u_{t\tau} = Y_t - \theta(\tau)^\top Z_t$. We have $E[\psi_\tau(u_{t\tau})|X_t] = 0$. Under appropriate assumptions on the moments of X_t and weak dependence of (X_t, u_t), we have

$$n^{-1/2} \sum_{t=1}^n Z_t \psi_\tau(u_{t\tau}) = \begin{bmatrix} n^{-1/2} \sum_{t=1}^n \psi_\tau(u_{t\tau}) \\ n^{-1/2} \sum_{t=1}^n X_t \psi_\tau(u_{t\tau}) \end{bmatrix} \Rightarrow N(0, \Sigma(\tau)),$$

where $\Sigma(\tau)$ is the long-run covariance matrix of $Z_t \psi_\tau(u_{t\tau})$, defined by

$$\Sigma(\tau) = \lim \left(n^{-1/2} \sum_{t=1}^n Z_t \psi_\tau(u_{t\tau}) \right) \left(n^{-1/2} \sum_{t=1}^n Z_t \psi_\tau(u_{t\tau}) \right) = \begin{bmatrix} \omega_\psi^2(\tau) & 0 \\ 0 & \Omega(\tau) \end{bmatrix}.$$

Under regularity assumptions, the quantile regression estimator (17.7) has asymptotic representation

$$\sqrt{n}(\hat{\theta}(\tau) - \theta(\tau)) = \frac{1}{f(F^{-1}(\tau))} \Sigma_z^{-1} S_{z\psi} + o_p(1),$$

where

$$\Sigma_z = \lim_{n \to \infty} \frac{1}{n} \sum_{t=1}^n Z_t Z_t^\top, \quad S_{z\psi} = \frac{1}{n^{1/2}} \sum_{t=1}^n Z_t \psi_\tau(u_{t\tau}).$$

As a result, for any fixed $\tau \in (0, 1)$,

$$\sqrt{n}(\hat{\theta}(\tau) - \theta(\tau)) \Rightarrow N\left(0, \frac{1}{f(F^{-1}(\tau))^2} \Sigma_z^{-1} \Sigma(\tau) \Sigma_z^{-1} \right).$$

The above results may be extended to the case of heteroskedastic data where other elements in $\theta(\tau)$ are also τ-dependent.

Statistical inference based on $\hat{\theta}(\tau)$ requires estimation of the covariance matrices Σ_z and $\Sigma(\tau)$. The matrix Σ_z can be easily estimated by its sample analog

$$\hat{\Sigma}_z = n^{-1} \sum_{t=1}^n Z_t Z_t^\top.$$

Let $\hat{u}_{t\tau} = Y_t - \hat{\theta}(\tau)^\top Z_t$. Following the heteroskedasticity and autocorrelation consistent estimation literature (see, for example, Andrews, 1991), $\Sigma(\tau)$ may be estimated by

$$\hat{\Sigma}(\tau) = \sum_{h=-M}^M k\left(\frac{h}{M} \right) \left[\frac{1}{n} \sum_{1 \leqslant t, t+h \leqslant n} Z_t \psi_\tau(\hat{u}_{t\tau}) Z_{t+h}^\top \psi_\tau(\hat{u}_{(t+h)\tau}) \right],$$

where $k(\cdot)$ is the lag window defined on $[-1, 1]$ with $k(0) = 1$, and M is the bandwidth parameter satisfying the property that $M \to \infty$ and $M/n \to 0$ as the sample size $n \to \infty$.

Portnoy (1991) studied asymptotic properties for regression quantiles with m-dependent errors. His analysis also allows for nonstationarity with a nonvanishing bias term.

The above quantile regression analysis can also be extended to the case with long-range dependent errors. Koul and Mukherjee (1994) considered linear model (17.6) when the errors are a function of Gaussian random variables that are stationary and long-range dependent so that

$$\text{Cov}(u_t, u_{t+h}) = h^{-\lambda} L(h), \quad \text{for some } 0 < \lambda < 1,$$

where $L(h)$ is positive for large h and slowly varying at infinity.

17.3 Quantile regressions with ARCH/GARCH errors

Autoregressive conditional heteroskedasticity (ARCH) and generalized ARCH (GARCH) models have proven to be highly successful in modeling financial data. Estimators of volatilities and quantiles based on ARCH and GARCH models are now widely used in finance applications. Koenker and Zhao (1996) studied quantile regression for linear ARCH models. They consider the following linear ARCH(p) process:

$$u_t = \sigma_t \cdot \varepsilon_t, \quad \sigma_t = \gamma_0 + \gamma_1 |u_{t-1}| + \cdots + \gamma_p |u_{t-p}|, \tag{17.8}$$

where $0 < \gamma_0 < \infty$, $\gamma_1, \ldots, \gamma_p \geq 0$, and ε_t are i.i.d.$(0, 1)$ random variables with pdf $f(\cdot)$ and cdf $F(\cdot)$. Let $Z_t = (1, |u_{t-1}|, \ldots, |u_{t-q}|)^\top$ and $\gamma(\tau) = (\gamma_0 F^{-1}(\tau), \gamma_1 F^{-1}(\tau), \ldots, \gamma_q F^{-1}(\tau))^\top$. Then the conditional quantiles of u_t are given by

$$Q_{u_t}(\tau | \mathcal{F}_{t-1}) = \gamma_0(\tau) + \gamma_1(\tau) |u_{t-1}| + \ldots + \gamma_p(\tau) |u_{t-p}| = \gamma(\tau)^\top Z_t,$$

and can be estimated by the following linear quantile regression of u_t on Z_t:

$$\min_{\gamma} \sum_t \rho_\tau (u_t - \gamma^\top Z_t), \tag{17.9}$$

where $\gamma = (\gamma_0, \gamma_1, \ldots, \gamma_q)^\top$.

The asymptotic behavior of the above quantile regression estimator is given by Koenker and Zhao (1996). Suppose that u_t is given by model (17.8), f is bounded and continuous, $f(F^{-1}(\tau)) > 0$ for any $0 < \tau < 1$. In addition, $E|u_t|^{2+\delta} < \infty$. Then the regression quantiles $\hat{\gamma}(\tau)$ of (17.9) have the Bahadur representation

$$\sqrt{n} (\hat{\gamma}(\tau) - \gamma(\tau)) = \frac{\Sigma_1^{-1}}{f(F^{-1}(\tau))} \frac{1}{\sqrt{n}} \sum_{t=1}^n Z_t^\top \psi_\tau(\varepsilon_{t\tau}) + o_p(1),$$

where $\Sigma_1 = E Z_t Z_t^\top / \sigma_t$ and $\varepsilon_{t\tau} = \varepsilon_t - F^{-1}(\tau)$. Consequently,

$$\sqrt{n} (\hat{\gamma}(\tau) - \gamma(\tau)) = N \left(0, \frac{\tau(1-\tau)}{f(F^{-1}(\tau))^2} \Sigma_1^{-1} \Sigma_0 \Sigma_1^{-1} \right), \quad \text{with } \Sigma_0 = E Z_t Z_t^\top.$$

In many applications, conditional heteroskedasticity is modeled on the residuals of a regression:

$$Y_t = \alpha^\top X_t + u_t, \tag{17.10}$$

where $X_t = (1, X_{1,t}, \ldots, X_{k,t})^\top$, $\alpha = (\alpha_0, \alpha_1, \ldots, \alpha_k)^\top$, and u_t is a linear ARCH(p) process given by model (17.8). A very popular model that is widely used in applications is the AR-ARCH model where $X_t = (1, Y_{t-1}, \ldots, Y_{t-k})^\top$. The conditional quantiles of Y_t are then given by

$$Q_{Y_t}(\tau|\mathcal{F}_{t-1}) = \alpha^\top X_t + \gamma(\tau)^\top Z_t. \tag{17.11}$$

We may consider a two-step procedure that estimates α in the first step, and then estimates $\gamma(\tau)$ based on the estimated residuals. The two-step procedure is usually less efficient because the preliminary estimation of α may affect the second-step estimation of $\gamma(\tau)$, but it is computationally much simpler and is widely used in empirical applications.

Koenker and Zhao (1996) studied the two-step estimation. Suppose that Y_t is given by (17.10) and (17.8), and the conditions of their Theorem 4.1 hold, $\hat{\alpha}$ is a root-n consistent estimator, and

$$\tilde{\gamma}(\tau) = \arg\min_\gamma \sum_t \rho_\tau(\hat{u}_t - \gamma^\top \hat{Z}_t),$$

where $\hat{Z}_t = (1, |\hat{u}_{t-1}|, \ldots, |\hat{u}_{t-q}|)^\top$, $\hat{u}_t = Y_t - \hat{\alpha}^\top X_t$. Then

$$\sqrt{n}\,(\tilde{\gamma}(\tau) - \gamma(\tau)) = \frac{\Sigma_1^{-1}}{f(F^{-1}(\tau))} \frac{1}{\sqrt{n}} \sum_{t=1}^n Z_t^\top \psi_\tau(\varepsilon_{t\tau}) + \Sigma_1^{-1} G_1 \sqrt{n}\,(\hat{\alpha} - \alpha) + o_p(1)$$

with $G_1 = \mathrm{E}(Z_t(X_t - B_t\gamma(\tau)))^\top/\sigma_t$, and $B_t = (0, \mathrm{sign}(u_{t-1})X_{t-1}, \ldots, \mathrm{sign}(u_{t-p})X_{t-p})$. If f is symmetric about zero, and $\alpha_0 = 0$, then $G_1 = 0$, and thus

$$\sqrt{n}\,(\tilde{\gamma}(\tau) - \gamma(\tau)) = N\left(0, \frac{\tau(1-\tau)}{f(F^{-1}(\tau))^2} \Sigma_1^{-1} \Sigma_0 \Sigma_1^{-1}\right).$$

Alternatively, in the regression setting (17.10), we may construct a *joint* estimation of α and ARCH/GARCH parameters based on a *nonlinear* quantile regression. Consider, for example, models (17.10) and (17.8). Denote by $Q_{u_t}(\tau|\mathcal{F}_{t-1})$ the conditional τ-quantile of u_t given $\mathcal{F}_{t-1}$, and by $Q_\varepsilon(\tau)$ the τ-quantile of ε_t, and assume that ε_t is independent of $\mathcal{F}_{t-1}$ for each t. Notice that, given (17.8),

$$Q_{u_t}(\tau|\mathcal{F}_{t-1}) = (\gamma_0 + \gamma_1|u_{t-1}| + \ldots + \gamma_p|u_{t-p}|)\,Q_\varepsilon(\tau),$$

and

$$u_t = Y_t - \alpha^\top X_t = Y_t - \alpha_0 - \alpha_1 X_{1t} - \cdots - \alpha_k X_{kt},$$

we have

$$Q_{Y_t}(\tau|\mathcal{F}_{t-1}) = g(\theta(\tau), X_t, Y_{t-1}, X_{t-1}, \ldots, Y_{t-p}, X_{t-p}),$$

where

$$g(\theta(\tau), X_t, Y_{t-1}, X_{t-1}, \ldots, Y_{t-p}, X_{t-p}) = \mu(\tau) + \sum_{j=1}^k \alpha_j X_{jt} + \sum_{i=1}^p \lambda_i(\tau)\left|Y_{t-i} - \alpha_0 - \sum_{j=1}^k \alpha_j X_{j,t-i}\right|,$$

with

$$\mu(\tau) = \alpha_0 + \gamma_0 Q_\varepsilon(\tau), \quad \lambda_i(\tau) = \gamma_i Q_\varepsilon(\tau), \quad \theta(\tau) = (\mu(\tau), \lambda_i(\tau), \alpha_j)^\top,$$

for $i = 1, \ldots, p$ and $j = 1, \ldots, k$. The parameter vector $\theta(\tau)$ contains two types of parameters: the local parameters $(\mu(\tau), \lambda_i(\tau), i = 1, \ldots, p)$ that vary with the quantile τ, and the global parameters $(\alpha_j, j = 0, \ldots, k)$ that remain the same across quantiles.

Given a random sample $\{(X_t, Y_t)\}_{t=1}^n$, we may estimate the parameter $\theta(\tau)$ based on the quantile regression

$$\hat{\theta}(\tau) = \arg\min_\theta \sum_{t=1}^n \rho_\tau(Y_t - g(\theta, X_t, Y_{t-1}, X_{t-1}, \ldots, Y_{t-p}, X_{t-p})). \tag{17.12}$$

Xiao and Zhao (2015) studied the joint estimation procedure where both the conditional mean and ARCH/GARCH errors are incorporated in the estimation. They established an asymptotic Bahadur representation and asymptotic normality for $\hat{\theta}(\tau)$. Monte Carlo experiments indicate that the joint estimation procedure is more efficient than the conventional two-step approach based on the estimated residuals.

Additional efficiency gain can be obtained via weighted quantile regression in the presence of conditional heteroskedasticity. In (17.8), notice that $u_t/\sigma_t = \varepsilon_t$ are i.i.d. and the dependence in $\{u_t\}$ is removed. Further efficiency gain can be achieved from a weighted quantile regression based on the joint model

$$\tilde{\theta}(\tau) = \arg\min_\theta \sum_{t=1}^n \frac{1}{\hat{\sigma}_t} \rho_\tau(Y_t - g(\theta, X_t, Y_{t-1}, X_{t-1}, \ldots, Y_{t-p}, X_{t-p})), \qquad (17.13)$$

where $\hat{\sigma}_t$ is an estimator of σ_t.

ARCH models are easier to estimate than but cannot parsimoniously capture the persistent influence of long past shocks as well as GARCH models. However, quantile regression on GARCH models is highly nonlinear and thus complicated to estimate. In particular, the quantile estimation problem in GARCH models corresponds to a restricted nonlinear quantile regression and conventional nonlinear quantile regression techniques are not directly applicable.

Consider, for example, the following linear GARCH(p, q) model:

$$
\begin{aligned}
u_t &= \sigma_t \cdot \varepsilon_t, & (17.14)\\
\sigma_t &= \beta_0 + \beta_1\sigma_{t-1} + \cdots + \beta_p\sigma_{t-p} + \gamma_1|u_{t-1}| + \cdots + \gamma_q|u_{t-q}|. & (17.15)
\end{aligned}
$$

Let $\mathcal{F}_{t-1}$ represent information up to time $t-1$. Then the τth conditional quantile of u_t is given by

$$Q_{u_t}(\tau|\mathcal{F}_{t-1}) = \theta(\tau)^\top Z_t, \qquad (17.16)$$

where $Z_t = (1, \sigma_{t-1}, \ldots, \sigma_{t-p}, |u_{t-1}|, \ldots, |u_{t-q}|)^\top$ and $\theta(\tau)^\top = (\beta_0, \beta_1, \ldots, \beta_p, \gamma_1, \ldots, \gamma_q)F^{-1}(\tau)$. Since Z_t contains σ_{t-k} $(k = 1, \ldots, q)$ which in turn depends on unknown parameters $\theta = (\beta_0, \beta_1, \ldots, \beta_p, \gamma_1, \ldots, \gamma_q)$, we may write Z_t as $Z_t(\theta)$ to emphasize the nonlinearity and its dependence on θ. If we use the nonlinear quantile regression

$$\min_\theta \sum_t \rho_\tau(u_t - \theta^\top Z_t(\theta)), \qquad (17.17)$$

for a fixed τ in isolation, a consistent estimate of θ cannot be obtained since it ignores the global dependence of the σ_{t-k} on the entire function $\theta(\cdot)$. If the dependence structure of u_t is characterized by (17.14) and (17.15), we can consider the following restricted quantile regression instead of (17.17):

$$
\left(\hat{\pi}, \hat{\theta}\right) =
\begin{cases}
\arg\min\limits_{\pi,\theta} \sum_i \sum_t \rho_{\tau_i}(u_t - \pi_i^\top Z_t(\theta)) \\[2mm]
\text{s.t. } \pi_i = \theta(\tau_i) = \theta F^{-1}(\tau_i).
\end{cases}
$$

Estimation of this global restricted nonlinear quantile regression is complicated. Xiao and Koenker (2009) propose a simpler two-stage estimator that both incorporates the global restrictions and focuses on the local approximation around the specified quantile. The proposed estimation consists of the following two steps. The first step considers a global estimation to incorporate the global dependence of the latent σ_{t-k} on θ. Then, using results from the first step, the second step focuses on the specified quantile to find the best local estimate for the conditional quantile. Let

$$A(L) = 1 - \beta_1 L - \cdots - \beta_p L^p, \quad B(L) = \gamma_1 + \cdots + \gamma_q L^{q-1}.$$

Under regularity assumptions ensuring that $A(L)$ is invertible, we obtain an ARCH(∞) representation for σ_t :

$$\sigma_t = a_0 + \sum_{j=1}^{\infty} a_j \, |u_{t-j}| \, . \tag{17.18}$$

For identification, we normalize $a_0 = 1$. Substituting the above ARCH(∞) representation into (17.14) and (17.15), we have

$$u_t = \left(a_0 + \sum_{j=1}^{\infty} a_j \, |u_{t-j}| \right) \varepsilon_t \tag{17.19}$$

and

$$Q_{u_t}(\tau|\mathcal{F}_{t-1}) = \alpha_0(\tau) + \sum_{j=1}^{\infty} \alpha_j(\tau) \, |u_{t-j}| \, ,$$

where $\alpha_j(\tau) = a_j Q_{\varepsilon_t}(\tau)$, $j = 0, 1, 2, \ldots$.

Let $m = m(n)$ be a truncation parameter. We may consider the following truncated quantile autoregression:

$$Q_{u_t}(\tau|\mathcal{F}_{t-1}) \approx a_0(\tau) + a_1(\tau) \, |u_{t-1}| + \cdots + a_m(\tau) \, |u_{t-m}| \, .$$

By choosing m suitably small relative to the sample size n, but large enough to avoid serious bias, we obtain a sieve approximation for the GARCH model.

One could estimate the conditional quantiles simply using a sieve approximation:

$$\check{Q}_{u_t}(\tau|\mathcal{F}_{t-1}) = \hat{a}_0(\tau) + \hat{a}_1(\tau) \, |u_{t-1}| + \cdots + \hat{a}_m(\tau) \, |u_{t-m}| \, ,$$

where $\hat{a}_j(\tau)$ are the quantile autoregression estimates. Under regularity assumptions,

$$\check{Q}_{u_t}(\tau|\mathcal{F}_{t-1}) = Q_{u_t}(\tau|\mathcal{F}_{t-1}) + O_p(m/\sqrt{n}).$$

However, Monte Carlo evidence indicates that the simple sieve approximation does not directly provide a good estimator for the GARCH model, but it serves as an adequate preliminary estimator. Since the first-step estimation focuses on the global model, it is desirable to use information on multiple quantiles in estimation. Combining information on multiple quantiles helps us to obtain a globally coherent estimate of the scale parameters.

Suppose that we estimate the mth-order quantile autoregression

$$\tilde{\alpha}(\tau) = \arg\min_{\alpha} \sum_{t=m+1}^{n} \rho_\tau \left(u_t - \alpha_0 - \sum_{j=1}^{m} \alpha_j \, |u_{t-j}| \right) \tag{17.20}$$

at quantiles $(\tau_1, \ldots, \tau_K)$, and obtain estimates $\tilde{\alpha}(\tau_k)$, $k = 1, \ldots, K$. Let $\tilde{a}_0 = 1$ in accordance with the identification assumption. Denote

$$\mathbf{a} = [a_1, \ldots, a_m, q_1, \ldots, q_K]^\top, \quad \tilde{\pi} = \left[\tilde{\alpha}(\tau_1)^\top, \ldots, \tilde{\alpha}(\tau_K)^\top \right]^\top,$$

where $q_k = Q_{\varepsilon_t}(\tau_k)$, and

$$\phi(\mathbf{a}) = g \otimes \alpha = [q_1, a_1 q_1, \ldots, a_m q_1, \ldots, q_K, a_1 q_K, \ldots, a_m q_K]^\top,$$

where $g = [q_1, \ldots, q_K]^\top$ and $\alpha = [1, a_1, a_2, \ldots, a_m]^\top$. We consider the following estimator for the vector $\mathbf{a}$ that combines information over the K quantile estimates based on the restrictions $\alpha_j(\tau) = a_j Q_{\varepsilon_t}(\tau)$:

$$\tilde{\mathbf{a}} = \arg \min_{\mathbf{a}} (\tilde{\boldsymbol{\pi}} - \phi(\mathbf{a}))^\top A_n (\tilde{\boldsymbol{\pi}} - \phi(\mathbf{a})), \tag{17.21}$$

where A_n is a $(K(m+1)) \times (K(m+1))$ positive definite matrix. Denoting $\tilde{\mathbf{a}} = (\tilde{a}_0, \ldots, \tilde{a}_m)$, σ_t can be estimated by

$$\tilde{\sigma}_t = \tilde{a}_0 + \sum_{j=1}^{m} \tilde{a}_j \left| u_{t-j} \right|.$$

In the second step, we perform a quantile regression,

$$\min_{\theta} \sum_t \rho_\tau (u_t - \theta^\top \tilde{Z}_t), \tag{17.22}$$

of u_t on $\tilde{Z}_t = (1, \tilde{\sigma}_{t-1}, \ldots, \tilde{\sigma}_{t-p}, |u_{t-1}|, \ldots, |u_{t-q}|)^\top$. The two-step estimator of $\theta(\tau)^\top = (\beta_0(\tau), \beta_1(\tau), \ldots, \beta_p(\tau), \gamma_1(\tau), \ldots, \gamma_q(\tau))$ is then given by the solution of (17.22), $\hat{\theta}(\tau)$, and the τth conditional quantile of u_t can be estimated by

$$\hat{Q}_{u_t}(\tau | \mathcal{F}_{t-1}) = \hat{\theta}(\tau)^\top \tilde{Z}_t.$$

Iteration can be applied to the above procedure for further improvement.

Let $\tilde{\alpha}(\tau)$ be the solution to (17.20). Then, under appropriate assumptions, we have

$$\|\tilde{\alpha}(\tau) - \alpha(\tau)\|^2 = O_p(m/n), \tag{17.23}$$

and for any $\lambda \in \mathcal{R}^{m+1}$,

$$\frac{\sqrt{n} \lambda^\top (\tilde{\alpha}(\tau) - \alpha(\tau))}{\sigma_\lambda} \Rightarrow N(0, 1),$$

where $\sigma_\lambda^2 = f_\varepsilon \left(F_\varepsilon^{-1}(\tau) \right)^{-2} \lambda^\top D_n^{-1} \Sigma_n(\tau) D_n^{-1} \lambda$, with

$$D_n = \left[\frac{1}{n} \sum_{t=m+1}^{n} \frac{x_t x_t^\top}{\sigma_t} \right], \quad \Sigma_n(\tau) = \frac{1}{n} \sum_{t=m+1}^{n} x_t x_t^\top \psi_\tau^2(u_{t\tau}),$$

in which $x_t = (1, |u_{t-1}|, \ldots, |u_{t-m}|)^\top$.

Define

$$G = \frac{\partial \phi(\mathbf{a})}{\partial \mathbf{a}^\top} \bigg|_{\mathbf{a}=\mathbf{a}_0} = \dot{\phi}(\mathbf{a}_0) = \left[g \otimes J_m \vdots I_K \otimes \alpha_0 \right], \quad g_0 = \begin{bmatrix} Q_{\varepsilon_t}(\tau_1) \\ \vdots \\ Q_{\varepsilon_t}(\tau_K) \end{bmatrix},$$

where g_0 and α_0 are the true values of vectors $g = [q_1, \ldots, q_K]^\top$ and $\alpha = [1, a_1, a_2, \ldots, a_m]^\top$, and

$$J_m = \begin{bmatrix} 0 & \cdots & 0 \\ 1 & \cdots & 0 \\ \vdots & \ddots & \vdots \\ 0 & \cdots & 1 \end{bmatrix}$$

is an $(m + 1) \times m$ matrix and I_K is a K-dimensional identity matrix. Under regularity assumptions, the minimum distance estimator $\tilde{a}$ solving (17.21) has asymptotic representation

$$\sqrt{n}(\hat{a} - a_0) = \left[G^\top A_n G\right]^{-1} G^\top A_n \sqrt{n}\,(\tilde{\pi} - \pi) + o_p(1),$$

where

$$\sqrt{n}(\tilde{\pi} - \pi) = -\frac{1}{\sqrt{n}} \sum_{t=m+1}^{n} \begin{bmatrix} D_n^{-1} x_t \frac{\psi_{\tau_1}(u_{t\tau_1})}{f_\varepsilon\left(F_\varepsilon^{-1}(\tau_1)\right)} \\ \vdots \\ D_n^{-1} x_t \frac{\psi_{\tau_m}(u_{t\tau_m})}{f_\varepsilon\left(F_\varepsilon^{-1}(\tau_m)\right)} \end{bmatrix} + o_p(1),$$

and the two-step estimator $\hat{\theta}(\tau)$ based on (17.22) has asymptotic representation

$$\sqrt{n}\left(\hat{\theta}(\tau) - \theta(\tau)\right) = -\frac{1}{f_\varepsilon\left(F_\varepsilon^{-1}(\tau)\right)}\Omega^{-1}\left\{\frac{1}{\sqrt{n}}\sum_t Z_t \psi_\tau(u_{t\tau})\right\} + \Omega^{-1}\Gamma\sqrt{n}\,(\tilde{a} - a) + o_p(1),$$

where $a = [a_1, a_2, \ldots, a_m]^\top$, $\Omega = E\left[Z_t Z_t^\top / \sigma_t\right]$, and

$$\Gamma = \sum_{k=1}^{p} \theta_k C_k, \quad C_k = E\left[\left(|u_{t-k-1}|, \ldots, |u_{t-k-m}|\right)\frac{Z_t}{\sigma_t}\right].$$

Quantile regression on other types of ARCH/GARCH models, for example, the quadratic ARCH or GARCH models, can also be analyzed based on nonlinear quantile regressions, although the implementations are different for different versions of ARCH/GARCH models.

Like the ARCH case, in applications we may also be interested in regression models with an error term characterized by a GARCH process. For example, one might be interested in the regression (17.10) when u_t is characterized by a GARCH(p, q) process given by, say, model (17.15). Again, we may consider a two-step procedure that estimates α in the first step, and then estimates the GARCH parameters based on the estimated residuals. And again, the preliminary estimation of the conditional mean generally affects the limiting distributions of these quantile regression estimates. Consequently, both the estimation efficiency and related inference are affected by the preliminary estimation. As in the ARCH model, the conditional mean parameters and conditional volatility can be estimated jointly. Weighted nonlinear regressions can also be conducted to further improve efficiency (see Xiao and Zhao, 2015).

Quantile regression can also provide robust estimators for volatility. Quantile regression based volatility estimators have better sampling performance than the traditional quasi-maximum likelihood estimators when the distribution of the underlying time series is non-normal. We illustrate this below for the case of a linear ARCH process, but the basic idea can be extended to various types of ARCH/GARCH models (see Xiao and Wan, 2012). Suppose that we consider the linear ARCH(p) process (17.8). We may first obtain a preliminary estimate of the standardized residual term ε_t via quantile regression (17.9), denoted by $\hat{\varepsilon}_t$. In particular, consider quantile regression of u_t on Z_t based on (17.9) and obtain

$$\hat{\gamma}(\tau) = \left(\hat{\gamma}_0(\tau), \hat{\gamma}_1(\tau), \ldots, \hat{\gamma}_p(\tau)\right).$$

Define $\sigma_t(\tau) = \gamma(\tau)^\top Z_t = \sigma_t F^{-1}(\tau)$ and $\varepsilon_t(\tau) = u_t/\sigma_t(\tau) = \varepsilon_t/F^{-1}(\tau)$, and let

$$\hat{\sigma}_t(\tau) = \hat{\gamma}(\tau)^\top Z_t, \quad \hat{\varepsilon}_t(\tau) = \frac{u_t}{\hat{\sigma}_t(\tau)} = \frac{u_t}{\hat{\gamma}(\tau)^\top Z_t}.$$

We can estimate ε_t by

$$\hat{\varepsilon}_t = \frac{1}{s_T(\tau)}\left(\hat{\varepsilon}_t(\tau) - \bar{\bar{\varepsilon}}(\tau)\right),$$

where

$$s_T(\tau)^2 = \frac{1}{T}\sum_{t=1}^{T}\left(\widehat{\varepsilon}_t(\tau) - \overline{\widehat{\varepsilon}}(\tau)\right)^2, \quad \overline{\widehat{\varepsilon}}(\tau) = \frac{1}{T}\sum_{t=1}^{T}\widehat{\varepsilon}_t(\tau) = \frac{1}{T}\sum_{t=1}^{T}\frac{u_t}{\widehat{\gamma}(\tau)^\top Z_t}.$$

Then we estimate the τth quantile of the distribution of ε, $F^{-1}(\tau)$, from the corresponding empirical quantile of $\{\widehat{\varepsilon}_t\}_{t=1}^{T}$, which is equivalent to solving the quantile regression

$$\widehat{F}^{-1}(\tau) = \widehat{\xi}(\tau) = \arg\min_{\xi}\sum_{t=1}^{T}\rho_\tau\left(\widehat{\varepsilon}_t - \xi\right). \tag{17.24}$$

The ARCH parameters $\gamma = (\gamma_0, \gamma_1, \ldots, \gamma_q)^\top$ are then estimated by

$$\widehat{\gamma} = \frac{\widehat{\gamma}(\tau)}{\widehat{\xi}(\tau)} = (\widehat{\gamma}_0, \widehat{\gamma}_1, \ldots, \widehat{\gamma}_p)^\top, \quad \text{where } \widehat{\gamma}_j = \frac{\widehat{\gamma}_j(\tau)}{\widehat{\xi}(\tau)}, \ j = 0, 1, \ldots, p, \tag{17.25}$$

and the conditional volatility σ_t is estimated by

$$\widehat{\sigma}_t = \widehat{\gamma}_0 + \widehat{\gamma}_1\left|u_{t-1}\right| + \cdots + \widehat{\gamma}_p\left|u_{t-p}\right|. \tag{17.26}$$

Under regularity conditions, the quantile estimator $\widehat{\xi}(\tau)$ of (17.24) has asymptotic representation

$$\sqrt{n}\left(\widehat{\xi}(\tau) - \xi(\tau)\right) = \frac{1}{\sqrt{T}}\frac{1}{f(F^{-1}(\tau))}\sum_{t=1}^{T}\psi_\tau(\varepsilon_{t\tau}) - \frac{1}{\sqrt{T}}\sum_{t=1}^{T}\varepsilon_t + o_p(1).$$

The ARCH parameter estimators $\widehat{\gamma}$ and volatility $\widehat{\sigma}_t$ given by (17.25) and (17.26) have asymptotic representations

$$\sqrt{T}\left(\widehat{\gamma} - \gamma\right) = \frac{1}{\sqrt{T}}\frac{1}{f(F^{-1}(\tau))F^{-1}(\tau)}\left[\sum_{t=1}^{n}\left(\Sigma_1^{-1}Z_t^\top - \gamma\right)\psi_\tau(\varepsilon_{t\tau})\right]$$

$$+ \frac{1}{\sqrt{T}}\frac{1}{F^{-1}(\tau)}\sum_{t=1}^{T}\gamma\varepsilon_t + o_p(1),$$

$$\sqrt{T}\left(\widehat{\sigma}_t - \sigma_t\right) = \frac{1}{\sqrt{T}}\frac{1}{f(F^{-1}(\tau))F^{-1}(\tau)}Z_t^\top\sum_{s=1}^{T}\left[\Sigma_1^{-1}Z_s - \gamma\right]\psi_\tau(\varepsilon_{s\tau})$$

$$+ \frac{1}{\sqrt{T}}\frac{1}{F^{-1}(\tau)}Z_t^\top\gamma\sum_{s=1}^{T}\varepsilon_s + o_p(1).$$

Similar (but more complicated) procedures can be applied to GARCH based models. In addition, more efficient estimators can be obtained by further combining information on multiple quantiles. Xiao and Wan (2012) studied two slightly different approaches to combining quantile information: combining quantile information based on a weighted average of these estimators obtained at different quantiles; and combining information on different τ directly based on the first-stage quantile regression (17.9), which delivers estimates $\widehat{\gamma}(\tau)$. Monte Carlo and empirical application results indicate that the quantile regression based volatility estimators outperform many present estimation techniques.

17.4   Quantile regressions with heavy-tailed errors

Quantile regression analysis can also be extended to time series regressions with infinite variances. This is an important topic for quantile regression since many financial time series have heavy tails and quantile regression based risk measures are calculated for these series (see, for example, Rachev and Mittnik, 2000; Fama, 1965; Mandelbrot, 1963; Taleb, 2007; Embrechts et al., 1997; Ibragimov, 2009; Ibragimov et al., 2009). For example, electricity prices can be subject to large spikes due to supply–demand imbalances that cannot be temporally mediated (Weron, 2008). There is also a literature on heavy-tail behavior in the distribution of wealth and income (see, for example, Ibragimov, 2009). Variance based risk measures may be nonrobust to large movements in series. The consequences of large crashes are enormous, and it is important to use quantile based measures that are robust to them.

Consider an autoregressive model of order p:

$$Y_t = \theta_0 + \theta_1 Y_{t-1} + \cdots + \theta_p Y_{t-p} + u_t,$$

where u_t is an i.i.d. sequence with zero mean and infinite variance. In particular, we assume that the distribution of Y_t is in the domain of attraction of a stable distribution with index $\alpha < 2$, so that $\mathrm{P}\left[|u_t| > u\right] = u^{-\alpha} L(u)$, where L is a slowly varying function at ∞.

The conditional quantile function of Y_t is again given by

$$Q_{Y_t}(\tau|\mathcal{F}_{t-1}) = \theta_0 + \theta_1 Y_{t-1} + \cdots + \theta_p Y_{t-p} + F^{-1}(\tau) = \theta(\tau)^\top X_t,$$

where $\theta_0(\tau) = \theta_0 + F_u^{-1}(\tau)$, $\theta(\tau) = (\theta_0(\tau), \theta_1, \ldots, \theta_p)^\top$, and $X_t = (1, Y_{t-1}, \ldots, Y_{t-p})^\top$. The τth conditional quantile of Y_t can be estimated based on a quantile regression of Y_t on $(1, Y_{t-1}, \ldots, Y_{t-p})^\top$:

$$\min_{\theta} \sum_{t=1}^{T} \rho_\tau \left(Y_t - X_t^\top \theta\right).$$

The convergence rates of the intercept and slope coefficients are different. For convenience of analysis, we denote

$$X_t = \begin{bmatrix} 1 \\ x_t \end{bmatrix}, \text{where } x_t = (Y_{t-1}, \ldots, Y_{t-p})^\top,$$

and

$$\hat{\theta}(\tau) = \begin{bmatrix} \hat{\theta}_0(\tau) \\ \hat{\theta}_x(\tau) \end{bmatrix}, \quad \hat{\theta}_x(\tau) = \begin{bmatrix} \hat{\theta}_1(\tau) \\ \vdots \\ \hat{\theta}_p(\tau) \end{bmatrix}.$$

In the presence of an infinite variance process, the convergence rate will be different from the case with finite variance and will be dependent on the index of the stable law, α. We introduce the standardization matrix

$$D_n = \operatorname{diag}\left[T^{1/2}, T^a, \ldots, T^a\right], \quad a = \frac{1}{\alpha},$$

and let $u_{t\tau} = Y_t - \theta\,(\tau)^\top X_t$,

$$\eta = \eta(\tau) = D_n\left(\theta - \theta(\tau)\right), \quad \hat{\eta}(\tau) = D_n\left(\hat{\theta}(\tau) - \theta(\tau)\right).$$

Then $Y_t - \theta^\top X_t = u_{t\tau} - \eta^\top \left[D_n^{-1} X_t \right]$. Define

$$Z_n(\eta) = \sum_t \left\{ \rho_\tau(u_{t\tau} - D_n^{-1} X_t^\top \eta) - \rho_\tau(u_{t\tau}) \right\}.$$

Then

$$Z_n(\eta) = Z_{0n}(\eta_0) + Z_{1n}(\eta_1, \ldots, \eta_p) + o_p(1),$$

where

$$Z_{0n}(\eta_0) = \sum_t \left\{ \rho_\tau(u_{t\tau} - n^{-1/2}\eta_0) - \rho_\tau(u_{t\tau}) \right\} \to Z_0(\eta_0)$$

and

$$Z_{1T}(\eta_1, \ldots, \eta_k) = \sum_t \left\{ \rho_\tau \left(u_{t\tau} - T^{-a} \sum_{j=1}^p \eta_j Y_{t-j} \right) - \rho_\tau(u_{t\tau}) \right\} \to Z_1(\eta_1, \ldots, \eta_p).$$

By an argument similar to that in Davis et al. (1992), since

$$Z_0(\eta_0) = -\eta_0 N(0, \tau(1-\tau)) + \frac{1}{2} f(F^{-1}(\tau))\eta_0^2,$$

we have

$$T^{1/2} \left(\widehat{\theta}_0(\tau) - \theta_0(\tau) \right) \Rightarrow \frac{1}{f(F^{-1}(\tau))} N(0, \tau(1-\tau)),$$

and the quantile regression estimator $\widehat{\theta}_x(\tau)$, has limiting behavior

$$T^{1/\alpha} \left(\widehat{\theta}_x(\tau) - \theta_x(\tau) \right) \Rightarrow \arg\min Z(v),$$

where $Z(v) = Z(\eta_1, \ldots, \eta_p)$ is the limiting process of

$$\sum_t \left\{ \rho_\tau(u_{t\tau} - T^{-a} \sum_{j=1}^p \eta_j Y_{t-j}) - \rho_\tau(u_{t\tau}) \right\}.$$

17.5 Quantile regression for nonstationary time series

Quantile regression can be applied to time series regressions with deterministic or stochastic trend.

17.5.1 Quantile regression for trending time series

Many economic time series contain a trend over time. Consider the trending regression model

$$y_t = \theta^\top d_t + u_t, \tag{17.27}$$

where y_t is the observed time series, the regressor vector d_t is a q-dimensional deterministic function of known form, say, $d_t = (1, t, \ldots, t^q)^\top$, and u_t is the stochastic component. Quantile regression can be performed on (17.27) via

$$\widehat{\theta}(\tau) = \arg\min_\theta \sum_{t=1}^n \rho_\tau \left(y_t - \theta^\top d_t \right). \tag{17.28}$$

Since d_t is a q-dimensional deterministic function, appropriate standardization is needed. We assume that there exists a scaling matrix G_n such that $G_n^{-1} d_{[nr]} \rightarrow d(r)$, as $n \rightarrow \infty$, uniformly in $r \in [0,1]$. For the leading case of a linear trend, $G_n = \text{diag}[1,n]$ and $d(r) = (1,r)^\top$. If x_t is a general pth-order polynomial trend, $G_n = \text{diag}[1,n,\ldots,n^p]$ and $d(r) = (1,r,\ldots,r^p)^\top$.

Let $\psi_\tau(u) = \tau - I(u < 0)$. The asymptotic behavior of the quantile regression estimator $\hat{\theta}(\tau)$ can be analyzed. If we assume that u_t is weakly dependent and satisfies appropriate regularity conditions that ensure a functional central limit theorem (in particular, let $u_{t\tau} = u_t - F_u^{-1}(\tau)$), the partial sum process of $\psi_\tau(u_{t\tau})$ satisfies an invariance principle:

$$\frac{1}{\sqrt{n}} \sum_{t=1}^{[nr]} \psi_\tau(u_{t\tau}) \Rightarrow B_\psi^\tau(r), \quad 0 \leqslant r \leqslant 1.$$

The random function $n^{-1/2} \sum_{t=1}^{[nr]} \psi_\tau(u_{t\tau})$ converges to a two-parameter Gaussian process $B_\psi^\tau(r) = B_\psi(\tau,r)$, which is partially a Brownian motion and partially a Brownian bridge in the sense that for fixed r, $B_\psi^\tau(r) = B_\psi(\tau,r)$ is a rescaled Brownian bridge, while for each τ, $n^{-1/2} \sum_{t=1}^{[nr]} \psi_\tau(u_{t\tau})$ converges weakly to a Brownian motion with variance $\tau(1-\tau)$. Thus, for each fixed pair (τ,r), $B_\psi^\tau(r) = B_\psi(\tau,r) \sim N(0,\tau(1-\tau)r)$.

The quantile regression estimator of trend coefficients has limit

$$\sqrt{n} G_n(\hat{\theta}(\tau) - \theta(\tau)) \Rightarrow \frac{1}{f_u(F_u^{-1}(\tau))} \left[\int d(r)d(r)^\top dr \right]^{-1} \int d(r) dB_\psi^\tau(r).$$

Inference procedures based on residuals from the above quantile regression can be constructed. For example, we may testing the null hypothesis of "trend stationarity" against the alternative of "unit root" by looking at the fluctuation in the residual processes (see Xiao, 2012).

17.5.2 Unit-root quantile regressions

A popular model in economic time series analysis is the autoregressive unit-root process, in which case the differenced time series is stationary (I(0)). Consider the widely used augmented Dickey–Fuller (ADF) regression model

$$Y_t = \alpha_1 Y_{t-1} + \sum_{j=1}^{q} \alpha_{j+1} \Delta Y_{t-j} + u_t, \tag{17.29}$$

where u_t is i.i.d.$(0,\sigma^2)$. Under assumptions that all the roots of $A(L) = 1 - \sum_{j=1}^{q} \alpha_{j+1} L^j$ lie outside the unit circle, if $\alpha_1 = 1$, Y_t contains a unit root; and if $|\alpha_1| < 1$, Y_t is stationary. If we denote the σ-field generated by $\{u_s, s \leqslant t\}$ by $\mathcal{F}_t$, then conditional on $\mathcal{F}_{t-1}$, the τth conditional quantile of Y_t is given by

$$Q_{Y_t}(\tau|\mathcal{F}_{t-1}) = Q_u(\tau) + \alpha_1 Y_{t-1} + \sum_{j=1}^{q} \alpha_{j+1} \Delta Y_{t-j}.$$

Let $\alpha_0(\tau) = Q_u(\tau)$, $\alpha_j(\tau) = \alpha_j$, $j = 1,\ldots,p$, $p = q+1$, and define

$$\alpha(\tau) = (\alpha_0(\tau), \alpha_1, \ldots, \alpha_{q+1}), \quad X_t = (1, Y_{t-1}, \Delta Y_{t-1}, \ldots, \Delta Y_{t-q})^\top;$$

we have $Q_{Y_t}(\tau|\mathcal{F}_{t-1}) = X_t^\top \alpha(\tau)$. The unit-root quantile autoregressive model can be estimated by

$$\min_\alpha \sum_{t=1}^{n} \rho_\tau(Y_t - X_t^\top \alpha).$$

Denote $w_t = \Delta Y_t$ and $u_{t\tau} = Y_t - X_t^\top \alpha(\tau)$. Under the unit-root hypothesis and other regularity assumptions,

$$n^{-1/2} \sum_{t=1}^{[nr]} (w_t, \psi_\tau(u_{t\tau}))^\top \Rightarrow (B_w(r), B_\psi^\tau(r))^\top = BM(0, \underline{\Sigma}(\tau)),$$

where

$$\underline{\Sigma}(\tau) = \begin{bmatrix} \sigma_w^2 & \sigma_{w\psi}(\tau) \\ \sigma_{w\psi}(\tau) & \sigma_\psi^2(\tau) \end{bmatrix}$$

is the long-run covariance matrix of the bivariate Brownian motion and can be written as $\Sigma_0(\tau) + \Sigma_1(\tau) + \Sigma_1^\top(\tau)$, where

$$\Sigma_0(\tau) = E[(w_t, \psi_\tau(u_{t\tau}))^\top (w_t, \psi_\tau(u_{t\tau}))], \quad \Sigma_1(\tau) = \sum_{s=2}^{\infty} E[(w_1, \psi_\tau(u_{1\tau}))^\top (w_s, \psi_\tau(u_{s\tau}))].$$

In addition, $n^{-1} \sum_{t=1}^{n} Y_{t-1} \psi_\tau(u_{t\tau}) \Rightarrow \int_0^1 B_w dB_\psi^\tau$. In particular, the random function $n^{-1/2} \sum_{t=1}^{[nr]} \psi_\tau(u_{t\tau})$ converges to a two-parameter process $B_\psi^\tau(r)$ that we discussed in the previous subsection.

Let $\hat{\alpha}(\tau) = (\hat{\alpha}_0(\tau), \hat{\alpha}_1, \ldots, \hat{\alpha}_p)^\top$ and $D_n = \mathrm{diag}(\sqrt{n}, n, \sqrt{n}, \ldots, \sqrt{n})$. The limiting distribution of $\hat{\alpha}(\tau)$ is given by Koenker and Xiao (2004). Let y_t be determined by (17.29). Under the unit-root assumption $\alpha_1 = 1$ and other regularity conditions,

$$D_n(\hat{\alpha}(\tau) - \alpha(\tau)) \Rightarrow \frac{1}{f(F^{-1}(\tau))} \begin{bmatrix} \int_0^1 \overline{B}_w \overline{B}_w^\top & 0_{2 \times q} \\ 0_{q \times 2} & \Omega_\Phi \end{bmatrix}^{-1} \begin{bmatrix} \int_0^1 \overline{B}_w dB_\psi^\tau \\ \Phi \end{bmatrix},$$

where $\overline{B}_w(r) = [1, B_w(r)]^\top$, $\Phi = [\Phi_1, \ldots, \Phi_q]^\top$ is a q-dimensional normal variate with covariance matrix $\tau(1-\tau)\Omega_\Phi$ where

$$\Omega_\Phi = \begin{bmatrix} \nu_0 & \cdots & \nu_{q-1} \\ \vdots & \ddots & \vdots \\ \nu_{q-1} & \cdots & \nu_0 \end{bmatrix}, \quad \nu_j = E[w_t w_{t-j}],$$

and Φ is independent with $\int_0^1 \overline{B}_w dB_\psi^\tau$.

As an immediate byproduct of the above result, the limiting distribution of $n(\hat{\alpha}_1(\tau) - 1)$ is invariant to the estimation of $\hat{\alpha}_j(\tau)$ $(j = 2, \ldots, p)$ and the lag length p. In particular,

$$n(\hat{\alpha}_1(\tau) - 1) \Rightarrow \frac{1}{f(F^{-1}(\tau))} \left[\int_0^1 \underline{B}_w^2 \right]^{-1} \int_0^1 \underline{B}_w dB_\psi^\tau, \tag{17.30}$$

where $\underline{B}_w(r) = B_w(r) - \int_0^1 B_w$ is a demeaned Brownian motion.

Inference based on the autoregression quantile process provides a robust approach to testing the unit-root hypothesis. Like the conventional ADF t-ratio test, we may consider the t-ratio statistic

$$t_n(\tau) = \frac{\widehat{f(F^{-1}(\tau))}}{\sqrt{\tau(1-\tau)}} \left(Y_{-1}^\top P_X Y_{-1}\right)^{1/2} (\hat{\alpha}_1(\tau) - 1),$$

where $\widehat{f(F^{-1}(\tau))}$ is a consistent estimator of $f(F^{-1}(\tau))$, Y_{-1} is the vector of lagged dependent variables (Y_{t-1}), and P_X is the projection matrix onto the space orthogonal to $X = (1, \Delta Y_{t-1}, \ldots, \Delta Y_{t-q})$. Under the unit-root hypothesis, we have

$$t_n(\tau) \Rightarrow t(\tau) = \frac{1}{\sqrt{\tau(1-\tau)}} \left[\int_0^1 \underline{B}_w^2 \right]^{-1/2} \int_0^1 \underline{B}_w dB_\psi^\tau. \tag{17.31}$$

At any fixed τ the test statistic $t_n(\tau)$ is simply the quantile regression counterpart of the well-known ADF t-ratio test for a unit root. The limiting distribution of $t_n(\tau)$ is nonstandard and depends on nuisance parameters $(\sigma_w^2, \sigma_{w\psi}(\tau))$ since B_w and B_ψ^τ are correlated Brownian motions.

The limiting distribution of $t_n(\tau)$ can be decomposed as a linear combination of two (independent) distributions, with weights determined by a long-run (zero-frequency) correlation coefficient that can be consistently estimated. Following Phillips and Hansen (1990), we have

$$\int_0^1 \underline{B}_w dB_\psi^\tau = \int \underline{B}_w dB_{\psi.w}^\tau + \lambda_{w\psi}(\tau) \int \underline{B}_w dB_w,$$

where $\lambda_{w\psi}(\tau) = \sigma_{w\psi}(\tau)/\sigma_w^2$ and $B_{\psi.w}^\tau$ is a Brownian motion with variance $\sigma_{\psi.w}^2(\tau) = \sigma_\psi^2(\tau) - \sigma_{w\psi}^2(\tau)/\sigma_w^2$, and is independent of $\underline{B}_w$. The limiting distribution of $t_n(\tau)$ can therefore be decomposed as

$$\frac{1}{\sqrt{\tau(1-\tau)}} \frac{\int \underline{B}_w dB_{\psi.w}^\tau}{\left(\int_0^1 \underline{B}_w^2\right)^{1/2}} + \frac{\lambda_{w\psi}(\tau)}{\sqrt{\tau(1-\tau)}} \frac{\int \underline{B}_w dB_w}{\left(\int_0^1 \underline{B}_w^2\right)^{1/2}}.$$

For convenience of exposition, we may rewrite the Brownian motions $B_w(r)$ and $B_{\psi.w}^\tau(r)$ as

$$B_w(r) = \sigma_w W_1(r), \quad B_{\psi.w}^\tau(r) = \sigma_{\psi.w}(\tau) W_2(r),$$

$$\underline{B}_w(r) = \sigma_w \underline{W}_1(r), \quad \underline{W}_1(r) = W_1(r) - \int_0^1 W_1(s)ds,$$

where $W_1(r)$ and $W_2(r)$ are standard Brownian motions and are independent of one another. Note that $\sigma_\psi^2(\tau) = \tau(1-\tau)$, and the limiting distribution of $t_n(\tau)$ can be written as

$$\delta \left(\int_0^1 \underline{W}_1^2\right)^{-1/2} \int_0^1 \underline{W}_1 dW_1 + \sqrt{1-\delta^2} N(0,1), \tag{17.32}$$

where

$$\delta = \delta(\tau) = \frac{\sigma_{w\psi}(\tau)}{\sigma_w \sigma_\psi(\tau)} = \frac{\sigma_{w\psi}(\tau)}{\sigma_w \sqrt{\tau(1-\tau)}}.$$

The above limiting distribution can be easily approximated using simulation methods. In fact, required critical values are tabulated in the literature and thus are available for use in applications.

Alternatively, we may consider a transformation of $t_n(\tau)$ that annihilates the nuisance parameter, and thereby provides a distributional-free form of inference. Hasan and Koenker (1997) consider rank-type tests based on regression rank scores in an ADF framework. A third option is to abandon the asymptotically distribution-free nature of tests and use critical values generated by resampling methods. One may also consider unit-root tests based on quantile autoregression over a range of (multiple) quantiles, targeting a somewhat broader class of alternatives than those considered in the OLS literature. See Koenker and Xiao (2004) for more discussions on this topic.

17.5.3 Quantile regression on cointegrated time series

Consider again the regression model (17.6). If X_t is a k-dimensional vector of integrated regressors and u_t is still mean-zero stationary (possibly correlated with X_t), it becomes the important cointegration regression model. In order to deal with endogeneity, we may use leads and lags of X_t (other methods may also be considered to deal with correlation between u_t and X_t). If we assume that u_t has representation

$$u_t = \sum_{j=-K}^{K} v_{t-j}^{\top} \Pi_j + \varepsilon_t, \tag{17.33}$$

where $v_t = \Delta X_t$, ε_t is a stationary process such that $E(v_{t-j}\varepsilon_t) = 0$, for any j, and

$$n^{-1/2} \sum_{t=1}^{[nr]} \begin{bmatrix} \psi_\tau(\varepsilon_{t\tau}) \\ v_t \end{bmatrix} \Rightarrow B(r) = \begin{bmatrix} B_\psi^*(r) \\ B_v(r) \end{bmatrix} = BM(0, \Omega^*),$$

the original cointegrating regression can be rewritten as

$$Y_t = \alpha + \beta^{\top} X_t + \sum_{j=-K}^{K} \Delta X_{t-j}^{\top} \Pi_j + \varepsilon_t.$$

If we denote the τth quantile of ε_t as $Q_\varepsilon(\tau)$, and let $\mathcal{G}_t = \sigma\{X_t, \Delta X_{t-j}, \forall j\}$, then, conditional on $\mathcal{G}_t$, the τth quantile of Y_t is given by

$$Q_{Y_t}(\tau|\mathcal{G}_t) = \alpha + \beta^{\top} X_t + \sum_{j=-K}^{K} \Delta X_{t-j}^{\top} \Pi_j + F_\varepsilon^{-1}(\tau),$$

where $F_\varepsilon(\cdot)$ is the cdf of ε_t. Let Z_t be the vector of regressors consisting of $z_t = (1, X_t)$ and $(\Delta X_{t-j}, j = -K, \ldots, K)$, $\Theta = (\alpha, \beta^{\top}, \Pi_{-K}^{\top}, \ldots, \Pi_K^{\top})^{\top}$, and

$$\Theta(\tau) = (\alpha(\tau), \beta^{\top}, \Pi_{-K}^{\top}, \ldots, \Pi_K^{\top})^{\top},$$

where $\alpha(\tau) = \alpha + F_\varepsilon^{-1}(\tau)$. Then we can rewrite the above regression as $Y_t = \Theta^{\top} Z_t + \varepsilon_t$, and

$$Q_{Y_t}(\tau|\mathcal{F}_t) = \Theta(\tau)^{\top} Z_t.$$

We now consider the following quantile cointegrating regression:

$$\hat{\Theta}(\tau) = \arg\min_{\theta} \sum_{t=1}^{n} \rho_\tau(Y_t - \Theta^{\top} Z_t). \tag{17.34}$$

Similar to case of the ADF regression, the components in $\hat{\Theta}(\tau)$ have different rates of convergence. Denote $G_n = \text{diag}(\sqrt{n}, n, \ldots, n, \sqrt{n}, \ldots, \sqrt{n})$. Conformable with $\Theta(\tau)$, we partition $\hat{\Theta}(\tau)$ as

$$\hat{\Theta}(\tau)^{\top} = \left[\hat{\alpha}(\tau), \hat{\beta}(\tau)^{\top}, \hat{\Pi}_{-K}(\tau)^{\top}, \ldots, \hat{\Pi}_K(\tau)^{\top} \right].$$

Under regularity assumptions,

$$G_n(\hat{\Theta}(\tau) - \Theta(\tau)) \Rightarrow \frac{1}{f_\varepsilon(F_\varepsilon^{-1}(\tau))} \begin{bmatrix} \int_0^1 \overline{B}_v \overline{B}_v^{\top} & 0 \\ 0 & \Gamma \end{bmatrix}^{-1} \begin{bmatrix} \int_0^1 \overline{B}_v dB_\psi^* \\ \Psi \end{bmatrix},$$

where $\overline{B}_v(r) = (1, B_v(r)^{\top})^{\top}$, $\Gamma = E(V_t V_t^{\top})$ with $V_t = (\Delta X_{t-K}^{\top}, \ldots, \Delta X_{t+K}^{\top})^{\top}$, and Ψ is a multivariate normal with dimension conformable with $(\Pi_{-K}(\tau)^{\top}, \ldots, \Pi_K(\tau)^{\top})^{\top}$. In particular,

$$n(\hat{\beta}(\tau) - \beta) \Rightarrow \frac{1}{f_\varepsilon(F_\varepsilon^{-1}(\tau))} \left[\int_0^1 \underline{B}_v \underline{B}_v^{\top} \right]^{-1} \int_0^1 \underline{B}_v dB_\psi^*,$$

where $\underline{B}_v(r) = B_v(r) - rB_v(1)$.

Consider the quantile regression residual

$$\varepsilon_{t\tau} = Y_t - Q_{Y_t}(\tau|\mathcal{F}_t) = Y_t - \Theta(\tau)^\top Z_t = \varepsilon_t - F_\varepsilon^{-1}(\tau).$$

We have $Q_{\varepsilon_{t\tau}}(\tau) = 0$, where $Q_{\varepsilon_{t\tau}}(\tau)$ signifies the τth quantile of $\varepsilon_{t\tau}$ and $E\psi_\tau(\varepsilon_{t\tau}) = 0$.

The cointegration relationship may be tested by directly looking at the fluctuation in the residual process $\varepsilon_{t\tau}$ from the quantile cointegrating regression. Consider the partial sum process

$$Y_n(r) = \frac{1}{\omega_\psi^* \sqrt{n}} \sum_{j=1}^{[nr]} \psi_\tau(\varepsilon_{j\tau}),$$

where ω_ψ^{*2} is the long-run variance of $\psi_\tau(\varepsilon_{j\tau})$. Under appropriate assumptions, the partial sum process follows an invariance principle and converges weakly to a standard Brownian motion $W(r)$. Choosing a continuous functional $h(\cdot)$ that measures the fluctuation of $Y_n(r)$, notice that $\psi_\tau(\varepsilon_{j\tau})$ is indicator-based, and a robust test for cointegration can be constructed based on $h(Y_n(r))$. By the continuous mapping theorem, under regularity conditions and the null of cointegration,

$$h(Y_n(r)) \Rightarrow h(W(r)).$$

In principle, any metric that measures the fluctuation in $Y_n(r)$ is a natural candidate for the functional h. The classical Kolmogorov–Smirnov type or Cramér–von Mises type measures are of particular interest. Under the alternative of no cointegration, the statistic diverges.

In practice, we estimate $\Theta(\tau)$ by $\widehat{\Theta}(\tau)$ using (17.34), and obtain the residuals

$$\widehat{\varepsilon}_{t\tau} = Y_t - \widehat{\Theta}(\tau)^\top Z_t.$$

A robust test for cointegration can then be constructed based on

$$\widehat{Y}_n(r) = \frac{1}{\widehat{\omega}_\psi^* \sqrt{n}} \sum_{j=1}^{[nr]} \psi_\tau(\widehat{\varepsilon}_{j\tau}),$$

where $\widehat{\omega}_\psi^{*2}$ is a consistent estimator of ω_ψ^{*2}. Under regularity assumptions and the hypothesis of cointegration,

$$\widehat{Y}_n(r) \Rightarrow \widetilde{W}(r) = W_1(r) - \left[\int_0^1 dW_1 \overline{W}_2^\top\right] \left[\int_0^1 \overline{W}_2 \overline{W}_2^\top\right]^{-1} \int_0^r \overline{W}_2(s),$$

where $\overline{W}_2(r) = (1, W_2(r)^\top)^\top$, and W_1 and W_2 are independent one- and k-dimensional standard Brownian motions. See Xiao (2012) for more discussion on robust tests for cointegration.

17.6   The QAR process

In classical autoregression (17.1), past information (Y_{t-j}) only affects the location of the conditional distribution of Y_t. In recent years, considerable research effort has been devoted to modifications of traditional constant coefficient dynamic models to incorporate a variety of heterogeneous innovation effects. An important motivation for such modifications is the introduction of asymmetries into model dynamics. It is widely acknowledged that many important economic variables may display asymmetric adjustment paths (e.g. Enders and Granger, 1998; Neftci, 1984). The observation that firms are more likely to increase than to reduce prices is a key feature of many macroeconomic models. Beaudry and Koop (1993) have argued that positive shocks to US gross domestic product are more persistent than negative shocks, indicating asymmetric business cycle dynamics over different quantiles of the innovation process. In addition, while it is generally recognized that output fluctuations are persistent, less persistent results are also found at longer horizons, suggesting some form of "local persistency."

Quantile regression provides an alternative way to study asymmetric dynamics and local persistency in time series. Let us look at a simple example based on a time series of (seasonally adjusted) US one-month interest rates from April 1971 to June 2002. This is a highly persistent series and the traditional ADF test cannot reject the null hypothesis of a unit root. Consider the following Dickey-Fuller type autoregression:

$$Y_t = \mu + \alpha Y_{t-1} + \sum_{j=1}^{q} \beta_j \Delta Y_{t-j} + u_t.$$

The OLS estimate of α is 0.96, and the traditional unit-root tests not reject the unit-root null hypothesis in this time series.

We next estimate the above regression at different quantiles in $(0, 1)$ based on

$$\min \sum_{t=1}^{n} \rho_\tau \left(Y_t - \mu - \alpha Y_{t-1} - \sum_{j=1}^{q} \beta_j \Delta Y_{t-j} \right).$$

The estimated values of α at different quantiles are reported in Figure 17.1, ranging from 0.74 at the 0.05th quantile to 1.16 at the 0.95th quantile.

The empirical example suggests that the dependence changes over different quantiles. In the lower quantiles, the series exhibits stationary behavior; around the center, Y_t looks like a unit-root process, but for large realizations of the innovation we even have some transient forms of explosive behavior. Thus, the model exhibits a form of asymmetric persistence in the sense that sequences of strongly positive innovations tend to generate a temporarily explosive path, while negative realizations of innovation induce mean reversion. On average, there is a strong persistency and the series looks like a unit-root process. Additional non-parametric analysis shows that past information in the series not only affects the location of the conditional distribution of the future interest rate, but also the shape of the conditional distribution.

In many empirical applications, the time series dynamics can be more complicated than a location (and/or scale) shift. In many time series applications, past information not only affects the location (and/or scale) of the conditional distribution of future values, but also the shape of the conditional distribution. See Koenker (2000) for an interesting example of a time series of daily temperature in Melbourne. It is quite clear that today's temperature not only affects the location and scale of the conditional distribution of tomorrow's temperature, but also the shape of the conditional distribution. As the value of the conditioning variable (Y_{t-1}) increases, the conditional distribution (of Y_t) becomes bimodal!

17.6.1   The linear QAR process

Quantile domain modeling methods provide a convenient way to address some of these problems. An important extension of the classical constant coefficient AR model is the

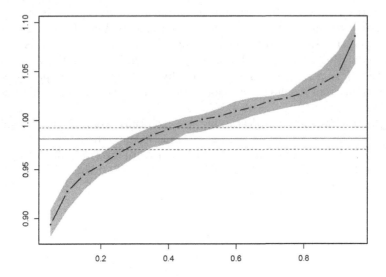

FIGURE 17.1
Estimated first-order quantile autoregression process for the one-month US Treasury bill
rate based on data from April 1971 to June 2002.

quantile autoregression model proposed by Koenker and Xiao (2006). Given a time series
$\{Y_t\}$, let $\mathcal{F}_t$ be the σ-field generated by $\{Y_s, s \leqslant t\}$. $\{Y_t\}$ is a pth-order QAR process if

$$Q_{Y_t}(\tau|\mathcal{F}_{t-1}) = \theta_0(\tau) + \theta_1(\tau)Y_{t-1} + \cdots + \theta_p(\tau)Y_{t-p}. \qquad (17.35)$$

This implies, of course, that the right-hand side of (17.35) is monotonically increasing in τ.
In the above QAR model, the autoregressive coefficients may be τ-dependent and thus can
vary over different quantiles of the conditional distribution. Consequently, the conditioning
variables not only shift the location of the distribution of Y_t, but also may alter the scale
and shape of the conditional distribution. QAR models play a useful role in expanding the
modeling territory of the classical autoregressive time series models, and the classical AR(p)
model can be viewed as a special case of QAR by setting $\theta_j(\tau)$ $(j = 1, \ldots, p)$ to constants.

The formulation in (17.35) reveals that the QAR model may be interpreted as a random
coefficient autoregressive model:

$$Y_t = \theta_0(U_t) + \theta_1(U_t)Y_{t-1} + \cdots + \theta_p(U_t)Y_{t-p}, \qquad (17.36)$$

where $\{U_t\}$ is a sequence of i.i.d. standard uniform random variables. In particular, by
allowing the autoregressive coefficients to vary over different quantiles of the conditional
distribution, the model can exhibit various forms of asymmetric dynamics and persistency,
including temporary nonstationary behavior coupled with long-term mean reversion of the
process. The long-run behavior of Y_t can be covariance stationary and satisfies a central limit
theorem, but some transient forms of explosive behavior can still be observed over shorter
periods. Denote $X_t = (1, Y_{t-1}, \ldots, Y_{t-p})^\top$ and $\theta(\tau) = (\theta_0(\tau), \theta_1(\tau), \ldots, \theta_p(\tau))^\top$. The QAR
model (17.35) can be estimated by the conventional quantile regression technique via (17.2).

Let $\mu_0 = E\theta_0(U_t)$, $u_t = \theta_0(U_t) - \mu_0$, and $\alpha_{j,t} = \theta_j(U_t)$, for $j = 1, \ldots, p$. We may
reformulate the QAR(p) model (17.36) in the more conventional random coefficient notation
as

$$Y_t = \mu_0 + \alpha_{1,t}Y_{t-1} + \cdots + \alpha_{p,t}Y_{t-p} + u_t. \qquad (17.37)$$

Let $A_{p-1,t} = [\alpha_{1,t}, \ldots, \alpha_{p-1,t}]$, denote by 0_{p-1} the $(p-1)$-dimensional vector of zeros, and define

$$A_t = \begin{bmatrix} A_{p-1,t} & \alpha_{p,t} \\ I_{p-1} & 0_{p-1} \end{bmatrix}, \quad \mathbf{V}_t = \begin{bmatrix} u_t \\ 0_{p-1} \end{bmatrix}.$$

then the QAR(p) process (17.37) can be expressed as a p-dimensional vector autoregression process of order 1:

$$\mathbf{Y}_t = \Gamma + A_t \mathbf{Y}_{t-1} + \mathbf{V}_t,$$

with

$$\Gamma = \begin{bmatrix} \mu_0 \\ 0_{p-1} \end{bmatrix}, \quad \mathbf{Y}_t = \begin{bmatrix} Y_t \\ \cdots \\ Y_{t-p+1} \end{bmatrix}.$$

Koenker and Xiao (2006) show that, under assumptions 1–3 stated below: (i) the QAR(p) process Y_t given by (17.37) is covariance stationary and satisfies a central limit theorem

$$\frac{1}{\sqrt{n}} \sum_{t=1}^{n} (Y_t - \mu_y) \Rightarrow N\left(0, \omega_y^2\right),$$

where $\mu_y = \mu_0/(1 - \sum_{j=1}^{p} \mu_j)$, $\omega_y^2 = \lim n^{-1} E[\sum_{t=1}^{n}(y_t - \mu_y)]^2$, and $\mu_j = E(\alpha_{j,t})$, $j = 1, \ldots, p$; (ii) the autoregression quantile process $\hat{\theta}(\tau)$ has the limiting representation

$$\Sigma^{-1/2} \sqrt{n}(\hat{\theta}(\tau) - \theta(\tau)) \Rightarrow B_k(\tau),$$

where $\Sigma = \Omega_1^{-1} \Omega_0 \Omega_1^{-1}$, $\Omega_1 = \lim n^{-1} \sum_{t=1}^{n} f_{t-1}[F_{t-1}^{-1}(\tau)] X_t X_t^{\top}$, $\Omega_0 = E\left(X_t X_t^{\top}\right)$, and $B_k(\tau)$ represents a k-dimensional standard Brownian bridge, $k = p + 1$. The assumptions are as follows:

1. $\{u_t\}$ are i.i.d. random variables with mean 0 and variance $\sigma^2 < \infty$. The distribution function of u_t, F, has a continuous density f with $f(u) > 0$ on $\mathcal{U} = \{u : 0 < F(u) < 1\}$.

2. Let $E(A_t \otimes A_t) = \Omega_A$. The eigenvalues of Ω_A have moduli less than unity.

3. Denote the conditional distribution function $P[y_t < \cdot | \mathcal{F}_{t-1}]$ by $F_{t-1}(\cdot)$ and its derivative by $f_{t-1}(\cdot)$. f_{t-1} is uniformly integrable on $\mathcal{U}$.

From the above result, we have, for fixed τ, that the limiting distribution of the QAR estimator is given by

$$\sqrt{n}(\hat{\theta}(\tau) - \theta(\tau)) \Rightarrow N(0, \tau(1-\tau)\Omega_1^{-1}\Omega_0\Omega_1^{-1}).$$

QAR models expand the modeling options for time series. They can capture systematic influences of conditioning variables on the location, scale, and shape of the conditional distribution of the response, and therefore significantly extend the classical constant coefficient linear time series models.

Given the QAR process (17.36), let $\theta_0 = E[\theta_0(U_t)]$, $\theta_1 = E[\theta_1(U_t)], \ldots, \theta_p = E[\theta_p(U_t)]$, and

$$V_t = \theta_0(U_t) - E\theta_0(U_t) + [\theta_1(U_t) - E\theta_1(U_t)] Y_{t-1} + \cdots + [\theta_p(U_t) - E\theta_p(U_t)] Y_{t-p}.$$

The QAR process can be rewritten as

$$Y_t = \theta_0 + \theta_1 Y_{t-1} + \cdots + \theta_p Y_{t-p} + V_t, \tag{17.38}$$

where V_t is martingale difference sequence. Apparently the QAR process is an AR process with conditional heteroskedasticity.

What is the difference between a QAR process and an AR process with conditional heteroskedastic errors? In short, the ARCH type model only focuses on the first two moments, while the QAR model goes beyond the second moment and allows for more flexible structure in higher moments. Both models allow for conditional heteroskedasticity and they are similar in the first two moments, but they can be quite different beyond conditional variance.

The classical time series analysis based on autocorrelations (and partial autocorrelations, etc.) only requires that the residuals are martingale difference sequences. As we show in (17.38), the autocovariance structure of the QAR process (17.36) is the same as that of a fixed coefficient AR(p) process with martingale difference sequence residuals. Thus, for two different QAR(p) processes,

$$Y_{i,t} = \theta_{i0}(U_t) + \theta_{i1}(U_t)Y_{i,t-1} + \cdots + \theta_{ip}(U_t)Y_{i,t-p}, \quad i = 1, 2,$$

if $E[\theta_{1j}(U_t)] = E[\theta_{2j}(U_t)]$, their autocorrelation structures are the same. Consequently, the classical time series analysis technique will identify QAR processes with different dependence structures as the same fixed coefficient AR(p) process. In this case, the QAR technique helps to reveal additional information that the classical time series analysis may ignore. In this sense, the QAR method provides a very useful complement to the classical analysis.

A simple high-level assumption that we made on the QAR process is monotonicity of the right-hand side of (17.35). The monotonicity of the conditional quantile functions usually imposes restrictions on the domain of the random variable Y_t unless Y_t is a traditional constant coefficient process. It requires that the domain of the random variables (or appropriately transformed versions) is bounded at least in one direction (say, nonnegative). Alternatively, we may consider quantile autoregression based on appropriately transformed time series (see the discussion later in this section).

If the monotonicity assumption does not hold, the quantile regression (17.2) can still be used in practice, but the regression quantiles $\hat{\theta}(\tau)$ should be analyzed under the assumption of misspecification, and the results need to be modified. In the case where the monotonicity assumption does not hold, apparently the conditional quantile function $Q_{Y_t}(\tau|\mathcal{F}_{t-1})$ cannot be written as a linear function of x_t, and the QAR estimator, $\hat{\theta}(\tau)$, will converge to some pseudo-parameter $\bar{\theta}(\tau)$ that minimizes some distance between $X_t^\top \theta$ and $Q_{Y_t}(\tau|\mathcal{F}_{t-1})$, that is,

$$\hat{\theta}(\tau) \to_p \bar{\theta}(\tau) = \arg\min_\theta E d\left(X_t^\top \theta, Q_{Y_t}(\tau|\mathcal{F}_{t-1})\right),$$

where the distance $d(\cdot, \cdot)$ is defined as $d\left(X_t^\top \theta, Q_{Y_t}(\tau|\mathcal{F}_{t-1})\right) = E\{(\delta - |\varepsilon_{t\tau}|)\, 1(|\varepsilon_{t\tau}| < \delta)|\mathcal{F}_{t-1}\}$, with $\delta(\theta, X_t) = |X_t^\top \theta - Q_{Y_t}(\tau|\mathcal{F}_{t-1})|$ and $\varepsilon_{t\tau} = Y_t - Q_{Y_t}(\tau|\mathcal{F}_{t-1})$. In this case, the quantile regression estimator $X_t^\top \hat{\theta}(\tau)$ provides a linear approximation for $Q_{Y_t}(\tau|\mathcal{F}_{t-1})$.

What is the distance (criterion) function $d(\cdot, \cdot)$? A good way to further investigate the distance function $d\left(x_t^\top \theta, Q_{y_t}(\tau|\mathcal{F}_{t-1})\right)$ is to look at the first-order condition of the corresponding population optimization problem (see Hallin and Werker, 2006). Let $F_t(\cdot)$ be the conditional distribution function of y_t given x_t. Then the first-order condition is

$$E\left\{\left(F_t(x_t^\top \theta) - \tau\right) x_t\right\} = 0. \tag{17.39}$$

If the conditional distribution function is linear over its domain of definition (which is the case foe a uniform distribution), the quantile regression estimation is then equivalent to a least-squares problem

$$\min_\theta E\left(F_t(x_t^\top \theta) - \tau\right)^2.$$

In this case the criterion function is a quadratic function.

In general, the criterion function $d\left(x_t^\top \theta, Q_{y_t}(\tau|\mathcal{F}_{t-1})\right)$ is a nonlinear function of $\left(x_t^\top \theta - Q_{y_t}(\tau|\mathcal{F}_{t-1})\right)$ and may not be simply written as a pure quadratic function of $\left(x_t^\top \theta - Q_{y_t}(\tau|\mathcal{F}_{t-1})\right)$. One could write it in the form of a "weighted" least-squares problem

$$\min_\theta \lambda(x_t, \theta)\left(F_t(x_t^\top \theta) - \tau\right)^2,$$

but the "weight" $\lambda(x, \theta)$ is a function of the unknow parameter θ, and thus the criterion is not exactly a quadratic function. If we denote the conditional density function of y_t given x_t by f_t, then the first-order condition (17.39) is the same as that of the (infeasible) density-weighted least-squares problem

$$\min_\theta \left(F_t(x_t^\top \theta) - \tau\right)^2 / f_t(x_t^\top \bar\theta(\tau))$$

using a weighting function $f_t(x_t^\top \bar\theta(\tau))$ (which depends on $\bar\theta(\tau)$ and is thus practically infeasible). The above optimization problem is infeasible but helps us in understanding the criterion function. Angrist et al. (2005) write the criterion function $d\left(x_t^\top \theta, Q_{y_t}(\tau|\mathcal{F}_{t-1})\right)$ in the form

$$w(x, \theta)\left(x_t^\top \theta - Q_{y_t}(\tau|\mathcal{F}_{t-1})\right)^2,$$

and call the weighting function $w(x, \theta)$ importance weights; see also their study of iterative approximation.

Of course, this pseudo-parameter $\bar\theta(\tau)$ will coincide with $\theta(\tau)$ when monotonicity and other regularity conditions hold. One can establish asymptotic normality of $\hat\theta(\tau)$ with the same rate around this target $\bar\theta(\tau)$. This is rather like the general theory of, say, AR(p) estimation under misspecification in time series literature (in which case the estimated AR parameters converge to some pseudo-parameters at rate root-n). Statistical inference can be conducted based on the new limiting distribution that accommodates the possible misspecification. Simulation based methods such as subsampling may be used to conduct statistical inference for the QAR models under misspecification.

Despite the possible crossing of quantile curves, the linear QAR model provides a convenient and useful local approximation to the conditional quantile function.

17.6.2 Nonlinear QAR models

The QAR process can be extended to nonlinear models. More complicated functional forms with nonlinearity can be considered for the conditional quantile function if we are interested in the global behavior of the time series. The absence of monotonicity implies that a more complicated functional form with nonlinearity is needed for $Q_{Y_t}(\tau|X_t)$. If the τth conditional quantile function of Y_t is given by

$$Q_{Y_t}(\tau|\mathcal{F}_{t-1}) = H(X_t; \theta(\tau)),$$

where X_t is the vector containing lagged Ys, we may estimate the vector of parameters $\theta(\tau)$ (and thus the conditional quantile of Y_t) by the nonlinear quantile regression

$$\min_\theta \sum_t \rho_\tau(Y_t - H(X_t, \theta)). \tag{17.40}$$

Let $\varepsilon_{t\tau} = y_t - H(x_t, \theta(\tau))$, $\dot{H}_\theta(x_t, \theta) = \partial H(x_t; \theta)/\partial\theta$. We assume that

$$V_n(\tau) = \frac{1}{n}\sum_t f_t(Q_{Y_t}(\tau|X_t))\dot{H}_\theta(X_t, \theta(\tau))\dot{H}_\theta(X_t, \theta(\tau))^\top \xrightarrow{P} V(\tau),$$

$$\Omega_n(\tau) = \frac{1}{n}\sum_t \dot{H}_\theta(X_t, \theta(\tau))\dot{H}_\theta(X_t, \theta(\tau))^\top \xrightarrow{P} \Omega(\tau),$$

and

$$\frac{1}{\sqrt{n}} \sum_t \dot{H}_\theta(x_t, \theta(\tau)) \psi_\tau(\varepsilon_{t\tau}) \Rightarrow N(0, \tau(1-\tau)\Omega(\tau)),$$

where $V(\tau)$ and $\Omega(\tau)$ are nonsingular. Then, under appropriate assumptions, the nonlinear QAR estimator $\hat{\theta}(\tau)$ defined as the solution to (17.40) is root-n consistent and

$$\sqrt{n}\left(\hat{\theta}(\tau) - \theta(\tau)\right) \Rightarrow N(0, \tau(1-\tau)V(\tau)^{-1}\Omega(\tau)V(\tau)^{-1}). \qquad (17.41)$$

In practice, one may employ parametric copula models to generate nonlinear-in-parameters QAR models (see, for example, Bouyé and Salmon, 2009; Chen et al., 2009). Copula-based Markov models provide a rich source of potential nonlinear dynamics describing temporal dependence and tail dependence. If we consider, for example, a first-order strictly stationary Markov process, $\{Y_t\}_{t=1}^n$, whose probabilistic properties are determined by the joint distribution of Y_{t-1} and Y_t, say, $G^*(y_{t-1}, y_t)$, and suppose that $G^*(y_{t-1}, y_t)$ has continuous marginal distribution function $F^*(\cdot)$, then, by Sklar's theorem, there exists a unique copula function $C^*(\cdot, \cdot)$ such that

$$G^*(y_{t-1}, y_t) \equiv C^*(F^*(y_{t-1}), F^*(y_t)),$$

where the copula function $C^*(\cdot, \cdot)$ is a bivariate probability distribution function with uniform marginals. Differentiating $C^*(u, v)$ with respect to u, and evaluating at $u = F^*(x)$, $v = F^*(y)$, we obtain the conditional distribution of Y_t given $Y_{t-1} = x$:

$$\mathrm{P}\left[Y_t < y | Y_{t-1} = x\right] = \left.\frac{\partial C^*(u, v)}{\partial u}\right|_{u=F^*(x), v=F^*(y)} \equiv C_1^*(F^*(x), F^*(y)).$$

For any $\tau \in (0, 1)$, solving $\tau = \mathrm{P}\left[Y_t < y | Y_{t-1} = x\right] \equiv C_1^*(F^*(x), F^*(y))$ for y (in terms of τ), we obtain the τth conditional quantile function of Y_t given $Y_{t-1} = x$:

$$Q_{Y_t}(\tau|x) = F^{*-1}(C_1^{*-1}(\tau; F^*(x))),$$

where $F^{*-1}(\cdot)$ signifies the inverse of $F^*(\cdot)$ and $C_1^{*-1}(\cdot; u)$ is the partial inverse of $C_1^*(u, v)$ with respect to $v = F^*(y_t)$.

In practice, neither the true copula function $C^*(\cdot, \cdot)$ nor the true marginal distribution function $F^*(\cdot)$ of $\{Y_t\}$ is known. If we model both parametrically by $C(\cdot, \cdot; \alpha)$ and $F(y; \beta)$, then the τth conditional quantile function of Y_t, $Q_{Y_t}(\tau|x)$, becomes a function of the unknown parameters α and β, that is,

$$Q_{Y_t}(\tau|x) = F^{-1}(C_1^{-1}(\tau; F(x, \beta), \alpha), \beta).$$

Denoting $\theta = (\alpha^\top, \beta^\top)^\top$ and $h(x, \alpha, \beta) \equiv C_1^{-1}(\tau; F(x, \beta), \alpha)$, we will write

$$Q_{Y_t}(\tau|x) = F^{-1}(h(x, \alpha, \beta), \beta) \equiv H(x; \theta). \qquad (17.42)$$

For example, considering the Clayton copula

$$C(u, v; \alpha) = [u^{-\alpha} + v^{-\alpha} - 1]^{-1/\alpha}, , \quad \text{where } \alpha > 0,$$

we can easily verify that the τth conditional quantile function of U_t given u_{t-1} is

$$Q_{U_t}(\tau|u_{t-1}) = [(\tau^{-\alpha/(1+\alpha)} - 1)u_{t-1}^{-\alpha} + 1]^{-1/\alpha},$$

See Bouyé and Salmon (2009) for additional examples of copula-based conditional quantile functions.

Although the quantile function specification in the above representation assumes the parameters to be identical across quantiles, we may permit the estimated parameters to vary with τ and thus extend the original copula-based QAR models to capture a wide range of systematic influences of conditioning variables on the conditional distribution of the response. By varying the choice of the copula specification we can induce a wide variety of nonlinear QAR(1) dependence, and the choice of the marginal enables us to consider a wide range of possible tail behavior as well. In many financial time series applications, the nature of the temporal dependence varies over the quantiles of the conditional distribution. Chen et al. (2009) studied asymptotic properties of the copula-based nonlinear quantile autoregression.

17.6.3 Quantile autoregression based on transformations

An alternative way to deal with monotonicity of the QAR process is to consider quantile autoregressions based on appropriately transformed time series. Consider the following (perhaps nonlinear) AR(1) model based on transformations on $F(Y_t)$ (notice that $F(Y_t)$ has bounded support):

$$H_1(F(Y_t)) = H_2(F(Y_{t-1})) + \sigma(F(Y_{t-1}))\varepsilon_t.$$

The conditional density of ε_t given that $F(Y_{t-1}) = u_{t-1}$ is

$$f_{\varepsilon|F(Y_{t-1})=y_{t-1}}(\varepsilon) = c(u_{t-1}, H_1^{-1}(H_2(u_{t-1}) + \sigma(u_{t-1})\varepsilon)) \div \frac{dH_1(H_2(u_{t-1}) + \sigma(u_{t-1})\varepsilon))}{d\varepsilon},$$

satisfying the condition that

$$H_2(u_{t-1}) = E\left[H_1(U_t)|\, U_{t-1} = u_{t-1}\right] = \int_0^1 H_1(u) \times c(u_{t-1}, u)du.$$

Consider the special case where $H_1(u) = u$,

$$U_t = H_2(U_{t-1}) + \sigma(U_{t-1})\varepsilon_t.$$

Let

$$\frac{\partial C(u_{t-1}, u)}{\partial u_{t-1}} = C_1(u_{t-1}, u),$$

$$H_2(u_{t-1}) = E\left[H_1(U_t)|\, U_{t-1} = u_{t-1}\right] = \int_0^1 uc(u_{t-1}, u)du = 1 - \int_0^1 C_1(u_{t-1}, u)du,$$

$$c(F_Y(Y_{t-1}), F_Y(Y_t)) = f_\varepsilon\left(\frac{Y_t - g(Y_{t-1}, \theta_1)}{\sigma(Y_{t-1}, \theta_2)}; \alpha\right) \frac{1}{f_Y(Y_t)}.$$

If Y_t is a stationary Markov process of order 1 characterized by a standard bivariate t_ν-copula function $C(\cdot, \cdot; \alpha)$ and marginal distribution function $F(\cdot)$, then let t_ν be the cdf of a t_ν random variable. The transformed variable $\{Z_t = t_\nu^{-1}(F(Y_t))\}$ is a Student t process that can be represented by

$$t_\nu^{-1}(F(Y_t)) = \rho t_\nu^{-1}(F(Y_{t-1})) + \sigma_\theta(F(Y_{t-1}))e_t,$$

where $e_t \sim t_{\nu+1}$ is independent of Y_{t-1}, and $\sigma_\theta(U_{t-1})$ is a known function of $t_\nu^{-1}(U_{t-1})$ and $\theta = (\nu, \rho)$. If the marginal distribution F is also t_ν then $t_\nu^{-1}(F(Y_t)) = Y_t$. In general, consider a random vector (Y_1, Y_2) with nonstandard bivariate t-distribution with degrees of freedom ν, mean vector μ, and covariance matrix Σ. Then the conditional distribution of Y_2 given Y_1 is also $t(\nu + 1)$ with conditional mean

$$\mu_2 + \Sigma_{12}^\top \Sigma_{11}^{-1}(Y_1 - \mu_1)$$

and conditional variance

$$\frac{\nu + (Y_1 - \mu_1)^\top \Sigma_{11}^{-1}(Y_1 - \mu_1)}{\nu + 1}(\Sigma_{22} - \Sigma_{12}^\top \Sigma_{11}^{-1}\Sigma_{21}).$$

Thus

$$Y_2 = \mu_2 + \Sigma_{12}^\top \Sigma_{11}^{-1}(Y_1 - \mu_1) + \sigma(Y_1)\varepsilon,$$

where

$$\sigma(Y_1) = \sqrt{\frac{\nu + (Y_1 - \mu_1)^\top \Sigma_{11}^{-1}(Y_1 - \mu_1)}{\nu + 1}(\Sigma_{22} - \Sigma_{12}^\top \Sigma_{11}^{-1}\Sigma_{21})}$$

and ε is a t-variate with $\nu + 1$ degrees of freedom. Thus,

$$Q_{Y_t}(\tau|Y_{t-1}) = \mu_2 + \Sigma_{12}^\top \Sigma_{11}^{-1}(Y_{t-1} - \mu_1) + \sigma(Y_1)F_\varepsilon(\tau).$$

In the special case when ε is a t-variate with 1 degree of freedom,

$$Q_{Y_t}(\tau|Y_{t-1}) = \theta_0(\tau) + \theta_1(\tau)Y_{t-1},$$

where

$$\theta_0(\tau) = \mu_2 - \Sigma_{12}^\top \Sigma_{11}^{-1}\mu_1 - \sqrt{\frac{\Sigma_{22} - \Sigma_{12}^\top \Sigma_{11}^{-1}\Sigma_{21}}{\Sigma_{11}}}\mu_1 F_\varepsilon(\tau),$$

$$\theta_1(\tau) = \Sigma_{12}^\top \Sigma_{11}^{-1} + \sqrt{\frac{\Sigma_{22} - \Sigma_{12}^\top \Sigma_{11}^{-1}\Sigma_{21}}{\Sigma_{11}}}F_\varepsilon(\tau).$$

17.7   Other dynamic quantile models

Quantile based methods provide a local approach to directly model the dynamics of a time series at a specified quantile. Consider again the linear GARCH model given by (17.14) and (17.15). Note that $\sigma_{t-j}F^{-1}(\tau) = Q_{u_{t-j}}(\tau|\mathcal{F}_{t-j-1})$, hence the conditional quantile $Q_{u_t}(\tau|\mathcal{F}_{t-1})$ has representation

$$Q_{u_t}(\tau|\mathcal{F}_{t-1}) = \beta_0^* + \sum_{i=1}^p \beta_i^* Q_{u_{t-i}}(\tau|\mathcal{F}_{t-i-1}) + \sum_{j=1}^q \gamma_j^* |u_{t-j}|, \qquad (17.43)$$

where $\beta_0^* = \beta_0(\tau) = \beta_0 F^{-1}(\tau)$, $\beta_i^* = \beta_i$, $i = 1,\ldots,p$, and $\gamma_j^* = \gamma_j(\tau) = \gamma_j F^{-1}(\tau)$, $j = 1,\ldots,q$. From (17.43) we can see an important feature of the linear GARCH model: conditional quantiles $Q_{u_t}(\tau|\mathcal{F}_{t-1})$ themselves follow an autoregression. This representation suggests that one may model the local dynamics or local correlation directly based on the conditional quantiles.

Engle and Manganelli (2004) propose the conditional autoregressive value at risk (CAViaR) specification for the τth conditional quantile of u_t:

$$Q_{u_t}(\tau|\mathcal{F}_{t-1}) = \beta_0 + \sum_{i=1}^p \beta_i Q_{u_{t-i}}(\tau|\mathcal{F}_{t-i-1}) + \sum_{j=1}^q \alpha_j \ell(X_{t-j}), \qquad (17.44)$$

where $X_{t-j} \in \mathcal{F}_{t-j}$, $\mathcal{F}_{t-j}$ being the information set at time $t - j$. A natural choice of X_{t-j} is the lagged u. When we choose $X_{t-j} = |u_{t-j}|$, we obtain (17.43). Engle and Manganelli (2004) discussed many choices of $\ell(X_{t-j})$ leading to different specifications of the CAViaR model.

Sim (2009) considered local models to study the asymmetric correlation of international stock returns. To study the correlations between the τ_Yth quantile of Y_t and the τ_Xth quantile of X_t, they consider the quantile dependence model

$$Q_{Y_t}(\tau_Y|\mathcal{F}_{t-1}) = h(Q_{X_t}(\tau_X|V_t), \beta(\tau_X, \tau_Y)). \tag{17.45}$$

Let $X_t = Y_{t-1}$, $\tau_X = \tau_Y = \tau$, $Q_{X_t}(\tau_X|V_t) = Q_{X_t}(\tau_X|\mathcal{F}_{t-2})$, and consider linear function of $h(\cdot)$. We obtain an autoregression model for the τth conditional quantile of Y_t:

$$Q_{y_t}(\tau|\mathcal{F}_{t-1}) = \beta_0 + \beta Q_{y_{t-1}}(\tau|\mathcal{F}_{t-2}).$$

Let $X_t = (Y_{t-1}, Z_{t-1})$ and $\tau_X = \tau_Y = \tau$. We have

$$Q_{Y_t}(\tau|\mathcal{F}_{t-1}) = h(Q_{Z_{t-1}}(\tau|\mathcal{F}_{t-2}), Q_{Y_{t-1}}(\tau|\mathcal{F}_{t-2}), \beta(\tau)),$$

where the τth conditional quantile of Y_t is affected by its own lagged value and lagged values of the conditional quantile of covariates.

Estimation of the CAViaR model is challenging. If we denote the vector of unknown parameters by θ, and, for simplicity, denote $Q_{u_t}(\tau|\mathcal{F}_{t-1})$ by $Q_t(\tau, \theta)$, then we may consider estimating θ by minimizing

$$RQ_n(\tau, \theta) = \sum_t \rho_\tau(Y_t - Q_t(\tau, \theta)), \tag{17.46}$$

where $Q_t(\tau, \theta) = \beta_0 + \sum_{i=1}^p \beta_i Q_{t-i}(\tau, \theta) + \sum_{j=1}^q \alpha_j \ell(X_{t-j})$. Since conditional quantiles enter the CAViaR regression model as regressors and are latent, conventional nonlinear quantile regression techniques are not directly applicable. De Rossi and Harvey (2009) studied an iterative Kalman filter method to calculate dynamic conditional quantiles that may be applied to calculate certain types of CAViaR model. In their model, the observed time series Y_t is described by the measurement equation

$$Y_t = \xi_t(\tau) + \varepsilon_t(\tau),$$

where $\xi_t(\tau) = Q_{Y_t}(\tau|\mathcal{F}_{t-1})$ is the state variable, and the disturbances $\varepsilon_t(\tau)$ are assumed to be serially independent and independent of $\xi_t(\tau)$. The dynamics of this system is characterized by the state transition equation based on $\xi_t(\tau)$. For example, if the conditional quantiles follow an autoregression, we have

$$\xi_t(\tau) = \beta \xi_{t-1}(\tau) + \eta_t(\tau).$$

Alternative forms of the state transition equations can be considered. The above state space model can then be estimated by iteratively applying an appropriate signal extraction algorithm.

Hsu (2010) studied estimating CAViaR with a Markov chain Monte Carlo (MCMC) method based on the Bayesian approach of Yu and Moyeed (2001). She considered the following asymmetric Laplace density as a working conditional density for the error term in the CAViaR model:

$$f(\varepsilon_{t\tau}|\mathcal{F}_{t-1}) = \frac{\tau(1-\tau)}{\sigma} \exp\left\{-\frac{1}{\sigma}\rho_\tau(\varepsilon_{t\tau})\right\},$$

where $\varepsilon_{t\tau} = Y_t - Q_t(\tau, \theta)$ and σ is a scale parameter. Then, given a data set of size n, the working likelihood is

$$f(\text{Data}|\theta, \sigma) = \left(\frac{\tau(1-\tau)}{\sigma}\right)^n \exp\left\{-\frac{1}{\sigma}RQ_n(\tau, \theta)\right\},$$

where $RQ_n(\tau, \theta)$ is given by (17.46). Hsu (2010) choose a flat prior for each coefficient in θ and the inverse gamma distribution $\text{IG}(\alpha_0, s_0)$ for σ; thus the joint prior for θ is given by

$$\pi(\theta, \sigma) \propto \frac{1}{\sigma^{\alpha_0+1}} \exp\left(-\frac{s_0}{\sigma}\right),$$

and the posterior for (θ, σ) is

$$f(\theta, \sigma|\text{Data}) \propto \frac{1}{\sigma^{\alpha_0+1}} \left(\frac{\tau(1-\tau)}{\sigma}\right)^n \exp\left\{-\frac{1}{\sigma}RQ_n(\tau, \theta) - \frac{s_0}{\sigma}\right\}.$$

Posterior inference on the CAViaR model can then be implemented. See Chapter 4 in this handbook for more discussion on Bayesian quantile methods.

Gourieroux and Jasiak (2008) proposed a dynamic additive quantile model based on a group of baseline quantile functions. In particular, let $Q_{Y_t}(\tau|\mathcal{F}_{t-1})$ be dependent on some unknown parameters θ and denote it by $Q_{Y_t|X_t}(\tau; \theta)$. They define a dynamic additive quantile model as

$$Q_{Y_t|X_t}(\tau; \theta) = \sum_{k=1}^{K} \rho_k(X_t, \alpha_k) Q_k(\tau, \beta_k) + \rho_0(X_t, \alpha_0),$$

where $Q_k(\tau, \beta_k)$ are baseline quantile functions with identical range and $\rho_k(X_t, \alpha_k)$ are positive functions of the past information. By construction, the quantile curves do not cross. Information based estimation methods are proposed by Gourieroux and Jasiak (2008) to estimate these models.

17.8   Quantile spectral analysis

Frequency domain analysis is an important approach in time series studies. The traditional spectral analysis of a time series $\{X_t\}_{t=1}^{n}$ is based on the spectral density, defined as the Fourier transform of the autocovariance function $\gamma_{xx}(h) = \text{Cov}(X_t, X_{t-h})$:

$$f_{xx}(\omega) = \frac{1}{2\pi} \sum_{h=-\infty}^{\infty} \gamma_{xx}(h)e^{-ih\omega}, \tag{17.47}$$

where $i = \sqrt{-1}$. By the Riesz–Fischer theorem, the spectral density function and the auto-covariance function contain the same information about the stochastic process. Being a weighted average of the autocovariance functions, the spectral density captures the correlation structure in $\{X_t\}_{t=1}^{n}$. If the temporal dependence in $\{X_t\}_{t=1}^{n}$ is accurately captured by the second moments (correlations), the spectral density provides a complete description of the dependence of the underlying time series. However, for nonlinear or non-Gaussian processes, the spectral density may miss some important information about the dependence structure in the underlying time series. It is well-known that an ARCH/GARCH process might be serially uncorrelated but dependent.

The existing literature (e.g. Granger, 2002) points out that the classical linear time series modeling based on the Gaussian distribution assumption clearly fails to explain the stylized facts observed in economic and financial data and that it is highly undesirable to perform various economic policy evaluations, financial forecasts, and risk managements based on such linear Gaussian models. Econometric analysis that ignores any of these characteristics may lead poor performance of the inference and applications. For these reasons, there has traditionally been a long-standing research effort in non-Gaussianity and nonlinearity. Higher-order spectra, which are the Fourier transforms of higher-order cumulants, have been proposed to capture some nonlinear dependencies. However, they usually still miss important information of dependence. Hong (1999) proposed spectral analysis based on the characteristic functions. Li (2008, 2012), Dette et al. (2015, 2016), Lee and Subba Rao (2011), and Hagemann (2013) studied quantile spectral analysis. These are important spectral tools in analyzing nonlinear, non-Gaussian time series.

17.8.1   Quantile cross-covariances and quantile spectrum

To go beyond the second-order moment serial features, we may consider the Laplace cross-covariance function defined by

$$r_{xx}(h, x_1, x_2) = \text{Cov}\left(1\left(X_t < x_1\right), 1\left(X_{t+h} < x_2\right)\right),$$

or, equivalently, the quantile cross-covariance function defined by

$$r^q_{xx}(h, \tau_1, \tau_2) = \text{Cov}\left(1\left(F_X\left(X_t\right) < \tau_1\right), 1\left(F_X\left(X_{t+h}\right) < \tau_2\right)\right).$$

The quantile (or Laplace) cross-covariance functions are important tools in analyzing serial dependence (see, for example, Hong, 1999). In comparison with the traditional covariance functions, the quantile (or Laplace) covariance functions do not require the existence of finite variance since the covariances of indicators $1\left(F_X\left(X_t\right) < \tau_1\right)$ (or $1\left(X_t < x_1\right)$) always exist for all lags. More importantly, by varying over the full range of quantiles, the quantile (or Laplace) cross-covariance functions provide a complete description of the bivariate distribution of (X_t, X_{t+h}), and thus capture dependence beyond the second-order moment features. These indicator-based covariance functions are able to account for sophisticated dependence features that covariance-based methods are unable to detect, such as nonlinearity, time irreversibility, tail dependence, and varying conditional skewness or kurtosis.

Just like the traditional spectral density, which is based on the Fourier transform of the autocovariance function, we may consider generalized versions of spectral density based on Fourier transforms of other covariance functions, such as the quantile (or Laplace) covariance functions. In particular, we may define the quantile spectral density based on the Fourier transform of the quantile covariance function

$$S^q_{xx}(\omega, \tau_1, \tau_2) = \frac{1}{2\pi} \sum_{h=-\infty}^{\infty} r^q_{xx}(h, \tau_1, \tau_2) e^{-ih\omega}.$$

When $\tau_1 = \tau_2 = \tau$, this is the level-crossing spectrum of X_t studied by Li (2012):

$$S^q_{xx}(\omega, \tau) = \frac{1}{2\pi} \sum_{h=-\infty}^{\infty} r^q_{xx}(h, \tau, \tau) e^{-ih\omega}.$$

The quantile spectral densities not only share many properties of the traditional spectral density functions in capturing serial correlation in the frequency domain, but also provide a much richer picture of the dependence in the underlying time series. Quantile spectral densities can detect hidden periodicity in quantiles. See, for example, Li (2008, 2012), Dette et al. (2015, 2016), Lee and Subba Rao (2011), and Hagemann (2013).

Quantile spectrum analysis can capture dependence beyond the conventional correlation analysis. Consider the QAR process (17.36) of Koenker and Xiao (2006),

$$
\begin{aligned}
Y_t &= \theta_0(U_t) + \theta_1(U_t)Y_{t-1} + \cdots + \theta_p(U_t)Y_{t-p} \\
&= \theta_0 + \theta_1 Y_{t-1} + \cdots + \theta_p Y_{t-p} + V_t,
\end{aligned}
$$

where $\theta_0 = \mathrm{E}\left[\theta_0(U_t)\right]$, $\theta_1 = \mathrm{E}\left[\theta_1(U_t)\right]$, $\ldots$, $\theta_p = \mathrm{E}\left[\theta_p(U_t)\right]$, and

$$
V_t = \theta_0(U_t) - \theta_0 + \sum_{j=1}^{p} \left[\theta_j(U_t) - \theta_j\right] Y_{t-j}
$$

is a martingale difference sequence. The classical spectral densities of QAR processes (17.36) with different $\theta_j(U_t)$, $j = 1, \ldots, p$, will be the same as the same fixed coefficient AR(p) process (17.38). In the case where $\mathrm{E}\left[\theta_1(U_t)\right] = \ldots = \mathrm{E}\left[\theta_p(U_t)\right] = 0$, the classical spectral density of Y_t will be flat, and therefore cannot reveal any information about cyclical or dependence features of Y_t. However, the quantile spectral density will not be flat over some quantiles (as long as $\theta_j(U_t)$ are not constant zeros), revealing additional information beyond the classical spectral density. See, for example, Hagemann (2013).

17.8.2   Quantile periodograms

Quantile spectral densities can be estimated based on their finite-sample analogs, the quantile periodograms. Given a time series $\{X_t\}_{t=1}^{n}$, the discrete Fourier transform (DFT) of the time series at frequency ω is defined as

$$
w_x(\omega) = \frac{1}{\sqrt{n}} \sum_{t=1}^{n} X_t e^{-it\omega},
$$

and the ordinary periodogram is defined as

$$
I_{xx}(\omega) = \frac{1}{n} \left| \sum_{t=1}^{n} X_t e^{-it\omega} \right|^2 .
$$

If we denote the sample covariance function by

$$
\widehat{\gamma}_{xx}(h) = \frac{1}{n} \sum_{1 \leqslant t, t+h \leqslant n} X_t X_{t+h},
$$

then the ordinary periodogram is simply a sample analog of the spectral density (multiplied by 2π) defined by (17.47):

$$
I_{xx}(\omega) = \sum_{h=-n+1}^{n-1} \left(\frac{1}{n} \sum_{1 \leqslant t, t+h \leqslant n} X_t X_{t+h} \right) e^{-ih\omega} = \sum_{h=-n+1}^{n-1} \widehat{\gamma}_{xX}(h) e^{-ih\omega}. \tag{17.48}
$$

In fact, if $\omega = 0$ (or, in general $\omega = 2\pi j$, where j is an integer), $I_{xx}(\omega)$ is asymptotically distributed as $2\pi f_{xx}(\omega)\chi_1^2$, where χ_1^2 is a chi-square random variables with 1 degree of freedom; if $\omega \neq 2\pi j$, then $I_{xx}(\omega)$ is asymptotically distributed as $\pi f_{xx}(\omega)\chi_2^2$, where χ_2^2 is a chi-square random variables with 2 degrees of freedom. Thus, the periodogram is an asymptotically unbiased (but inconsistent) estimator of the spectral density in the sense

$$
\mathrm{E}\left[I_{xx}(\omega)\right] \to 2\pi f_{xx}(\omega), \quad \text{as } n \to \infty.
$$

The above results can be extended to quantile spectral analysis. In particular, similarly to (17.48), let $\hat{r}_{xx}^q(h, \tau_1, \tau_2)$ be the sample covariance function of $(1\left[X_t < F_X^{-1}(\tau_1)\right],$ $1\left[X_{t+h} < F_X^{-1}(\tau_2)\right])$. We may construct the quantile periodogram as follows:

$$L_{xx}(\omega, \tau_1, \tau_2) = \sum_{h=-n+1}^{n-1} \hat{r}_{xx}^q(h, \tau_1, \tau_2) e^{-ih\omega}.$$

The above quantile periodogram is completely parallel to the ordinary periodogram, and shares some properties similar to the ordinary periodogram. In particular, the quantile periodogram is an asymptotically unbiased estimator of the quantile spectral density in the sense

$$\mathrm{E}\left(L_{xx}(\omega, \tau_1, \tau_2)\right) \to 2\pi \cdot S_{xx}^q(\omega, \tau_1, \tau_2), \quad \text{as } n \to \infty.$$

More specifically,

$$L_{xx}(\omega, \tau, \tau) \Rightarrow \pi \cdot S_{xx}^q(\omega, \tau, \tau) \cdot \chi_2^2, \quad \text{if } \tau_1 = \tau_2 = \tau,$$

and

$$L_{xx}(\omega, \tau_1, \tau_2) \Rightarrow \pi \cdot Z_1^\top J Z_2, \quad \text{if } \tau_1 \neq \tau_2,$$

where

$$Z_1 \sim N\left(0, S_{xx}^q(\omega, \tau_1, \tau_1) I_2\right), \quad Z_2 \sim N\left(0, S_{xx}^q(\omega, \tau_2, \tau_2) I_2\right),$$

and

$$\mathrm{Cov}\left(Z_1, Z_2\right) = \begin{bmatrix} \mathrm{Re}S_{xx}^q(\omega, \tau_1, \tau_2) & \mathrm{Im}S_{xx}^q(\omega, \tau_1, \tau_2) \\ -\mathrm{Im}S_{xx}^q(\omega, \tau_1, \tau_2) & \mathrm{Re}S_{xx}^q(\omega, \tau_1, \tau_2) \end{bmatrix},$$

where $\mathrm{Re}S_{xx}^q(\omega, \tau_1, \tau_2)$ and $\mathrm{Im}S_{xx}^q(\omega, \tau_1, \tau_2)$ are the real and imaginary parts of $S_{xx}^q(\omega, \tau_1, \tau_2)$. When we focus on one quantile, $\tau_1 = \tau_2 = \tau$, this is basically the quantile periodogram studied by Hagemann (2013).

17.8.3 Relationship to quantile regression on harmonic regressors

Periodograms are closely connected to projections of a time series $\{X_t\}_{t=1}^n$ onto the harmonic basis. If we consider the fundamental frequencies (or the Fourier frequencies) $\omega_j = \omega_{j,n} = 2\pi j/n$, for some integer j, the ordinary periodogram can be written as

$$I_{xx}(\omega_j) = \frac{1}{4} n \left\| \hat{\beta}(\omega_j) \right\|^2 = \frac{n}{4} \cdot \hat{\beta}(\omega_j)^\top J \hat{\beta}(\omega_j),$$

where $\hat{\beta}(\omega_j)$ is the OLS regression solution of X_t on the harmonic regressors $Z_t(\omega_j) = [\cos(t\omega_j), \sin(t\omega_j)]^\top$:

$$\hat{\beta}(\omega_j) = \arg\min \sum_{t=1}^n \left(X_t - \alpha - \beta^\top Z_t(\omega_j)\right)^2,$$

and

$$J = \begin{bmatrix} 1 & i \\ -i & 1 \end{bmatrix}.$$

Li (2008, 2012) and Dette et al. (2015) considered the L_1 projection of X_t on the harmonic basis, and introduced the Laplace and quantile periodograms based on L_1 regression of X_t on $Z_t(\omega_j)$. In particular, Li (2008) considered the LAD regression

$$\tilde{\beta}(\omega_j) = \arg\min \sum_{t=1}^n \left|X_t - \alpha - \beta^\top Z_t(\omega_j)\right|,$$

and defined the (Laplace) periodogram based on $\widetilde{\beta}(\omega_j)$,

$$\ell_{xx}(\omega_j) = \frac{1}{4}n\left\|\widetilde{\beta}(\omega_j)\right\|^2 = \frac{n}{4}\cdot\widetilde{\beta}(\omega_j)^\top J\widetilde{\beta}(\omega_j).$$

The Laplace periodogram $L_{xx}(\omega_j)$ corresponds to the zero cross-spectrum which is the spectral density based on $\mathrm{Cov}\left(1\left(X_t < 0\right), 1\left(X_{t+h} < 0\right)\right)$.

The above idea can be extended to quantile regression. Li (2012) considered the τth quantile regression

$$\widetilde{\beta}_\tau(\omega_j) = \arg\min\sum_{t=1}^{n}\rho_\tau\left(X_t - \alpha - \beta^\top Z_t(\omega_j)\right),$$

and defined the quantile periodogram

$$\ell_{xx}(\omega_j, \tau) = \frac{1}{4}n\left\|\widetilde{\beta}_\tau(\omega_j)\right\|^2 = \frac{n}{4}\cdot\widetilde{\beta}_\tau(\omega_j)^\top J\widetilde{\beta}_\tau(\omega_j).$$

Dette et al. (2015) defined a more general version of the quantile periodogram based on the (τ_1, τ_2)th quantiles:

$$\ell_{xx}(\omega_j, \tau_1, \tau_2) = \frac{n}{4}\widetilde{\beta}_{\tau_1}(\omega_j)^\top J\widetilde{\beta}_{\tau_2}(\omega_j).$$

Periodograms based on harmonic regression projections are asymptotically equivalent to periodograms based on sample analogs of the spectral density. In particular, it can be shown that, under appropriate assumptions,

$$\frac{n}{4}\widetilde{\beta}_{\tau_1}(\omega_j)^\top J\widetilde{\beta}_{\tau_2}(\omega_j) = \frac{1}{f(F^{-1}(\tau_1))f(F^{-1}(\tau_2))}\sum_{h=-n+1}^{n-1}\widehat{r}_{xx}^q(h, \tau_1, \tau_2)e^{-ih\omega_j} + o_p(1),$$

thus

$$\ell_{xx}(\omega_j, \tau_1, \tau_2) = \frac{1}{f(F^{-1}(\tau_1))f(F^{-1}(\tau_2))}L_{xx}(\omega_j, \tau_1, \tau_2) + o_p(1).$$

Consider the standardized version of the quantile spectral density,

$$s_{xx}^q(\omega, \tau_1, \tau_2) = \frac{S_{xx}^q(\omega, \tau_1, \tau_2)}{f(F_X^{-1}(\tau_1))f(F_X^{-1}(\tau_2))},$$

where $f(\cdot)$ is the density of X. Then $(2\pi)^{-1}\ell_{xx}(\omega, \tau_1, \tau_2)$ are sample analogs of $s_{xx}^q(\omega, \tau_1, \tau_2)$, and $\mathrm{E}(\ell_{xx}(\omega, \tau_1, \tau_2)) \to 2\pi\cdot s_{xx}^q(\omega, \tau_1, \tau_2)$, as $n \to \infty$. Dette et al. (2015) show that

$$\ell_{xx}(\omega, \tau, \tau) \Rightarrow \pi\cdot s_{xx}^q(\omega, \tau, \tau)\cdot\chi_2^2, \quad \text{if } \tau_1 = \tau_2 = \tau,$$

and

$$\ell_{xx}(\omega, \tau_1, \tau_2) \Rightarrow \pi\cdot z_1^\top J z_2, \quad \text{if } \tau_1 \neq \tau_2,$$

where

$$z_1 \sim N\left(0, s_{xx}^q(\omega, \tau_1, \tau_1)I_2\right), \quad z_2 \sim N\left(0, s_{xx}^q(\omega, \tau_2, \tau_2)I_2\right),$$

and

$$\mathrm{Cov}(z_1, z_2) = \begin{bmatrix} \mathrm{Res}_{xx}^q(\omega, \tau_1, \tau_2) & \mathrm{Ims}_{xx}^q(\omega, \tau_1, \tau_2) \\ -\mathrm{Ims}_{xx}^q(\omega, \tau_1, \tau_2) & \mathrm{Res}_{xx}^q(\omega, \tau_1, \tau_2) \end{bmatrix},$$

where $\mathrm{Res}_{xx}^q(\omega, \tau_1, \tau_2)$ and $\mathrm{Ims}_{xx}^q(\omega, \tau_1, \tau_2)$ are the real and imaginary parts of $s_{xx}^q(\omega, \tau_1, \tau_2)$.

17.8.4 Estimation of quantile spectral density

The quantile periodogram shares properties similar to the ordinary periodogram, and these properties facilitate nonparametric estimation of the quantile spectral density. In particular, the quantile spectral density can be estimated based on either a smoothed quantile periodogram around the specified frequency, or a truncated summation of the sample covariance functions of $\tau - 1\left(X_t < F_X^{-1}(\tau)\right)$.

Kernel estimator. We may estimate the quantile spectral density based on a truncated summation of the sample covariances. Since

$$S_{xx}^q(\omega, \tau_1, \tau_2) = \frac{1}{2\pi} \sum_{h=-\infty}^{\infty} r_{xx}^q(h, \tau_1, \tau_2) e^{-ih\omega}$$

is a weighted summation of the quantile covariance function $r_{xx}^q(h, \tau_1, \tau_2)$, we may estimate the quantile spectrum based on the kernel average of the sample quantile covariance function

$$\hat{r}_{xx}^q(h, \tau_1, \tau_2) = \frac{1}{n} \sum_{1 \leq t, t+h \leq n} \left(\tau_1 - 1\left(X_t < \hat{Q}_X(\tau_1)\right)\right)\left(\tau_2 - 1\left(X_{t+h} < \hat{Q}_X(\tau_2)\right)\right),$$

where $\hat{Q}_X(\tau_j)$ are sample quantiles of X_t. In particular, we construct the quantile spectrum estimator

$$\hat{S}_{xx}^q(\omega, \tau_1, \tau_2) = \frac{1}{2\pi} \sum_{h=-n+1}^{n-1} k\left(\frac{h}{M}\right) \hat{r}_{xx}^q(h, \tau_1, \tau_2) e^{-ih\omega},$$

where $k(\cdot)$ is an appropriate kernel function and M is a bandwidth parameter that goes to ∞ slower than n. Under regularity conditions, $\hat{S}_{xx}^q(\omega, \tau_1, \tau_2) \to S_{xx}^q(\omega, \tau_1, \tau_2)$ as $n \to \infty$.

Smoothed periodogram. Alternatively, the quantile spectral density can be constructed based on the smoothed periodograms. Let

$$L_{xx}(\lambda_s, \tau_1, \tau_2) = \sum_{h=-n+1}^{n-1} \hat{r}_{xx}^q(h, \tau_1, \tau_2) e^{-ih\lambda_j}.$$

We are interested in estimating $S_{xx}^q(\omega, \tau_1, \tau_2)$. Consider a frequency band around ω,

$$B(\omega) = \left\{\lambda_s = 2\pi s/n : \omega - \frac{\pi}{2M} \leq \lambda_s \leq \omega + \frac{\pi}{2M}\right\}$$

where, again, M is a bandwidth parameter that goes to ∞ slower than n. By definition, the length of $B(\omega)$ is π/M. Denote the number of fundamental frequencies $\lambda_s = 2\pi s/n$ in the frequency band $B(\lambda)$ by m. Then

$$m = \frac{n}{2M},$$

that is, $2mM = n$. We may construct the following smoothed quantile periodograms to estimate the quantile spectral density:

$$\hat{S}_{xx}^q(\omega, \tau_1, \tau_2) = \frac{1}{2\pi m} \sum_{\lambda_s \in B(\omega)} K(\lambda_s - \omega) L_{xx}(\lambda_s, \tau_1, \tau_2),$$

where $K(\cdot)$ is the spectral window such that $\{K(\lambda_s - \omega) : \lambda_s \in B(\omega)\}$ is a sequence of weights satisfying

$$K(\lambda_s - \omega) = K(\omega - \lambda_s), \quad \sum_{\lambda_s \in B(\omega)} K(\lambda_s - \omega) = 1.$$

Under appropriate regularity conditions, $\widehat{S}_{xx}^{q}(\omega, \tau_1, \tau_2)$ is a consistent estimator for $S_{xx}^{q}(\omega, \tau_1, \tau_2)$.

These two approaches are closely related to each other. Given a spectral window $K(\cdot)$, if we consider

$$k\left(\frac{h}{M}\right) = \frac{1}{m}\sum_{\lambda_j \in B} K(\lambda_j - \omega)e^{-ih(\lambda_j - \omega)},$$

then the kernel estimator for the quantile spectral density based on $k(\cdot)$ can be written as a smoothed periodogram using spectral window $K(\cdot)$:

$$\begin{aligned}
\widehat{S}_{xx}^{q}(\omega, \tau_1, \tau_2) &= \frac{1}{2\pi}\sum_{h=-n+1}^{n-1} k\left(\frac{h}{M}\right)\widehat{r}_{xx}^{q}(h, \tau_1, \tau_2)e^{-ih\omega} \\
&= \frac{1}{2\pi}\sum_{h=-n+1}^{n-1}\frac{1}{m}\sum_{\lambda_j \in B} K(\lambda_j - \omega)e^{-ih\lambda_j}\widehat{r}_{xx}^{q}(h, \tau_1, \tau_2) \\
&= \frac{1}{2\pi}\frac{1}{m}\sum_{\lambda_j \in B} K(\lambda_j - \omega)L_{xx}(\lambda_j, \tau_1, \tau_2).
\end{aligned}$$

17.9 Quantile regression based forecasting

Dynamic quantile regression models offer a natural approach for interval forecasting. Given the time series observations $\{Y_t\}_{t=1}^{T}$, and a dynamic quantile regression model

$$Q_{Y_t}(\tau|\mathcal{F}_{t-1}) = g(X_t, \theta(\tau)), \tag{17.49}$$

where $X_t = (1, Y_{t-1}, \ldots, Y_{t-p})^{\top}$, we consider out-of-sample prediction based on the available observations. In the special case $g(X_t, \theta(\tau)) = X_t^{\top}\theta(\tau)$, this is the QAR model (17.35). If the parameters $\theta(\tau)$ are known, the interval

$$[g(X_{T+1}, \theta(\alpha/2)), \ g(X_{T+1}, \theta(1 - \alpha/2))]$$

is an exact $(1 - \alpha)$-level interval forecast of Y_{T+1}. In practice, we do not know $\theta(\tau)$ and have to use a quantile regression estimator $\widehat{\theta}(\tau)$ in the above construction. Following Zhou and Portnoy (1996), we may use the modified interval forecast

$$\left[g(X_{T+1}, \widehat{\theta}(\alpha/2 - h_T)), \ g(X_{T+1}, \widehat{\theta}(1 - \alpha/2 + h_T))\right],$$

where $h_T \to 0$, to account for the uncertainty from the preliminary quantile regression estimation $\widehat{\theta}(\tau)$.

To generate an p-step interval forecast for Y_{T+p}, notice that the one-step-ahead forecast conditional distribution of Y_{T+1} can be obtained from

$$\widehat{Y}_{T+1} = g(X_{T+1}, \widehat{\theta}(U)),$$

where U are random draws from a uniform distribution $U[0, 1]$. Let U_1^* be a draw from $U[0, 1]$. Then a draw from the one-step-ahead forecast distribution of Y_{T+1} is given by

$$\widehat{Y}_{T+1}^{*} = g(X_{T+1}, \widehat{\theta}(U_1^*)).$$

Next, let $\widetilde{X}_{T+2} = (1, \widehat{Y}_{T+1}^*, Y_T, \ldots, Y_{T-p+2})^\top$, and $U_2^* \sim U[0,1]$. Then a draw from the two-step ahead forecast distribution of Y_{T+2} is given by

$$\widehat{Y}_{T+2}^* = g(\widetilde{X}_{T+2}, \widehat{\theta}(U_2^*)).$$

At step s, let $\widetilde{X}_{T+s} = (1, \widehat{Y}_{T+s-1}^*, \ldots, \widehat{Y}_{T+s-p}^*)^\top$ (where $\widehat{Y}_j^* = Y_j$ if $j \leqslant T$) and $U_s^* \sim U[0,1]$. Then we can obtain a draw from the forecast

$$\widehat{Y}_{T+s}^* = g(\widetilde{X}_{T+s}, \widehat{\theta}(U_s^*)).$$

Applying the above sampling procedure recursively, we obtain a sample path of forecast

$$\left(\widehat{Y}_{T+1}^*, \widehat{Y}_{T+2}^*, \ldots, \widehat{Y}_{T+p}^*\right).$$

Repeating this process R times, a forecast of the conditional distribution of Y_{T+p} can then be approximated based on an ensemble of such sample paths $\left\{(\widehat{Y}_{T+1}^{(r)}, \ldots, \widehat{Y}_{T+p}^{(r)})\right\}_{r=1}^R$, and an p-step $(1-\alpha)$-level interval forecast can be constructed based on the sample quantiles of $\left\{\widehat{Y}_{T+p}^{(r)}\right\}_{r=1}^R$.

Other types of models, such as the ARCH model, or a combination of AR structure in the mean equation and ARCH error, may be forecasted in the same way; see Granger et al. (1989) and Koenker and Zhao (1996) for a discussion based on linear ARCH models.

Yu and Moyeed (2001) proposed a Bayesian solution to the quantile regression problem via the likelihood of a skewed Laplace distribution. If the density of a random variable takes the form

$$f(\varepsilon) = \tau(1-\tau)\exp\left\{-\rho_\tau(\varepsilon)\right\},$$

the density is called an asymmetric Laplace density. Consider, say, an autoregression model. If the error term u_t has probability density function proportional to $\exp\left\{-\rho_\tau(u)\right\}$, the associated maximum likelihood estimation is equivalent to minimizing the check function of a quantile regression. The likelihood connection of quantile regression facilitates extracting the posterior distributions of unknown parameters via the MCMC method, and provides a convenient way of incorporating parameter uncertainty into quantile predictive inference. Given a specified quantile model, Bayesian quantile forecasting in this direction can be obtained.

17.10   Conclusion

Quantile time series analysis is a fast-growing research subject. Several important topics are not included in this chapter, some of which are covered by other chapters in this handbook. For example, time series quantile regression has important applications in finance; extreme quantiles are also widely used in risk management; and dynamic panel models are now attracting increasing research attention (see related chapters in this handbook). In addition, as a natural way of studying the conditional distribution, quantile regression offers a variety of techniques for making inferences about conditional quantile/distribution functions. For example, it can provide a useful approach in testing for changes in distribution or conditional distribution. It also offers a convenient way in comparing distributional properties among different samples. For semiparametric and nonparametric time series quantile regressions, see, for example, Xiao (2009) and references therein. A systematic description of quantile regression is given by Koenker (2005).

Bibliography

D. W. K. Andrews. Heteroskedasticity and autocorrelation consistent covariance matrix estimation. *Econometrica*, 59:817–858, 1991.

J. Angrist, V. Chernozhukov, and I. Fernández-Val. Quantile regression under misspecification, with an application to the US wage structure. *Econometrica*, 74:539–563, 2005.

P. Beaudry and G. Koop. Do recessions permanently change output? *Journal of Monetary Economics*, 31:149–163, 1993.

E. Bouyé and M. Salmon. Dynamic copula quantile regressions and tail area dynamic dependence in Forex markets. *European Journal of Finance*, 15(7-8):721–750, 2009.

X. Chen, R. Koenker, and Z. Xiao. Copula-based nonlinear quantile autoregression. *Econometrics Journal*, 12:50–67, 2009.

R. Davis, K. Knight, and J. Liu. M-estimation for autoregressions with infinite variance. *Stochastic Processes and Their Applications*, 40:145–180, 1992.

G. De Rossi and A. Harvey. Quantiles, expectiles and splines. *Journal of Econometrics*, 152:179–185, 2009.

H. Dette, M. Hallin, T. Kley, and S. Volgushev. Of copulas, quantiles, ranks and spectra: An l_1-approach to spectral analysis. *Bernoulli*, 21:781–831, 2015.

H. Dette, M. Hallin, T. Kley, and S. Volgushev. Quantile spectral processes: Asymptotic analysis and inference. *Bernoulli*, 22:1770–1807, 2016.

P. Embrechts, C. Klüppelberg, and T. Mikosch. *Modelling Extremal Events for Insurance and Finance*. Springer, Berlin, 1997.

W. Enders and C. Granger. Unit root tests and asymmetric adjustment with an example using the term structure of interest rates. *Journal of Business and Economic Statistics*, 16:304–311, 1998.

R. Engle and S. Manganelli. CAViaR: Conditional autoregressive value at risk by regression quantiles. *Journal of Business and Economic Statistics*, 22:367–381, 2004.

E. Fama. Portfolio analysis in a stable Paretian market. *Management Science*, 11:404–419, 1965.

C. Gourieroux and J. Jasiak. Dynamic quantile models. *Journal of Econometrics*, 147:198 – 205, 2008.

C. Granger. Time series concept for conditional distributions. Technical Report, University of California at San Diego, 2002.

C. Granger, H. White, and M. Kamstra. Interval forecasting: An analysis based on ARCH-quantile estimators. *Journal of Econometrics*, 40:87–96, 1989.

A. Hagemann. Robust spectral analysis. UIUC Technical Report, 2013.

M. Hallin and B. Werker. Discussion of quantile autoregression. *Journal of the American Statistical Association*, 101:475, 2006.

M. N. Hasan and R. Koenker. Robust rank tests of the unit root hypothesis. *Econometrica*, 65:133–161, 1997.

Y. Hong. Hypothesis testing in time series via the empirical characteristic function: A generalized spectral density approach. *Journal of the American Statistical Association*, 94:1201–1220, 1999.

Y.-H. Hsu. *Applications of quantile regression to estimation and detection of some tail characteristics*. PhD thesis, University of Illinois, 2010.

R. Ibragimov. Portfolio diversification and value at risk under thick-tailedness. *Quantitative Finance*, 9:565–580, 2009.

R. Ibragimov, D. Jaffee, and J. Walden. Nondiversification traps in catastrophe insurance markets. *Review of Financial Studies*, 22:959–993, 2009.

R. Koenker. Galton, Edgeworth, Frisch and prospects for quantile regression in econometrics. *Journal of Econometrics*, 95:347–374, 2000.

R. Koenker. *Quantile Regression*. Cambridge University Press, Cambridge, 2005.

R. Koenker and G. Bassett. Regression quantiles. *Econometrica*, 46:33–49, 1978.

R. Koenker and Z. Xiao. Unit root quantile regression inference. *Journal of the American Statistical Association*, 99:775–787, 2004.

R. Koenker and Z. Xiao. Quantile autoregression. *Journal of the American Statistical Association*, 101:980–1006, 2006.

R. Koenker and Q. Zhao. Conditional quantile estimation and inference for ARCH models. *Econometric Theory*, 12:793–813, 1996.

H. Koul and K. Mukherjee. Regression quantiles and related processes under long range dependent errors. *Journal of Multivariate Analysis*, 51:318–337, 1994.

J. Lee and S. Subba Rao. The quantile spectral density and comparison based tests for nonlinear time series. arXiv:1112.2759, 2011.

T.-H. Li. Laplace periodogram for time series analysis. *Journal of the American Statistical Association*, 103:757–768, 2008.

T.-H. Li. Quantile periodograms. *Journal of the American Statistical Association*, 107: 765–776, 2012.

B. Mandelbrot. The variation of certain speculative prices. *Journal of Business*, 3:394–411, 1963.

S. Neftci. Are economic time series asymmetric over the business cycle? *Journal of Political Economy*, 92:307–328, 1984.

P. C. B. Phillips and B. E. Hansen. Statistical inference in instrumental variables regression with I(1) processes. *Review of Economic Studies*, 57:99–125, 1990.

S. Portnoy. Asymptotic behavior of regression quantiles in non-stationary dependent cases. *Journal of Multivariate Analysis*, 38:100–113, 1991.

S. Rachev and S. Mittnik. *Stable Paretian Models in Finance*. Wiley, New York, 2000.

N. Sim. *Modeling quantile dependence*. PhD thesis, Boston College, 2009.

N. N. Taleb. *The Black Swan: The Impact of the Highly Improbable.* Penguin, London, 2007.

R. Weron. Heavy-tails and regime-switching in electricity prices. *Mathematical Methods of Operations Research*, 69:457–473, 2008.

Z Xiao. Quantile cointegrating regression. *Journal of Econometrics*, 150:248–260, 2009.

Z. Xiao. Robust inference in nonstationary time series models. *Journal of Econometrics*, 169:211–223, 2012.

Z. Xiao and R. Koenker. Conditional quantile estimation and inference for GARCH models. *Journal of the American Statistical Association*, 104:1696–1712, 2009.

Z. Xiao and C. Wan. Robust estimation of conditional volatility. Working Paper, 2012.

Z. Xiao and Z. Zhao. Efficient quantile estimation of regressions with garch errors. Working Paper, 2015.

K. Yu and R. A. Moyeed. Bayesian quantile regression. *Statistics & Probability Letters*, 54: 437–447, 2001.

K. Q. Zhou and S. L. Portnoy. Direct use of regression quantiles to construct confidence sets in linear models. *Annals of Statistics*, 24:287–306, 1996.

18

Extremal Quantile Regression

Victor Chernozhukov

MIT, Cambridge, Massachusetts, USA

Iván Fernández-Val

Boston University, Boston, Massachusetts, USA

Tetsuya Kaji

MIT, Cambridge, Massachusetts, USA

CONTENTS

18.1   Introduction

In 1895, the Italian econometrician Vilfredo Pareto discovered that the power law describes well the tails of income and wealth data. This simple observation stimulated further applications of the power law to economic data, including Zipf (1949), Mandelbrot (1963), Fama (1965), Praetz (1972), Sen (1973), and Longin (1996), among many others. It also led to a theory to analyze the properties of the tails of the distributions, so-called extreme value (EV) theory, which was developed by Gnedenko (1943) and de Haan (1970). Jansen and de Vries (1991) applied this theory to analyze the tail properties of US financial returns and concluded that the 1987 market crash was not an outlier; rather, it was a rare event whose magnitude could have been predicted by prior data. This work stimulated numerous other studies that rigorously documented the tail properties of economic data (Embrechts et al., 1997).

Chernozhukov (2005) extended the EV theory to develop extreme quantile regression models in the tails, and to analyze the properties of the Koenker and Bassett (1978) quantile regression estimator, called *extremal quantile regression*. This work builds especially on Feigin and Resnick (1994) and Knight (2001), who studied the most extreme, frontier regression case in the location model. Related results for the frontier case – the regression frontier estimators – were developed by Smith (1994), Chernozhukov (1998), Jurečková (1999), and Portnoy and Jurečková (1999). The work of Portnoy and Koenker (1989) and Gutenbrunner et al. (1993) implicitly contained some results on extending the normal approximations to intermediate order regression quantiles (moderately extreme quantiles) in location models. The complete theory for intermediate order regression quantiles was developed in Chernozhukov (2005). Jurečková (2016) recently characterized properties of averaged extreme regression quantiles.

In this chapter we review the theory of extremal quantile regression. We start by introducing the general setup used throughout the chapter. Let Y be a continuous response variable of interest with distribution function $F_Y(y) = \mathrm{P}(Y \leqslant y)$. The marginal τ-quantile of Y is the left-inverse of $y \mapsto F_Y(y)$ at τ, that is, $Q_Y(\tau) := \inf\{y : F_Y(y) \geqslant \tau\}$ for some $\tau \in (0,1)$. Let X be a d_x-dimensional vector of covariates related to Y, F_X be the distribution function of X, and $F_Y(y|x) = \mathrm{P}(Y \leqslant y|X = x)$ be the conditional distribution function of Y given $X = x$. The conditional τ-quantile of Y given $X = x$ is the left-inverse of $y \mapsto F_Y(y|x)$ at τ, that is, $Q_Y(\tau|x) := \inf\{y : F_Y(y|x) \geqslant \tau\}$ for some $\tau \in (0,1)$. We refer to $x \mapsto Q_Y(\tau|x)$ as the τ-quantile regression (τ-QR) function. This function measures the effect of X on Y, both at the center and at the tails of the outcome distribution. A marginal or conditional τ-quantile is *extremal* whenever the probability index τ is either close to 0 or close to 1. Without loss of generality, we focus the discussion on τ close to 0.

The analysis of the properties of the estimators of extremal quantiles relies on EV theory. This theory uses sequences of quantile indexes $\{\tau_T\}_{T=1}^{\infty}$ that change with the sample size T. Let $\tau_T T$ be the order of the τ_T-quantile. A sequence of quantile index and sample size pairs $\{\tau_T, T\}_{T=1}^{\infty}$ is said to be an *extreme order* sequence if $\tau_T \searrow 0$ and $\tau_T T \to k \in (0, \infty)$ as $T \to \infty$; an *intermediate order* sequence if $\tau_T \searrow 0$ and $\tau_T T \to \infty$ as $T \to \infty$; and a *central order* sequence if τ_T is fixed as $T \to \infty$. In this chapter we show that each of these sequences produces different asymptotic approximation to the distribution of the quantile regression estimators. The extreme order sequence leads to an EV law in large samples, whereas the intermediate and central sequences lead to normal laws. The EV law provides a better approximation to the extremal quantile regression estimators.

We conclude this introductory section with a review of some applications of extremal quantile regression to economics and finance.

Example 1 (Conditional value-at-risk) *Value-at-risk (VaR) analysis seeks to forecast or explain low quantiles of future portfolio returns of an institution, Y, using current information, X (Chernozhukov and Umantsev, 2001; Engle and Manganelli, 2004). Typically, the extremal τ-QR functions $x \mapsto Q_Y(\tau|x)$, for $\tau = 0.01$ and $\tau = 0.05$, are of interest. The VaR is a risk measure commonly used in real-life financial management, insurance, and actuarial science (Embrechts et al., 1997). We provide an empirical example of VaR in Section 18.4.*

Example 2 (Determinants of birthweights) *In health economics, we may be interested in how smoking, absence of prenatal care, and other maternal behavior during pregnancy, X, affect infant birthweights, Y (Abrevaya, 2001). Very low birthweights are connected with subsequent health problems and therefore extremal quantile regression can help identify factors to improve adult health outcomes. Chernozhukov and Fernández-Val (2011) provide an empirical study of the determinants of extreme birthweights.*

Example 3 (Probabilistic production frontiers) *An important application to industrial organization is the determination of efficiency or production frontiers, pioneered by Aigner and Chu (1968), Timmer (1971), and Aigner et al. (1976). Given the cost of production and possibly other factors, X, we are interested in the highest production levels, Y, that only a small fraction of firms, the most efficient firms, can attain. These (nearly) efficient production levels can be formally described by the extremal τ-QR function $x \mapsto Q_Y(\tau|x)$ for $\tau \in [1 - \varepsilon, 1)$ and $\varepsilon > 0$, so that only an ε-fraction of firms produce $Q_Y(\tau|X)$ or more.*

Example 4 (Approximate reservation rules) *In labor economics, Flinn and Heckman (1982) proposed a job search model with approximate reservation rules. The reservation rule measures the wage level, Y, below which a worker with characteristics, X, accepts a job with small probability ε, and can by described by the extremal τ-QR $x \mapsto Q_Y(\tau|x)$ for $\tau \in (0, \varepsilon]$.*

Example 5 (Approximate (S, s)-rules) *The (S, s)-adjustment models arise as an optimal policy in many economic models (Arrow et al., 1951). For example, the capital stock, Y, of a firm with characteristics X is adjusted sharply up to the level $S(X)$ once it has depreciated below some low level $s(X)$ with probability close to one, $1-\varepsilon$. This conditional (S, s)-rule can be described by the extremal τ-QR functions $x \mapsto Q_Y(\tau|x)$ and $x \mapsto Q_Y(1 - \tau|x)$ for $\tau \in (0, \varepsilon]$.*

Example 6 (Structural auction models) *Consider a first-price procurement auction where bidders hold independent valuations. Donald and Paarsch (2002) modeled the winning bid, Y, as $Y = c(Z)\beta(N) + \varepsilon$, where $c(Z)$ is the efficient cost function that depends on the bid characteristics, Z, $\beta(N) \geqslant 1$ is a mark-up that approaches 1 as the number of bidders N approaches infinity, and the disturbance ε captures small bidding mistakes independent of Z and N. By construction, the structural function $(z, n) \mapsto c(z)\beta(n)$ corresponds to the extremal quantile regression function $x \mapsto Q_Y(\tau|x)$ for $x = (z, n)$ and $\tau \in (0, P(\varepsilon \leqslant 0)]$.*

Example 7 (Other recent applications) *Following the pioneering work of Powell (1984), Altonji et al. (2012) applied extremal quantile regression to estimate extensive margins of demand functions with corner solutions. D'Haultfœuille et al. (2015) used extremal quantile regression to deal with endogenous sample selection under the assumption that there is no selection at very high quantiles of the response variable Y conditional on covariates X. They applied this approach to the estimation of the black wage gap for young males in the USA. Zhang (2015) employed extremal quantile regression methods to estimate tail quantile treatment effects under a selection on observables assumption.*

Notation. The symbol $\to_d$ denotes convergence in law. For two real numbers a, b, $a \ll b$ means that a is much less than b. More notation will be introduced when it is first used.

Outline. The rest of the chapter is organized as follows. Section 18.2 reviews models for marginal and conditional extreme quantiles. Section 18.3 describes estimation and inference methods for extreme quantile models. Section 18.4 presents two empirical applications of extremal quantile regression to conditional VaR and financial contagion.

18.2 Extreme quantile models

This section reviews typical modeling assumptions in extremal quantile regression. They embody Pareto conditions on the tails of the distribution of the response variable Y and linear specifications for the τ-QR function $x \mapsto Q_Y(\tau|x)$.

18.2.1 Pareto-type and regularly varying tails

The theory for extremal quantiles often assumes that the tails of the distribution of Y exhibit Pareto-type behavior, meaning that the tails decay approximately as a power function, or, more formally, a regularly varying function. The Pareto-type tails encompass a rich variety of tail behaviors, from thick- to thin-tailed distributions, and from bounded- to unbounded-support distributions.

Define the variable U by $U := Y$ if the lower end-point of the support of Y is $-\infty$ and by $U := Y - Q_Y(0)$ if the lower end-point of the support of Y is finite. In words, U is a shifted copy of Y whose support ends at either $-\infty$ or 0. The assumption that the random variable U exhibits a *Pareto-type tail* is stated by the following two equivalent conditions:[1]

$$Q_U(\tau) \sim L(\tau) \cdot \tau^{-\xi} \qquad \text{as } \tau \searrow 0, \tag{18.1}$$

$$F_U(u) \sim \bar{L}(u) \cdot u^{-1/\xi} \qquad \text{as } u \searrow Q_U(0), \tag{18.2}$$

for some $\xi \neq 0$, where $\tau \mapsto L(\tau)$ is a nonparametric slowly varying function at 0, and $u \mapsto \bar{L}(u)$ is a nonparametric slowly varying function at $Q_U(0)$.[2] The leading examples of slowly varying functions are the constant function and the logarithmic function. The number ξ as defined in (18.1) or (18.2) is called the *extreme value (EV) index* or the *tail index*.

The absolute value of ξ measures the heavy-tailedness of the distribution. The support of a Pareto-type tailed distribution necessarily has a finite lower bound if $\xi < 0$ and an infinite lower bound if $\xi > 0$. Distributions with $\xi > 0$ include stable, Pareto, and Student's t, among many others. For example, the t-distribution with ν degrees of freedom has $\xi = 1/\nu$ and exhibits a wide range of tail behaviors. In particular, setting $\nu = 1$ yields the Cauchy distribution which has heavy tails with $\xi = 1$, while setting $\nu = 30$ gives a distribution that has light tails with $\xi = 1/30$ and is very close to the normal distribution. On the other hand, distributions with $\xi < 0$ include the uniform, exponential, and Weibull, among many others.

---

[1]$a \sim b$ means that $a/b \to 1$ with an appropriate notion of limit.
[2]A function $z \mapsto f(z)$ is said to be *slowly varying* at z_0 if $\lim_{z \searrow z_0} f(z)/f(mz) = 1$ for every $m > 0$.

The assumption of Pareto-type tails can be equivalently cast in terms of a regular variation assumption, as is commonly done in EV theory. A distribution function $u \mapsto F_U(u)$ is said to be *regularly varying* at $u = Q_U(0)$ with index of regular variation $-1/\xi$ if

$$\lim_{y \searrow Q_U(0)} F_U(ym)/F_U(y) = m^{-1/\xi}, \quad \text{for every } m > 0.$$

This condition is equivalent to the regular variation of the quantile function $\tau \mapsto Q_U(\tau)$ at $\tau = 0$ with index $-\xi$,

$$\lim_{\tau \searrow 0} Q_U(\tau m)/Q_U(\tau) = m^{-\xi}, \quad \text{for every } m > 0.$$

The case of $\xi = 0$ corresponds to the class of rapidly varying distribution functions. Such distribution functions have exponentially light tails, with the normal and exponential distributions being the chief examples. For the sake of simplicity, we omit this case from our discussion. Note, however, that since the limit distribution of the main statistics is continuous in ξ, the inference theory for $\xi = 0$ is included by taking $\xi \to 0$.

18.2.2 Extremal quantile regression models

The most common model for the QR function is the linear-in-parameters specification

$$Q_Y(\tau|x) = B(x)'\beta(\tau), \quad \text{for all } \tau \in (0, \eta] \text{ and some } \eta \in (0,1), \tag{18.3}$$

and for every $x \in \mathbf{X}$, where $\mathbf{X}$ is the support of X. This linear functional form not only provides computational convenience but also has good approximation properties. Thus, the set $B(x)$ can include transformations of x such as polynomials, splines, indicators, or interactions such that $x \mapsto B(x)'\beta(\tau)$ is close to $x \mapsto Q_Y(\tau|x)$. In what follows, without loss of generality we lighten the notation by using x instead of $B(x)$. We also assume that the d_x-dimensional vector x contains a constant as the first element, has compact support $\mathbf{X}$, and satisfies the regularity conditions stated in Chernozhukov and Fernández-Val (2011, Assumption 3). Compactness is needed to ensure the continuity and robustness of the mapping from extreme events in Y to the extremal QR statistics. Even if $\mathbf{X}$ is not compact, we can select the data for which X belongs to a compact region.

The main additional assumption for extremal QR is that Y, transformed by some auxiliary regression line $X'\beta_e$, has Pareto-type tails. More precisely, together with (18.3), it assumes that there exists an auxiliary regression parameter $\beta_e \in \mathbb{R}^{d_x}$ such that the disturbance $V := Y - X'\beta_e$ has lower end-point $s = 0$ or $s = -\infty$ almost surely (a.s.), and its conditional quantile function $Q_V(\tau|x)$ satisfies the tail equivalence relationship

$$Q_V(\tau|x) \sim x'\gamma \cdot Q_U(\tau), \quad \text{as } \tau \searrow 0 \text{ uniformly in } x \in \mathbf{X} \subseteq \mathbb{R}^{d_x}, \tag{18.4}$$

for some quantile function $Q_U(\tau)$ that exhibits a Pareto-type tail (18.1) with EV index ξ, and some vector parameter γ such that $E[X]'\gamma = 1$ and $X'\gamma > 0$ a.s.

Condition (18.4) imposes a location–scale shift model. This model is more general than the standard location shift model that replaces $x'\gamma$ by a constant, because it permits the conditional heteroskedasticity that is common in economic applications. Moreover, condition (18.4) only affects the far tails, and therefore allows covariates to affect extremal and central quantiles very differently. Even at the tails, the local effect of the covariates is approximately given by $\beta(\tau) \approx \beta_e + \gamma Q_U(\tau)$, which can be heterogeneous across extremal quantiles.

Existence and Pareto-type behavior of the conditional quantile density function are also often imposed as a technical assumption that facilitates the derivation of inference results. Accordingly, we will assume that the conditional quantile density function $\partial Q_V(\tau|x)/\partial\tau$ exists and satisfies the tail equivalence relationship

$$\partial Q_V(\tau|x)/\partial\tau \sim x'\gamma \cdot \partial Q_U(\tau)/\partial\tau, \quad \text{as } \tau \searrow 0$$

uniformly in $x \in \mathbf{X}$, where $\partial Q_U(\tau)/\partial\tau$ exhibits Pareto-type tails as $\tau \searrow 0$ with EV index $\xi + 1$.

18.3   Estimation and inference methods

This section reviews estimation and inference methods for extremal quantile regression. Estimation is based on the Koenker and Bassett (1978) quantile regression estimator. We consider both analytical and resampling methods. These methods are introduced in the univariate case of marginal quantiles and then extended to the multivariate or regression case of conditional quantiles. We start by imposing some general sampling conditions.

18.3.1   Sampling conditions

We assume that we have a sample of (Y, X) of size T that is either independent and identically distributed (i.i.d.) or stationary and weakly dependent, with extreme events satisfying a nonclustering condition. In particular, the sequence $\{(Y_t, X_t)\}_{t=1}^T$ is assumed to form a stationary, strongly mixing process with a geometric mixing rate, which satisfies the condition that curbs clustering of extreme events (Chernozhukov and Fernández-Val, 2011). The assumption of mixing dependence is standard in econometrics (White, 2001). The nonclustering condition is of Meyer (1973) type and states that the probability of two extreme events co-occurring at nearby dates is much lower than the probability of just one extreme event. For example, it assumes that a large market crash is not likely to be immediately followed by another large crash. This assumption is convenient because it leads to limit distributions of extremal quantile regression estimators as if independent sampling had taken place. The plausibility of the nonclustering assumption is an empirical matter.

18.3.2   Univariave case: Marginal quantiles

The analog estimator of the marginal τ-quantile is the sample τ-quantile,

$$\hat{Q}_Y(\tau) = Y_{(\lfloor \tau T \rfloor)},$$

where $Y_{(s)}$ is the sth order statistic of $(Y_1, \ldots, Y_T)$, and $\lfloor z \rfloor$ denotes the integer part of z. The sample τ-quantile can also be computed as a solution to an optimization program,

$$\hat{Q}_Y(\tau) \in \mathrm{argmin}_{\beta \in \mathbb{R}} \sum_{t=1}^T \rho_\tau(Y_t - \beta),$$

where $\rho_\tau(u) := (\tau - 1\{u < 0\})u$ is the asymmetric absolute deviation function of Fox and Rubin (1964).

We review the asymptotic behavior of the sample quantiles under extreme and intermediate order sequences, and describe inference methods for extremal marginal quantiles.

18.3.2.1 Extreme order approximation

Recall the following classical result from EV theory on the limit distribution of extremal order statistics: as $T \to \infty$, for any integer $k \geq 1$ such that $k_T := \tau_T T \to k$,

$$\hat{Z}_T(k_T) = A_T(\hat{Q}_Y(\tau_T) - Q_Y(\tau_T)) \to_d \hat{Z}_\infty(k) = \Gamma_k^{-\xi} - k^{-\xi}, \tag{18.5}$$

where

$$A_T = 1/Q_U(1/T), \quad \Gamma_k = \mathcal{E}_1 + \cdots + \mathcal{E}_k. \tag{18.6}$$

The variable U is defined in Section 18.2.1 and $\{\mathcal{E}_1, \mathcal{E}_2, \ldots\}$ is an i.i.d. sequence of standard exponential variables. We call $\hat{Z}_T(k_T)$ the *canonically normalized quantile (CN-Q) statistic* because it depends on the canonical scaling constant A_T. The result (18.5) was obtained by Gnedenko (1943) for i.i.d. sequences of random variables. It continues to hold for stationary weakly dependent series, provided the probability of extreme events occurring in clusters is negligible relative to the probability of a single extreme event (Meyer, 1973). Leadbetter et al. (1983) extended the result to more general time series processes.

The limit in (18.5) provides an EV distribution as an approximation to the finite-sample distribution of $\hat{Q}_Y(\tau_T)$, given the scaling constant A_T. The limit distribution is characterized by the EV index ξ and the variables Γ_k. The index ξ is usually unknown but can be estimated by one of the methods described below. The variables Γ_k are *gamma* random variables. The limit variable $\hat{Z}_\infty(k)$ is therefore a transformation of gamma variables, which has finite mean if $\xi < 1$ and has finite moments of up to order $1/\xi$ if $\xi > 0$. Moreover, the limit EV distribution is not symmetric, predicting significant median asymptotic bias in $\hat{Q}_Y(\tau_T)$ with respect to $Q_Y(\tau_T)$ that motivates the use of the median bias correction techniques discussed below.

The classical result (18.5) is often not feasible for inference on $Q_Y(\tau_T)$ because the constant A_T is unknown and generally cannot be estimated consistently (Bertail et al., 2004). One way to deal with this problem is to add strong parametric assumptions on the nonparametric, slowly varying function $\tau \mapsto L(\tau)$ in equation (18.1) in order to estimate A_T consistently. This approach is discussed in Section 18.3.2.3. An alternative is to consider the *self-normalized quantile (SN-Q) statistic*

$$Z_T(k_T) := \mathcal{A}_T(\hat{Q}_Y(\tau_T) - Q_Y(\tau_T)), \quad \mathcal{A}_T := \frac{\sqrt{k_T}}{\hat{Q}_Y(m\tau_T) - \hat{Q}_Y(\tau_T)}, \tag{18.7}$$

for $m > 1$ such that mk is an integer. For example, $m = p/k_T + 1 = p/k + 1 + o(1)$ for some spacing parameter $p \geq 1$ (e.g. $p = 5$). The scaling factor $\mathcal{A}_T$ is completely a function of data and therefore feasible, avoiding the need for the consistent estimation of A_T. Chernozhukov and Fernández-Val (2011) showed that as $T \to \infty$, for any integer $k \geq 1$ such that $k_T := \tau_T T \to k$,

$$Z_T(k_T) \to_d Z_\infty(k) := \frac{\sqrt{k}(\Gamma_k^{-\xi} - k^{-\xi})}{\Gamma_{mk}^{-\xi} - \Gamma_k^{-\xi}}. \tag{18.8}$$

The limit distribution in (18.8) only depends on the EV index ξ, and its quantiles can be easily obtained by the resampling methods described in Section 18.3.2.5.

18.3.2.2 Intermediate order approximation

Dekkers and de Haan (1989) show that as $\tau \searrow 0$ and $k_T \to \infty$, and under further regularity conditions,

$$Z_T(k_T) := A_T(\hat{Q}_Y(\tau_T) - Q_Y(\tau_T)) \to_d \mathcal{N}\left(0, \frac{\xi^2}{(m^\xi - 1)^2}\right), \tag{18.9}$$

where $\mathcal{A}_T$ is defined as in (18.7). This result yields a normal approximation to the finite-sample distribution of $\widehat{Q}_Y(\tau)$. Note that this normal approximation holds only when $k_T \to \infty$, while the EV approximation (18.8) holds not only when $k_T \to k$ but also when $k_T \to \infty$ because the EV distribution converges to the normal distribution as $k \to \infty$. In finite samples, we may interpret the condition $k_T \to \infty$ as requiring that $k_T \geqslant 30$.

Figure 18.1, taken from Chernozhukov and Fernández-Val (2011), shows that the EV distribution provides a better approximation to the distribution of extremal sample quantiles than the normal distribution when $k_T < 30$. It plots the quantiles of these distributions against the quantiles of the finite-sample distribution of the sample τ-quantile for $T = 200$ and $\tau \in \{0.025, 0.2, 0.3\}$. If either the EV or the normal distributions were to coincide with the exact distribution, then their quantiles would fall on the 45-degree line shown by the solid line. When the order τT is 5 or 40, the quantiles of the EV distribution are very close to the 45-degree line, and in fact are much closer to this line than the quantiles of the normal distribution. Only for the case where the order τT becomes 60 do the quantiles of the EV and normal distributions become comparably close to the 45-degree line.

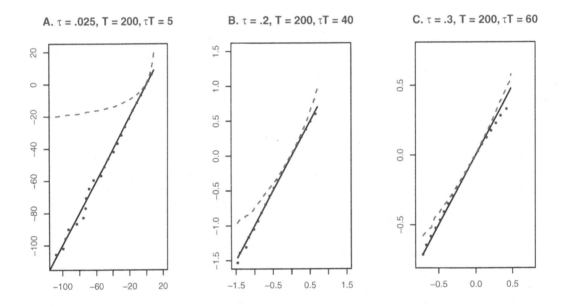

FIGURE 18.1
Quantiles of the distribution of the sample quantiles versus quantiles of EV and normal approximations. The figure is based on a design where Y follows a Cauchy distribution. The solid line shows the quantiles of the exact distribution of the sample τ-quantile with $\tau \in \{0.025, 0.2, 0.3\}$, obtained from 10,000 simulations. The dashed line shows the quantiles of the normal approximation. The dotted line shows the quantiles of the EV approximation.

18.3.2.3   Estimation of ξ

Some inference methods for extremal marginal quantiles require a consistent estimator of the EV index ξ. We describe two well-known estimators of ξ. The first, due to Pickands

(1975), relies on the ratio of sample quantile spacings:

$$\hat{\xi}_P := \frac{-1}{\ln 2} \ln \left(\frac{\hat{Q}_Y(4\tilde{\tau}_T) - \hat{Q}_Y(2\tilde{\tau}_T)}{\hat{Q}_Y(2\tilde{\tau}_T) - \hat{Q}_Y(\tilde{\tau}_T)} \right). \tag{18.10}$$

Under further regularity conditions, for $\tilde{\tau}_T \searrow 0$ and $\tilde{\tau}_T T \to \infty$ as $T \to \infty$, we have

$$\sqrt{\tilde{\tau}_T T}(\hat{\xi}_P - \xi) \to_d \mathcal{N}\left(0, \frac{\xi^2(2^{2\xi+1}+1)}{[2(2^\xi-1)\ln 2]^2}\right)$$

The second estimator, developed by Hill (1975), is the moment estimator,

$$\hat{\xi}_H := -\frac{\sum_{t=1}^{T} 1\{Y_t < \hat{Q}_Y(\tilde{\tau}_T)\} \ln(Y_t/\hat{Q}_Y(\tilde{\tau}_T))}{\tilde{\tau}_T \sum_{t=1}^{T} 1\{Y_t < \hat{Q}_Y(\tilde{\tau}_T)\}}, \tag{18.11}$$

which is applicable when $\xi > 0$ and $\hat{Q}_Y(\tilde{\tau}_T) < 0$. This estimator is motivated by the maximum likelihood method that fits an exact power law to the tail data. Under further regularity conditions, for $\tilde{\tau}_T \searrow 0$ and $\tilde{\tau}_T T \to \infty$ as $T \to \infty$,

$$\sqrt{\tilde{\tau}_T T}(\hat{\xi}_H - \xi) \to_d \mathcal{N}(0, \xi^2).$$

The previous limit results can be used to construct the confidence intervals and median bias corrections for ξ. We give an example of these confidence intervals and corrections in Section 18.4.

Embrechts et al. (1997) provide methods for choosing $\tilde{\tau}_T$. There is a variance–bias trade-off, as the variance of the estimator decreases, but the bias increases, as $\tilde{\tau}_T$ increases. Another view on the choice of $\tilde{\tau}_T$ is that the statistical models are approximations, but not literal descriptions of the data. In practice, the dependence of $\hat{\xi}$ on the threshold $\tilde{\tau}_T$ reflects that power laws with different values of ξ would fit some tail regions better than others. Therefore, if the interest lies in making the inference on $Q_Y(\tau_T)$ for a particular τ_T, it seems reasonable to use $\hat{\xi}$ constructed using $\tilde{\tau}_T = \tau_T$ or the closest $\tilde{\tau}_T$ to τ_T subject to $\tilde{\tau}_T T \geqslant 30$. This condition ensures a sufficient number of observations to estimate ξ.

18.3.2.4 Estimation of A_T

To use the CN-Q statistic for inference, we need to estimate the scaling constant A_T defined in (18.6). This requires additional strong restrictions on the underlying model. For instance, assume that the slowly varying function $\tau \mapsto L(\tau)$ is just a constant L, that is, as $\tau \searrow 0$,

$$1/Q_U(\tau) = L \cdot \tau^\xi \cdot (1 + \delta(\tau)), \quad \text{for some } L \in \mathbb{R} \text{ and } \delta(\tau) \to 0. \tag{18.12}$$

Then, we can estimate L by

$$\hat{L} := \frac{\hat{Q}_Y(2\tilde{\tau}_T) - \hat{Q}_Y(\tilde{\tau}_T)}{(2^{-\hat{\xi}} - 1) \cdot \tilde{\tau}_T^{-\hat{\xi}}}, \tag{18.13}$$

where $\hat{\xi}$ is either the Pickands (18.10) or Hill (18.11) estimator, and $\tilde{\tau}_T$ can be chosen using the same methods as in the estimation of ξ. Then the estimator of A_T is

$$\hat{A}_T := \hat{L} \cdot T^{-\hat{\xi}}. \tag{18.14}$$

18.3.2.5 Computing quantiles of the limit extreme value distributions

The inference and bias corrections for extremal quantiles are based on the EV approxima-tions given in (18.5) and (18.8), with an estimator in place of the EV index if needed. In practice, it is convenient to compute the quantiles of the EV distributions using simulation or resampling methods, instead of an analytical method. Here we illustrate two such meth-ods: the extremal bootstrap and extremal subsampling. The bootstrap method relies on simulation, whereas the subsampling method relies on drawing subsamples from the orig-inal sample. Subsampling has the advantages that it does not require the estimation of ξ and is consistent under general conditions (e.g. subsampling does not require i.i.d. data). Nevertheless, the bootstrap is more accurate than subsampling when a stronger set of as-sumptions holds. It should be noted here that the empirical or nonparametric bootstrap is not consistent for extremal quantiles (Bickel and Freedman, 1981).

The extremal bootstrap is based on simulating samples from the distribution with the same tail behavior as Y. Consider the random variable,[3]

$$Y^* = \frac{\mathcal{E}^{-\xi} - 1}{-\xi}, \quad \mathcal{E} \sim \text{Exponential}(1). \tag{18.15}$$

This variable has quantile function

$$Q_{Y*}(\tau) = \frac{[-\ln(1 - \tau)]^{-\xi} - 1}{-\xi}, \tag{18.16}$$

which satisfies condition (18.2) because $Q_{Y*}(\tau) - 1/\xi \sim \tau^{-\xi}/\xi$. The extremal bootstrap estimates the distribution of $Z_T(k_T) = \mathcal{A}_T(\hat{Q}_Y(\tau_T) - Q_Y(\tau_T))$ by the distribution of $Z_T^*(k_T) = \mathcal{A}_T(\hat{Q}_{Y*}(\tau_T) - Q_{Y*}(\tau_T))$ obtained by simulation. This approximation reproduces both the EV limit (18.8) under extreme order sequences and the normal limit (18.9) under intermediate order sequences. Algorithm 1 describes the implementation of this method.

Algorithm 1 (Extremal bootstrap) *(1) Choose the quantile index of interest τ_T and the number of simulations S (e.g., $S = 500$). (2) For each $s \in \{1, \ldots, S\}$, draw an i.i.d. sequence $\{Y_{1,s}^*, \ldots, Y_{T,s}^*\}$ from the random variable Y^* defined in (18.15), replacing ξ by the estimator (18.10) or (18.11). (3) For each $s \in \{1, \ldots, S\}$, compute the statistic $Z_{T,s}^*(k_T) = \mathcal{A}_T(\hat{Q}_{Y*}(\tau_T) - Q_{Y*}(\tau_T))$, where $\hat{Q}_{Y*}(\tau_T)$ is the sample τ_T-quantile in the bootstrap sample $\{Y_{1,s}^*, \ldots, Y_{T,s}^*\}$ and $Q_{Y*}(\tau_T)$ is defined as in (18.16), replacing ξ by the same estimator as in step (2). (4) Estimate the quantiles of $Z_T(k_T) = \mathcal{A}_T(\hat{Q}_Y(\tau_T) - Q_Y(\tau_T))$ by the sample quantiles of $\{Z_{T,s}^*(k_T)\}_{s=1}^S$.*

The extremal bootstrap can be also applied to estimate the distribution of other statis-tics, including the estimators of the EV index in (18.10) or (18.11) and the extrapolation estimators of Section 18.3.2.7.

We now describe an extremal subsampling method to estimate the distributions of $Z_T(k_T)$ and $\hat{Z}_T(k_T)$ developed by Chernozhukov and Fernández-Val (2011). It is based on drawing subsamples of size $b < T$ from $(Y_1, \ldots, Y_T)$ such that $b \to \infty$ and $b/T \to 0$ as $T \to \infty$, and computing the subsampling version of the SN statistic $Z_T(k_T)$ as

$$Z_{b,T}^*(k_T) := \mathcal{A}_{b,T}(\hat{Q}_Y^b(\tau_b) - \hat{Q}_Y(\tau_b)), \quad \mathcal{A}_{b,T} := \frac{\sqrt{\tau_b b}}{\hat{Q}_Y^b(m\tau_b) - \hat{Q}_Y^b(\tau_b)}, \tag{18.17}$$

---

[3]The variable Y^* follows the generalized extreme value distribution, which nests the Frechet, Weibull, and Gumbell distributions. There are other possibilities, for example, the standard exponential distribution can be replaced by the standard uniform distribution, in which case Y follows the generalized Pareto distribution.

where $\widehat{Q}_Y^b(\tau)$ is the sample τ-quantile in the subsample of size b, and $\tau_b := (\tau_T T)/b$. Similarly, the subsampling version of the CN-Q statistic $\widehat{Z}_T(k_T)$ is

$$\widehat{Z}_{b,T}^*(k_T) := \widehat{A}_b(\widehat{Q}_Y^b(\tau_b) - \widehat{Q}_Y(\tau_b)), \qquad (18.18)$$

where $\widehat{A}_b$ is a consistent estimator of A_b. For example, under the parametric restrictions specified in (18.12), we can set $\widehat{A}_b = \widehat{L}b^{-\widehat{\xi}}$ for $\widehat{L}$ as in (18.13) and $\widehat{\xi}$ one of the estimators in (18.10) or (18.11).

The distributions of $Z_{b,T}^*(k_T)$ and $\widehat{Z}_{b,T}^*(k_T)$ over all the subsamples estimate the distributions of $Z_T(k_T)$ and $\widehat{Z}_T(k_T)$, respectively. The number of possible subsamples depends on the structure of serial dependence in the data, but it can be very large. In practice, the distributions over all subsamples are approximated by the distributions over a smaller number S of randomly chosen subsamples, such that $S \to \infty$ as $T \to \infty$ (Politis et al., 1999, Chapter 2.5). Politis et al. (1999) and Bertail et al. (2004) provide methods for choosing the subsample size b. Algorithm 2 describes the implementation of extremal subsampling.

Algorithm 2 (Extremal subsampling) *(1) Choose the quantile index of interest τ_T, the subsample size b (e.g., $b = \lfloor 50 + \sqrt{T} \rfloor$), and the number of subsamples S (e.g., $S = 500$).[4] (2) If the data have serial dependence, draw S subsamples from $(Y_1, \ldots, Y_T)$ of size b of the form $(Y_{1,s}^*, \ldots, Y_{b,s}^*) = (Y_k, \ldots, Y_{k+b+1})$ where $k \in \{1, \ldots, T - b + 1\}$. If the data are independent, $(Y_{1,s}^*, \ldots, Y_{b,s}^*)$ can be drawn as a random subsample of size b from $(Y_1, \ldots, Y_T)$ without replacement. (3) For each $s \in \{1, \ldots, S\}$, compute $Z_{b,s}^*(k_T)$ or $\widehat{Z}_{b,s}^*(k_T)$, applying (18.17) or (18.18) to the subsample $(Y_{1,s}^*, \ldots, Y_{b,s}^*)$. (4) Estimate the quantiles of $Z_T(k_T)$ or $\widehat{Z}_T(k_T)$ by the sample quantiles of $\{Z_{b,s}^*(k_T)\}_{s=1}^S$ or $\{\widehat{Z}_{b,s}^*(k_T)\}_{s=1}^S$.*

Extremal subsampling differs from conventional subsampling, which is inconsistent for extremal quantiles. This difference appears more sharply in the case of $\widehat{Z}_T(k_T)$. Here, conventional subsampling would recenter the subsampling version of the statistic by the estimator in the full sample $\widehat{Q}_Y(\tau_T)$. Recentering in this way requires $A_b/A_T \to 0$ for consistency (see Politis et al., 1999, Theorem 2.2.1), but $A_b/A_T \to \infty$ when $\xi > 0$. Thus, when $\xi > 0$ the extremal sample quantiles $\widehat{Q}_Y(\tau_T)$ diverge, rendering conventional subsampling inconsistent. In contrast, extremal subsampling uses $\widehat{Q}_Y(\tau_b)$ for recentering. This sample quantile may diverge, but because it is an intermediate order quantile if $b/T \to 0$, the speed of its divergence is strictly slower than that of A_T. Hence, extremal subsampling exploits the special structure of the order statistics to do the recentering.

18.3.2.6 Median bias correction and confidence intervals

Chernozhukov and Fernández-Val (2011) construct asymptotically median-unbiased estimators and $(1 - \alpha)$-confidence intervals (CIs) for $Q_Y(\tau_T)$ based on the SN-Q statistic as

$$\widehat{Q}_Y(\tau) - \frac{\widehat{c}_{1/2}}{\mathcal{A}_T} \quad \text{and} \quad \left[\widehat{Q}_Y(\tau) - \frac{\widehat{c}_{1-\alpha/2}}{\mathcal{A}_T}, \ \widehat{Q}_Y(\tau) - \frac{\widehat{c}_{\alpha/2}}{\mathcal{A}_T}\right],$$

where $\widehat{c}_p$ is a consistent estimator of the p-quantile of $Z_T(k_T)$ that can be obtained using

---

[4]Chernozhukov and Fernández-Val (2005) suggest the choice $b = \lfloor m + T^{1/c} \rfloor$ with $c \geq 2$ and $m > 0$, to guarantee that the minimal subsample size is m.

Algorithm 1 or 2. Chernozhukov and Fernández-Val (2011) also construct asymptotically median-unbiased estimators and $(1 - \alpha)$-CIs for $Q_Y(\tau_T)$ based on the CN-Q statistic as

$$\hat{Q}_Y(\tau) - \frac{\tilde{c}_{1/2}}{\hat{A}_T} \quad \text{and} \quad \left[\hat{Q}_Y(\tau) - \frac{\tilde{c}_{1-\alpha/2}}{\hat{A}_T}, \; \hat{Q}_Y(\tau) - \frac{\tilde{c}_{\alpha/2}}{\hat{A}_T} \right],$$

where $\tilde{c}_p$ is a consistent estimator of the p-quantile of $\hat{Z}_T(k_T)$ that can be obtained using Algorithm 2, and $\hat{A}_T$ is a consistent estimator of A_T such as (18.14).

18.3.2.7   Extrapolation estimator for very extremes

Sample τ-quantiles can be very inaccurate estimators of marginal τ-quantiles when τT is very small, say $\tau T < 1$. For such very extremal cases we can construct more precise estimators using the assumptions on the behavior of the tails. In particular, we can estimate less extreme quantiles reliably, and extrapolate them to the quantile of interest using the tail assumptions.

Dekkers and de Haan (1989) developed the extrapolation estimator

$$\tilde{Q}_Y(\tau_T) = \hat{Q}_Y(\tilde{\tau}_T) + \frac{(\tau_T/\tilde{\tau}_T)^{-\hat{\xi}} - 1}{2^{-\hat{\xi}} - 1} \left[\hat{Q}_Y(2\tilde{\tau}_T) - \hat{Q}_Y(\tilde{\tau}_T) \right], \tag{18.19}$$

where $\tau_T \ll \tilde{\tau}_T$ and $\hat{\xi}$ is a consistent estimator of ξ such as (18.10) or (18.11). Then, for $\tilde{\tau}_T T \to \tilde{k}$ and $\tau_T T \to k$ with $\tilde{k} > k$,

$$\frac{\tilde{Q}_Y(\tau_T) - Q_Y(\tau_T)}{\hat{Q}_Y(\tilde{\tau}_T) - \hat{Q}_Y(2\tilde{\tau}_T)} \to_d \frac{(\tilde{k}/k)\xi - 2^{-\xi}}{1 - 2^{-\xi}} + \frac{1 - (\Gamma_{\tilde{k}}/k)\xi}{e^{\xi \mathcal{E}_{\tilde{k}}} - 1},$$

where $\mathcal{E}_{\tilde{k}}$ and $\Gamma_{\tilde{k}}$ are independent, $\Gamma_{\tilde{k}}$ has a standard gamma distribution with shape parameter $(2\tilde{k} + 1)$, and $\mathcal{E}_{\tilde{k}} \sim \sum_{j=\tilde{k}+1}^{2\tilde{k}} Z_j/j$ with $Z_1, Z_2, \dots$ i.i.d. standard exponential. He et al. (2016) proposed the closely related estimator

$$\breve{Q}_Y(\tau_T) = \hat{Q}_Y(\tilde{\tau}_T) + \frac{(\tau_T/\tilde{\tau}_T)^{-\hat{\xi}} - 1}{2^{\hat{\xi}} - 1} \left[\hat{Q}_Y(\tilde{\tau}_T/2) - \hat{Q}_Y(\tilde{\tau}_T) \right]. \tag{18.20}$$

Under some regularity conditions, they show that for $\tau_T/\tilde{\tau}_T \to 0$ as $T \to \infty$, this estimator converges to a normal distribution jointly with the EV index estimator $\hat{\xi}$.

The estimators in (18.19) and (18.20) have good properties provided that the quantities on the right-hand side are well estimated, which in turn requires that $\tilde{\tau}_T T$ be large, and that the Pareto-type tail model be a good approximation.

18.3.3   Multivariate case: Conditional quantiles

The τ-QR estimator of the conditional τ-quantile $Q_Y(\tau|x) = x'\beta(\tau)$ is

$$\hat{Q}_Y(\tau|x) := x'\hat{\beta}(\tau), \quad \hat{\beta}(\tau) \in \operatorname{argmin}_{\beta \in \mathbb{R}^d} \sum_{t=1}^{T} \rho_\tau(Y_t - X_t'\beta). \tag{18.21}$$

This estimator was introduced by Laplace in 1818 (see Laplace, 1995) for the median case, and extended by Koenker and Bassett (1978) to include other quantiles and regressors.

In this subsection, we review the asymptotic behavior of the QR estimator under extreme and intermediate order sequences, and describe inference methods for extremal quantile regression. The analysis for the multivariate case parallels the analysis for the univariate case in Section 18.3.2.

18.3.3.1 Extreme order approximation

Consider the *canonically normalized quantile regression (CN-QR) statistic*

$$\widehat{Z}_T(k_T) := A_T(\widehat{\beta}(\tau_T) - \beta(\tau_T)) \quad \text{for } A_T := 1/Q_U(1/T), \tag{18.22}$$

and the *self-normalized quantile regression (SN-QR) statistic*

$$Z_T(k_T) := \mathcal{A}_T(\widehat{\beta}(\tau_T) - \beta(\tau_T)), \quad \text{for } \mathcal{A}_T := \frac{\sqrt{k_T}}{\bar{X}_T'(\widehat{\beta}(m\tau_T) - \widehat{\beta}(\tau_T))}, \tag{18.23}$$

where $\bar{X}_T := T^{-1} \sum_{t=1}^{T} X_t$ and m is a real number such that $k(m-1) > d_x$ for $k_T = \tau_T T \to k$. For example, $m = (d_x + p)/k_T + 1 = (d_x + p)/k + 1 + o(1)$ where $p \geqslant 1$ is a spacing parameter (e.g. $p = 5$). The CN-QR statistic is generally infeasible for inference because it depends on the unknown canonical normalization constant A_T. This constant can only be estimated consistently under strong parametric assumptions, which will be discussed in Section 18.3.3.4. The SN-QR statistic is always feasible because it uses a normalization that only depends on the data.

Chernozhukov and Fernández-Val (2011) show that for $k_T \to k > 0$ as $T \to \infty$,

$$\widehat{Z}_T(k_T) \to_d \widehat{Z}_\infty(k), \tag{18.24}$$

where for $\chi = 1$ if $\xi < 0$ and $\chi = -1$ if $\xi > 0$,

$$\widehat{Z}_\infty(k) := \chi \cdot \mathrm{argmin}_{z \in \mathbb{R}^{d_x}} \left[-kE[X]'z + \sum_{t=1}^{\infty} \{\mathcal{X}_t'z - \chi(\Gamma_t^{-\xi} - k^{-\xi})\mathcal{X}_t'\gamma\}_+ \right], \tag{18.25}$$

in which $\{\mathcal{X}_1, \mathcal{X}_2, \dots\}$ is an i.i.d. sequence with distribution F_X; $\{\Gamma_1, \Gamma_2, \dots\} := \{\mathcal{E}_1, \mathcal{E}_1 + \mathcal{E}_2, \dots\}$; $\{\mathcal{E}_1, \mathcal{E}_2, \dots\}$ is an i.i.d. sequence of standard exponential variables that is independent of $\{\mathcal{X}_1, \mathcal{X}_2, \dots\}$; and $\{y\}_+ := \max(0, y)$. Furthermore,

$$Z_T(k_T) \to_d Z_\infty(k) := \frac{\sqrt{k}\widehat{Z}_\infty(k)}{E[X]'(\widehat{Z}_\infty(mk) - \widehat{Z}_\infty(k)) + \chi \cdot (m^{-\xi} - 1)k^{-\xi}}. \tag{18.26}$$

The limit EV distributions are more complicated than in the univariate case, but they share some common features. First, they depend crucially on the gamma variables Γ_t, are not necessarily centered at zero, and can have a significant first-order asymptotic median bias. Second, as mentioned above, the limit distribution of the CN-QR statistic in equation (18.25) is generally infeasible for inference due to the difficulty in consistently estimating the scaling constant A_T.

Remark 6 (Very extreme order quantiles) *Feigin and Resnick (1994), Smith (1994), Chernozhukov (1998), Portnoy and Jurečková (1999), and Knight (2001) derived related results for canonically normalized linear programming or frontier regression estimators under very extreme order sequences where $\tau_T T \searrow 0$ as $T \to \infty$.*

18.3.3.2 Intermediate order approximation

Chernozhukov (2005) shows that for $\tau_T \searrow 0$ and $k_T \to \infty$ as $T \to \infty$,

$$Z_T(k_T) = \mathcal{A}_T(\widehat{\beta}(\tau_T) - \beta(\tau_T)) \to_d \mathcal{N}\left(0, E[XX']^{-1}\frac{\xi^2}{(m^{-\xi} - 1)^2}\right), \tag{18.27}$$

where $\mathcal{A}_T$ is defined as in (18.23). As in the univariate case, this normal approximation provides a less accurate approximation to the distribution of the extremal quantile regression than the EV approximation when $k_T \nrightarrow \infty$. The condition $k_T \to \infty$ can be interpreted in finite samples as requiring that $k_T/d_x \geq 30$, where k_T/d_x is a dimension-adjusted order of the quantile explained in Section 18.3.4.

18.3.3.3 Estimation of ξ and γ

Some inference methods for extremal quantile regression require consistent estimators of the EV index ξ and the scale parameter γ. The regression analog of the Pickands estimator is

$$\widehat{\xi}_P := \frac{-1}{\ln 2} \ln \left(\frac{\bar{X}_T'(\widehat{\beta}(4\tilde{\tau}_T) - \widehat{\beta}(2\tilde{\tau}_T))}{\bar{X}_T'(\widehat{\beta}(2\tilde{\tau}_T) - \widehat{\beta}(\tilde{\tau}_T))} \right). \tag{18.28}$$

This estimator is consistent if $\tilde{\tau}_T T \to \infty$ and $\tilde{\tau}_T \searrow 0$ as $T \to \infty$. Under additional regularity conditions, for $\tilde{\tau}_T \searrow 0$ and $\tilde{\tau}_T T \to \infty$ as $T \to \infty$,

$$\sqrt{\tilde{\tau}_T T}(\widehat{\xi}_P - \xi) \to_d \mathcal{N}\left(0, \frac{\xi^2(2^{2\xi+1} + 1)}{[2(2^\xi - 1)\ln 2]^2}\right). \tag{18.29}$$

The regression analog of the Hill estimator is

$$\widehat{\xi}_H := -\frac{\sum_{t=1}^T 1\{Y_t < X_t'\widehat{\beta}(\tilde{\tau}_T)\} \ln(Y_t/X_t'\widehat{\beta}(\tilde{\tau}_T))}{\tilde{\tau}_T \sum_{t=1}^T 1\{Y_t < X_t'\widehat{\beta}(\tilde{\tau}_T)\}}, \tag{18.30}$$

which is applicable when $\xi > 0$ and $X_t\widehat{\beta}(\tilde{\tau}_T) < 0$. Under further regularity conditions, for $\tilde{\tau}_T \searrow 0$ and $\tilde{\tau}_T T \to \infty$,

$$\sqrt{\tilde{\tau}_T T}(\widehat{\xi}_H - \xi) \to_d \mathcal{N}(0, \xi^2). \tag{18.31}$$

These limit results can be used to construct CIs for ξ. The scale parameter γ can be estimated by

$$\widehat{\gamma} = \frac{\widehat{\beta}(2\tilde{\tau}_T) - \widehat{\beta}(\tilde{\tau}_T)}{\bar{X}_T'(\widehat{\beta}(2\tilde{\tau}_T) - \widehat{\beta}(\tilde{\tau}_T))}, \tag{18.32}$$

which is consistent if $\tilde{\tau}_T T \to \infty$ and $\tilde{\tau}_T \searrow 0$ as $T \to \infty$.

The choice of $\tilde{\tau}_T$ is similar to the univariate case in Section 18.3.2.3. This time, however, one needs to take into account the multivariate nature of the problem. For example, if the interest lies in making the inference on $\beta(\tau_T)$ for a particular τ_T, it is reasonable to set $\tilde{\tau}_T$ equal to the closest value to τ_T such that $\tilde{\tau}_T T/d_x \geq 30$. We again refer the reader to Section 18.3.4 for a discussion on the difference in the choice of $\tilde{\tau}_T$ between the univariate and multivariate cases.

18.3.3.4 Estimation of A_T

To use the CN-QR statistic for inference, we need to estimate the scaling constant A_T defined in (18.22). This requires strong restrictions and an additional estimation procedure. For example, assume that the nonparametric slowly varying component $L(\tau)$ of A_T is replaced by a constant L, that is, as $\tau \searrow 0$,

$$1/Q_U(\tau) = L \cdot \tau^\xi \cdot (1 + \delta(\tau)), \quad \text{for some } L \in \mathbb{R} \text{ and } \delta(\tau) \to 0. \tag{18.33}$$

Then we can estimate the constant L by

$$\widehat{L} := \frac{\bar{X}'_T(\widehat{\beta}(2\tilde{\tau}_T) - \widehat{\beta}(\tilde{\tau}_T))}{(2^{-\widehat{\xi}} - 1) \cdot \tilde{\tau}_T^{-\widehat{\xi}}}, \tag{18.34}$$

where $\widehat{\xi}$ is either the Pickands (18.28) or Hill (18.30) estimator. Thus, the scaling constant A_T is estimated by

$$\widehat{A}_T := \widehat{L} T^{-\widehat{\xi}}.$$

18.3.3.5 Computing quantiles of the limit extreme value distributions

We consider inference and asymptotically median-unbiased estimation for linear functions of the coefficient vector $\beta(\tau)$, $\psi'\beta(\tau)$, for some nonzero vector $\psi \in \mathbb{R}^{d_x}$, based on the EV approximations $\psi'\widehat{Z}_\infty(k)$ and $\psi'Z_\infty(k)$ from (18.26). We describe three methods to compute critical values of the limit EV distributions: analytical computation, extremal bootstrap, and extremal subsampling. The analytical and bootstrap methods require estimation of the EV index ξ and the scale parameter γ. Subsampling applies under more general conditions than the other methods, and hence we would recommend its use. However, the analytical and bootstrap methods can be more accurate than subsampling if the data satisfy a stronger set of assumptions.

The analytical computation method is based directly on the limit distributions (18.24) and (18.26), replacing ξ and γ by consistent estimators. Define the d_x-dimensional random vector

$$\widehat{Z}^*_\infty(k) = \widehat{\chi} \cdot \text{argmin}_{z \in \mathbb{R}^{d_x}} \left[-k\bar{X}'_T z + \sum_{t=1}^{\infty} \{\mathcal{X}'_t z - \widehat{\chi}(\Gamma_t^{-\widehat{\xi}} - k^{-\widehat{\xi}})\mathcal{X}'_t\widehat{\gamma}\}_+ \right], \tag{18.35}$$

where $\widehat{\chi} = 1$ if $\widehat{\xi} < 0$ and $\widehat{\chi} = -1$ if $\widehat{\xi} > 0$, $\widehat{\xi}$ is an estimator of ξ such as (18.28) or (18.30), $\widehat{\gamma}$ is an estimator of γ such as (18.32), $\{\Gamma_1, \Gamma_2, \dots\} = \{\mathcal{E}_1, \mathcal{E}_1 + \mathcal{E}_2, \dots\}$, $\{\mathcal{E}_1, \mathcal{E}_2, \dots\}$ is an i.i.d. sequence of standard exponential variables, and $\{\mathcal{X}_1, \mathcal{X}_2, \dots\}$ is an i.i.d. sequence independent of $\{\mathcal{E}_1, \mathcal{E}_2, \dots\}$ with distribution function $\widehat{F}_X$, where $\widehat{F}_X$ is any smooth consistent estimator of F_X, for example, a smoothed empirical distribution function of the sample $(X_1, \dots, X_T)$. Also, let

$$Z^*_\infty(k) = \frac{\sqrt{k}\widehat{Z}^*_\infty(k)}{\bar{X}'_T(\widehat{Z}^*_\infty(mk) - \widehat{Z}^*_\infty(k)) + \widehat{\chi}(m^{-\widehat{\xi}} - 1)k^{-\widehat{\xi}}}.$$

The quantiles of $\psi'\widehat{Z}_\infty(k)$ and $\psi'Z_\infty(k)$ are estimated by the corresponding quantiles of $\psi'\widehat{Z}^*_\infty(k)$ and $\psi'Z^*_\infty(k)$, respectively. In practice, these quantiles can only be evaluated numerically via the following algorithm.

Algorithm 3 (QR analytical computation) *(1) Choose the quantile index of interest τ_T and the number of simulations S (e.g., $S = 200$). (2) For each $s \in \{1, \dots, S\}$, draw an i.i.d. sequence $\{\widehat{Z}^*_{\infty,s}(k), \dots, \widehat{Z}^*_{\infty,s}(k)\}$ from the random vector $Z^*_\infty(k)$ defined in (18.35) with $k = \tau_T T$ and the infinite summation truncated at some finite value M (e.g. $M = T$). (3) For each $s \in \{1, \dots, S\}$, compute $Z^*_{\infty,s}(k) = \sqrt{k}\widehat{Z}^*_{\infty,s}(k)/(\bar{X}'_T(\widehat{Z}^*_{\infty,s}(mk) - \widehat{Z}^*_{\infty,s}(k)) + \widehat{\chi}(m^{-\widehat{\xi}} - 1)k^{-\widehat{\xi}})$. (4) Estimate the quantiles of $\psi'\widehat{Z}_T(k_T)$ and $\psi'Z_T(k_T)$ by the sample quantiles of $\{\psi'\widehat{Z}^*_{T,s}(k_T)\}_{s=1}^S$ and $\{\psi'Z^*_{T,s}(k_T)\}_{s=1}^S$.*

The extremal bootstrap is computationally less demanding than the analytical methods. It is based on simulating samples from a random variable with the same tail behavior as $(Y_1, \ldots, Y_T)$. Consider the bootstrap sample $\{(Y_1^*, X_1), \ldots, (Y_T, {}^* X_T)\}$, where

$$Y_t^* = \frac{\mathcal{E}_t^{-\xi} - 1}{-\xi} X_t' \gamma, \quad \mathcal{E}_t \sim \text{i.i.d. Exponential}(1), \tag{18.36}$$

and $\{X_1, \ldots, X_T\}$ is a fixed set of observed regressors from the data. The variable Y_t^* has conditional quantile function

$$Q_{Y_t^*}(\tau|x) = x' \beta^*(\tau), \quad \beta^*(\tau) = \frac{[-\ln(1-\tau)]^{-\xi} - 1}{-\xi} \gamma. \tag{18.37}$$

The extremal bootstrap approximates the distribution of $Z_T(k_T) = \mathcal{A}_T(\hat{\beta}(\tau) - \beta(\tau))$ by the distribution of $Z_T^*(k_T) = \mathcal{A}_T(\hat{\beta}^*(\tau) - \beta^*(\tau))$, where $\hat{\beta}^*(\tau)$ is the τ-QR estimator in the bootstrap sample. This approximation reproduces both the EV limit (18.26) under extreme value sequences, and the normal limit (18.27) under intermediate order sequences. The distribution of $Z_T^*(k_T)$ can be obtained by simulation using the following algorithm:

Algorithm 4 (QR extremal bootstrap) *(1) Choose the quantile index of interest τ_T and the number of simulations S (e.g., $S = 500$). (2) For each $s \in \{1, \ldots, S\}$, draw a bootstrap sample $\{(Y_{1,s}^*, X_1), \ldots, (Y_{T,s}^*, X_T)\}$ from the random vector $\{(Y_1^*, X_1), \ldots, (Y_T^*, X_T)\}$ defined in (18.36), replacing ξ by the estimator (18.28) or (18.30) and γ by the estimator (18.32). (3) For each $s \in \{1, \ldots, S\}$, compute the statistic $Z_{T,s}^*(k_T) = \mathcal{A}_T(\hat{\beta}_s^*(\tau_T) - \beta^*(\tau_T))$, where $\hat{\beta}_s^*(\tau_T)$ is the τ_T-QR in the bootstrap sample $\{(Y_{1,s}^*, X_1), \ldots, (Y_{T,s}^*, X_T)\}$ and $\beta^*(\tau_T)$ is defined as in (18.37), replacing ξ and γ by the same estimators as in step (2). (4) Estimate the quantiles of $\psi' Z_T(k_T)$ by the sample quantiles of $\{\psi' Z_{T,s}^*(k_T)\}_{s=1}^{S}$.*

Chernozhukov and Fernández-Val (2011) developed an extremal subsampling method to estimate the distributions of $\hat{Z}_T(k_T)$ and $Z_T(k_T)$. It is based on drawing subsamples of size $b < T$ from $\{(X_t, Y_t)\}_{t=1}^{T}$ such that $b \to \infty$ and $b/T \to 0$ as $T \to \infty$, and computing the subsampling version of the SN-QR statistic as

$$Z_{b,T}^*(k_T) := \mathcal{A}_{b,T}(\hat{\beta}_b(\tau_b) - \hat{\beta}(\tau_b)), \quad \mathcal{A}_{b,T} := \frac{\sqrt{\tau_b b}}{\bar{X}_{b,T}'[\hat{\beta}_b(m\tau_b) - \hat{\beta}_b(\tau_b)]}, \tag{18.38}$$

where $m = (d_x + p)/(\tau_T T)$ for some *spacing parameter* $p \geqslant 1$ (e.g. $p = 5$), $\hat{\beta}_b(\tau)$ is the τ-QR estimator in the subsample of size b, $\bar{X}_{b,T}$ is the sample mean of the regressors in the subsample, and $\tau_b := (\tau_T T)/b$.[5] Similarly, the subsampling version of the CN-QR statistic $\hat{Z}_T(k_T)$ is

$$\hat{Z}_{b,T}^*(k_T) := \hat{A}_b(\hat{\beta}_b(\tau_b) - \hat{\beta}(\tau_b)), \tag{18.39}$$

where $\hat{A}_b$ is a consistent estimator for A_b. For example, $\hat{A}_b = \hat{L} b^{-\hat{\xi}}$, for $\hat{L}$ given by (18.34), and $\hat{\xi}$ is the estimator of ξ given in (18.28) or (18.30).

As in the univariate case, the distributions of $Z_{b,T}^*(k_T)$ and $\hat{Z}_{b,T}^*(k_T)$ over all the possible

---

[5]In practice, it is reasonable to use the following finite-sample adjustment to τ_b: $\tau_b = \min\{(\tau_T T)/b, 0.2\}$ if $\tau_T < 0.2$, and $\tau_b = \tau_T$ if $\tau_T \geqslant 0.2$. The idea is that τ_T is adjusted to be nonextremal if $\tau_T > 0.2$, and the subsampling procedure reverts to central order inference. The truncation of τ_b by 0.2 is a finite-sample adjustment that restricts the key statistics $Z_{b,T}^*(k_T)$ to be extremal in subsamples. These finite-sample adjustments do not affect the asymptotic arguments.

subsamples estimate the distributions of $Z_T(k_T)$ and $\hat{Z}_T(k_T)$, respectively. These distributions can be obtained by simulation using the following algorithm:

Algorithm 5 (QR extremal subsampling) *(1) Choose the quantile index of interest τ_T, the subsample size b (e.g., $b = \lfloor 50 + \sqrt{T} \rfloor$), and the number of subsamples S (e.g., $S = 500$). (2) If the data have serial dependence, draw S subsamples from $\{(Y_t, X_t)\}_{t=1}^T$ of size b, $\{(Y_{t,s}^*, X_{t,s}^*)\}_{t=1}^b$, of the form $(Y_{t,s}^*, X_{t,s}^*) = (Y_{t+k}, X_{t+k})$ where $k \in \{1, \dots, T - b + 1\}$. If the data are independent, $\{(Y_{t,s}^*, X_{t,s}^*)\}_{t=1}^b$ can be drawn as a random subsample of size b from $\{(Y_t, X_t)\}_{t=1}^T$ without replacement. (3) For each $s \in \{1, \dots, S\}$, compute $Z_{b,s}^*(k_T)$ or $\hat{Z}_{b,s}^*(k_T)$, applying (18.38) or (18.39) to the subsample $\{(Y_{t,s}^*, X_{t,s}^*)\}_{t=1}^b$. (4) Estimate the quantiles of $\psi' Z_T(k_T)$ or $\psi' \hat{Z}_T(k_T)$ by the sample quantiles of $\{\psi' Z_{b,s}^*(k_T)\}_{s=1}^S$ or $\{\psi' \hat{Z}_{b,s}^*(k_T)\}_{s=1}^S$.*

The comments in Section 18.3.2.5 on the choice of subsample size, number of simulations, and differences with conventional subsampling also apply to the regression case.

18.3.3.6 Median bias correction and confidence intervals

Chernozhukov and Fernández-Val (2011) construct asymptotically median-unbiased estimators and $(1 - \alpha)$-CIs for $\psi' \beta(\tau)$ based on the SN-QR statistic as

$$
\psi' \hat{\beta}(\tau) - \frac{\hat{c}_{1/2}}{\mathcal{A}_T} \quad \text{and} \quad \left[\psi' \hat{\beta}(\tau) - \frac{\hat{c}_{1-\alpha/2}}{\mathcal{A}_T}, \; \psi' \hat{\beta}(\tau) - \frac{\hat{c}_{\alpha/2}}{\mathcal{A}_T} \right],
$$

where $\hat{c}_p$ is a consistent estimator of the p-quantile c_α of $Z_T(k_T)$ that can be obtained using Algorithm 3, 4, or 5. Chernozhukov and Fernández-Val (2011) also construct asymptotically median-unbiased estimators and $(1 - \alpha)$-CIs for $\psi' \beta(\tau)$ based on the CN-QR statistic as

$$
\psi' \hat{\beta}(\tau) - \frac{\tilde{c}_{1/2}}{\mathcal{A}_T} \quad \text{and} \quad \left[\psi' \hat{\beta}(\tau) - \frac{\tilde{c}_{1-\alpha/2}}{\mathcal{A}_T}, \; \psi' \hat{\beta}(\tau) - \frac{\tilde{c}_{\alpha/2}}{\mathcal{A}_T} \right],
$$

where $\tilde{c}_p$ is a consistent estimator of the p-quantile of $\hat{Z}_T(k_T)$ that can be obtained using Algorithm 3 or 5.

18.3.3.7 Extrapolation estimator for very extremes

The τ-QR estimators can be very inaccurate when $\tau T / d_x$ is very small, say $\tau T / d_x < 1$. We can construct extrapolation estimators for these cases that use the assumptions on the behavior of the tails. By analogy with the univariate case,

$$
\tilde{\beta}(\tau_T) = \hat{\beta}(\tilde{\tau}_T) + \frac{(\tau_T / \tilde{\tau}_T)^{-\hat{\xi}} - 1}{2^{-\hat{\xi}} - 1} \left[\hat{\beta}(2\tilde{\tau}_T) - \hat{\beta}(\tilde{\tau}_T) \right] \tag{18.40}
$$

or

$$
\breve{\beta}(\tau_T) = \hat{\beta}(\tilde{\tau}_T) + \frac{(\tau_T / \tilde{\tau}_T)^{-\hat{\xi}} - 1}{2^{\hat{\xi}} - 1} \left[\hat{\beta}(\tilde{\tau}_T / 2) - \hat{\beta}(\tilde{\tau}_T) \right], \tag{18.41}
$$

where $\tau_T \ll \tilde{\tau}_T$, and $\hat{\xi}$ is the Pickands (18.28) or Hill (18.30) estimator of ξ. He et al. (2016) derived the joint asymptotic distribution of $(\breve{\beta}(\tau_T), \hat{\xi}_P)$. Wang et al. (2012) developed other extrapolation estimators for heavy-tailed distributions with $\xi > 0$.

The estimators in (18.40) and (18.41) have good properties provided that the quantities on the right-hand side are well estimated, which in turn requires that $\tilde{\tau}_T T/d_x$ be large, and that the Pareto-type tail model be a good approximation. To construct the confidence interval for $\beta(\tau_T)$ based on extrapolation, we can apply extremal subsampling to the statistic

$$\tilde{\mathcal{A}}_T[\tilde{\beta}(\tau_T) - \beta(\tau_T)], \quad \tilde{\mathcal{A}}_T = \frac{\sqrt{\tilde{\tau}_T T}}{\bar{X}'_T(\hat{\beta}(m\tilde{\tau}_T) - \hat{\beta}(\tilde{\tau}_T))}.$$

For the estimator (18.41), we can also use analytical methods based on the asymptotic distribution given in He et al. (2016, Corollary 3.4).

18.3.4 Extreme value versus normal inference

Chernozhukov and Fernández-Val (2011) provided a simple rule of thumb for the application of EV inference. Recall that the order of a sample τ_T-quantile from a sample of size T is $\tau_T T$ (rounded to the next integer). This order plays a crucial role in determining the quality of the EV or normal approximations. Indeed, the former requires $\tau_T T \to k$, whereas the latter requires $\tau_T T \to \infty$. In the regression case, in addition to the order of the quantile, we need to take into account d_x, the dimension of X. As an example, consider the case where all d_x covariates are indicators that divide equally the sample into subsamples of size T/d_x. Then each of the components of the τ_T-QR estimator will correspond to a sample quantile of order $\tau_T T/d_x$. We may therefore think of $\tau_T T/d_x$ as a dimension-adjusted order for quantile regression.

A common simple rule for the application of the normal is that the sample size is greater than 30. This suggests that we should use extremal inference whenever $\tau_T T/d_x \lesssim 30$. This simple rule may or may not be conservative. For example, when regressors are continuous, the computational experiments in Chernozhukov and Fernández-Val (2011) show that normal inference performs as well as EV inference provided that $\tau_T T/d_x \gtrsim 15$ to 20, which suggests using EV inference when $\tau_T T/d_x \lesssim 15$ to 20 for this case. On the other hand, if we have an indicator in X equal to 1 only for 2% of the sample, then the coefficient of this indicator behaves as a sample quantile of order $0.02\tau_T T = \tau_T T/50$, which would motivate using EV inference when $\tau_T T/50 \lesssim 15$ to 20 in this case. This rule is far more conservative than the original simple rule when $d_x \lll 50$. Overall, it seems prudent to use both EV and normal inference methods in most cases, with the idea that the discrepancies between the two can indicate extreme situations.

18.4 Empirical applications

We consider two applications of extremal quantile regression: conditional value-at-risk and financial contagion. We implement the empirical analysis in R with Koenker's (2016) quantreg package and the code from Chernozhukov and Du (2008) and Chernozhukov and Fernández-Val (2011). The data are obtained from Yahoo! Finance.[6]

---

[6]The data set and the code are available online at Fernández-Val's website: http://sites.bu.edu/ivanf/research/.

18.4.1   Value-at-risk prediction

We revisit the problem of forecasting the conditional value-at-risk of a financial institution posed by Chernozhukov and Umantsev (2001) with more recent methodology. The response variable Y_t is the daily return of Citigroup stock, and the covariates X_{1t}, X_{2t}, and X_{3t} are the lagged daily returns of Citigroup stock (C), the Dow Jones Industrial Index (DJI), and the Dow Jones US Financial Index (DJUSFN), respectively. The lagged own return captures dynamics, the DJI is a measure of overall market return, and the DJUSFN is a measure of market return in the financial sector. We estimate quantiles of Y_t conditional on $X_t = (1, X_{1t}^+, X_{1t}^-, X_{2t}^+, X_{2t}^-, X_{3t}^+, X_{3t}^-)$ with $x^+ = \max\{x, 0\}$ and $x^- = -\min\{x, 0\}$. There are 1,738 daily observations in the sample covering the period from January 1, 2009 to November 30, 2015.

Figure 18.2 plots the QR estimates $\hat{\beta}(\tau)$ along with 90% pointwise CIs. The solid lines represent the extremal CIs and the dashed lines the normal CIs. The extremal CIs are computed by the extremal subsampling method described in Algorithm 5 with subsample size $b = \lfloor 50 + \sqrt{1{,}738} \rfloor = 91$ and $S = 500$ simulations. We use the SN-QR statistic with spacing parameter $p = 5$. The normal CIs are based on the normal approximation with standard errors computed by the method proposed by Powell (1991).[7] Figure 18.3 plots the median bias-corrected QR estimates along with 90% pointwise CIs for the lower tail (note that due to the median bias correction, the coefficient estimates are slightly different from those in Figure 18.2). The bias correction is also implemented using extremal subsampling with the same specifications.

We focus the discussion on the impact of downward movements of the explanatory variables (the C lag X_{1t}^-, the DJI lag X_{2t}^-, and the DJUSFN lag X_{3t}^-) on the extreme risk, that is, on the low conditional quantiles of the Citigroup stock return. To interpret the results, it is helpful to keep in mind that if the covariates were completely irrelevant (i.e. independent of the response), then their coefficients would be equal to 0 uniformly over τ, except for the constant term. The intercept would coincide with the unconditional quantile of Citigroup daily return. Another general remark is that we would expect the estimates and CIs to be more volatile at the tails than at the center due to data sparsity. Figures 18.2 and 18.3 show that most of the coefficients are insignificant throughout the distribution, which confirms the expected unpredictability of the stock returns. However, we do find that the coefficient on Citigroup's lagged return X_{1t}^- is significantly different from 0 in the extreme low quantiles (see the upper right-hand figure in Figure 18.3). This suggests that from 2009 to 2015, a past drop in the stock price of Citigroup has significantly pushed down the extreme low quantiles of the current stock price. Informally speaking, the negative return on the stock price induced the risk of a further negative outcome in the near future.

Comparing the CIs produced by the extremal inference and the normal inference, Figure 18.2 shows that they closely match in the central region, while Figures 18.3 reveals that the normal CIs are often narrower than the extremal CIs in the tails, especially for $\tau < 0.05$. As briefly mentioned in Section 18.3.2.2, the extremal CIs coincide with the normal CIs when the situation is nonextremal. Therefore, this discrepancy indicates that the normal CIs on the tails substantially underestimate the sampling variation and hence might lead to substantial undercoverage in the CIs.

We next characterize the tail properties of the model. Table 18.1 reports the estimates of the EV index ξ obtained by the Hill estimator in (18.30), together with bias-corrected estimates and 90% CIs based on (18.31), which were obtained using the QR extremal

---

[7] We used the command `summary.rq` with the option `ker` in the quantreg package to compute the standard errors.

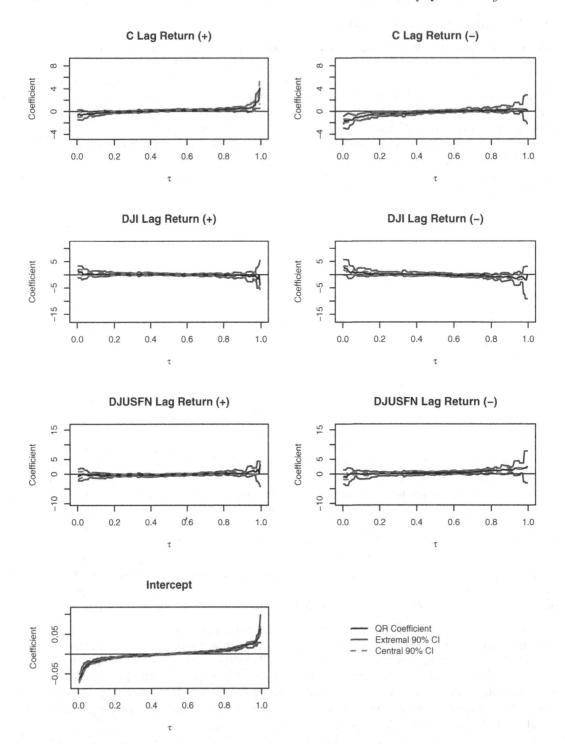

FIGURE 18.2
Value-at-risk: QR coefficient estimates and 90% pointwise CIs. The response variable is the daily Citigroup return from January 1, 2009 to November 30, 2015. The solid lines depict extremal CIs and the dashed lines depict normal CIs.

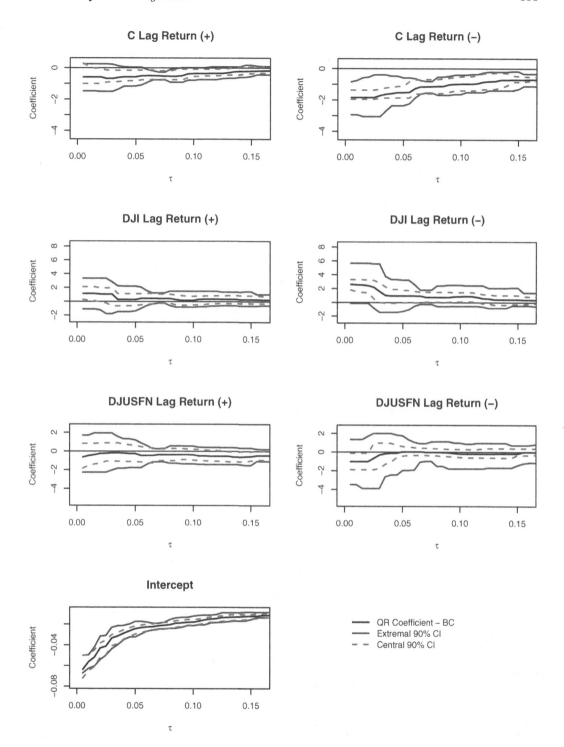

FIGURE 18.3
Value-at-risk: bias-corrected QR coefficient estimates and 90% pointwise CIs for low quantiles. The response variable is the daily Citigroup return from January 1, 2009 to November 30, 2015. The solid lines depict extremal CIs and the dashed lines depict normal CIs.

bootstrap of Algorithm 4 with $S = 500$ applied to the Hill estimator. The bias-corrected estimates of ξ are relatively stable even at the extreme tails. They are greater than 0, confirming that the distribution of stock returns has a much thicker lower tail than the normal distribution. It is noteworthy that none of these estimates were used to produce Figues 18.2 and 18.3 because they were obtained by the extremal subsampling method applied to the SN-QR statistic.

TABLE 18.1

Value-at-Risk: Hill estimation results for the EV index

	Estimate	Bias-Corrected Estimate	90% Confidence Interval
$\tau = 0.01$	0.330	0.311	[0.164, 0.441]
$\tau = 0.05$	0.427	0.350	[0.250, 0.461]
$\tau = 0.1$	0.447	0.293	[0.215, 0.369]

Having characterized the EV index, we can now estimate the very extreme quantiles using extrapolation methods. We set $\hat{\xi}$ to be the estimate with $\tau = 0.05$, and compute the extrapolation estimator (18.40) for $\tau = 0.005$, 0.001, and 0.0001 in Table 18.2. For comparison purposes, the first column reports the τ-QR estimates for $\tau = 0.005$ obtained from (18.21). This estimator cannot be calculated for the other quantile indexes considered. We find some discrepancies between the two estimators especially for the coefficients of the negative lags at $\tau = 0.005$. Figure 18.4 plots the predicted values for the conditional 0.005-quantiles in the second half of 2015 obtained from the QR and extrapolation estimators. The standard QR fit uses sample data that contains few observations on the extreme events, while the extrapolated fit uses the tail model and reliably estimated conditional 0.05-quantile coefficients to predict the magnitude of such events. The quality of this prediction clearly depends on whether the tails model is accurate.

TABLE 18.2

Value-at-risk: Extrapolation estimators for the quantile regression

Variable	Regression estimate $\tau = 0.005$	Extrapolation estimate $\tau = 0.005$	$\tau = 0.001$	$\tau = 0.0001$
Intercept	−0.066	−0.067	−0.122	−0.274
C lag return (+)	−0.646	−1.530	−2.888	−6.642
C lag return (−)	−2.179	−4.806	−9.117	−21.033
DJI lag return (+)	1.583	0.630	1.167	2.652
DJI lag return (−)	3.283	2.425	4.194	9.085
DJUSFN lag return (+)	−0.742	0.558	1.597	4.470
DJUSFN lag return (−)	−1.004	1.107	2.547	6.526

18.4.2 Contagion of financial risk

We consider an application to contagion of financial risk between commercial banks. The response variable Y_t is the daily return of Citigroup stock (C), and the covariates X_{1t}, X_{2t}, and X_{3t} are the contemporaneous daily returns of the stocks of other banks, namely, Bank of America (BAC), J.P.Morgan Chase & Co. (JPM), and Wells Fargo & Co. (WFC). As in the previous subsection, we estimate the quantiles of Y_t conditional

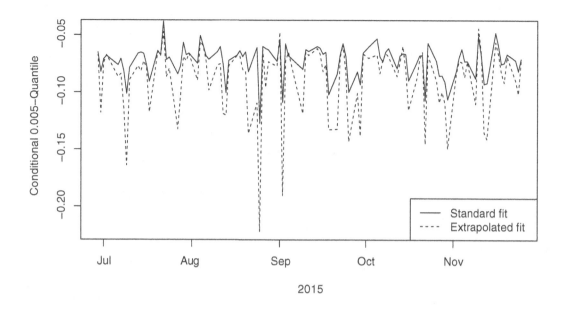

FIGURE 18.4

Value-at-risk: Extrapolation versus standard estimates of conditional quantiles of the daily Citigroup return between July 1, 2015 and November 30, 2015.

on $X_t = (1, X_{1t}^+, X_{1t}^-, X_{2t}^+, X_{2t}^-, X_{3t}^+, X_{3t}^-)$ using 1,738 daily observations covering the period from January 1, 2009 to November 30, 2015.

Figure 18.5 plots the QR estimates $\widehat{\beta}(\tau)$ along with 90% pointwise CIs. The solid lines represent the extremal CIs and the dashed lines the normal CIs. The extremal CIs are computed by the extremal subsampling method described in Algorithm 5 with subsample size $b = \lfloor 50 + \sqrt{1,738} \rfloor = 91$ and $S = 500$ simulations. We use the SN-QR statistic with spacing parameter $p = 5$. The normal CIs are based on the normal approximation with standard errors computed by the method proposed by Powell (1991). Figure 18.6 plots the median bias-corrected QR estimates along with 90% pointwise CIs for the lower tail. The bias correction is also implemented using extremal subsampling with the same specifications.

We find a significant effect of Bank of America's risk on Citigroup's risk. Observe that the coefficient of BAC (+) is positive and that of BAC (−) is negative across most of the quantiles. This tells that BAC and C hold similar portfolios and that there might be a direct contagion of BAC's risk to C's risk (negative return of BAC is likely to cause negative return of C). Similar observation holds for JPM's risk onto C's risk. However, there is no such contagion effect of WFC's risk onto C's. In Figure 18.6 we see that the negative return of Bank of America stock has a large effect on the extreme low quantile of C's return, while its positive return has no significant effect. This indicates that Bank of America's risk has an asymmetric and large impact on its competitor. As in the value-at-risk application, we find that the normal and extremal CIs are similar in the central region, while the normal CIs are narrower than the extremal CIs in the tails, especially for $\tau < 0.05$.

Table 18.3 reports the estimates of the EV index ξ obtained by the Hill estimator (18.30), together with bias-corrected estimates and 90% CIs based on (18.31), which were obtained

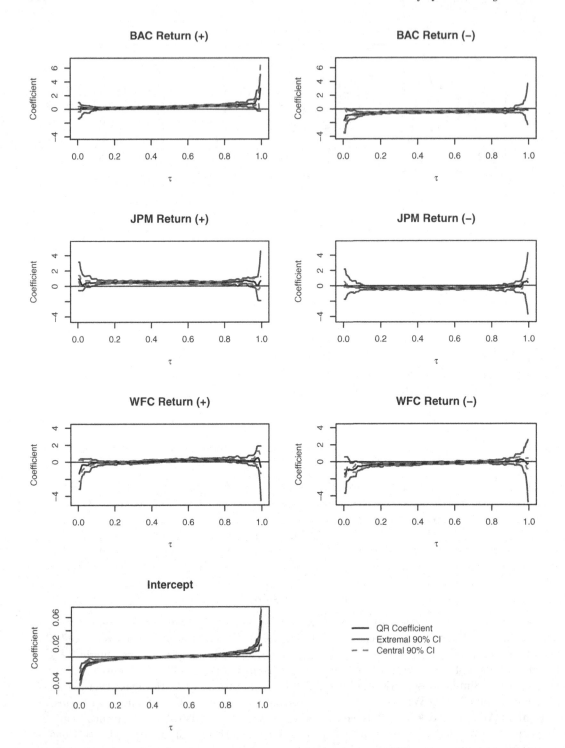

FIGURE 18.5
Financial contagion: QR coefficient estimates and 90% pointwise CIs. The response variable is the daily Citigroup return from January 1, 2009 to November 30, 2015. The solid lines depict extremal CIs and the dashed lines depict normal CIs.

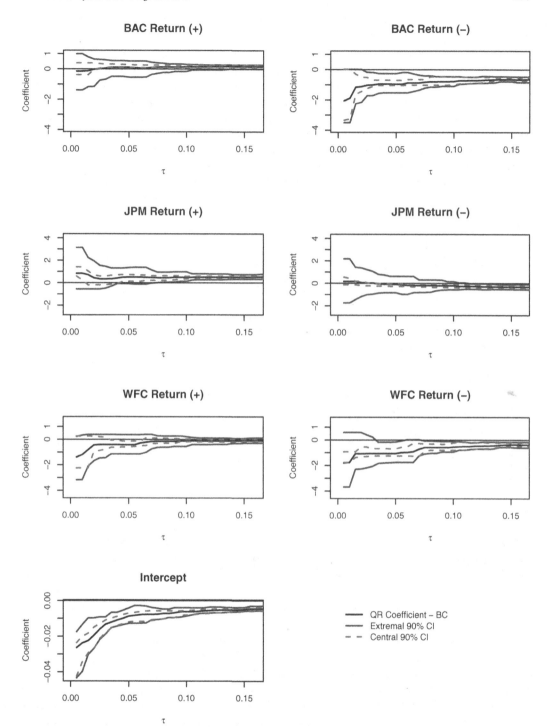

FIGURE 18.6

Financial contagion: Bias-corrected QR coefficient estimates and 90% pointwise CIs for low quantiles. The response variable is the daily Citigroup return from January 1, 2009 to November 30, 2015. The solid lines depict extremal CIs and the dashed lines depict normal CIs.

using the QR extremal bootstrap of Algorithm 4 with $S = 500$ applied to the Hill estimator. Again we find estimates significantly greater than 0, confirming that stock returns have thick lower tails relative to the normal distribution. Table 18.4 shows the estimates of the QR coefficients for very low quantiles obtained from QR and the extrapolation estimator (18.40) with $\widehat{\xi} = 0.263$, the estimate from Table 18.3 for $\tau = 0.05$. The largest difference between the regression and extrapolation estimates occurs for the WFC return. Here we find a large negative coefficient for the return ($-$) that indicates that there might be contagion of financial risk from WFC to C at very low quantiles. Figure 18.7 contrasts the predicted values for the conditional 0.005-quantiles in the second half of 2015 obtained from the QR and extrapolation estimators. Overall, the two methods produce similar estimates, although the extrapolated estimator predicts deeper troughs in the quantiles.

TABLE 18.3
Financial contagion: Hill estimation results for the EV index

	Estimate	Bias-Corrected Estimate	90% Confidence Interval
$\tau = 0.01$	0.263	0.255	[0.086, 0.461]
$\tau = 0.05$	0.646	0.611	[0.468, 1.000]
$\tau = 0.1$	0.500	0.357	[0.276, 0.441]

TABLE 18.4
Financial contagion: Extrapolation estimators for the quantile regression

Variable	Regression estimate	Extrapolation estimate		
	$\tau = 0.005$	$\tau = 0.005$	$\tau = 0.001$	$\tau = 0.0001$
Intercept	−0.035	−0.022	−0.037	−0.074
BAC return (+)	0.042	0.203	0.227	0.285
BAC return (−)	−1.657	−1.487	−2.202	−3.928
JPM return (+)	0.931	0.328	0.205	−0.091
JPM return (−)	−0.149	0.165	0.542	1.451
WFC return (+)	−1.350	−1.775	−3.437	−7.449
WFC return (−)	−1.352	−3.279	−5.960	−12.430

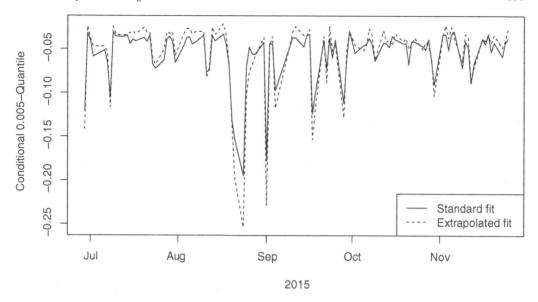

FIGURE 18.7
Financial contagion: Extrapolation versus standard estimates of conditional quantiles of the daily Citigroup return between July 1, 2015 and November 30, 2015.

Bibliography

J. Abrevaya. The effect of demographics and maternal behavior on the distribution of birth outcomes. *Empirical Economics*, 26(1):247–259, 2001.

D. J. Aigner and S. F. Chu. On estimating the industry production function. *American Economic Review*, 58:826–839, 1968.

D. J. Aigner, T. Amemiya, and D. J. Poirier. On the estimation of production frontiers: Maximum likelihood estimation of the parameters of a discontinuous density function. *International Economic Review*, 17(2):377–396, 1976.

J. G. Altonji, H. Ichimura, and T. Otsu. Estimating derivatives in nonseparable models with limited dependent variables. *Econometrica*, 80(4):1701–1719, 2012.

K. J. Arrow, T. Harris, and J. Marschak. Optimal inventory policy. *Econometrica*, 19(3): 205–272, 1951.

P. Bertail, C. Haefke, D. N. Politis, and H. White. A subsampling approach to estimating the distribution of diverging extreme statistics with applications to assessing financial market risks. *Journal of Econometrics*, 120(2):295–326, 2004.

P. Bickel and D. Freedman. Some asymptotic theory for the bootstrap. *Annals of Statistics*, 9:1196–1217, 1981.

V. Chernozhukov. Nonparametric extreme regression quantiles. Working paper, Standord Univ. Presented at Princeton Econometrics Seminar, 1998.

V. Chernozhukov. Extremal quantile regression. *Annals of Statistics*, 33(2):806–839, 2005.

V. Chernozhukov and S. Du. Extremal quantiles and value-at-risk. In S. N. Durlauf and

L. E. Blume, editors, *The New Palgrave Dictionary of Economics*. Palgrave Macmillan, Basingstoke, 2008.

V. Chernozhukov and I. Fernández-Val. Subsampling inference on quantile regression processes. *Indian Journal of Statistics*, 67:253–276, 2005.

V. Chernozhukov and I. Fernández-Val. Inference for extremal conditional quantile models, with an application to market and birthweight risks. *Review of Economic Studies*, 78: 559–589, 2011.

V. Chernozhukov and L. Umantsev. Conditional value-at-risk: Aspects of modeling and estimation. *Empirical Economics*, 26(1):271–293, 2001.

L. de Haan. *On Regular Variation and Its Applications to the Weak Convergence*. Mathematical Centre Tract 32. Mathematical Centre, Amsterdam, 1970.

A. Dekkers and L. de Haan. On the estimation of the extreme-value index and large quantile estimation. *Annals of Statistics*, 17(4):1795–1832, 1989.

X. D'Haultfœuille, A. Maurel, and Y. Zhang. Extremal quantile regressions for selection models and the black-white wage gap. Working Paper, 2015.

S. G. Donald and H. J. Paarsch. Superconsistent estimation and inference in structural econometric models using extreme order statistics. *Journal of Econometrics*, 109(2):305–340, 2002.

P. Embrechts, C. Klüppelberg, and T. Mikosch. *Modelling Extremal Events for Insurance and Finance*. Springer-Verlag, Berlin, 1997.

R. F. Engle and S. Manganelli. CAViaR: Conditional autoregressive value at risk by regression quantiles. *Journal of Business and Economic Statistics*, 22(4):367–381, 2004.

E. F. Fama. The behavior of stock market prices. *Journal of Business*, 38:34–105, 1965.

P. D. Feigin and S. I. Resnick. Limit distributions for linear programming time series estimators. *Stochastic Processes and Their Applications*, 51:135–165, 1994.

C. J. Flinn and J. J. Heckman. New methods for analyzing structural models of labor force dynamics. *Journal of Econometrics*, 18(1):115–168, 1982.

M. Fox and H. Rubin. Admissibility of quantile estimates of a single location parameter. *Annals of Mathematical Statistics*, 35:1019–1030, 1964.

B. Gnedenko. Sur la distribution limité du terme d' une série alétoire. *Annals of Mathematics*, 44:423–453, 1943.

C. Gutenbrunner, J. Jurečková, R. Koenker, and S. Portnoy. Tests of linear hypotheses based on regression rank scores. *Journal of Nonparametric Statistics*, 2(4):307–331, 1993.

F. He, Y. Cheng, and T. Tong. Estimation of extreme conditional quantiles through an extrapolation of intermediate regression quantiles. *Statistics and Probability Letters*, 113: 30–37, 2016.

B. M. Hill. A simple general approach to inference about the tail of a distribution. *Annals of Statistics*, 3(5):1163–1174, 1975.

D. W. Jansen and C. G. de Vries. On the frequency of large stock returns: Putting booms and busts into perspective. *Review of Economics and Statistics*, 73:18–24, 1991.

J. Jurečková. Regression rank-scores tests against heavy-tailed alternatives. *Bernoulli*, 5 (4):659–676, 1999.

J. Jurečková. Averaged extreme regression quantile. *Extremes*, 19:41–49, 2016.

K. Knight. Limiting distributions of linear programming estimators. *Extremes*, 4(2):87–103, 2001.

R. Koenker. *quantreg: Quantile Regression*, 2016. URL https://CRAN.R-project.org/ package=quantreg. R package version 5.21.

R. Koenker and G. Bassett. Regression quantiles. *Econometrica*, 46:33–50, 1978.

P.-S. Laplace. *Théorie analytique des probabilités*. Éditions Jacques Gabay, Paris, 1995.

M. R. Leadbetter, G. Lindgren, and H. Rootzén. *Extremes and Related Properties of Random Sequences and Processes*. Springer-Verlag, New York, 1983.

F. M. Longin. The asymptotic distribution of extreme stock market returns. *Journal of Business*, 69(3):383–408, 1996.

M. Mandelbrot. The variation of certain speculative prices. *Journal of Business*, 36:394–419, 1963.

R. M. Meyer. A Poisson-type limit theorem for mixing sequences of dependent "rare" events. *Annals of Probability*, 1:480–483, 1973.

J. Pickands, III. Statistical inference using extreme order statistics. *Annals of Statistics*, 3: 119–131, 1975.

D. N. Politis, J. P. Romano, and M. Wolf. *Subsampling*. Springer-Verlag, New York, 1999.

S. Portnoy and J. Jurečková. On extreme regression quantiles. *Extremes*, 2(3):227–243, 1999.

S. Portnoy and R. Koenker. Adaptive *L*-estimation for linear models. *Annals of Statistics*, 17(1):362–381, 1989.

J. L. Powell. Least absolute deviations estimation for the censored regression model. *Journal of Econometrics*, 25:303–325, 1984.

J. L. Powell. Estimation of monotonic regression models under quantile restrictions. In W. A. Barnett, J. Powell, and G. E. Tauchen, editors, *Nonparametric and Semiparametric Methods in Econometrics: Proceedings of the Fifth International Symposium in Economic Theory and Econometrics*, pages 357–384. Cambridge University Press, Cambridge, 1991.

V. Praetz. The distribution of share price changes. *Journal of Business*, 45(1):49–55, 1972.

A. Sen. *On Economic Inequality*. Clarendon Press, Oxford, 1973.

R. L. Smith. Nonregular regression. *Biometrika*, 81(1):173–183, 1994.

C. P. Timmer. Using a probabilistic frontier production function to measure technical efficiency. *Journal of Political Economy*, 79:776–794, 1971.

H. Wang, D. Li, and X. He. Estimation of high conditional quantiles for heavy-tailed distributions. *Journal of the American Statistical Association*, 107:1453–1464, 2012.

H. White. *Asymptotic Theory for Econometricians.* Academic Press, New York, revised edition, 2001.

Y. Zhang. Extremal quantile treatment effects. Job Market Paper, 2015.

G. Zipf. *Human Behavior and the Principle of Last Effort.* Addison-Wesley, Cambridge, MA, 1949.

19

Quantile Regression Methods for Longitudinal Data

Antonio F. Galvao

University of Arizona, Tucson, Arizona, USA

Kengo Kato

University of Tokyo, Tokyo, Japan

CONTENTS

19.1   Introduction

Since the seminal work of Koenker and Bassett (1978), quantile regression (QR) has been a pivotal statistical technique; it offers an easy-to-implement method to estimate conditional quantiles, and by estimating several different conditional quantiles, one is able to make inference on the effects of the regressors on the entire conditional distribution. The properties of the QR method are well established for cross-sectional models; see the landmark monograph by Koenker (2005) for references and discussion.

One of the "future topics" noted in Koenker (2005) is the analysis of the QR method for longitudinal or panel data. At the time of publication of Koenker (2005), there were only a few papers on that subject, an important example being Koenker (2004). Koenker (2004) introduced a general approach to the estimation of QR models for longitudinal data where individual effects are treated as pure location shift parameters common to all quantiles and may be subject to shrinkage toward a common value.

Since Koenker (2004), there has been a growing literature on QR methods for panel data. This chapter attempts to provide a progress report on the current state of this topic. We shall mainly concentrate on the most basic additive linear QR model, but also will describe other extensions. We hope this chapter will appeal to both econometricians and empirical economists interested in the use of QR for panel data.

A conceptually simple way of applying the QR method to panel data is to model the conditional quantile of the response variable given the regressors and (scalar) individual effect as the addition of the individual effect and a linear function of the regressors (see the model described in equation (19.1) below), and then apply fixed effects (FE) estimation by treating the individual effects as parameters to be estimated. However, as we will see later, this FE estimator for this panel QR model suffers from the incidental parameters problem (Neyman and Scott, 1948), and fails to be consistent when the number of time periods is fixed. The incidental parameters problem arises because the number of parameters to be estimated is proportional to n, and so if T, the number of time periods, is fixed, then the number of observations available for estimation is comparable to the number of parameters, which prevents consistent estimation of the common parameter.

To overcome this drawback, it has become standard in the literature to employ "large-T" asymptotics where n and T jointly go to infinity. Kato et al. (2012) formally derived the asymptotic properties of the FE-QR estimator under large time series asymptotics and established sufficient conditions for consistency and asymptotic normality of the estimator. Unfortunately, existing sufficient conditions under which the asymptotic bias of the FE-QR estimator is negligible require $T \gg n$. We shall mention that the nondifferentiability of the QR objective function (check function) significantly complicates the asymptotic analysis of the FE-QR estimator. The analysis of the FE estimator in a general nonlinear panel data model relies on higher-order asymptotic theory (see Hahn and Newey, 2004; Hahn and Kuersteiner, 2011). However, the higher-order asymptotic theory for QR is nonstandard, and Kato et al. (2012) could not derive the analytic expression for the incidental parameters bias of the FE-QR estimator. Given these difficulties, Galvao and Kato (2016) analyzed the FE estimator defined by a minimizer of an objective function which uses kernels to smoothly approximate the nondifferentiable quantile regression objective function, which they call the fixed effects smoothed quantile regression (FE-SQR) estimator, and showed that the FE-SQR estimator has a limiting normal distribution with a bias in the mean when n/T goes to some constant. Importantly, by using the smoothing technique, they were able to derive the analytic expression for the incidental parameters bias. Galvao and Kato (2016) further proposed a one-step bias correction to the fixed effects estimator based on the analytic bias formula. These results are essentially parallel to those obtained in the study of general nonlinear panel data models such as Hahn and Newey (2004) and Hahn and Kuersteiner (2011), but are not covered by those general results. In the following sections we will begin with a review of the main results of Kato et al. (2012) and Galvao and Kato (2016).

There are some alternative estimation methods beyond mere FE estimation (but still within the FE framework). They include penalized estimation (Koenker, 2004; Lamarche, 2010), minimum-distance estimation (Galvao and Wang, 2015), and two-step estimation (Canay, 2011). We will also review these alternative methods.

The aforementioned papers assume that the parameter of interest (typically the common parameter) is point-identified. Rosen (2012) studied the identifying power of conditional quantile restrictions in short panels with fixed effects. He showed that if the τ-quantile of the error term is assumed to be 0 conditional only on the regressors but not on the individual effect, the parameter of interest cannot even be (nontrivially) set-identified. By imposing an extra weak conditional independence assumption on the error terms, he showed that the parameter can be set-identified and provided a characterization of the sharp identified set.

Abrevaya and Dahl (2008) introduced yet another alternative approach to FE-QR that employs the "correlated random effects" (CRE) model of Chamberlain (1982). The CRE-QR is an attempt to develop a class of QR methods for short panels, but the unobserved effect needs to be specified as a function of observables. Abrevaya and Dahl (2008) assumed the unobservable individual effect to be a linear function of the observable regressors plus an error term. Along these lines, Arellano and Bonhomme (2016) developed nonlinear estimation for QR with short panels. Their framework covers static and dynamic autoregressive models, models with general predetermined regressors, and models with multiple individual effects. We discuss the CRE-QR methods in Section 19.4.

Finally, in Section 19.5, we describe a few extensions of QR methods for longitudinal data to models with endogeneity (Galvao, 2011; Harding and Lamarche, 2009), censoring (Wang and Fygenson, 2009; Galvao et al., 2013), group-level treatments (Chetverikov et al., 2016), and semiparametric QR models for longitudinal data (Wei and He, 2006).

19.2 Panel quantile regression model

In the panel data setting, we observe pairs of response variables and regressor vectors $(y_{it}, \mathbf{x}_{it})$, $i = 1, \ldots, n$, $t = 1, \ldots, T$, where y_{it} is scalar while $\mathbf{x}_{it}$ is p-dimensional, and i denotes the index for individuals while t denotes the time index. In addition to these variables, suppose that there are unobservable individual-specific variables η_i, $i = 1, \ldots, n$, that represent individual characteristics such as ability or firm's managerial quality which we would include in the analysis if they were observable.

In the mean regression case, the primary focus is on the estimation of the partial effect of $\mathbf{x}_{it}$ on the conditional mean $\mathrm{E}[y_{it} \mid \mathbf{x}_{it}, \eta_i]$, and typically we assume that $\mathrm{E}[y_{it} \mid \mathbf{x}_{it}, \eta_i]$ is additive in η_i and $\mathbf{x}_{it}$, and linear in $\mathbf{x}_{it}$. Then there are a number of estimation methods for the coefficient vector on $\mathbf{x}_{it}$ depending on exogeneity assumptions. In contrast, in QR, we are interested in the conditional quantile of y_{it} given $(\mathbf{x}_{it}, \eta_i)$, and try to estimate the partial effect of $\mathbf{x}_{it}$ on a conditional quantile. Let $Q_{y_{it}}(\tau \mid \mathbf{x}_{it}, \eta_i)$ denote the conditional τ-quantile of y_{it} given $(\mathbf{x}_{it}, \eta_i)$, where $\tau \in (0, 1)$ is a quantile index. We shall first focus on the following basic model for $Q_{y_{it}}(\tau \mid \mathbf{x}_{it}, \eta_i)$:

$$Q_{y_{it}}(\tau \mid \mathbf{x}_{it}, \eta_i) = \alpha_i(\tau, \eta_i) + \mathbf{x}_{it}'\boldsymbol{\beta}_0(\tau), \tag{19.1}$$

where $\alpha_i(\tau, \eta_i)$ is some (unknown) function of τ and η_i, and $\boldsymbol{\beta}_0(\tau)$ is the p-dimensional vector of unknown parameters.[1] Then we are interested in estimation and inference on the common parameter $\boldsymbol{\beta}_0(\tau)$.

Model (19.1) can be written as

$$y_{it} = \alpha_i(\tau, \eta_i) + \mathbf{x}_{it}'\boldsymbol{\beta}_0(\tau) + u_{it}(\tau), \tag{19.2}$$

where $u_{it}(\tau)$ has zero conditional τ-quantile given $(\mathbf{x}_{it}, \eta_i)$. Under standard assumptions, the conditional quantile restriction is equivalent to

$$\mathrm{P}(u_{it}(\tau) \leqslant 0 \mid \mathbf{x}_{it}, \eta_i) = \tau. \tag{19.3}$$

---

[1]Instead of the linear quantile regression model considered here, it is possible to consider a parametric nonlinear quantile model with unobserved heterogeneity that entered in an explicit, parametric known way. The analysis of nonlinear quantile regression for panel data should be analogous to the linear case discussed below, but the literature lacks a formal treatment.

Given model (19.2), one might be tempted to difference out α_i as in the mean regression case. However, as with most nonlinear panel models with unobserved effects, differencing or other transformations to remove unobserved heterogeneity are not available for panel QR models. In fact, even if we strengthen the restriction (19.3) to the strict exogeneity type restriction $P(u_{it}(\tau) \leqslant 0 \mid \mathbf{x}_{i1}, \ldots, \mathbf{x}_{iT}, \eta_i) = \tau$, the differencing out does not work in the QR case essentially because the quantile restriction is a nonlinear restriction. For example, let us take the difference $y_{it} - y_{i,t-1}$; then we formally have

$$y_{it} - y_{i,t-1} = (\mathbf{x}_{it} - \mathbf{x}_{i,t-1})'\boldsymbol{\beta}_0(\tau) + \underbrace{u_{it}(\tau) - u_{i,t-1}(\tau)}_{=v_{it}(\tau)}.$$

The problem is that the term $v_{it}(\tau)$ does not satisfy the desired conditional quantile restriction. This intrinsic difficulty has been recognized by Abrevaya and Dahl (2008), among others, and was clarified by Koenker and Hallock (2000). They remarked that "Quantiles of convolutions of random variables are rather intractable objects, and preliminary differencing strategies familiar from Gaussian models have sometimes unanticipated effects" (p. 19).

19.3   Fixed effects estimation

In this section, we discuss fixed effects quantile regression estimation, namely, we make no assumption on the joint distribution of $\mathbf{x}_{it}$ and η_i, and treat $\alpha_i(\tau, \eta_i)$ as parameters to be estimated. The latter can be done by conditioning on $\eta_i = \eta_{i0}$ and letting $\alpha_{i0}(\tau) = \alpha_i(\tau, \eta_{i0})$. In addition, we fix $\tau \in (0, 1)$ and suppress the dependence on τ to simplify the notation, giving $\alpha_{i0} = \alpha_{i0}(\tau)$ and $\boldsymbol{\beta}_0 = \boldsymbol{\beta}_0(\tau)$.

19.3.1   FE-QR estimator

We first consider the FE estimator studied in Kato et al. (2012):

$$(\hat{\boldsymbol{\alpha}}, \hat{\boldsymbol{\beta}}) := \arg\min_{\boldsymbol{\alpha}, \boldsymbol{\beta}} \frac{1}{nT} \sum_{i=1}^{n} \sum_{t=1}^{T} \rho_\tau(y_{it} - \alpha_i - \mathbf{x}_{it}'\boldsymbol{\beta}), \tag{19.4}$$

where $\boldsymbol{\alpha} := (\alpha_1, \ldots, \alpha_n)'$ and $\rho_\tau(u) := \{\tau - I(u \leqslant 0)\}u$ is the check function (Koenker and Bassett, 1978). We call $\hat{\boldsymbol{\beta}}$ the FE-QR estimator of $\boldsymbol{\beta}_0$.

We observe that the FE-QR estimator $\hat{\boldsymbol{\beta}}$ in general fails to be consistent when T is fixed. To fix the idea, let us consider the following data-generating process:

$$y_{it} = 1 + d_{it} + (1 + d_{it})\epsilon_{it}, \quad i = 1, \ldots, n, \ t = 1, 2,$$

where (ϵ_{it}, d_{it}) is independent and identically distributed (i.i.d.) across i and t, $\{\epsilon_{it}\}$ and $\{d_{it}\}$ are independent, $P(d_{it} = 0) = P(d_{it} = 1) = 1/2$, and $\epsilon_{it} \sim N(0, 1)$. Then the pair (y_{it}, d_{it}) satisfies the conditional quantile restriction

$$Q_{y_{it}}(\tau \mid d_{it}) = \{1 + \Phi^{-1}(\tau)\} + \{1 + \Phi^{-1}(\tau)\}d_{it} =: \alpha_{i0} + \beta_0 d_{it},$$

where $\Phi(\cdot)$ is the distribution function of the standard normal distribution. Then (Graham et al., 2009, Theorem 1) shows that the FE-QR estimate $\hat{\beta}$ satisfies

$$\sum_{i=1}^{n} |(y_{i2} - y_{i1}) - \hat{\beta}(d_{i2} - d_{i1})| = \min_{\beta} \sum_{i=1}^{n} |(y_{i2} - y_{i1}) - \beta(d_{i2} - d_{i1})|.$$

That is, $\widehat{\beta}$ is the least absolute deviation estimate when $y_{i2} - y_{i1}$ is a response variable and $d_{i2} - d_{i1}$ is a regressor. Further, according to (Angrist et al., 2006, Theorem 3), as $n \to \infty$, $\widehat{\beta}$ converges in probability to the unique β^* that solves

$$\mathrm{E}[\{I(y_{i2} - y_{i1} \leqslant \beta^*(d_{i2} - d_{i1})) - 1/2\}(d_{i2} - d_{i1})] = 0.$$

If $d_{i2} = 1$ and $d_{i1} = 0$, then $y_{i2} - y_{i1} = 1 + 2\epsilon_{i2} - \epsilon_{i1} \sim N(1, 5)$, and so $\mathrm{E}[I(y_{i2} - y_{i1} \leqslant \beta^*(d_{i2}-d_{i1})) \mid d_{i2} = 1, d_{i1} = 0] = \Phi((\beta^*-1)/\sqrt{5})$. On the other hand, if $d_{i2} = 0$ and $d_{i1} = 1$, then $y_{i2} - y_{i1} = -1 + \epsilon_{i2} - 2\epsilon_{i1} \sim N(-1, 5)$, and so $\mathrm{E}[I(y_{i2} - y_{i1} \leqslant \beta^*(d_{i2} - d_{i1})) \mid d_{i2} = 0, d_{i1} = 1] = 1 - \Phi((\beta^* - 1)/\sqrt{5})$. Thus we have

$$\mathrm{E}[\{I(y_{i2} - y_{i1} \leqslant \beta^*(d_{i2} - d_{i1})) - 1/2\}(d_{i2} - d_{i1})] = \frac{1}{2}\Phi\left(\frac{\beta^* - 1}{\sqrt{5}}\right) - \frac{1}{4},$$

which is equal to 0 only when $\beta^* = 1$, so that $\widehat{\beta}$ is inconsistent unless $\tau = 1/2$. The absolute bias of $\widehat{\beta}$ is $|\Phi^{-1}(\tau)|$, which diverges as τ goes to 0 or 1.

Kato et al. (2012) formally derived the asymptotic properties of the FE-QR estimator under the asymptotic framework where n and T jointly go to infinity. We refer the reader to that paper for the precise regularity conditions. Here we informally state their main results.

Suppose that $(y_{it}, \mathbf{x}_{it})$ are independent across i, and stationary and weakly dependent across t where the distribution of $(y_{it}, \mathbf{x}_{it})$ may depend on i. Let $f_i(u \mid \mathbf{x})$ denote the conditional density of u_{it} ($= u_{it}(\tau)$) given $\mathbf{x}_{it} = \mathbf{x}$, and let $f_i(u)$ denote the marginal density of u_{it}. In addition, let $\boldsymbol{\gamma}_i = \mathrm{E}[f_i(0 \mid \mathbf{x}_{it})\mathbf{x}_{it}]/f_i(0)$, and let V_{ni} denote the covariance matrix of the term $T^{-1/2}\sum_{t=1}^{T}\{\tau - I(u_{it} \leqslant 0)\}(\mathbf{x}_{it} - \boldsymbol{\gamma}_i)$. Then under suitable regularity conditions, if $(\log n)^2/T \to 0$, then $\widehat{\boldsymbol{\beta}}$ is consistent, that is, $\widehat{\boldsymbol{\beta}} \xrightarrow{p} \boldsymbol{\beta}_0$, and if $n^2(\log n)^3/T \to 0$, then

$$\sqrt{nT}(\widehat{\boldsymbol{\beta}} - \boldsymbol{\beta}_0) \xrightarrow{d} N(\mathbf{0}, \Gamma^{-1}V\Gamma^{-1}),$$

where $\Gamma = \lim_n n^{-1}\sum_{i=1}^{n}\mathrm{E}[f_i(0 \mid \mathbf{x}_{it})\mathbf{x}_{it}(\mathbf{x}_{it} - \boldsymbol{\gamma}_i)']$ and $V = \lim_n n^{-1}\sum_{i=1}^{n}V_{ni}$ (assume those limits exist, and the matrix Γ is invertible). Kato et al. (2012) further proposed consistent estimates of the matrices Γ and V.

Unfortunately, the condition on T for mean-zero asymptotic normality for the FE-QR estimator is more stringent than the standard one (i.e., $n/T \to 0$) found in the nonlinear panel data literature. This comes from the fact that the asymptotic analysis of FE estimators in general nonlinear panel data models relies on higher-order asymptotic theory, but higher-order asymptotic theory is nonstandard in the QR case because of the nondifferentiability of the check function – hence the proofs in Kato et al. (2012) require different techniques than those used in Hahn and Newey (2004) and Hahn and Kuersteiner (2011). It is an open question whether the above condition on T can be significantly relaxed.

A more important question would be whether it is possible to obtain the analytic bias (of order $O(T^{-1})$) for the FE-QR estimator, which seems to challenging. One of the difficulties is that in a smooth nonlinear panel data model, the second-order bias of each estimate of the individual effect, which is of order $O(T^{-1})$, contributes to the incidental parameters bias. However, in general, the second-order bias of the QR estimate is not uniquely determined as the QR problem generally does not have a unique solution in finite samples, and the solution of the quantile estimation problem is a set. Consider a simple example. Let $X_1, \ldots, X_T$ be i.i.d. uniform random variables on $(0, 1)$, and suppose that $T\tau$ is integral. Then

$$\arg\min \sum_{t=1}^{T} \rho_\tau(X_t - \tau) = [X_{(T\tau)}, X_{(T\tau+1)}]$$

where $X_{(1)} < \cdots < X_{(T)}$ are order statistics. It is a simple exercise to show that $\mathrm{E}[X_{(T\tau+1)} - X_{(T\tau)}] = 1/(T+1)$, and so in this example the bias of order $O(T^{-1})$ is not uniquely determined for the QR estimate.

19.3.2 FE-SQR estimator

In the previous subsection, we discussed fundamental technical difficulties that arise in the formal analysis of the FE-QR estimator. Galvao and Kato (2016) studied the asymptotic properties of the estimator defined by a minimizer of an objective function which uses kernels to smoothly approximate the nondifferentiable QR objective function (which they call the FE-SQR estimator); importantly, by using the smoothing technique, they were able to derive an analytic expression for the incidental parameters bias.

Smoothing the check function was employed in Horowitz (1998). He studied the bootstrap refinement for inference in QR models, which, however, requires higher-order asymptotic theory. To bypass the difficulty in higher-order asymptotic theory for QR, he used a smoothing technique. Galvao and Kato (2016) adopt the basic idea from Horowitz (1998) by smoothing over $I(y_{it} \leqslant \alpha_i + \mathbf{x}'_{it}\boldsymbol{\beta})$ by using a kernel function. Let $K(\cdot)$ be a kernel function and $G(\cdot)$ be the survival function of $K(\cdot)$,

$$\int_{-\infty}^{\infty} K(u)du = 1, \quad G(u) := \int_{u}^{\infty} K(v)dv,$$

where we do not require $K(\cdot)$ to be nonnegative (we will indeed use higher-order kernels). Let $\{h_n\}$ be a sequence of positive numbers (bandwidths) such that $h_n \to 0$ as $n \to \infty$ and write $G_{h_n}(\cdot) = G(\cdot/h_n)$. Note that $G_{h_n}(y_{it} - \alpha_i - \mathbf{x}'_{it}\boldsymbol{\beta})$ is a smooth approximation to $I(y_{it} \leqslant \alpha_i + \mathbf{x}'_{it}\boldsymbol{\beta})^2$. Then we consider the estimator

$$(\widehat{\boldsymbol{\alpha}}, \widehat{\boldsymbol{\beta}}) := \arg\min_{(\boldsymbol{\alpha},\boldsymbol{\beta})\in\mathcal{A}^n\times\mathcal{B}} \left[\frac{1}{nT}\sum_{i=1}^{n}\sum_{t=1}^{T}(y_{it} - \alpha_i - \mathbf{x}'_{it}\boldsymbol{\beta})\{\tau - G_{h_n}(y_{it} - \alpha_i - \mathbf{x}'_{it}\boldsymbol{\beta})\} \right], \quad (19.5)$$

where $\mathcal{A}$ is a compact subset of $\mathbb{R}$, $\mathcal{A}^n$ is the product of n copies of $\mathcal{A}$, and $\mathcal{B}$ is a compact subset of $\mathbb{R}^p$ (the compactness of the set $\mathcal{A}^n \times \mathcal{B}$ is required to ensure the existence of $(\widehat{\boldsymbol{\alpha}}, \widehat{\boldsymbol{\beta}})$). We call $\widehat{\boldsymbol{\beta}}$ the FE-SQR estimator of $\boldsymbol{\beta}_0$.

Galvao and Kato (2016) provided conditions under which the FE-SQR estimator is consistent and has a limiting normal distribution with a bias in the mean when n and T grow at the same rate. In particular, assuming that $n/T \to \rho$ for some $\rho > 0$, and under some regularity conditions,

$$\sqrt{nT}(\widehat{\boldsymbol{\beta}} - \boldsymbol{\beta}_0) \xrightarrow{d} N(\sqrt{\rho}\mathbf{b}, \Gamma^{-1}V\Gamma^{-1}), \quad (19.6)$$

where the bias term is given by

$$\mathbf{b} := \Gamma^{-1}\left[\lim_{n\to\infty}\left\{\frac{1}{n}\sum_{i=1}^{n} s_i\left(\omega_{ni}^{(1)}\boldsymbol{\gamma}_i - \omega_{ni}^{(2)} + \frac{s_i\omega_{ni}^{(3)}\boldsymbol{\nu}_i}{2}\right)\right\}\right], \quad (19.7)$$

and where $s_i := 1/f_i(0), \boldsymbol{\gamma}_i := s_i\mathrm{E}[f_i(0\mid\mathbf{x}_{i1})\mathbf{x}_{i1}], \boldsymbol{\nu}_i := f'_i(0)\boldsymbol{\gamma}_i - \mathrm{E}[f'_i(0\mid\mathbf{x}_{i1})\mathbf{x}_{i1}]$, and the

---

[2]In Galvao and Kato (2016), we formally think of $T = T_n$ as a function of n and assume that $T_n \to \infty$ as $n \to \infty$; the bandwidth is chosen as a function of n and T, but since T is a function of n, it is denoted just as a function of n. In any case since the dependence of T on n is arbitrary, this assumption covers the case where n and T jointly go to infinity.

matrices V and Γ are the same as in the previous section. Further,

$$
\omega_{ni}^{(1)} := \sum_{1 \leqslant |j| \leqslant T-1} (1 - |j|/T) \left\{ \tau f_i(0) - \int_{-\infty}^{0} f_{i,j}(0, u) du \right\},
$$

$$
\omega_{ni}^{(2)} := \sum_{1 \leqslant |j| \leqslant T-1} (1 - |j|/T) \left\{ \tau \mathrm{E}[f_i(0 \mid \mathbf{x}_{i1})\mathbf{x}_{i1}] - \mathrm{E}\left[\mathbf{x}_{i1} \int_{-\infty}^{0} f_{i,j}(0, u \mid \mathbf{x}_{i1}, \mathbf{x}_{i,1+j}) du \right] \right\},
$$

$$
\omega_{ni}^{(3)} := \sum_{|j| \leqslant T-1} (1 - |j|/T) \, \mathrm{Cov}\{I(u_{i1} \leqslant 0), I(u_{i,1+j} \leqslant 0)\},
$$

where $f_{i,j}(u_1, u_{1+j})$ denotes the joint density of $(u_{i1}, u_{i,1+j})$, and $f_{i,j}(u_1, u_{1+j} \mid \mathbf{x}_{i1}, \mathbf{x}_{i,1+j})$ denotes the joint conditional density of $(u_{i1}, u_{i,1+j})$ given $(\mathbf{x}_{i1}, \mathbf{x}_{i,1+j})$.

19.3.2.1   Bias correction: Analytical method

The problem of the limiting distribution of $\sqrt{nT}(\hat{\boldsymbol{\beta}} - \boldsymbol{\beta}_0)$ not being centered at zero is that the usual confidence intervals based on the asymptotic approximation will be incorrect. In particular, even if $\mathbf{b}$ is small, the asymptotic bias can be of moderate size when the ratio n/T is large. Here, we shall consider the bias correction to the FE-SQR estimator.

We first consider a one-step bias correction based on the analytic form of the asymptotic bias. Put $\hat{u}_{it} := y_{it} - \hat{\alpha}_i - \mathbf{x}_{it}'\hat{\boldsymbol{\beta}}$. The terms $f_i := f_i(0), s_i, \gamma_i, \boldsymbol{\nu}_i$ and Γ can be estimated by

$$
\hat{f}_i := \frac{1}{T}\sum_{t=1}^{T} K_{h_n}(\hat{u}_{it}), \quad \hat{s}_i := \frac{1}{\hat{f}_i}, \quad \hat{\gamma}_i := \frac{\hat{s}_i}{T}\sum_{t=1}^{T} K_{h_n}(\hat{u}_{it})\mathbf{x}_{it},
$$

$$
\hat{\boldsymbol{\nu}}_i := \frac{1}{Th_n^2}\sum_{t=1}^{T} K^{(1)}(\hat{u}_{it}/h_n)(\mathbf{x}_{it} - \hat{\gamma}_i), \quad \hat{\Gamma}_n := \frac{1}{nT}\sum_{i=1}^{n}\sum_{t=1}^{T} K_{h_n}(\hat{u}_{it})\mathbf{x}_{it}(\mathbf{x}_{it}' - \hat{\gamma}_i'),
$$

where $K^{(1)}(u) = dK(u)/du$. The estimation of the terms $\omega_{ni}^{(1)}, \omega_{ni}^{(2)}$ and $\omega_{ni}^{(3)}$ is a more delicate issue, since it reduces to the estimation of long-run covariances. As in Hahn and Kuersteiner (2011), we make use of a truncation strategy. Define

$$
\phi_i(j) := \int_{-\infty}^{0} f_{i,j}(0, u) du,
$$

$$
\varphi_i(j) := \mathrm{E}\left[\mathbf{x}_{i1} \int_{-\infty}^{0} f_{i,j}(0, u | \mathbf{x}_{i1}, \mathbf{x}_{i,1+j}) du \right],
$$

$$
\varrho_i(j) := \mathrm{E}[I(u_{i1} \leqslant 0)I(u_{i,1+j} \leqslant 0)].
$$

Since $\phi_i(j) \approx \mathrm{E}[K_{h_n}(u_{i1})I(u_{i,1+j} \leqslant 0)]$, it can be estimated by

$$
\hat{\phi}_i(j) := \frac{1}{T}\sum_{t=\max\{1, -j+1\}}^{\min\{T, T-j\}} K_{h_n}(\hat{u}_{it})I(\hat{u}_{i,t+j} \leqslant 0).
$$

Similarly, $\varphi_i(j)$ can be estimated by

$$
\hat{\varphi}_i(j) := \frac{1}{T}\sum_{t=\max\{1, -j+1\}}^{\min\{T, T-j\}} K_{h_n}(\hat{u}_{it})I(\hat{u}_{i,t+j} \leqslant 0)\mathbf{x}_{it}.
$$

The term $\varrho_i(j)$ can be estimated by its sample analog:

$$
\hat{\varrho}_i(j) := \frac{1}{T}\sum_{t=\max\{1, -j+1\}}^{\min\{T, T-j\}} I(\hat{u}_{it} \leqslant 0)I(\hat{u}_{i,t+j} \leqslant 0).
$$

Take a sequence m_n such that $m_n \to \infty$ sufficiently slowly. Then $\omega_{ni}^{(1)}, \omega_{ni}^{(2)}$ and $\omega_{ni}^{(3)}$ can be estimated by

$$\hat{\omega}_{ni}^{(1)} := \sum_{1 \le |j| \le m_n} (1 - |j|/T) \{ \tau \hat{f}_i - \hat{\phi}_i(j) \},$$

$$\hat{\omega}_{ni}^{(2)} := \sum_{1 \le |j| \le m_n} (1 - |j|/T) \{ \tau \hat{f}_i \hat{\gamma}_i - \hat{\varphi}_i(j) \},$$

$$\hat{\omega}_{ni}^{(3)} := \tau(1 - \tau) + \sum_{1 \le |j| \le m_n} (1 - |j|/T) \{ -\tau^2 + \hat{\varrho}_i(j) \}.$$

The bias term $\mathbf{b}$ is thus estimated by

$$\hat{\mathbf{b}} := \hat{\Gamma}_n^{-1} \left\{ \frac{1}{n} \sum_{i=1}^n \hat{s}_i \left(\hat{\omega}_{ni}^{(1)} \hat{\gamma}_i - \hat{\omega}_{ni}^{(2)} + \frac{\hat{s}_i \hat{\omega}_{ni}^{(3)} \hat{\nu}_i}{2} \right) \right\}.$$

We define the one-step bias-corrected estimator by $\hat{\beta}^1 := \hat{\beta} - \hat{\mathbf{b}}/T$. In practice, there is no need to compute the the terms $\tau \hat{f}_i$ and $\tau \hat{f}_i \hat{\gamma}_i$ in $\hat{\omega}_{ni}^{(1)}$ and $\hat{\omega}_{ni}^{(2)}$, respectively, as they are canceled out by the difference $\hat{\omega}_{ni}^{(1)} \hat{\gamma}_i - \hat{\omega}^{(2)}$. In addition, there is no need to use the same kernel and the same bandwidth to estimate β_0 and $\mathbf{b}$.

Galvao and Kato (2016) showed that the bias-corrected estimator, $\hat{\beta}^1$, has the limiting normal distribution with mean zero and the same covariance matrix as $\hat{\beta}$:

$$\sqrt{nT}(\hat{\beta}^1 - \beta_0) \xrightarrow{d} N(\mathbf{0}, \Gamma^{-1} V \Gamma^{-1}),$$

when $m_n \to \infty$ such that $m_n^2 (\log n)/(T h_n^2) \to 0$.

19.3.2.2 Bias correction: Jackknife

Galvao and Kato (2016) also considered the half-panel jackknife method originally proposed by Dhaene and Jochmans (2015), which is an automatic way of removing the bias of $\hat{\beta}$. Suppose for a moment that T is even. Partition $\{1, \dots, T\}$ into two subsets, $S_1 := \{1, \dots, T/2\}$ and $S_2 := \{T/2 + 1, \dots, T\}$. Let $\hat{\beta}_{S_l}$ be the FE-SQR estimate based on the data $\{(y_{it}, \mathbf{x}_{it}), 1 \le i \le n, t \in S_l\}$ for $l = 1, 2$. The half-panel jackknife estimator is defined as $\hat{\beta}_{1/2} := 2\hat{\beta} - \bar{\beta}_{1/2}$, where $\bar{\beta}_{1/2} := (\hat{\beta}_{S_1} + \hat{\beta}_{S_2})/2$. For simplicity, suppose for a moment that we use the same bandwidth to construct $\hat{\beta}$ and $\hat{\beta}_{S_l}$ ($l = 1, 2$). Then, from the asymptotic representation of the FE-SQR estimator, it can be shown that when $n/T \to \rho$ and under some regularity conditions,

$$\sqrt{nT}(\hat{\beta}_{1/2} - \beta_0) \xrightarrow{d} N(\mathbf{0}, \Gamma^{-1} V \Gamma^{-1}).$$

The half-panel jackknife estimator is attractive in both theoretical and practical senses. In fact, its validity does not require any extra condition. It is also preferable from a practical point of view since it does not require the nonparametric estimation of the bias term and at the same time is easy to implement.

Practically computing the FE-SQR, the bias-corrected estimator, and their corresponding asymptotic variance–covariance matrices requires choosing kernels and bandwidths. Existing theory provides little guidance on how to do this for statistical inferences, and there are no current good recommendations for making these choices. Nevertheless, Galvao and Kato (2016) provide numerical experiments using a fourth-order kernel,

$K(u) = \frac{105}{64}(1 - 5u^2 + 7u^4 - 3u^6)I(|u| \leqslant 1)$, together with bandwidth $h_{1n} = c_1 s_1 (nT)^{-1/7}$ to construct the FE-SQR estimate, where c_1 is some constant and s_1 is the sample standard deviation of errors of the standard FE-QR, and $h_{2n} = c_2 s_2 T^{-1/5}$ to construct the bias estimate, where c_2 is some constant and s_2 is the sample standard deviation of errors of the FE-SQR. These choices are only for simplicity but provide good small-sample results.

19.3.3 Alternative FE approaches

19.3.3.1 Shrinkage

Koenker (2004) proposed a penalized estimation method where the individual effects are treated as pure location shift parameters common to all quantiles and subject to the ℓ_1 penalty. It is well known that the optimal estimator for the random effects Gaussian model involves shrinking the individual effects toward a common value. When there is an intercept in the model this common value can be taken to be the conditional central tendency of the response at a point determined by the centering of the other covariates. In the QR model this would be some corresponding conditional quantile of the response.

Shrinkage for QR longitudinal data may be advantageous in controlling the variability introduced by the large number of estimated individual-specific parameters. In addition, by shrinking one is effectively reducing the number of parameters to be estimated. For the quantile loss function, it is convenient to consider the ℓ_1 penalty, in place of the conventional Gaussian penalty. This choice maintains the linear programming form of the problem and also preserves the sparsity of the resulting design matrix.[3]

Specifically, Koenker (2004) considered the model

$$Q_{y_{it}}(\tau \mid x_{it}, \alpha_i) = \alpha_i + \mathbf{x}'_{it}\boldsymbol{\beta}(\tau), \quad i = 1, \ldots, n, \ t = 1, \ldots, T, \tag{19.8}$$

where y_{it} is the outcome of interest, $\mathbf{x}_{it}$ are the regressors, α_i is the individual effect, and $\tau \in (0,1)$ is a quantile index. Model (19.8) differs from that considered in Kato et al. (2012) since in the latter the individual effects are allowed to depend on the quantile, while in (19.8) they are not. Koenker (2004) proposed the following penalized estimator $(\widehat{\boldsymbol{\alpha}}(\lambda), \widehat{\boldsymbol{\beta}}(\lambda, \tau))$:

$$(\widehat{\boldsymbol{\alpha}}(\lambda), \widehat{\boldsymbol{\beta}}(\lambda, \tau_1), \ldots, \widehat{\boldsymbol{\beta}}(\lambda, \tau_q)) := \arg\min \frac{1}{nT} \sum_{k=1}^{q} \sum_{i=1}^{n} \sum_{t=1}^{T} w_k \rho_\tau (y_{it} - \alpha_i - \mathbf{x}'_{it}\boldsymbol{\beta}(\tau_k)) + \lambda \sum_{i=1}^{n} |\alpha_i|,$$

where $w_1, \ldots, w_q$ are nonnegative weights, and $\lambda \geqslant 0$ is the penalty level. This penalized estimation procedure depends on the tuning parameter λ. Lamarche (2010) proposes methods to select the penalty level. The tuning parameter affects the asymptotic variance, and thus in practice, one can select the penalty by minimizing the trace of the estimated asymptotic variance matrix.

19.3.3.2 Minimum distance

Galvao and Wang (2015) proposed a minimum-distance (MD) QR estimator for panels with FEs. The feasible MD-QR estimator is defined as

$$\widehat{\boldsymbol{\beta}}_{\text{MD}} = \left(\sum_{i=1}^{n} \widehat{V}_i^{-1} \right)^{-1} \sum_{i=1}^{n} \widehat{V}_i^{-1} \widehat{\boldsymbol{\beta}}_i, \tag{19.9}$$

---

[3]The ℓ_1-penalization or "lasso" was proposed by Tibshirani (1996) for mean regression. Belloni and Chernozhukov (2011) studied the properties of the ℓ_1-penalized quantile regression in detail.

where $\widehat{\beta}_i$ is the QR estimate for β_0 using the ith time series data only, and $\widehat{V}_i$ is a consistent estimator of V_i, which is the (asymptotic) covariance matrix of $\widehat{\beta}_i$ for each individual.

The feasible two-step estimator can be obtained by first computing QR estimates $\widehat{\beta}_i$ for all individuals, together with consistent estimates for their asymptotic covariance matrices $\widehat{V}_i$, and then the estimates are substituted into (19.9).

Building on Kato et al. (2012), Galvao and Wang (2015) showed asymptotic normality of the MD-QR estimator under sequential and simultaneous asymptotics. Specifically, under suitable regularity conditions, we have that

$$\sqrt{nT}(\widehat{\beta}_{\mathrm{MD}} - \beta_0) \xrightarrow{d} N(0, V),$$

where $V := \lim_{n\to\infty} \left(n^{-1}\sum_{i=1}^{n} V_i^{-1}\right)^{-1}$. For the simultaneous limits, the requirements on the sample growth are similar to those in Kato et al. (2012), that is, $(T, n) \to \infty$ and $\frac{n^2(\log n)}{T}\left|\log\frac{(\log n)^{0.5}}{T^{0.5}}\right|^2 \to 0$.

Practical inference in FE-QR models has been a source of investigation as well. Yoon and Galvao (2015) develop cluster robust inference methods for standard panel QR models with FEs, allowing arbitrary temporal correlation structure within each individual. They build on the asymptotic results derived in Kato et al. (2012) and propose a clustered covariance matrix estimator, which is the quantile analog to that of Arellano (1987), and also suggest two cluster robust tests. The first considers a variant of the quantile rank score test, while the second considers a conventional Wald test for linear restrictions.

19.3.3.3   Two-step estimation of Canay (2011)

Canay (2011) suggested a variant of a two-step estimation procedure assuming that the individual effects are independent of τ. Canay (2011) begins with a random coefficient model

$$y_{it} = \mathbf{x}_{it}'\boldsymbol{\theta}(u_{it}) + \alpha_i,$$

where $(y_{it}, \mathbf{x}_{it})$ are observable and (u_{it}, α_i) are unobservable. Further, suppose that the map $\tau \mapsto \mathbf{x}_{it}'\boldsymbol{\theta}(\tau)$ is strictly increasing in $\tau \in (0, 1)$, and u_{it} has uniform distribution on $[0, 1]$ independent of $(\mathbf{x}_{i1}, \ldots, \mathbf{x}_{iT}, \alpha_i)$. Then for each $\tau \in (0, 1)$,

$$\mathrm{P}(y_{it} \leqslant \mathbf{x}_{it}'\boldsymbol{\theta}(\tau) + \alpha_i \mid \mathbf{x}_i, \alpha_i) = \tau,$$

where $\mathbf{x}_i = (\mathbf{x}_{i1}, \ldots, \mathbf{x}_{iT})$. Canay (2011) considered the auxiliary equation

$$y_{it} = \mathbf{x}_{it}'\boldsymbol{\theta}_\mu + \alpha_i + v_{it}, \quad \mathrm{E}[v_{it} \mid \mathbf{x}_i, \alpha_i] = 0,$$

where $\theta_\mu = \mathrm{E}[\theta(u_{it})]$ and $v_{it} = \mathbf{x}_{it}'(\boldsymbol{\theta}(u_{it}) - \boldsymbol{\theta}_\mu)$. From this equation, given a consistent estimator of θ_μ, one is able to construct an estimator of α_i. The second step is then to estimate $\boldsymbol{\theta}(\tau)$ by applying standard QR to $y_{it} - \widehat{\alpha}_i$ on $\mathbf{x}_{it}$. Formally, this two-step estimation procedure can be described as follows:

Step 1. Let $\widehat{\boldsymbol{\theta}}_\mu$ be an estimator of $\boldsymbol{\theta}_\mu$, and define $\widehat{\alpha}_i = T^{-1}\sum_{t=1}^{T}(y_{it} - \mathbf{x}_{it}'\widehat{\boldsymbol{\theta}}_\mu)$.

Step 2. Let $\widehat{y}_{it} = y_{it} - \widehat{\alpha}_i$, and define the two-step estimator $\widehat{\boldsymbol{\theta}}(\tau)$ by

$$\widehat{\boldsymbol{\theta}}(\tau) := \arg\min_{\boldsymbol{\theta}} \frac{1}{nT}\sum_{i=1}^{n}\sum_{t=1}^{T} \rho_\tau(\widehat{y}_{it} - \mathbf{x}_{it}'\boldsymbol{\theta}).$$

19.4 Correlated random effects

From the basic model (19.1), several different types of panel data models arise from the assumptions concerning the unobservable α_i. In Section 19.3 we discussed the fixed effects model, where we imposed no restriction on the joint distribution between the unobserved individual effect α_i and the observable regressors $\mathbf{x}_{it}$. However, despite its flexibility, the validity of asymptotic approximations of the distributions of the estimators requires $T \to \infty$ at some rates. In this section we discuss alternative solutions focused upon models that allow for dependence between α_i and $\mathbf{x}_{it}$ in a more restrictive form, but are able to allow for short panels.

The correlated random effects (CRE) approach provides an alternative solution to the incidental parameters problem in panel QR models. Abrevaya and Dahl (2008) suggested a CRE model similar to that of Chamberlain (1982, 1984). An important distinction between the standard CRE and FE models is that once modeling the unobserved individual heterogeneity, one is able to circumvent the incidental parameters problem and restrict the time dimension to be fixed.

Abrevaya and Dahl (2008) considered the CRE-QR model with the unobservable α_i as a linear projection onto the observables plus a disturbance. The baseline model is

$$y_{it} = \alpha_i + \mathbf{x}'_{it}\boldsymbol{\beta} + u_{it},$$

while the relationship between α_i and $\mathbf{x}_i = (\mathbf{x}_{i1}, \ldots, \mathbf{x}_{iT})$ is described (in general) by

$$\alpha_i = \phi(\mathbf{x}_i) + v_i, \quad \mathrm{E}[v_i \mid \mathbf{x}_i] = 0.$$

In this model the number of time periods $T \geqslant 2$ for each cross-sectional unit is assumed to be fixed, whereas the cross-sectional dimension $n \to \infty$. The random variables $\{(\mathbf{x}_{i1}, \ldots, \mathbf{x}_{iT}, u_{i1}, \ldots, u_{iT}, \alpha_i)\}_{i=1}^n$ are assumed to be i.i.d. draws from their underlying distributions. The observed data are $\{(y_{i1}, \ldots, y_{iT}, \mathbf{x}_{i1}, \ldots, \mathbf{x}_{iT})\}_{i=1}^n$.

For each $\tau \in (0,1)$, the conditional τth quantile of y_{it} is

$$Q_{y_{it}}(\tau \mid \mathbf{x}_{it}) = \mathbf{x}'_{it}\boldsymbol{\beta} + \phi(\mathbf{x}_i) + Q_{v_i + u_{it}}(\tau \mid \mathbf{x}_i). \tag{19.10}$$

Then Abrevaya and Dahl (2008) made the following assumptions: first, that v_i is independent of $\mathbf{x}_i$; and second, that $Q_{u_{it}}(\tau \mid \mathbf{x}_i, v_i) = Q_{u_{it}}(\tau \mid \mathbf{x}_{it})$. These assumptions imply that $Q_{v_i + u_{it}}(\tau \mid \mathbf{x}_i) = Q_{v_i + u_{it}}(\tau \mid \mathbf{x}_{it}) \equiv g_{\tau,t}(\mathbf{x}_{it})$, and hence the conditional quantile function can be written as

$$Q_{y_{it}}(\tau \mid \mathbf{x}_{it}) = \mathbf{x}'_{it}\boldsymbol{\beta} + g_{\tau,t}(\mathbf{x}_{it}) + \phi(\mathbf{x}_i). \tag{19.11}$$

Abrevaya and Dahl (2008) specifically assumed that the functions $g_{\tau,t}(\mathbf{x}_{it})$ and $\phi(\mathbf{x}_i)$ are linear and estimate the coefficients by applying linear QR. Since the number of parameters is fixed in this setup, the incidental parameters problem does not arise.

Arellano and Bonhomme (2016) introduced an alternative class of QR estimators for short panels. The model has a correlated random effects flavor. The outcomes y_{it} are specified as a function of covariates $\mathbf{x}_{it}$ and latent heterogeneity α_i,

$$y_{it} = \sum_{k=1}^{K_1} \beta_k(U_{it}) g_k(\mathbf{x}_{it}, \alpha_i),$$

and similarly the dependence of α_i on covariates $\mathbf{x}_i = (\mathbf{x}'_{i1}, \ldots, \mathbf{x}'_{iT})'$ is given by

$$\alpha_i = \sum_{k=1}^{K_2} \delta_k(V_i) h_k(\mathbf{x}_i),$$

where $U_{i1}, \ldots, U_{iT}, V_i$ are independent uniform random variables, and the gs and hs belong to some family of functions. Outcomes y_{it} and heterogeneity α_i are monotone in U_{it} and V_i, respectively, so the last two equations are models of conditional quantile functions. The second equation characterizes the correlated random effects model. The specifications of the two equations above allow for flexible patterns of interactions between covariates and heterogeneity at various quantiles, and although stated for the static case and a scalar unobserved effect, the framework allows for dynamics and multi-dimensional latent components. Arellano and Bonhomme (2016) are able to allow for multi-dimensional latent components because they enter as conditioning variables in the model, and moreover, identification and estimation rely on integrating these unobservables out.

The main econometric challenge is that there are no data on heterogeneity α_i. As α_i is not observed, Arellano and Bonhomme (2016) use an iterative approach for estimation by construct some imputations. The resulting algorithm is a variant of the expectation-maximization algorithm, sometimes referred to as "stochastic EM." Arellano and Bonhomme (2016) characterized the asymptotic distribution of the sequential method-of-moments estimators.

19.5 Extensions

19.5.1 Endogeneity

Endogeneity is an important concern in economics. Chernozhukov and Hansen (2006) developed an instrumental variables (IV) estimator for quantile regression. Galvao (2011) and Harding and Lamarche (2009) apply the Chernozhukov–Hansen approach to panel data.

Consider the setting in Section 19.2, but suppose that we now have endogenous variables d_{it} in addition to exogenous variables x_{it}, and we have a vector of instruments w_{it} such that

$$P(y_{it} \leqslant \alpha_i(\tau, \eta_i) + d_{it}'\mu(\tau) + x_{it}'\beta(\tau) \mid x_{it}, w_{it}) = \tau.$$

As usual, let τ be fixed, and omit the dependence on τ. Ignoring endogeneity in d_{it} will lead to inconsistent estimation of the parameters, and so simple application of FE-QR is not recommended. The Chernozhukov–Hansen approach adapted to estimation of the above model is described as follows. First, for each fixed μ, run QR regression:

$$(\widehat{\alpha}(\mu), \widehat{\beta}(\mu), \gamma(\mu)) = \arg \min_{\alpha, \beta, \gamma} \sum_{i=1}^{n} \sum_{t=1}^{T} \rho_\tau(y_{it} - \alpha_i - d_{it}'\mu - x_{it}'\beta - w_{it}'\gamma).$$

Then estimate μ by

$$\widehat{\mu} = \arg \min_{\mu} \|\widehat{\gamma}(\mu)\|.$$

The final estimator for (α, β, μ) is given by $(\widehat{\alpha}(\widehat{\mu}), \widehat{\beta}(\widehat{\mu}), \widehat{\mu})$. See Galvao (2011) and Harding and Lamarche (2009) for details.

19.5.2 Censoring

Censored observations are common in applied work. There is a growing literature on QR methods for panel data with censored observations. Wang and Fygenson (2009) developed inference procedures for censored QR models with longitudinal data where some of the measurements are censored by fixed constants. In the analysis, they examine and account for both the effects of fixed censoring in the dependent variable and the longitudinal nature of the observations. Wang and Fygenson (2009) considered a random effects model where the random effect is part of the error term, and proposed a rank score test for large-sample inference on a subset of the covariates, and importantly, they account for the intra-subject dependence induced by the random effects term.

Galvao et al. (2013) studied estimation of a panel QR model with fixed effects and censoring. Let y_{it}^* denote the potentially left censored tth response of the ith individual and let $y_{it} = \max(C_{it}, y_{it}^*)$ be its corresponding observed value, where C_{it} is a known censoring point. Suppose that y_{it}^* is independent of the censoring point C_{it} conditionally on covariates, $\mathbf{x}_{it}$, and an individual effect, η_i. Given a quantile index $\tau \in (0,1)$, consider as in Section 19.2 the model

$$y_{it}^* = \alpha_i(\tau, \eta_i) + \mathbf{x}_{it}'\boldsymbol{\beta}_0(\tau) + u_{it}(\tau), \quad i = 1, \ldots, n, \ t = 1, \ldots, T,$$

where $\mathbf{x}_{it}$ is a $p \times 1$ vector of regressors, $\boldsymbol{\beta}_0(\tau)$ is a $p \times 1$ vector of parameters, and $u_{it}(\tau)$ is the error term whose τth conditional quantile is zero. Then the conditional quantile function of y_{it}^* is described as

$$Q_{y_{it}^*}(\tau \mid \mathbf{x}_{it}, \eta_i) = \alpha_i(\tau, \eta_i) + \mathbf{x}_{it}'\boldsymbol{\beta}_0(\tau). \tag{19.12}$$

By equivariance of the quantile function to monotone transformation, we arrive at a version of the Powell (1986) censored QR model:

$$Q_{y_{it}}(\tau \mid \mathbf{x}_{it}, \eta_i, C_{it}) = \max\{C_{it}, \alpha_i(\tau, \eta_i) + \mathbf{x}_{it}'\boldsymbol{\beta}_0(\tau)\}.$$

In what follows, we condition on $\eta_i = \eta_{i0}$, and let $\alpha_{i0}(\tau) = \alpha_i(\tau, \eta_{i0})$. In addition, we fix $\tau \in (0,1)$ and suppress the dependence on τ.

An FE-QR estimator $(\widehat{\boldsymbol{\alpha}}, \widehat{\boldsymbol{\beta}})$ could be defined by a minimizer of the following objective function:

$$Q_{1,n}(\boldsymbol{\alpha}, \boldsymbol{\beta}) = \frac{1}{nT} \sum_{i=1}^{n} \sum_{t=1}^{T} \rho_\tau(y_{it} - \max(C_{it}, \alpha_i + \mathbf{x}_{it}'\boldsymbol{\beta})), \tag{19.13}$$

where $\boldsymbol{\alpha} := (\alpha_1, \ldots, \alpha_n)'$. Despite its intuitive appeal, this estimator has not become popular in empirical research due to its computational difficulty. But it can be shown that a minimizer of (19.13) is asymptotically equivalent to a minimizer of

$$Q_{2,n}(\boldsymbol{\alpha}, \boldsymbol{\beta}) = \frac{1}{nT} \sum_{i=1}^{n} \sum_{t=1}^{T} \rho_\tau(y_{it} - \alpha_i - \mathbf{x}_{it}'\boldsymbol{\beta})1(\alpha_{i0} + \mathbf{x}_{it}'\boldsymbol{\beta}_0 > C_{it}). \tag{19.14}$$

Let $\delta_{it} = 1(y_{it}^* > C_{it})$. Because $\pi_{i0}(\mathbf{x}_{it}, \alpha_{i0}, C_{it}) := \mathrm{P}(\delta_{it} = 1 \mid \mathbf{x}_{it}, \eta_{i0}, C_{it}) = \mathrm{P}(u_{it} > -\alpha_{i0} - \mathbf{x}_{it}'\boldsymbol{\beta}_0 + C_{it} \mid \mathbf{x}_{it}, \eta_{i0}, C_{it})$ and $\mathrm{P}(u_{it} > 0 \mid \mathbf{x}_{it}, \eta_{i0}, C_{it}) = 1 - \tau$, and noting that the restriction set selects those observations (i,t) where the conditional quantile line is above the censoring point C_{it}, the objective function (19.14) is (asymptotically) equivalent to

$$Q_{3,n}(\boldsymbol{\alpha}, \boldsymbol{\beta}) = \frac{1}{nT} \sum_{i=1}^{n} \sum_{t=1}^{T} \rho_\tau(y_{it} - \alpha_i - \mathbf{x}_{it}'\boldsymbol{\beta})1(\pi_{i0}(\mathbf{x}_{it}, \alpha_{i0}, C_{it}) > 1 - \tau). \tag{19.15}$$

These observations suggest estimating $\boldsymbol{\beta}_0$ by applying FE-QR to the subset $\{(i,t) : \pi_{i0}(\mathbf{x}_{it}, \alpha_{i0}, C_{it}) > 1 - \tau\}$, including all the observations, even censored ones, for which the true τth conditional quantile is above the censoring point C_{it}. However, in applications, the true propensity score function $\pi_{i0}(\cdot)$ is unknown. Thus we would first estimate $\pi_{i0}(\cdot)$, only using the values of δ_{it} and regressors. From this step, the fitted function would be used to determine the observations to be included in panel QR. Thus the estimation consists of the following two steps:

1. Estimate $\pi_{i0}(\mathbf{x}_{it}, \alpha_{i0}, C_{it})$ by using either a parametric or nonparametric regression method for binary data, and denote the estimated conditional probability by $\widehat{\pi}_i(\mathbf{x}_{it}, \alpha_{i0}, C_{it})$. Determine the informative subset $J_T = \{(i,t) : \widehat{\pi}_i(\mathbf{x}_{it}, \alpha_{i0}, C_{it}) > 1 - \tau + c_n\}$, where c_n is a pre-specified small positive value with $c_n \to 0$ as $n \to \infty$.

2. Then $\boldsymbol{\theta}_0 = (\boldsymbol{\alpha}_0', \boldsymbol{\beta}_0')'$ can be estimated by applying FE-QR to the subset J_T, that is, $\widehat{\boldsymbol{\theta}} = (\widehat{\boldsymbol{\alpha}}', \widehat{\boldsymbol{\beta}}')'$ is a minimizer of the objective function

$$Q_n(\boldsymbol{\alpha}, \boldsymbol{\beta}, \widehat{\pi}) = \frac{1}{nT} \sum_{i=1}^{n} \sum_{t=1}^{T} \rho_\tau (y_{it} - \alpha_i - \mathbf{x}_{it}'\boldsymbol{\beta}) 1\left(\widehat{\pi}_i(\mathbf{x}_{it}, \alpha_{i0}, C_{it}) > 1 - \tau + c_n\right). \tag{19.16}$$

In numerical experiments Galvao et al. (2013) suggest $c_n = (nT)^{-1/5}\tau$. In addition, Galvao et al. (2013) derived the asymptotic properties of the proposed two-step estimator when n and T jointly go to infinity, building upon Kato et al. (2012). Thus, inference procedures impose a large T requirement with a restriction that T grows at most polynomially in n.

Consistent estimation of $\pi_{i0}(\mathbf{x}_{it}, \alpha_{i0}, C_{it})$ is simple once we assume a parametric model: $\pi_{i0}(\mathbf{x}_{it}, \alpha_{i0}, C_{it}) = p(\dot{\mathbf{x}}_{it}'\gamma_0)$, where $p(\cdot)$ is a link function and $\dot{\mathbf{x}}_{it}$ is the vector consisting of $\mathbf{z}_{it}$ and $\mathbf{x}_{it}$, with $\mathbf{z}_{it}$ an n-dimensional indicator variable for the individual effect. When the parametric form of the true propensity score is unknown, then one can obtain a consistent estimator of π_{i0} by applying nonparametric or semiparametric methods. Although the nonparametric methods for the first stage are attractive, they are practical only in low dimensions, have slow convergence rates, and might not allow for categorical data. Thus Galvao et al. (2013) use parametric regression to estimate the conditional censoring probability, and consider a three-step estimator for the censored QR model.

The three-step estimation consists of the following steps. The first step selects the sample $J_0 = \{(i,t) : p(\dot{\mathbf{x}}_{it}'\widehat{\gamma}) > 1 - \tau + d\}$, where d is strictly between 0 and τ and $p(\cdot)$ is a parametric link function, for instance a logit function. The goal of the first step is to select some, and not necessarily the largest, subset of observations where $\pi_{i0}(\mathbf{x}_{it}, \alpha_{i0}, C_{it}) > 1 - \tau$, that is, where the quantile line $\alpha_{i0} + \mathbf{x}_{it}'\boldsymbol{\beta}_0$ is above the censoring point C_{it}. The second step applies FE-QR to the subset J_0, selecting the subset $J_1 = \{(i,t) : \widehat{\alpha}_i^0 + \mathbf{x}_{it}'\widehat{\boldsymbol{\beta}}^0 > \delta_{nT} + C_{it}\}$, where δ_{nT} is a small positive number such that $\delta_{nT} \downarrow 0$ and $\sqrt{nT} \times \delta_{nT}$ is bounded, and $\widehat{\boldsymbol{\theta}}^0 = (\widehat{\boldsymbol{\alpha}}^{0'}, \widehat{\boldsymbol{\beta}}^{0'})'$ is the second-stage estimator of $\boldsymbol{\theta}_0 = (\boldsymbol{\alpha}_0', \boldsymbol{\beta}_0')'$. Lastly, we take a third step by solving an FE-QR problem on the subset J_1 if $J_0 \subset J_1$, which is denoted by $\widehat{\boldsymbol{\theta}}^1 = (\widehat{\boldsymbol{\alpha}}^{1'}, \widehat{\boldsymbol{\beta}}^{1'})'$.

Naturally, the two-step and three-step procedures have advantages and disadvantages when applied to panel data. On the one hand, in practice, the parametric form of the propensity score is unknown and estimation might be subject to misspecification. On the other hand, the nonparametric two-step estimator requires relatively larger T and additional assumptions that control the degree of smoothness.

19.5.3 Group-level treatments

Chetverikov et al. (2016) developed a methodology for estimating the distributional effects of an endogenous treatment that varies at the group level when there are group-level unobservables, providing a quantile extension of Hausman and Taylor (1981). They proposed a grouped instrumental variables quantile regression estimator and established inference procedures.

We study the following model for the response variable y_{ig} of individual i in group g:

$$y_{ig} = \mathbf{z}_{ig}'\boldsymbol{\gamma}(u_{ig}) + \mathbf{x}_g'\boldsymbol{\beta}(u_{ig}) + \epsilon(u_{ig}, \eta_g),$$

where $\mathbf{z}_{ig}$ and $\mathbf{x}_g$ are d_z- and d_x-vectors of individual- and group-level observable covariates ($\mathbf{x}_g$ contains the constant), η_g is a vector of group-level unobservable covariates, and u_{ig} is a scalar random variable representing individual heterogeneity. To turn this model into a quantile regression, they assume that for any given value of $(\mathbf{z}_{ig}, \mathbf{x}_g, \eta_g)$ on its domain, y_{ig} is increasing in u_{ig}, and u_{ig} is distributed uniformly on $[0,1]$. Given these assumptions, the conditional τth quantile of y_{ig} given $(\mathbf{z}_{ig}, \mathbf{x}_g, \eta_g)$ is

$$Q_{y_{ig}}(\tau \mid \mathbf{z}_{ig}, \mathbf{x}_g, \eta_g) = \mathbf{z}_{ig}'\boldsymbol{\gamma}(\tau) + \mathbf{x}_g'\boldsymbol{\beta}(\tau) + \epsilon(\tau, \eta_g).$$

We have data on G groups with N_g individuals within group g, $g = 1, \ldots, G$.

In this model, $\mathbf{z}_{ig}'\boldsymbol{\gamma}(\tau)$ is the individual effect, and $\mathbf{x}_g'\boldsymbol{\beta}(\tau) + \epsilon(\tau, \eta_g)$ is the group effect. So arbitrary dependence between these effects is allowed. The unobservable component $\epsilon(\tau, \eta_g)$ is modeled as a general nonparametric function, and arbitrary nonlinear effects of the group-level unobservable covariates are also allowed.

The main purpose is to estimate $\boldsymbol{\beta}(\tau)$, and to accomplish this they assume that they have instrumental variables, $\mathbf{w}_g$, that satisfy the condition

$$\mathrm{E}[\epsilon(\tau, \eta_g) \mid \mathbf{w}_g] = 0.$$

Given such IVs and the condition $\sqrt{G} \max_{g=1,\ldots,G}\{(\log N_g)/N_g\} \to 0$, Chetverikov et al. (2016) showed that

$$\sqrt{G}(\hat{\boldsymbol{\beta}}(\cdot) - \boldsymbol{\beta}(\cdot)) \Rightarrow \mathbb{G}(\cdot) \quad \text{in } \ell^\infty(\mathcal{U})$$

where $\mathcal{U}$ is a closed interval in $(0,1)$, and $\mathbb{G}(\cdot)$ is a centered Gaussian process with continuous paths. They propose the following two-step estimation strategy:

1. Estimate $\boldsymbol{\gamma}_g(\tau)$ and $\alpha_g(\tau) = \mathbf{x}_g'\boldsymbol{\beta}(\tau) + \epsilon(\tau, \eta_g)$ using QR:

$$(\hat{\boldsymbol{\gamma}}_g(\tau), \hat{\alpha}_g(\tau)) = \arg \min_{h,a} \sum_{i=1}^{N_g} \rho_\tau(y_{ig} - \mathbf{z}_{ig}'\boldsymbol{h} - a).$$

2. Apply IV regression to $\hat{\alpha}_g(\tau)$ on $\mathbf{x}_g$ using $\mathbf{w}_g$ as an instrument to obtain an estimator $\hat{\boldsymbol{\beta}}(\tau)$ of $\boldsymbol{\beta}(\tau)$, that is, run IV estimation to obtain

$$\alpha_g(\tau) = \mathbf{x}_g'\beta(u) + \epsilon(u, \eta_g).$$

19.5.4   Semiparametric QR for longitudinal data

Wei and He (2006) studied an alternative model for longitudinal QR and propose a global semiparametric QR model. The model includes a nonparametric intercept function of the current measurement time, and can be estimated from longitudinal reference data with irregular measurement times and with some level of robustness against outliers, and it is also flexible for including covariate information.

Suppose that one has n subjects, and the ith subject has m_i measurements at times $t_{i,1}, t_{i,2}, \ldots, t_{i,m_i}$, which are not necessarily evenly spaced. Denote by $Y_{i,j}$ the jth measurement of the ith subject, and $D_{i,j,k} = t_{i,j} - t_{i,j-k}$ is the time distance between the jth and $(j-k)$th measurements. For a given and fixed $\tau \in (0,1)$, consider the following model for $Y_{i,j}$:

$$Y_{i,j} = g_0(t_{i,j}) + \sum_{k=1}^p (a_k + b_k D_{i,j,k}) Y_{i,j-k} + \mathbf{X}_{i,j}'\boldsymbol{\gamma}_0 + e_{i,j}, \tag{19.17}$$

where $i = 1, \ldots, n$, $j = p+1, \ldots, m_i$, $\mathbf{X}_{i,j} = (\mathbf{X}_{i,j,1}, \ldots, \mathbf{X}_{i,j,l})'$ consists of l covariates for the ith subject at time $t_{i,j}$, and $e_{i,j}$ is a random variable whose τth quantile, given the growth path up to the $(j-1)$th measurement and $\mathbf{X}_{i,j}$, is zero. The conditional quantile of $Y_{i,j}$ given the p prior measurements and the covariates $\mathbf{X}_{i,j}$ includes g_0 as a nonparametric intercept function of the current measurement time, an autoregressive function of $Y_{i,j-1}, \ldots, Y_{i,j-p}$ whose coefficients are linear functions of measurement time distances $D_{i,j,k}$, and a linear function of the covariate $\mathbf{X}_{i,j}$.

In general, the subjects take measurements at irregular time intervals. It is natural to assume that the dependence between two measurements varies with their measurement time distance. This motivates the choice of the autoregressive coefficients as functions of $D_{i,j,k}$. If all the subjects take measurements at the fixed time intervals, the coefficients simplify to constants.

In (19.17), $g_0(t)$ is a nonparametric function with a certain degree of smoothness. Suppose that the range of interest in time is $t \in [t_L, t_U]$. Different smoothing methods may be used to estimate g_0, but Wei and He (2006) used regression splines. Specifically, we approximate $g_0(t)$ by a linear combination of B-spline basis functions. Given a set of knots, we denote by $\pi(t) = (\pi_1(t), \pi_2(t), \ldots, \pi_{k_n}(t))'$ the set of k_n B-spline basis functions. Let $\pi_{i,j} = \pi(t_{i,j})$ and $\tilde{\mathbf{X}}_{i,j} = (Y_{i,j-1}, D_{i,j,1}Y_{i,j-1}, \ldots, Y_{i,j-p}, D_{i,j,p}Y_{i,j-p}, \mathbf{X}'_{i,j})'$; then they estimate the τth conditional quantile function of $Y_{i,j}$ by $\pi'_{i,j}\hat{\boldsymbol{\alpha}}_n + \tilde{\mathbf{X}}'_{i,j}\hat{\boldsymbol{\beta}}_n$, where $\hat{\boldsymbol{\alpha}}_n$ and $\hat{\boldsymbol{\beta}}_n$ are obtained by minimizing

$$\sum_{i=1}^{n} \sum_{j=p+1}^{m_i} \rho_\tau(Y_{i,j} - \pi'_{i,j}\boldsymbol{\alpha} - \tilde{\mathbf{X}}'_{i,j}\boldsymbol{\beta}).$$

Wei and He (2006) showed, under some regularity conditions, $\sqrt{n}$-asymptotic normality of $\hat{\boldsymbol{\beta}}_n$.

19.6 Conclusion

In this chapter, we have reviewed the literature on quantile regression models for panel data. We focused on linear models with unobserved individual effects. We described alternative approaches to estimating fixed effects models and briefly presented the corresponding asymptotic theory to carry inference.

There are other important developments in quantile regression for panel data. An example is nonseparable panel models with time homogeneity proposed by Chernozhukov et al. (2013, 2015).

Bibliography

J. Abrevaya and C. M. Dahl. The effects of birth inputs on birthweight: Evidence from quantile estimation on panel data. *Journal of Business and Economic Statistics*, 26: 379–397, 2008.

J. Angrist, V. Chernozhukov, and I. Fernández-Val. Quantile regression under misspecification, with an application to the U.S. wage structure. *Econometrica*, 74:539–563, 2006.

M. Arellano. Computing robust standard errors for within-groups estimators. *Oxford Bulletin of Economics and Statistics*, 49:431–434, 1987.

M. Arellano and S. Bonhomme. Nonlinear panel data estimation via quantile regressions. *Econometrics Journal*, 19(3):C61–C94, 2016.

A. Belloni and V. Chernozhukov. ℓ_1-penalized quantile regression in high-dimensional sparse models. *Annals of Statistics*, 39:82–130, 2011.

I. A. Canay. A simple approach to quantile regression for panel data. *Econometrics Journal*, 14:368–386, 2011.

G. Chamberlain. Multivariate regression models for panel data. *Journal of Econometrics*, 18:5–46, 1982.

G. Chamberlain. Panel data. In Z. Griliches and M. D. Intriligator, editors, *Handbook of Econometrics*, pages 1248–1313. North-Holland, Amsterdam, 1984.

V. Chernozhukov and C. Hansen. Instrumental quantile regression inference for structural and treatment effects models. *Journal of Econometrics*, 132:491–525, 2006.

V. Chernozhukov, I. Fernández-Val, J. Hahn, and W. Newey. Average and quantile effects in nonseparable panel modelss. *Econometrica*, 81:535–580, 2013.

V. Chernozhukov, I. Fernández-Val, S. Hoderlein, H. Holzmann, and W. Newey. Nonparametric identification in panels using quantiles. *Journal of Econometrics*, 188:378–392, 2015.

D. Chetverikov, B. Larsen, and C. Palmer. IV quantile regression for group-level treatments, with an application to the effects of trade on the distribution of wage. *Econometrica*, 84: 809–833, 2016.

G. Dhaene and K. Jochmans. Split-panel jackknife estimation of fixed-effect models. *Review of Economic Studies*, 82:991–1030, 2015.

A. F. Galvao. Quantile regression for dynamic panel data with fixed effects. *Journal of Econometrics*, 164:142–157, 2011.

A. F. Galvao and K. Kato. Smoothed quantile regression for panel data. *Journal of Econometrics*, 193:92–112, 2016.

A. F. Galvao and L. Wang. Efficient minimum distance estimator for quantile regression fixed effects panel data. *Journal of Multivariate Analysis*, 133:1–26, 2015.

A. F. Galvao, C. Lamarche, and L. Lima. Estimation of censored quantile regression for panel data with fixed effects. *Journal of the American Statistical Association*, 108:1075–1089, 2013.

B. S. Graham, J. Hahn, and J. L. Powell. The incidental parameter problem in a non-differentiable panel data model. *Economics Letters*, 105:181–182, 2009.

J. Hahn and G. M. Kuersteiner. Bias reduction for dynamic nonlinear panel models with fixed effects. *Econometric Theory*, 27:1152–1191, 2011.

J. Hahn and W. Newey. Jackknife and analytical bias reduction for nonlinear panel models. *Econometrica*, 72:1295–1319, 2004.

M. Harding and C. Lamarche. A quantile regression approach for estimating panel data models using instrumental variables. *Economics Letters*, 104:133–135, 2009.

J. Hausman and W. Taylor. Panel data and unobservable individual effects. *Econometrica*, 49:1377–1398, 1981.

J. L. Horowitz. Bootstrap methods for median regression models. *Econometrica*, 66:1327–1351, 1998.

K. Kato, A. F. Galvao, and G. Montes-Rojas. Asymptotics for panel quantile regression models with individual effects. *Journal of Econometrics*, 170:76–91, 2012.

R. Koenker. Quantile regression for longitudinal data. *Journal of Multivariate Analysis*, 91:74–89, 2004.

R. Koenker. *Quantile Regression*. Cambridge University Press, Cambridge, 2005.

R. Koenker and Gilbert W. Bassett. Regression quantiles. *Econometrica*, 46:33–49, 1978.

R. Koenker and K. Hallock. Quantile regression: An introduction. Manuscript, University of Illinois at Urbana-Champaign, 2000.

C. Lamarche. Robust penalized quantile regression estimation for panel data. *Journal of Econometrics*, 157:396–408, 2010.

J. Neyman and E. L. Scott. Consistent estimates based on partially consistent observations. *Econometrica*, 16:1–32, 1948.

J. L. Powell. Censored regression quantiles. *Journal of Econometrics*, 32:143–155, 1986.

A. Rosen. Set identification via quantile restrictions in short panels. *Journal of Econometrics*, 166:127–137, 2012.

R. Tibshirani. Regression shrinkage and selection via the lasso. *Journal of the Royal Statistical Society, Series B*, 58:267–288, 1996.

H. J. Wang and M. Fygenson. Inference for censored quantile regression models in longitudinal studies. *Annals of Statistics*, 37:756–781, 2009.

Y. Wei and X. He. Conditional growth charts. *Annals of Statistics*, 34:2069–2097, 2006.

J. Yoon and A. F. Galvao. Robust inference for panel quantile regression models with individual fixed effects and serial correlation. Working paper, 2015.

20

Quantile Regression Applications in Finance

Oliver Linton

Cambridge University, Cambridge, UK

Zhijie Xiao

Boston College, Chesnut Hill, Massachusetts, USA

CONTENTS

20.1 Introduction

Distributional information, such as variance, skewness, value-at-risk, expected shortfall (or conditional versions of these) of returns and other economic variables are very important for investment and decision-making in finance. Using variance as a measure for risk, Markowitz (1952) proposed the mean-variance efficient portfolio that minimizes variance for a given expected return. Even nowadays, the mean-variance portfolio construction is widely used. An important and very popular model that takes some account of distributional information is the ARCH/GARCH model (Engle, 1982; Bollerslev, 1986) where the role of conditional volatility, in addition to the conditional mean, is emphasized. An observed process can be serially uncorrelated but the conditional variances can have significant serial correlation. Since Engle (1982), volatility has played an important role in finance.

Despite the large number of financial applications using information about the mean and variance, other distributional information can also play an important role in financial applications. First of all, stock returns and other financial variables may not be normally distributed. Second, variance may not be a satisfactory measure for risk. Third, empirical and experimental evidence indicate that when people evaluate risk or opportunity, they often depart from the predictions of expected utility. Investors may have different attitudes/preferences on financial securities with different distributions, even when the means and variances are the same across these securities. In particular, researchers have found that the information contained in the left tail of the return distribution is important in risk management and portfolio construction. Left tail measures such as expected shortfall (ES) and value-at-risk (VaR) are now widely used in financial applications (see, for example, Artzner et al., 1999; Bassett et al., 2004). In addition to left tail information, the right tail of return distributions also contains important information and affects investors' behavior in financial markets. Investors not only look at the left tail distribution of returns to control risk, but also consider the right tail distributional information for opportunity (see Xiao, 2014).

Research in finance in the last twenty years has devoted a lot of effort to developing models that can capture investors' behavior more accurately. A leading example is the "cumulative prospect theory" proposed by Tversky and Kahneman (1992). A prospect investor is affected by the whole distribution of the returns process (see, for example, Barberis and Huang, 2008).

In order to use distributional information in financial data one needs an appropriate tool. Quantile regression provides a convenient way of estimating the conditional distribution. The *conditional quantile function* of Y given X is the inverse of the corresponding conditional distribution function, that is,

$$Q_Y(\tau|X) = F_Y^{-1}(\tau|X) = \inf\{y : F_Y(y|X) \geqslant \tau\}.$$

By definition, the conditional quantile function provides a complete description of the whole conditional distribution. The τth conditional quantile function of Y given X can be estimated by appropriate quantile regressions, and, as a byproduct, the conditional density of Y can also be estimated by the difference quotients (numerical derivatives) thereof. Combining these methods with additional restrictions or further efforts, quantile regression can be used in various applications in financial data analysis.

In addition to directly estimating the conditional distributions, quantile regression based methods also provide an important complementary way to study the relationship between variables in financial markets. Distributional or quantile dependence is a well-known empirical feature in finance, and there is a large literature in finance on the study of directional predictability in stock prices. Financial analysts and portfolio managers who make use of quantitative methods typically focus much of their attention on assessing the effectiveness of particular factors for investment purposes. Quantitative investors frequently analyze factor performance using regression based on the familiar least-squares approach. This is effective for understanding the central tendency within a data set, but will often be less useful for assessing the behavior of data points close to the upper or lower extremes within a population. However, from the perspective of active investors or risk managers, the data points at the extremes may be precisely the ones of greatest interest. For such applications, quantile regression can play an important role as an extension of the conventional ordinary least-squares (OLS) method in directing investment strategies.

In this chapter we discuss certain important quantile regression applications in finance. In particular, left tail measures and risk management based on quantile regression are discussed in Section 20.2. The importance of upper quantile information in financial markets is discussed in Section 20.3. Quantile regression applications in portfolio allocation are discussed in Section 20.4. Stochastic dominance applications are considered in Section 20.5. Quantile dependence and the quantilogram are studied in Section 20.6.

20.2 Quantile regression in risk management

The left tails of the distributions of financial variables contain important information on the downside risk present in securities. Left tail quantiles and expected values of the left tail distribution are easily interpretable measures of risk that summarize information regarding the distribution of potential losses. In requiring publicly traded firms to report risk exposure, the Securities and Exchange Commission (SEC) lists these measures of the left tail distribution as disclosure methods "expressing the potential loss in future earnings, fair values, or cash flows from market movements over a selected period of time and with a selected likelihood of occurrence." As a method of estimating the conditional quantiles, quantile regression offers an important approach to estimating left tail measures, such as VaR and ES, that are widely used in risk management.

20.2.1 Value-at-risk

The value-at-risk, as mandated in many current regulatory contexts, is the loss in market value of a security over a given time horizon that is exceeded with probability τ, where τ is often set at a small number, say 0.01 or 0.05. For a time series of returns on an asset, $\{r_t\}_{t=1}^{n}$, the τ (or $100\tau\%$) VaR at time t, denoted by VaR_t, is simply the (negative) τth conditional quantile of returns defined by

$$P\left(r_t < -VaR_t | \mathcal{F}_{t-1}\right) = \tau, \tag{41.1}$$

where $\mathcal{F}_{t-1}$ denotes the information set at time $t-1$, including past values of returns and possibly the value of some covariates X_t.

VaR is a popular tool in the measurement and management of financial risk. This popularity is spurred both by the need of various institutions to manage risk and by government regulations (see Dowd, 2000; Saunders, 1999).

Estimation of VaR has attracted much attention from researchers. Many existing methods of VaR estimation in economics and finance are based on the assumption that financial returns have normal (or conditional normal) distributions (usually with ARCH or GARCH effects). Under the assumption of a conditionally normal returns distribution, the estimation of conditional quantiles is equivalent to estimating the conditional volatility of returns. The massive literature on volatility modeling offers a rich source of parametric methods of this type.

Suppose that the time series of returns is modelled by

$$r_t = \mu_t + \sigma_t \varepsilon_t,$$

where r_t is the return of an asset at time t, and $\mu_t, \sigma_t \in \mathcal{F}_{t-1}$. The random variables ε_t are martingale difference sequences. The conditional value-at-risk of r_t given $\mathcal{F}_{t-1}$ is

$$VaR_t(\tau) = \mu_t + \sigma_t Q_\varepsilon(\tau),$$

where $Q_\varepsilon(\tau)$ denotes the unconditional VaR of the error term ε_t. Assuming conditional normality, the 5% VaR at time t can be computed as

$$VaR_t(0.05) = \mu_t + 1.65\sigma_t,$$

where μ_t and σ_t are the conditional mean and conditional volatility for r_t.

Based on the assumption of a conditionally normal stock returns distribution, the estimation of VaR is equivalent to estimating the conditional volatility of returns. One important practical approach to estimating VaR is the RiskMetrics method (Morgan and Reuters, 1996) with its simple and pragmatic method of modeling the conditional volatility. The forecast for time t variance in the RiskMetrics method is a weighted average of the previous forecast, using weight λ, and of the latest squared innovation, using weight $1 - \lambda$,

$$\sigma_t^2 = \lambda \sigma_{t-1}^2 + (1 - \lambda) r_{t-1}^2, \tag{41.2}$$

where the parameter λ is called the decay factor $(1 > \lambda > 0)$. Conceptually λ should be estimated using a maximum likelihood approach. RiskMetrics simply set it at 0.94 for daily data and 0.97 for monthly data.

More generally, one may model the conditional volatility using the ARCH process, which assumes a general autoregressive structure in volatility,

$$\sigma_t^2 = \omega_0 + \sum_{j=1}^{p} \omega_j \epsilon_{t-j}^2,$$

or the GARCH process,

$$\sigma_t^2 = \omega_0 + \sum_{i=1}^{q} \lambda_i \sigma_{t-i}^2 + \sum_{j=1}^{p} \omega_j \epsilon_{t-j}^2.$$

There is extensive empirical evidence supporting the use of ARCH and GARCH models in conditional volatility estimation. Bollerslev et al. (1992) provide a nice overview of the subject. Sarma et al. (2000) showed that at the 5% level, the AR(1)–GARCH(1,1) model is a preferred model under the conditional normality assumption. The AR(1)–GARCH(1,1) model is specified as

$$\begin{aligned} r_{t+1} &= a_0 + a_1 r_t + \epsilon_{t+1}, \quad \epsilon_{t+1} | \mathcal{F}_t \sim N(0, \sigma_t^2), \\ \sigma_t^2 &= \omega_0 + \omega_1 \sigma_{t-1}^2 + \omega_2 \epsilon_{t-1}^2. \end{aligned} \tag{20.1}$$

The conditional mean equation is modelled as an AR(1) process to account for the weakly autoregressive behavior of daily returns.

However, there is accumulating evidence that financial time series and returns distributions are *not* well approximated by Gaussian models. In particular, it is frequently found that market returns display negative skewness and very large kurtosis. Extreme realizations of returns can adversely affect the performance of estimation and inference designed for Gaussian conditions; this is particularly true of ARCH and GARCH models whose estimation of variances is very sensitive to large innovations. For this reason, research attention has recently shifted toward the development of more robust estimators of conditional quantiles.

Quantile regression is well suited to estimating VaR. Value-at-risk is a conditional quantile by definition. This concept is intimately linked to quantile regression estimation. Suppose that we have a random sample $\{r_1, \ldots, r_n\}$ of returns and that the τth conditional quantile of r_t given $X_t \in \mathcal{F}_{t-1}$ is

$$Q_{r_t}(\tau | X_t) = g(X_t, \beta(\tau)), \tag{20.2}$$

where g is a known function. Then we may consider the quantile regression

$$\widehat{\beta}(\tau) = \arg \min_{\beta \in \mathbb{R}^k} \sum_{t=1}^{n} \rho_\tau (Y_t - g(X_t, \beta)),$$

where $\rho_\tau(u) = u(\tau - I(u < 0))$, and the τth conditional quantile of r_t can be estimated by

$$\widehat{Q}_{r_t}(\tau | X_t) = g(X_t, \widehat{\beta}(\tau)).$$

As a special case, the unconditional quantile can be estimated by setting $X_t = 1$ and $g(X_t, \beta) = \beta$.

One may couple quantile regression with different types of popular volatility models in finance (ARCH, GARCH, etc.) in estimating VaR. Koenker and Zhao (1996) studied quantile regression for linear ARCH models. They consider the linear ARCH(p) process

$$u_t = \sigma_t \cdot \varepsilon_t, \quad \sigma_t = \gamma_0 + \gamma_1 |u_{t-1}| + \cdots + \gamma_p |u_{t-p}|, \tag{20.3}$$

where $0 < \gamma_0 < \infty$, $\gamma_1, \ldots, \gamma_p \geq 0$, and ε_t are independent and identically distributed (i.i.d.) $(0,1)$ random variables with pdf $f(\cdot)$ and cdf $F(\cdot)$. Let $\gamma(\tau) = (\gamma_0 F^{-1}(\tau), \gamma_1 F^{-1}(\tau), \ldots, \gamma_q F^{-1}(\tau))^\top$ and $Z_t = (1, |u_{t-1}|, \ldots, |u_{t-q}|)^\top$. The τ conditional quantile of u_t is given by

$$Q_{u_t}(\tau|\mathcal{F}_{t-1}) = \gamma_0(\tau) + \gamma_1(\tau)|u_{t-1}| + \ldots + \gamma_p(\tau)|u_{t-p}| = \gamma(\tau)^\top Z_t,$$

which can be estimated by the following linear quantile regression of u_t on Z_t:

$$\min_\gamma \sum_t \rho_\tau(u_t - \gamma^\top Z_t), \tag{20.4}$$

where $\gamma = (\gamma_0, \gamma_1, \ldots, \gamma_q)^\top$. The asymptotic behavior of the above quantile regression estimator is analyzed in Koenker and Zhao (1996).

Xiao and Koenker (2009) studied quantile regression estimation of the linear GARCH(p,q) model:

$$u_t = \sigma_t \cdot \varepsilon_t, \tag{20.5}$$
$$\sigma_t = \beta_0 + \beta_1 \sigma_{t-1} + \cdots + \beta_p \sigma_{t-p} + \gamma_1 |u_{t-1}| + \cdots + \gamma_q |u_{t-q}|. \tag{20.6}$$

Corresponding to the above linear GARCH model, the τth conditional quantile of u_t given information $\mathcal{F}_{t-1}$ up to time $t-1$ is given by

$$Q_{u_t}(\tau|\mathcal{F}_{t-1}) = \theta(\tau)^\top Z_t, \tag{20.7}$$

where

$$Z_t^\top = (1, \sigma_{t-1}, \ldots, \sigma_{t-p}, |u_{t-1}|, \ldots, |u_{t-q}|),$$
$$\theta(\tau)^\top = (\beta_0, \ldots, \beta_p, \gamma_1, \ldots, \gamma_q) F^{-1}(\tau).$$

Since Z_t contains σ_{t-k} ($k = 1, \ldots, q$), which in turn depends on unknown parameters $\theta = (\beta_0, \beta_1, \ldots, \beta_p, \gamma_1, \ldots, \gamma_q)$ and past data, quantile regression based on (20.7) is a highly nonlinear optimization problem. Xiao and Koenker (2009) propose a simpler two-stage estimator. Let

$$A(L) = 1 - \beta_1 L - \cdots - \beta_p L^p, \quad B(L) = \gamma_1 + \cdots + \gamma_q L^{q-1}.$$

Under regularity assumptions ensuring that $A(L)$ is invertible, we obtain an ARCH(∞) representation for σ_t:

$$\sigma_t = a_0 + \sum_{j=1}^\infty a_j |u_{t-j}|. \tag{20.8}$$

Thus

$$Q_{u_t}(\tau|\mathcal{F}_{t-1}) = \alpha_0(\tau) + \sum_{j=1}^\infty \alpha_j(\tau)|u_{t-j}|,$$

where $\alpha_j(\tau) = a_j Q_{\varepsilon_t}(\tau)$, $j = 0, 1, 2, \ldots$. For identification, we normalize $a_0 = 1$.

We may consider a truncated quantile autoregression:

$$Q_{u_t}(\tau|\mathcal{F}_{t-1}) \approx a_0(\tau) + a_1(\tau)|u_{t-1}| + \cdots + a_m(\tau)|u_{t-m}|.$$

By choosing m suitably small relative to the sample size n, but large enough to avoid serious bias, we obtain a sieve approximation for the GARCH model. Suppose that we estimate the above truncated quantile autoregression

$$\tilde{\alpha}(\tau) = \arg\min_{\alpha} \sum_{t=m+1}^{n} \rho_\tau \left(u_t - \alpha_0 - \sum_{j=1}^{m} \alpha_j |u_{t-j}| \right) \tag{20.9}$$

at quantiles $\{\tau_1, \ldots, \tau_K\}$, and obtain estimates $\tilde{\alpha}(\tau_k)$, $k = 1, \ldots, K$. Let $\tilde{a}_0 = 1$ in accordance with the identification assumption. Denote

$$\mathbf{a} = [a_1, \ldots, a_m, q_1, \ldots, q_K]^\top, \quad \tilde{\boldsymbol{\pi}} = \left[\tilde{\alpha}(\tau_1)^\top, \ldots, \tilde{\alpha}(\tau_K)^\top \right]^\top,$$

where $q_k = Q_{\varepsilon_t}(\tau_k)$, and $\phi(\mathbf{a}) = g \otimes \alpha$, with $g = [q_1, \ldots, q_K]^\top$ and $\alpha = [1, a_1, a_2, \ldots, a_m]^\top$. Xiao and Koenker (2009) suggest estimating $\mathbf{a}$ by combining information over the K quantile estimates via the following minimum distance method:

$$\tilde{\mathbf{a}} = \arg\min_{\mathbf{a}} \left(\tilde{\boldsymbol{\pi}} - \phi(\mathbf{a}) \right)^\top A_n \left(\tilde{\boldsymbol{\pi}} - \phi(\mathbf{a}) \right), \tag{20.10}$$

and σ_t can be estimated based on $\tilde{\mathbf{a}}$: $\tilde{\sigma}_t = \tilde{a}_0 + \sum_{j=1}^{m} \tilde{a}_j |u_{t-j}|$. The conditional quantile of u_t can then be estimated by a quantile regression of u_t on

$$\tilde{Z}_t = (1, \tilde{\sigma}_{t-1}, \ldots, \tilde{\sigma}_{t-p}, |u_{t-1}|, \ldots, |u_{t-q}|)^\top.$$

Quantile regression can also be applied to other types of ARCH/GARCH models. See, for example, Lee and Noh (2013) and Zheng et al. (2016) for quantile regression on quadratic GARCH models.

Other popularly used quantile models includes the conditional autoregressive value-at-risk (CAViaR) model of Engle and Manganelli (2004). The CAViaR model assumes that the τth conditional quantile of r_t itself follows an autoregression

$$Q_{r_t}(\tau|\mathcal{F}_{t-1}) = \beta_0 + \sum_{i=1}^{p} \beta_i Q_{r_{t-i}}(\tau|\mathcal{F}_{t-i-1}) + \sum_{j=1}^{q} \alpha_j \ell(X_{t-j}), \tag{20.11}$$

where $X_{t-j} \in \mathcal{F}_{t-j}$, and $\mathcal{F}_{t-j}$ is the information set at time $t - j$. This representation suggests that one may model the local dynamics or local correlation directly based on the conditional quantiles. Engle and Manganelli (2004) discussed many choices of $\ell(X_{t-j})$, leading to different specifications of the CAViaR model.

A natural choice of X_{t-j} is the lagged r. When we choose $X_{t-j} = |r_{t-j}|$, we obtain

$$Q_{r_t}(\tau|\mathcal{F}_{t-1}) = \beta_0^* + \sum_{i=1}^{p} \beta_i^* Q_{r_{t-i}}(\tau|\mathcal{F}_{t-i-1}) + \sum_{j=1}^{q} \gamma_j^* |r_{t-j}|. \tag{20.12}$$

Estimation of the CAViaR model is challenging. If we denote the vector of unknown parameters by θ, and, for simplicity, denote $Q_{r_t}(\tau|\mathcal{F}_{t-1})$ by $Q_t(\tau, \theta)$, then we may consider estimating θ by minimizing the objective function

$$RQ_n(\tau, \theta) = \sum_t \rho_\tau(Y_t - Q_t(\tau, \theta)), \tag{20.13}$$

where $Q_t(\tau, \theta) = \beta_0 + \sum_{i=1}^{p} \beta_i Q_{t-i}(\tau, \theta) + \sum_{j=1}^{q} \alpha_j \ell(X_{t-j})$.

Since conditional quantiles enter the CAViaR regression model as regressors and they are latent, conventional nonlinear quantile regression techniques are not directly applicable. Because the linear GARCH process has a CAViaR$(p; q)$ representation, in this sense, the two-step estimation procedure proposed in Xiao and Koenker (2009) provides a method of estimating a class of CAViaR models. De Rossi and Harvey (2009) studied an iterative Kalman filter method to calculate dynamic conditional quantiles that may be applied to calculate certain types of CAViaR models. In their model, the observed time series Y_t is described by the measurement equation

$$Y_t = \xi_t(\tau) + \varepsilon_t(\tau),$$

where $\xi_t(\tau) = Q_{Y_t}(\tau|\mathcal{F}_{t-1})$ is the state variable, and the disturbances $\varepsilon_t(\tau)$ are assumed to be serially independent and independent of $\xi_t(\tau)$. The dynamics of this system is characterized by the state transition equation based on $\xi_t(\tau)$. For example, if the conditional quantiles follow an autoregression, we have

$$\xi_t(\tau) = \beta \xi_{t-1}(\tau) + \eta_t(\tau).$$

Alternative forms of the state transition equations can be considered. The above state space model can then be estimated by iteratively applying an appropriate signal extraction algorithm.

Chernozhukov and Hong (2003) studied the Laplace-type estimators using Markov chain Monte Carlo (MCMC) methods, which can be applied to CAViaR models. Hsu (2010) also studied the estimation of CAViaR by an MCMC method. Following Yu and Moyeed (2001), she considered the asymmetric Laplace density as a working conditional density for the error term in the CAViaR model

$$f(\varepsilon_{t\tau}|\mathcal{F}_{t-1}) = \frac{\tau(1-\tau)}{\sigma} \exp\left\{-\frac{1}{\sigma}\rho_\tau(\varepsilon_{t\tau})\right\},$$

where $\varepsilon_{t\tau} = Y_t - Q_t(\tau, \theta)$. Then, given a data set of size n, the working likelihood is

$$f(\text{Data}|\theta, \sigma) = \left(\frac{\tau(1-\tau)}{\sigma}\right)^n \exp\left\{-\frac{1}{\sigma}RQ_n(\tau, \theta)\right\}$$

where $RQ_n(\tau, \theta)$ is given by (20.13). Posterior inference on the CAViaR model can then be implemented with appropriate chosen prior. Hsu (2010) choose a flat prior for each coefficient in θ and the inverse gamma distribution IG(α_0, s_0) for σ.

There is also a growing interest in nonparametric estimation of conditional quantiles. In general, if we do not assume a parametric form for the conditional quantile function $Q_{r_t}(\tau|X_t)$, we may consider the nonparametric quantile regression estimation. For example, the local linear nonparametric quantile regression estimator is defined as

$$(\widehat{a}, \widehat{b}) = \operatorname*{argmin}_{a,b} \sum_{t=1}^{n} \rho_\tau\{r_t - a - b(X_t - x)\} K\left(\frac{X_t - x}{h}\right). \tag{20.14}$$

The solution value for a in the above optimization, $\widehat{a} = \widehat{a}(\tau, x)$, provides a nonparametric estimate for $Q_{r_t}(\tau|x)$, while $\widehat{b} = \widehat{b}(\tau, x)$ provides a nonparametric estimate for $\partial Q_{r_t}(\tau|x)/\partial x$.

Nonparametric methods can perform poorly in the presence of many covariates, due to the curse of dimensionality. Other approaches to estimating VaR include the hybrid method and methods based on extreme value theory (see Chernozhukov, 2005). For additional research on quantile regression estimation of VaR, see Chernozhukov and Umantsev (2001) and Chernozhukov and Fernández-Val (2011).

20.2.2   Expected shortfall

As an intuitive measure of risk, VaR has become a standard measure of market risk in the financial services sector and has gained considerable popularity in asset allocations. However, despite its popularity, VaR as a risk measure has also been criticized. An important criticism to VaR is that it is not a "coherent" risk measure.[1]

Following the axiomatic approach, Artzner et al. (1999) define a coherent risk measurement from a regulator's point of view. Their starting point is that although we all have an intuitive sense of what financial risk entails, it is difficult to give a good assessment of financial risk unless we specify what a measurement of financial risk actually means.

Definition 2 *A mapping $\rho = \rho_0 : \chi \to \mathbb{R} \cup \{+\infty\}$ is called a* coherent *measurement of risk if it satisfies the following conditions for all $X, Y \in \chi$.*
- *Monotonicity: if $X \leqslant Y$, then $\rho(X) \geqslant \rho(Y)$.*
- *Translation invariance: if $a \in \mathbb{R}$, then $\rho(X + a) = \rho(X) - a$.*
- *Positive homogeneity: if $\lambda \geqslant 0$, then $\rho(\lambda X) = \lambda \rho(X)$.*
- *Subadditivity: $\rho(X + Y) \leqslant \rho(X) + \rho(Y)$.*

The four axioms have clear meaning in the finance context: monotonicity says that the risk increases when the return decreases; a risk measurement satisfies the translation invariance axiom if adding alpha dollars of capital to an asset reduces the risk measure by alpha dollars; positive homogeneity says that the risk exposure of a financial position grows linearly as the size of the position increases; and subadditivity specifies that the risk of the sum never exceeds the sum of the risks, which is closely related to the concept of risk diversification in a portfolio of risky assets.

In the mean-variance based approach, the standard deviation (variance) is used as a measure of risk. According to the definition of Artzner et al., such a measure is ruled out as a coherent risk measurement by the monotonicity requirement.

VaR is not subadditive. That means that the VaR of a portfolio may be greater than the sum of individual VaRs of these assets. As a result, managing risk by VaR may fail to stimulate diversification. In addition, VaR based risk management only focuses on controlling the probability of loss, rather than its magnitude (Basak and Shapiro, 2001). However, there are limitations for diversification when losses may be large. For extremely heavy-tailed risks with unbounded distributions support (assets with infinite mean), diversification may increase VaR, and generally it is difficult to construct an appropriate risk measure for such distributions. See,for example, Ibragimov and Walden (2007).

While VaR does not fall into the category of coherent risk measurement as not being subadditive, expected shortfall, a coherent risk measurement, has been suggested as an alternative (remedy) for VaR based risk measurement. ES is defined as the expected loss exceeding VaR. More specifically, the ES at level τ is the expected loss of portfolio value given that a loss is occurring at or below the τth quantile:

$$ES_\tau = \mathrm{E}(r_t | r_t < Q_{r_t}(\tau)),$$

where $Q_{r_t}(\tau)$ is the τth quantile of r_t. And the τth *conditional* ES of r_t is defined as

$$ES_\tau(x) = \mathrm{E}[r_t | r_t \leqslant Q_{r_t}(\tau | X_t), X_t = x], \qquad (20.15)$$

where $Q_{r_t}(\tau | X_t)$ is the τth conditional quantile of r_t. Note that

$$ES_\tau(x) = \int_0^1 w_\tau(s) Q_{r_t}(s | X_t = x) ds,$$

---

[1]Despite the popularity of coherent risk metrics, there are limitations to coherent risk measures (see, for example, Ibragimov and Walden, 2007; Ibragimov et al., 2009; Dhaene et al., 2008).

where $w_\tau(s) = 1(s \leqslant \tau)/\tau$, which shows the connection with VaR. This is included in the more general class of spectral risk measures in which $w_\tau(s)$ is any nonnegative, non-increasing, right-continuous, integrable function defined on $[0,1]$ such that $\int_0^1 w_\tau(s)ds = 1$. All members of this class are coherent risk measures according to the above axioms.

Unlike VaR, which is insensitive to the magnitude of loss beyond a certain percentile, ES weights large losses by their magnitude. ES is also called expected tail loss or conditional VaR in the literature. Bassett et al. (2004) show that a coherent regular risk measure must be a *pessimistic* measure, which accentuates the implicit likelihood of the least favorable outcomes and depresses that of the most favorable ones. In particular, they demonstrate that the τ-level ES is simply (the negative value of) a Choquet expected return when the probabilities of the τ least favorable outcomes are inflated and the $1 - \alpha$ proportion of most favorable outcomes are discounted entirely. In addition, Rockafellar and Uryasev (2000, 2002) highlight the mathematical properties of ES and suggest ES as a more suitable measure of market risk for the purpose of portfolio optimization.[2] The recent revision of the main international banking regulations included the recommendation that ES be used in place of VaR (see Basel Committee on Banking Supervision, 1999).

ES can be estimated based on quantile regressions. Since the estimation of ES requires knowledge about the corresponding quantile, we may estimate the τth quantile of r_t, $Q_{r_t}(\tau)$, by the τth sample quantile of $\{r_1, \ldots, r_n\}$, say, $\hat{Q}_r(\tau)$. Then we may construct an unsmoothed estimator of ES_τ (Chen, 2008),

$$\widehat{ES}_\tau = \frac{1}{\tau n} \sum_{t=1}^{n} r_t 1\left(r_t \leqslant \hat{Q}_r(\tau)\right),$$

or a smoothed estimator (Scaillet, 2004),

$$\widetilde{ES}_\tau = \frac{1}{\tau n} \sum_{t=1}^{n} r_t \mathcal{K}_h\left(r_t - \hat{Q}_r(\tau)\right),$$

where $\mathcal{K}_h(\cdot) = \mathcal{K}(\cdot/h)$, and $\mathcal{K}$ is a twice continuously differentiable survivor function with compact support satisfying $\mathcal{K}(x) = 1$ for $x \leqslant -1$ and $\mathcal{K}(x) = 0$ for $x \geqslant 1$, while h is a bandwidth.

The conditional ES can be estimated similarly. In this case, we consider an appropriate quantile regression (parametric or nonparametric) to estimate the conditional quantile $Q_{r_t}(\tau|X_t)$. Notice that

$$\begin{aligned} ES_\tau(x) &= \mathrm{E}\left[r_t 1\left(r_t < Q_{r_t}(\tau|X_t)\right)| X_t = x\right] \\ &= \mathrm{E}\left(r|X = x\right) - \frac{1}{\tau}\mathrm{E}\left[\rho_\tau\left(r - Q_r(\tau|X)\right)| X = x\right]. \end{aligned}$$

We can estimate the conditional expected shortfall $ES_\tau(x)$ by replacing the conditional expectations with corresponding local sample averages:

$$\widehat{ES_\tau(x)} = \frac{1}{\hat{f}_X(x)} \frac{1}{nh^d} \sum_{t=1}^{n} K\left(\frac{X_t - x}{h}\right)\left[r_t - \frac{1}{\tau}\rho_\tau\left(r_t - \hat{Q}_{r_t}(\tau|X_t)\right)\right],$$

where $\hat{f}_X(x) = n^{-1}h^{-d}\sum_{t=1}^{n} K\left(\frac{X_t - x}{h}\right)$ and $h = h(n)$ is the bandwidth parameter that we use for estimating ES.

---

[2] Again, this does not apply to assets with extremely heavy-tailed distributions (assets with infinite means). Generally it is difficult to construct an appropriate risk measure for such distributions. See, for example, Ibragimov and Walden (2007).

Under regularity conditions, consistency and asymptotic normality of these estimators can be derived, and statistical inference of ES can be constructed based on appropriate estimates of the variance.

Linton and Xiao (2013) studied estimation and inference regarding ES for time series with infinite variance. In particular, they consider a random sample $\{r_1, \ldots, r_n\}$ from a stationary and mixing time series with regularly varying tail probabilities, with index of regular variation[3] of r_t, $\theta \in (1, 2)$, so that the expected shortfall exists, but r_t has infinite variance. Linton and Xiao (2013) show that, under regularity conditions of (1) strong mixing that controls the dependence, (2) continuity of the distribution, (3) a tail balance condition, and (4) an anti-clustering condition (see, for example, Bartkiewicz et al., 2011, for discussions on these conditions for stable processes), both the unsmoothed estimator $\widehat{ES}_\tau$ and the smoothed estimator $\widetilde{ES}_\tau$ have limiting distributions given by

$$n^{(\theta-1)/\theta}(\widehat{ES}_\tau - ES_\tau) \Longrightarrow \frac{1}{\tau}S,$$

and

$$T^{(\theta-1)/\theta}(\widetilde{ES}_\tau - ES_\tau) \Longrightarrow \frac{1}{\tau}S,$$

where S is a stable random variable with characteristic function

$$\exp\left[-|t|^\theta \frac{\Gamma(2-\theta)}{1-\theta}\{(c_+ + c_-)\cos(\pi\theta/2) - i\mathrm{sign}(t)(c_+ - c_-)\sin(\pi\theta/2)\}\right].$$

The rate of convergence of the estimators is determined by the tail thickness parameter θ, and the limiting distribution S is in the stable class with parameters (θ, c_+, c_-) depending on the tail thickness of the time series (θ) and on the dependence structure and skewness (c_+, c_-), which makes inference complicated. Linton and Xiao (2013) propose a subsampling procedure to carry out statistical inference. First, the parameter θ can be estimated consistently under weak dependence conditions by many methods (see, for example, Hill, 2010; Embrechts et al., 1999). Let $\hat{\theta}$ be a consistent estimator of θ. Given the random sample $\{r_t, t = 1, \ldots, n\}$, we consider subsamples of size M,

$$\{r_t, \ldots, r_{t+M-1}\}, \quad t = 1, \ldots, n - M + 1,$$

and estimate the expected shortfall based on subsamples by

$$\widehat{ES}_\tau(M, t) = \frac{1}{\tau M} \sum_{s=0}^{M-1} r_{t+s} 1\left(r_{t+s} \leq \hat{Q}_{r,t}(\tau)\right),$$

where $\hat{Q}_{r,t}(\tau)$ is the corresponding estimator based on the subsample $\{r_t, \ldots, r_{t+M-1}\}$.[4]

We approximate the sampling distribution of $n^{(\theta-1)/\theta}\left(\widehat{ES}_\tau - ES_\tau\right)$, denoted by $\hat{F}_n(y)$, by

$$\hat{F}_{n,M}(y) = \frac{1}{n-M+1} \sum_{t=1}^{n-M+1} 1\left(M^{(\hat{\theta}-1)/\hat{\theta}}\left[\widehat{ES}_\tau(M, t) - \widehat{ES}_\tau\right] \leq y\right).$$

Under regularity conditions (see Linton and Xiao, 2013), $\hat{F}_{n,M}(y) \xrightarrow{P} F(y)$, as $n \to \infty$, where $F(y)$ is the limiting distribution function of $n^{(\theta-1)/\theta}(\widehat{ES}_\tau - ES_\tau)$. The proposed subsampling method is "robust" in the sense that it is also consistent even in the case of finite variance, where normal asymptotics prevail. A similar subsampling procedure can be constructed for the smoothed estimator.

---

[3]That is, $\mathrm{P}[|Y_t| > x] = x^{-\theta}L(x)$, where L is a slowly varying function at ∞.

[4]Since that the preliminary estimation does not affect the limiting distribution of estimators of the ES, one can also use other quantile estimators.

20.3 Upper quantile information and financial markets

Not only are lower quantile measures such as expected shortfall and value-at-risk useful in financial applications, the *upper quantile* of returns distributions also contains important information and affects investors' behavior in financial markets (Xiao, 2014). Investors not only look at the left tail of returns to control risk, but also consider the right tail information for opportunities: people buy insurance as well as playing the lottery. While people try to control risk by purchasing insurance, they also try to capture the small chance of making a fortune by playing lotteries – although different people may have different attitudes toward risk and opportunity. They are interested in the information contained in *both* the left tail and the right tail of returns distributions.

In the last few decades, researchers in economics and finance have devoted a lot of effort to developing models that can capture investors' behavior more accurately. One important model along these lines is the Tversky and Kahneman (1992) "cumulative prospect theory." The behavior (say, portfolio choice) in financial markets of an investor characterized by the prospect theory is affected not only by the mean and variance of the asset return, but also by the distribution of return process. In particular, an investor will exhibit particular sensitivity to both the left tail and right tail distribution of the return, and the sensitivity is asymmetric in the two tails. Barberis and Huang (2008) study the implications of the cumulative prospect theory and show that a prospect investor overweights the tail probabilities and holds lottery-like securities.

In order to capture information contained in the right tail distribution of financial variables such as returns, appropriate measures summarizing the right tail distributional information are needed. Just like the role of moments in summarizing the information about a distribution, right tail moments are natural measures capturing the right tail distributional property. Xiao (2014) proposes using the *right tail mean* (RTM) as a measurement of opportunity. Let Y be the return of a security. The RTM is simply the mean of a gain exceeding a specified upper quantile of the return distribution. Let τ be an upper quantile, say $\tau = 95\%$, and denote the τth quantile of the return distribution by $Q_Y(\tau)$, that is, $P(Y < Q_Y(\tau)) = \tau$. Then the τth RTM of Y is defined as follows:[5]

$$M(\tau) = E[Y|Y \geqslant Q_Y(\tau)].$$

The RTM can also be extended to conditional RTM since, in many financial applications, investors look at the *conditional* distribution of returns given current information. Let Y_t be the return of an asset at time t, and denote the τth conditional quantile of Y_t as $Q_{Y_t}(\tau|X_t)$, that is, $P(Y_t < Q_{Y_t}(\tau|X_t)|X_t) = \tau$, where X_t is a vector containing the information available at time t. Then the τth conditional right tail mean is defined as

$$M(\tau, x) = E[Y_t|Y_t \geqslant Q_{Y_t}(\tau|X_t), X_t = x].$$

Xiao (2014) studied some basic properties of the RTM. In particular, for real-valued random variables $Y \in X$, denote the associated RTM by $\mu(Y)$. Then $\mu(Y)$ has the following properties:

1. *Monotonicity*: for any $Y_1, Y_2 \in \mathcal{X}$, if $Y_1 \geqslant Y_2$, then $\mu(Y_1) \geqslant \mu(Y_2)$.
2. *Subadditivity*: for any $Y_1, Y_2 \in \mathcal{X}$, $Y_1 + Y_2 \in \mathcal{X}$, and $\mu(Y_1 + Y_2) \leqslant \mu(Y_1) + \mu(Y_2)$.
3. *Linear homogeneity*: for any $\lambda \geqslant 0$, and all $Y \in \mathcal{X}$, $\mu(\lambda Y) = \lambda \mu(Y)$.

---

[5]Since this is the right tail counterpart of the expected shortfall (ES), we may call the RTM the "expected windfall" as in Wan and Xiao (2014).

4.   *Translation invariance*: for any $a \in \mathbb{R}$, and all $Y \in \mathcal{X}$, $\mu(Y + a) = \mu(Y) + a$.

The monotonicity property of the RTM indicates that it is consistent with stochastic dominance: if a security's distribution dominates another security's distribution, its opportunity measure should be larger – opportunity increases when the return increases. Subadditivity of the RTM is a property of no extra synergy. This means that a merger does not bring extra opportunity. In practice, although the optimal level of diversification (measured by the rules of mean-variance portfolio theory) exceeds 300 stocks (see, for example, Campbell et al., 2001; Statman, 2004), it is well known that the average investor holds rather fewer stocks. This is because although diversification reduces downside risk, it also reduces the upside opportunity. Investors not only are sensitive to the downside protection, but also care about the upside potential. Such a property is also reflected in the corporate focus in practice. It is found in the literature on mergers that marginally profitable projects (whose risks are high) merge to survive a period of distress but, if profitability improves, divestiture occurs.[6] Linear homogeneity says that the opportunity of a financial position grows linearly as the size of the position increases. The last property of translation invariance indicates that adding (subtracting) a sure amount a to the portfolio simply increases (decreases) the opportunity measure by a. The opportunity of a risk-free asset should be the same as the certain payoff provided by the risk-free asset.

In general, one may consider right tail moments. For example, one may consider the right tail variance (RTV), the variance of return exceeding the $100\tau\%$-level quantile:

$$RTV(\tau) = E\{[Y - M(\tau)]^2 | Y \geqslant Q_Y(\tau)\},$$

and the conditional right tail variance, defined as

$$RTV(\tau, x) = E\{[Y_t - M(\tau, x)]^2 | Y_t \geqslant Q_{Y_t}(\tau|X_t), X_t = x\}.$$

Simulation experiments show that a prospect investor who is characterized by the "cumulative prospect theory" of Tversky and Kahneman (1992) have different preferences over securities with identical means and variances but different tail distributions. The prospect investor's valuation of stocks with different distributions increases as the RTM increases.

Similar to the left tail measures, estimation of the RTM (and other empirical studies using the RTM) can be conducted based on quantile regression methods. Let Y be a real-valued random variable with $E(Y) < \infty$. Then the RTM is given by

$$M(\tau) = E(Y) + \frac{1}{1-\tau} E[\rho_\tau (Y - Q_Y(\tau))],$$

and the conditional RTM is given by

$$M(\tau, x) = M(\tau, x) = E[Y|Y \geqslant Q_\tau(Y|X), X = x]$$
$$= E(Y|X = x) + \frac{1}{1-\tau} E[\rho_\tau (Y - g(X, \beta(\tau)))| X = x].$$

Given a random sample $\{Y_t, t = 1, \ldots, n\}$ on Y, we can estimate the RTM $M(\tau)$ by replacing the expectations by corresponding sample averages.

Wan and Xiao (2014) studied the role of right tail information in cross-sectional regression of returns. The asset pricing theory of incomplete markets predicts a positive idiosyncratic volatility effect: investors should command a higher expected return for bearing higher idiosyncratic risks. In the presence of costly information, Merton (1987) provides a

---

[6]This does not apply to extremely heavy-tailed assets, as noted earlier and discussed further by Ibragimov and Walden (2007).

theoretical framework where firms with high idiosyncratic volatility require high expected returns to compensate investors for holding imperfectly diversified portfolios. Some behavioral models also predict that stocks with higher idiosyncratic volatility should earn higher expected returns.

However, empirical studies have yielded mixed results on idiosyncratic volatility and there has been a lively debate on the role of idiosyncratic risk in determining cross-sectional stock returns. Ang et al. (2006) find that stocks with high idiosyncratic volatility in the current month subsequently earn abnormally low returns the following month. Moreover, Ang et al. (2009) show that this negative idiosyncratic volatility effect is robust after controlling for other risk factors and occurs internationally across 23 developed markets. Since this negative idiosyncratic volatility effect is inconsistent with most asset pricing theories, These findings are referred to as an "idiosyncratic volatility puzzle," and have attracted much research attention.

Wan and Xiao (2014) argue that many volatile stocks tend to be small, and low-price stocks generally have more positively skewed returns than large-cap high price stocks. According to the equilibrium model of Barberis and Huang (2008), the securities capable of delivering windfall profits may be overpriced and consequently expected to earn low returns. This might help to explain why high idiosyncratic volatility stocks have low returns if those stocks are expected to realize windfall profits.

In order to take this effect into account, Wan and Xiao (2014) consider the RTM as a measurement to quantify the upside potential of stock return distributions, and use the RTM to identify stocks having a small chance of "making it big." By including the RTM in the cross-sectional regression of stock returns, they show that the RTM has significant predictive power for expected returns. This negative coefficient on the RTM is consistent with the prediction of Barberis and Huang (2008), since a large RTM value would indicate a similar stock return distribution to that of a lottery. More importantly, after controlling for the RTM, the cross-sectional regression reveals a positive relationship between conditional idiosyncratic volatility and expected returns, which is consistent with the predictions of most finance theories.

20.4 Quantile regression and portfolio allocation

Quantile regression has important applications in portfolio construction. The question of how to allocate wealth among alternative assets is crucial to all investors. Using variance as a measure for risk, Markowitz (1952) proposed that investors should choose the portfolio that offers the smallest variance for a given level of the expected return. For more than five decades, the mean-variance analysis has served as the standard procedure for portfolio management. The mean-variance optimal portfolio, which is fully determined by the mean and variance of portfolio returns, maximizes the expected utility if asset returns are normally distributed, or if investors have quadratic utility. However, strong empirical evidence against the normality assumption of asset returns has been reported and numerous empirical analyses have shown that financial time series tend to be skewed and heavy-tailed. Furthermore, variance is not a satisfactory measure for risk. Moreover, a quadratic utility function cannot fully capture investors' different attitudes toward gains and losses as it assumes that investors are as averse to upside gains as they are to downside losses. In practice, however, investors are loss-averse and care mainly about the losses caused by extreme downside movements while upside gains should not be penalized.

For this reason, researchers and practitioners have shifted their attention from variance to downside-risk measures, in particular VaR and ES, in financial applications. Mean-risk portfolio construction has also been proposed in the literature. Using an appropriate measure for risk, investors may consider choosing the portfolio that offers the smallest risk for a given level of the expected return. Recall that Rockafellar and Uryasev (2000, 2002) suggest ES as a suitable measure of market risk for the purpose of portfolio optimization.

20.4.1   The mean-ES portfolio construction

Bassett et al. (2004) studied mean-ES portfolio allocation and Choquet expected utility maximization via quantile regression. Let q_τ be the τth quantile of the returns distribution. The mean-ES approach corresponds to a simple truncated utility function,

$$u(R) = \begin{cases} R/\tau, & \text{if } R \leqslant q_\tau, \\ 0, & \text{otherwise.} \end{cases} \qquad (20.16)$$

In the mean-ES setting, investors' expected utility is the ES of the portfolio return distribution corresponding to the τth quantile:

$$Eu(R) = \int_{-\infty}^{+\infty} u(R)dF(R) = \frac{1}{\tau} \int_{-\infty}^{q_\tau} RdF(R).$$

Bassett et al. define the τ-risk of R as

$$\varrho_\tau(R) = -\frac{1}{\tau} \int_{-\infty}^{q_\tau} RdF(R),$$

where q_τ is the τth quantile of the return distribution, and show that empirical strategies for minimizing the τ-risk lead to methods of quantile regression. They show that

$$\min_\theta E\rho_\tau(R - \theta) = \alpha(\mu + \varrho_\tau(R)), \qquad (20.17)$$

where μ is the mean.

Consider an investment decision over L underlying assets with random returns $r = (r_1, \ldots, r_L)^\top$. If we construct a portfolio by choosing portfolio weights $w = (w_1, \ldots, w_L)^\top$, $\sum_{i=1}^L w_i = 1$, the portfolio return rate R is equal to $w^\top r$. Denote the mean of the portfolio by $\mu(w^\top r)$. The optimal portfolio choice for an mean-ES investor corresponds to

$$\begin{cases} \min_w & \varrho_\tau(w^\top r) \\ \text{s.t.} & \mu(w^\top r) = \mu_0, \quad \sum_{i=1}^L w_i = 1. \end{cases}$$

In practice, given a sample of n observations of the assets' return (say, $\{r_t, \ t = 1, \ldots, n\}$, $r_t = (r_{t1}, \ldots, r_{tL})^\top$), using (20.17) and replacing the expectations by their sample analogs, the portfolio construction problem can be expressed in terms of the quantile regression

$$\begin{cases} \min_{w,\theta} & \sum_{t=1}^n \rho_\tau(w^\top r_t - \theta) \\ \text{s.t.} & \frac{1}{n} \sum_{t=1}^n w^\top r_t = \mu_0, \quad \sum_{i=1}^L w_i = 1. \end{cases}$$

20.4.2   The multi-quantile portfolio construction

Since the ES corresponds to a simplified truncated utility function, one way to allocate assets more appropriately is to include additional information such as other quantiles or moments of the return distribution to portfolio selection. For example, Roman et al. (2007) consider both the first and second moments in addition to the ES in portfolio construction, while many others consider higher moments. However, incorporating additional information of other quantiles or moments brings considerable technical difficulty into portfolio construction. Even imposing one more restriction in addition to the mean return could substantially complicate the implementation.

Bassett et al. (2004) proposed a combined quantile approach of portfolio construction that takes into account of information over multiple quantiles of the return distribution. The multi-quantile approach is quite general, numerically desirable, and can also be used to approximate a wide range of nonlinear utility functions. One can use the utility specification as a guideline in choosing the market risk measure and in solving the portfolio allocation problem. In particular, they consider the utility maximization problem for an investor, whose preference is characterized by a piecewise linear utility function with reference points that capture the degrees of downside risk aversion. The utility function is

$$U(R) = \begin{cases} a_1 R + b_1, & \text{if } R \in (-\infty, q_1], \\ \vdots & \vdots \\ a_i R + b_i, & \text{if } R \in (q_{i-1}, q_i], \\ \vdots & \vdots \\ a_{M+1} R + b_{M+1}, & \text{if } R \in (q_M, +\infty), \end{cases} \tag{20.18}$$

where R is the portfolio return, and M is the total number of reference points (kinks). Suppose the portfolio return R has a distribution function, $F(\cdot)$. Then, in (20.18), the ith reference point, q_i, corresponds to the α_ith quantile of the portfolio return distribution, $F_R^{-1}(\alpha_i)$. Set α_0 to 0 and α_{M+1} to 1, so that q_0 and q_{M+1} correspond to negative and positive infinity, respectively. The slope of the ith line segment, a_i, indicates the degree of loss aversion in the corresponding range of returns. Bassett et al. assume that $0 \leqslant a_{i+1} \leqslant a_i$ for all $i \in [1, M]$, which implies a diminishing marginal utility as the level of return increases. As a conventional risk management practice, the value of α_i is pre-specified to capture investors' attitude change toward $100\alpha_i\%$ least favorable returns compared with the more favorable ones.

The utility function (20.18) is closely related to the rank dependent utility studied by Quiggin (1982), Schmeidler (1989), Segal (1984), and Yaari (1987). The piecewise linear structure in (20.18) distorts the probabilities according to the rank ordering of the outcomes. Like the rank-dependent expected utility model, it overweights extreme outcomes, corresponding to a transformation of the cumulative probability distribution function. The central idea of rank-dependent weightings was incorporated by Daniel Kahneman and Amos Tversky into prospect theory, resulting in the cumulative prospect theory of Tversky and Kahneman (1992). See also Rostek (2010), de Lara Resende and Wu (2010), and Basili and Chateauneuf (2011) for discussions on quantile utility functions.

Consider portfolio selection motivated by the utility function (20.18). Notice that the $100\alpha\%$ ES of R is

$$\varrho_\alpha(R) = -\frac{1}{\alpha} \int_{-\infty}^{q_\alpha} R \, dF(R).$$

Denoting the expected portfolio return by μ_R, and letting $\lambda_i = a_i - a_{i+1}$, the expected utility of (20.18) is given by

$$\mathrm{E}[U(R)] = -\sum_{i=1}^{M} \lambda_i \alpha_i \varrho_{\alpha_i}(R) + \delta, \tag{20.19}$$

where $\delta = a_{M+1}\mu_R - \sum_{i=1}^{M}(b_{i+1} - b_i)\alpha_i - b_{M+1}$. Thus, given an average portfolio return, the utility maximization problem is equivalent to the minimization of a weighted summation of ESs, namely $\sum_{i=1}^{M}(a_i - a_{i+1})\alpha_i \varrho_{\alpha_i}(R)$. The weights used to combine ESs at different quantiles, $(a_i - a_{i+1})\alpha_i$, are jointly determined by the location of a reference point (i.e. the α_ith quantile of the return distribution) and the changes in marginal utility at the kink, $a_i - a_{i+1}$.

Consider an investment decision over L underlying assets with random returns $r = (r_1, \ldots, r_L)^\top$. The investment portfolio is constructed by choosing portfolio weights $\omega = (\omega_1, \ldots, \omega_L)^\top$ that satisfy the constraint $\sum_{i=1}^{L} \omega_i = 1$. The portfolio return R is equal to $\omega^\top r$. Then the optimal portfolio choice for an investor with utility function (20.18) corresponds to the maximization problem

$$\begin{cases} \max_\omega & \mathrm{E}[U(\omega^\top r)] \\ \text{s.t.} & \omega^\top \iota = 1, \end{cases} \tag{P1}$$

where $\iota = (1, \ldots 1)^\top$.

By (20.19), the portfolio selection problem (P1) can be rewritten as

$$\begin{cases} \min_\omega & \left[\sum_{i=1}^{M} \lambda_i \alpha_i \varrho_{\alpha_i}(\omega^\top r) - a_{M+1}\mu_R \right] \\ \text{s.t.} & \omega^\top \iota = 1, \end{cases} \tag{P1'}$$

where the constant term, $\sum_{i=1}^{M}(b_{i+1} - b_i)\alpha_i$, is dropped from the expected utility expression. Note that, using the result of Bassett et al. (2004), the value of ES can be obtained from the optimization problem

$$\varrho_\alpha(X) = \frac{1}{\alpha}\min_\xi E\rho_\alpha(X - \xi) - \mu_X.$$

Therefore, the utility maximization problem (P1') has a quantile regression representation. In particular, the optimal portfolio selection based on the utility function (20.18) is determined by the quantile regression problem

$$\min_{\omega, \, \xi_i, i=1,\ldots,M} \left[\sum_{i=1}^{M} \lambda_i E\rho_{\alpha_i}(\omega^\top r - \xi_i) - \left(\sum_{i=1}^{M} \lambda_i \alpha_i + a_{M+1} \right) \mu_{\omega^\top r} \right].$$

In practice, the returns distribution is unknown and thus $E\rho_{\alpha_i}(\omega^\top r - \xi_i)$ is unknown. Suppose there is a sample of N observations from the distribution of assets returns (say, r_t where $t = 1, \ldots, n$). Replacing the expectations with the sample analogs, the optimal portfolio weights can be determined by solving the minimization problem

$$\min_{\omega, \, \xi_i, i=1,\ldots,M} \left[\sum_{i=1}^{M} \lambda_i \sum_{t=1}^{n} \rho_{\alpha_i}(\omega^\top r_t - \xi_i) - \left(\sum_{i=1}^{M} \lambda_i \alpha_i + a_{M+1} \right) \sum_{t=1}^{n} \omega^\top r_t \right]. \tag{20.20}$$

The solution to the above optimization problem gives the optimal portfolio weights that maximize investors' expected utility specified in (20.18). The weighted summation of ESs, referred to as the combined ES, is a pessimistic risk measure that reflects investors' preference over the entire distribution of portfolio returns. The statistical properties of the portfolio weights and financial applications are considered in Wan and Xiao (2013).

20.5   Stochastic dominance and quantile regression

Stochastic dominance finds applications in many areas. In finance it is used to assess portfolio diversification, capital structure, bankruptcy risk, and price bounds on options. In reinsurance coverage, the insured use it to select the best coverage option, while the insurers use it to assess whether the options are consistently priced. Partial *strong* orders are based on specific utility (loss) functions. By their very nature, strong orders do not command consensus. In contrast, *uniform* order relations such as stochastic dominance (SD) rankings can produce universal assessments based on the expected utility paradigm and its mathematical regularity conditions. These relations are defined over relatively large classes of utility functions.

The following definitions will be useful. Let X_1 and X_2 be two variables (incomes, returns/prospects) at two different points in time, or for different regions or countries, or with or without a program (treatment). Let $\mathcal{U}_1$ denote the class of all von Neumann–Morgenstern type utility functions, u, such that $u' \geqslant 0$ (increasing). Also, let $\mathcal{U}_2$ denote the class of all utility functions in $\mathcal{U}_1$ for which $u'' \leqslant 0$ (strict concavity).

Definition 3 X_1 *first-order stochastically dominates* X_2, *written* $X_1 \geq_{\mathrm{FSD}} X_2$, *if and only if:*

(1) $E[u(X_1)] \geqslant E[u(X_2)]$ *for all* $u \in \mathcal{U}_1$, *with strict inequality for some* u; *or*

(2) $F_{X_1}(x) \leqslant F_{X_2}(x)$ *for all* x, *with strict inequality for some* x; *or*

(3) $Q_{X_1}(\alpha) \geqslant Q_{X_2}(\alpha)$ *for all* $\alpha \in [0,1]$, *with strict inequality for some* α.

Definition 4 X_1 *second-order stochastically dominates* X_2, *written* $X_1 \geq_{\mathrm{SSD}} X_2$, *if and only if:*

(1) $E[u(X_1)] \geqslant E[u(X_2)]$ *for all* $u \in \mathcal{U}_2$, *with strict inequality for some* u; *or*

(2) $\int_{-\infty}^{x} F_{X_1}(t)dt \leqslant \int_{-\infty}^{x} F_{X_2}(t)dt$ *for all* x, *with strict inequality for some* x; *or*

(3) $\int_{-\infty}^{x} Q_{X_1}(t)dt \geqslant \int_{-\infty}^{x} Q_{X_2}(t)dt$, *for all* $\alpha \in [0,1]$, *with strict inequality for some* α.

Statistical tests of stochastic dominance hypotheses can be equivalently based on cdfs or quantile functions (see Linton et al., 2005; Chernozhukov and Fernández-Val, 2005; Ng et al., 2011).

As an estimation method for the conditional density, quantile regression provides a natural approach to testing for stochastic dominance. Consider two populations (X_1 and X_2) with distribution functions F_1 and F_2, and quantile functions Q_1 and Q_2, respectively. Suppose that the two distributions have means α_1 and α_2, respectively. We can write $X_1 = \alpha_1 + U$, and $X_2 = \alpha_2 + V$, where U and V are zero-mean random variables with distribution functions F_u and F_v, respectively. The distribution function of X_1 is then given by $F_1(x) = F_u(x - \alpha_1)$ and the distribution function of X_2 is given by $F_2(x) = F_v(x - \alpha_2)$. X_1 strictly stochastically dominates X_2 at first order ($Y_1 >_1 Y_2$) if and only if $Q_1(\tau) \geqslant Q_2(\tau)$ for all $\tau \in (0,1)$ and there exists a nonempty interval $T \subset (0,1)$ such that $Q_1(\tau) > Q_2(\tau)$ for any $\tau \in T$. Suppose that $\{x_{11}, \ldots, x_{1n_1}\}$ and $\{x_{21}, \ldots, x_{2n_2}\}$ are i.i.d. realizations of X_1 and X_2. To test stochastic dominance via quantile regression, we may combine these two observations and define:

$$x_t = \begin{cases} x_{1t} & \text{for } t = 1, \ldots, n_1, \\ x_{2,t-n_1} & \text{for } t = n_1 + 1, \ldots, n, \end{cases} \tag{20.21}$$

where $n = n_1 + n_2$. Corresponding to the combined sample $\{x_t\}$, we introduce a dummy variable D_t defined as follows:

$$D_t = \begin{cases} 1 & \text{for } t = 1, \ldots, n_1, \\ 0 & \text{for } t = n_1 + 1, \ldots, n. \end{cases} \tag{20.22}$$

Thus,

$$x_t = \alpha + \beta D_t + w_t = z_t^\top \theta + w_t, \tag{20.23}$$

where $z_t = (1, D_t)^\top$, $\theta = (\alpha, \beta)^\top$, $w = v_t + (u_t - v_t)D_t$, $\alpha = \alpha_2$ is the mean of the second distribution, and $\beta = \alpha_1 - \alpha_2$ gives the difference between the two distributions.

The conditional quantile function of x_t given D_t can be written as

$$Q_{x_t}(\tau | D_t) = \alpha(\tau) + \beta(\tau)D_t = z_t^\top \theta(\tau), \tag{20.24}$$

where $\alpha(\tau) = \alpha + Q_v(\tau)$, $\beta(\tau) = \beta + Q_u(\tau) - Q_v(\tau)$, and $\theta(\tau) = (\alpha(\tau), \beta(\tau))^\top$. In particular,

$$Q_{x_t}(\tau | D_t = 1) = \alpha_1 + Q_u(\tau) = Q_1(\tau)$$

and

$$Q_{x_t}(\tau | D_t = 0) = \alpha_2 + Q_v(\tau) = Q_2(\tau).$$

The hypothesis that $X_1 \succ_1 X_2$ can then be expressed as

$$Q_{x_t}(\tau | D_t = 1) \geqslant Q_{x_t}(\tau | D_t = 0), \quad \text{for } \tau \in (0, 1),$$

with strict inequality on at least a nonempty interval.

It is easy to see that the regression quantile process, $\beta(\tau)$, of the dummy variable D_t measures the distributional difference between the two groups (assets) so that

$$\beta(\tau) = Q_{x_t}(\tau | D_t = 1) - Q_{x_t}(\tau | D_t = 0) = Q_1(\tau) - Q_2(\tau).$$

Hence, the FSD can be further reformulated as shown in the quantile regression model (20.24) based on the pooled sample,

$$\beta(\tau) \geqslant 0, \quad \text{for all } \tau,$$

with strict inequality on at least a nonempty interval. As a result, we can construct statistical tests for FSD based on the quantile regression

$$\min_{\alpha, \beta} \sum_{t=1}^n \rho_\tau(x_t - \alpha - \beta D_t), \tag{20.25}$$

where $\rho_\tau(u) = u(\tau - I(u < 0))$ is the "check function."

In the case of FSD, $X_1 \succ_1 X_2$, we have $Q_1(\tau) \geqslant Q_2(\tau)$ for all $\tau \in (0, 1)$, which is equivalent to $\beta(\tau) \geqslant 0$ for all $\tau \in (0, 1)$, with strict inequality on at least a nonempty interval. Thus a testing procedure for stochastic dominance can be constructed based on $\hat{\beta}(\tau)$. Ng et al. (2011) proposed one-sided Kolmogorov–Smirnov statistics along these lines. Higher-order stochastic dominance can be tested based on integrals of $\hat{\beta}(\tau)$ or, more precisely, integrals of $\hat{V}(\tau)$.

Quantile regression methods can be applied to many other related applications. So far we have been comparing the distribution or quantile function of two given risky outcomes, whereas in practice one may want to combine the risky outcomes with riskless assets to form a portfolio. Quantile methods are easily extended to the case of diversification between risky asset and riskless assets. Consider the portfolios

$$Y_j(\omega) = \omega X_j + (1 - \omega)r,$$

where $\omega > 0$ is the portfolio weight, while r is the risk-free rate of return. It follows that

$$Q_{Y_j(\omega)}(\alpha) = \omega Q_{X_j}(\alpha) + (1 - \omega)r.$$

Even when neither F_{X_1} nor F_{X_2} dominates the other by FSD, in some cases we find that for every portfolio $Y_2(\omega)$ we can find a portfolio $Y_1(\omega')$ that dominates it, which property we call FSDR (Levy, 2016).

Definition 5 F_{X_1} *dominates* F_{X_2} *by FSDR if and only if*

$$\inf_{0 \leqslant \alpha \leqslant F_{X_1}(r)} \frac{Q_{X_2}(\alpha) - r}{Q_{X_1}(\alpha) - r} \geqslant \sup_{F_{X_1}(r) \leqslant \alpha \leqslant 1} \frac{Q_{X_2}(\alpha) - r}{Q_{X_1}(\alpha) - r}.$$

This condition is related to the Sharpe (1964) and Lintner (1965) condition: $(\mathrm{E}X_1 - r)/\sigma_1 > (\mathrm{E}X_2 - r)/\sigma_2$, which ensures that asset 1 will be preferred to asset 2 by the mean-variance rule in the presence of a riskless asset. Levy (2016, Theorem 4.5) generalizes this to second-order dominance (SSDR).

Definition 6 F_1 *dominates* F_2 *by SSDR if and only if*

$$\inf_{0 \leqslant \alpha \leqslant \alpha_0} \frac{\int_0^\alpha (Q_{X_2}(s) - r)\, ds}{\int_0^\alpha (Q_{X_1}(s) - r)\, ds} \geqslant \sup_{\alpha_0 \leqslant \alpha \leqslant 1} \frac{\int_0^\alpha (Q_{X_2}(s) - r)\, ds}{\int_0^\alpha (Q_{X_1}(s) - r)\, ds},$$

where α_0 solves the equation

$$r = \frac{1}{\alpha_0} \int_0^{\alpha_0} Q_{X_1}(s)\, ds.$$

These conditions are special to the quantile formulation of the problem. Given observations on X_1 and X_2, $Q_{X_i}(s)$ can be estimated by sample quantiles and thus inference procedures can be constructed based on sample analogs of these quantities.

20.6 Quantile dependence

Quantile regression provides a convenient tool for studying dependence in distributions. For risk management and many other financial applications, understanding how the correlation may vary is a very important endeavor.[7] The body of evidence we now have tells a remarkably consistent story – that the correlation tends to be larger when markets are bearish, but bull markets are much less dependent (see, for example Erb et al., 1994; Longin and Solnik, 2001; Hu, 2006; Sim and Xiao, 2009). For the investor who diversifies across international stock markets, hoping to reduce the correlation in her portfolio, the strengthened asset dependence during bear markets implies that the benefits of diversification are diminished especially when they are needed most. Moreover, there is unfavorable evidence suggesting

---

[7]See Ang and Chen (2002) for an example of asymmetric correlation between US portfolio returns and US aggregate returns.

that the welfare losses are significant if potential variations in correlation are overlooked
when financial conditions shift (Ang and Bekaert, 2002).

In economics and finance, policy-makers and portfolio managers are particularly concerned with tail events and their impact. For example, a popular measure of systematic risk
of a financial institution is to look at the lower quantile (VaR), and the VaR of a particular
institution or market is usually correlated with the VaR of other institutions or markets.
Quantile-based models provide convenient tools for studying the correlation of the conditional quantiles of a market or firm with the conditional quantiles of another market or
institution.

Quantile dependence is also closely related to contagion in financial applications. Financial contagion refers to the spread of market disturbances (mostly on the downside) from
one country to another, a process observed through co-movements in exchange rates, stock
prices, sovereign spreads, and capital flows. Financial contagion can be a potential risk for
countries trying to integrate their financial system with international financial markets and
institutions. It helps explain an economic crisis extending across neighboring countries, or
even regions.

Connected to the problem of quantile dependence, there is a large literature in empirical finance attempting to find predictability in the direction of stock prices, based on the
statistics of signs and ranks. Cowles and Jones (1937) proposed a statistic for testing market efficiency based on the frequencies of up movements relative to down movements for
i.i.d. data. Adrian and Brunnermeier (2016) propose a measure for systemic risk, defined as
the difference between the conditional value-at-risk (CoVaR) of the financial system conditional on an institution being in distress and the CoVaR conditional on the median state of
the institution. The proposed index measures what happens to the system's VaR when one
particular institution is under financial stress. Christoffersen and Diebold (2006) have investigated the predictability of signed returns under more general sampling schemes where the
returns are conditionally normal and allowed for time-varying volatility. Linton and Whang
(2009) propose a quantile-based diagnostic statistic for measuring the extent of directional
predictability based on a sample correlation. In particular, they test the predictability of
the direction of stock returns relative to a threshold corresponding to a particular quantile
of the distribution. Han et al. (2015) further extended this idea to study cross-dependence
of different time series.

20.6.1 Directional predictability via the quantilogram

Suppose that random variables $\{Y_t\}_{t=1}^n$ are from a stationary process, and denote the τth
quantile of Y_t by $Q_Y(\tau)$, $0 < \tau < 1$. Let $\psi_\tau(y) = \tau - 1(y < 0)$. Then $E\psi_\tau(Y_t - Q_Y(\tau)) = 0$.
Linton and Whang (2009) consider the conditional quantiles. Let $\mathcal{F}_t = \sigma(Y_t, Y_{t-1}, \ldots)$. They
consider the null hypothesis that the τth conditional quantiles are time invariant,

$$H_0 : E\left[\psi_\tau(Y_t - Q_Y(\tau))| \mathcal{F}_{t-1}\right] = 0, \quad \text{almost surely,} \tag{20.26}$$

and call $\{Y_t, \mathcal{F}_t\}_{t=1}^n$ a quantilegale. Under this hypothesis, if you are below the unconditional
τ-quantile today, the chance is no more than τ that you will be below it tomorrow. In the
absence of this property there is obviously some predictability in the $\psi_\tau(Y_t - Q_Y(\tau))$.

Linton and Whang (2009) argue that it is important to consider quantiles instead of the
mean. First, note that there is a sort of fair bet aspect to (20.26) that is relevant to investors:
if you are positioned on τ-type investments, you want to know whether the best current
guess of outcome is always still at the end of the distribution or more or less likely given
current information. Second, there is a powerful statistical reason for preferring quantiles
over moments, robustness: empirical evidence indicates that high-frequency data especially

may not have moments beyond the second and maybe not even that many. Third, quantiles also have a role to play in decision-making.

Linton and Whang (2009) define the quantilogram as

$$\rho_\tau (k) = \frac{\mathrm{E}\left[\psi_\tau \left(Y_t - Q_Y (\tau)\right)\psi_\tau \left(Y_{t+k} - Q_Y (\tau)\right)\right]}{\mathrm{E}\left[\psi_\tau \left(Y_t - Q_Y (\tau)\right)^2\right]}, \quad k = 1, 2, 3, \ldots.$$

Under the null hypothesis (20.26), $\mathrm{E}[\psi_\tau (Y_t - Q_Y (\tau))\psi_\tau (Y_{t+k} - Q_Y (\tau))] = 0$, for all k. Therefore, $\rho_\tau (k)$ is zero for all k. Under the alternative hypothesis $\rho_\tau (k)$ can take a variety of shapes across τ and k. The quantilogram is robust to the nonexistence of moments: no moments are required of Y_t for the existence of $\rho_\tau (k)$ or for the distribution theory of estimators of it. This means it can be used in circumstances where the usual correlogram is suspect.

The quantilogram can be computed using the empirical counterpart of $\rho_\tau (k)$. We first estimate $Q_Y (\tau)$ by the sample quantile $\hat{Q}_Y (\tau)$:

$$\hat{Q}_Y (\tau) = \arg\min_{\beta \in \mathbb{R}} \sum_{t=1}^{n} \rho_\tau (Y_t - \beta).$$

Then we estimate the quantilogram by

$$\hat{\rho}_\tau (k) = \frac{\sum_{t=1}^{n-k}\left[\psi_\tau \left(Y_t - \hat{Q}_Y (\tau)\right)\psi_\tau \left(Y_{t+k} - \hat{Q}_Y (\tau)\right)\right]}{\sqrt{\sum_{t=1}^{n-k}\left[\psi_\tau \left(Y_t - \hat{Q}_Y (\tau)\right)^2\right]}\sqrt{\sum_{t=1}^{n-k}\left[\psi_\tau \left(Y_{t+k} - \hat{Q}_Y (\tau)\right)^2\right]}}.$$

Linton and Whang show that under H_0, and regularity conditions,

$$\begin{bmatrix} \sqrt{n}\hat{\rho}_\tau (1) \\ \vdots \\ \sqrt{n}\hat{\rho}_\tau (p) \end{bmatrix} \Rightarrow N(0, V),$$

where $V = (V_\tau (i, j))$, $i, j = 1, 2, \ldots, p$,

$$V_\tau (i, j) = \begin{cases} 1 + \left(\dfrac{\tau(1-\tau)\mathrm{E}\sigma_{t+j}^{-1}\psi_\tau (Y_t - Q_Y (\tau))}{\left[\mathrm{E}\sigma_t^{-1}\right]^2 \mathrm{E}\left[\psi_\tau (Y_t - Q_Y (\tau))^2\right]}\right)^2, & i = j, \\[4mm] \dfrac{\mathrm{E}\sigma_{t+j}^{-1}\psi_\tau (Y_t - Q_Y (\tau))\mathrm{E}\sigma_{t+k}^{-1}\psi_\tau (Y_t - Q_Y (\tau))}{\left[\mathrm{E}\sigma_t^{-1}\right]^2 \mathrm{E}\left[\psi_\tau (Y_t - Q_Y (\tau))^2\right]}, & i \neq j. \end{cases}$$

The asymptotic normality facilitates statistical inference based on the quantilogram. A Box–Pierce–Ljung type test statistic can be constructed,

$$Q = n \sum_{h=1}^{p} \hat{\rho}_\tau (h)^2,$$

which can be compared with critical values from a $\chi^2(p)$ distribution.

The idea of quantile dependence can be extended to different time series to study the joint dependence structure. Let $Y_t = (Y_{1t}, Y_{2t})$ and $X_t = (X_{1t}^\top, X_{2t}^\top)^\top$, consider stationary random vectors $\{Y_t, X_t\}_{t=1}^{n}$, and denote the τ_ith conditional quantile of Y_{it} given X_{it} by $Q_{Y_{it}} (\tau_i | X_{it})$, $0 < \tau_i < 1$. Then the cross-quantilogram is defined (Han et al., 2015) as

$$\rho_{\tau_1, \tau_2} (k) = \frac{\mathrm{E}\left[\psi_{\tau_1} \left(Y_{1t} - Q_{Y_{1t}} (\tau_1 | X_{1t})\right)\psi_\tau \left(Y_{2,t+k} - Q_{Y_{2,t+k}} (\tau_2 | X_{2,t+k})\right)\right]}{\sqrt{\mathrm{E}\left[\psi_{\tau_1} \left(Y_{1t} - Q_{Y_{1t}} (\tau_1 | X_{1t})\right)^2\right]}\sqrt{\mathrm{E}\left[\psi_\tau \left(Y_{2,t+k} - Q_{Y_{2,t+k}} (\tau_2 | X_{2,t+k})\right)^2\right]}}.$$

The cross-quantilogram captures dependence between the two series at different quantile levels. It has properties similar to the quantilogram; in particular, it is robust to the nonexistence of moments.

The cross-quantilogram can be computed using the empirical counterpart of $\rho_{\tau_1,\tau_2}(k)$. We first estimate $Q_{Y_{it}}(\tau_i|X_{it})$ via a quantile regression

$$\hat{\beta}(\tau) = \arg\min_{\beta \in \mathbb{R}} \sum_{t=1}^{n} \rho_\tau(Y_{it} - g(X_{it}, \beta)).$$

Let $\hat{Q}_{Y_{it}}(\tau_i|X_{it}) = g(X_{it}, \hat{\beta}(\tau))$. Then we estimate the cross-quantilogram by

$$\hat{\rho}_{\tau_1,\tau_2}(k) = \frac{\sum_{t=1}^{n-k}\left[\psi_{\tau_1}\left(Y_{1t} - \hat{Q}_{Y_{1t}}(\tau_1|X_{1t})\right)\psi_{\tau_2}\left(Y_{2,t+k} - \hat{Q}_{Y_{2,t+k}}(\tau_2|X_{2t})\right)\right]}{\sqrt{\sum_{t=1}^{n-k}\left[\psi_{\tau_1}\left(Y_{1t} - \hat{Q}_{Y_{1t}}(\tau_1|X_{1t})\right)^2\right]}\sqrt{\sum_{t=1}^{n-k}\left[\psi_{\tau_2}\left(Y_{2,t+k} - \hat{Q}_{Y_{2,t+k}}(\tau_2|X_{2t})\right)^2\right]}}.$$

Han et al. derived the asymptotic properties of the estimated cross-quantilograms and proposed inference methods based on it.

20.6.2 Causality in quantiles

Another related application of dependence in quantiles is Granger causality. Following the definition of Granger (1969, 1980), we say that the random variable X does not Granger-cause the random variable Y if

$$F_{Y_t}(y|X_{t-j}, Y_{t-j}, j \geqslant 1) = F_{Y_t}(y|Y_{t-j}, j \geqslant 1), \quad \forall y, \tag{20.27}$$

where $F_{Y_t}(y|\mathcal{F})$ is the conditional distribution of Y_t given $\mathcal{F}$, or equivalently,

$$Q_{Y_t}(\tau|X_{t-j}, Y_{t-j}, j \geqslant 1) = Q_{Y_t}(\tau|Y_{t-j}, j \geqslant 1), \quad \forall \tau,$$

where $Q_{Y_t}(\tau|\mathcal{F})$ is the conditional quantile function of Y_t given $\mathcal{F}$. Granger noncausality implies that the past information on X does not affect the conditional distribution of Y_t. The variable X is said to Granger-cause Y when (20.27) fails to hold.

If we assume that the quantile function of Y_t conditional on $\{X_{t-j}, Y_{t-j}, j \geqslant 1\}$ is given by, say,

$$Q_{Y_t}(\tau|X_{t-j}, Y_{t-j}, j \geqslant 1) = \alpha(\tau) + \sum_{j=1}^{p} \beta_j(\tau) Y_{t-j} + \sum_{j=1}^{q} \gamma_j(\tau) X_{t-j},$$

then the null hypothesis of no Granger causality in distribution corresponds to

$$H_0 : \gamma_j(\tau) = 0, \quad \forall \tau \in [0,1], \ j = 1, \ldots, q,$$

and can be tested based on quantile regression estimators $\hat{\gamma}_j(\tau)$, $j = 1, \ldots, q$:

$$\left(\hat{\alpha}(\tau), \hat{\beta}(\tau), \hat{\gamma}(\tau)\right) = \arg\min_{\alpha,\beta,\gamma} \sum_t \rho_\tau\left(Y_t - \alpha - \sum_{j=1}^{p} \beta_j Y_{t-j} - \sum_{j=1}^{q} \gamma_j X_{t-j}\right).$$

In particular, regression Wald processes, or other similar statistics, can be constructed using results of Koenker and Machado (1999). See Koenker and Xiao (2002), Chernozhukov and Fernández-Val (2005), Lee and Yang (2006), and Chuang et al. (2009) for related studies.

20.7 Concluding remarks

Quantile regression has a wide range of applications in finance. In addition to the selected topics covered in this chapter, we mention here density forecasting, factor analysis for long–shorting, and cross-sectional analysis of stock returns.

Quantile regression can be applied to cross-sectional analysis of stock returns (see, for example, Barnes and Hughes, 2002; Gowlland et al., 2009). Quantile regression allows us to model the performance of firms or portfolios that under- or overperform in the sense that the conditional mean under- or overpredicts the firm's return. Barnes and Hughes apply quantile regression to modeling returns and testing the capital asset pricing model. Although many OLS-based researchers have found conflicting or inconclusive evidence regarding the importance of beta in explaining the cross-section of returns (around the center of the distribution, beta "flops around" either side of zero and is nonsignificant), they find that beta is strongly significant though with contrasting signs.

Quantile regression also provides very important information in analyzing factor effectiveness in investment. Most quantitative analysts back-test individual factors on a univariate basis, in order to check whether investing on the basis of that factor would have been associated with positive or negative returns in the past. If a particular factor would have been effective at identifying future outperformers and/or future underperformers, then it may have value as part of a broader multi-factor model, which can be used to rank individual stocks in terms of their overall attractiveness in terms of projected returns. There are two commonly used methods for assessing factor effectiveness: regression-based analysis and synthetic factor-mimicking portfolios. Regression-based analysis is widely used by academic researchers and by many practitioners, partly because it can readily be extended to a multivariate context. Many practitioners also make use of factor-mimicking portfolios because these impose less structure on the data and consequently offer greater robustness in the presence of outliers and similar problems. However, most users of factor-mimicking portfolios tend to focus on factor effectiveness at a univariate level. Quantile regression can provide a more complete picture of factor effectiveness than the conventional OLS methods, while also being more easily adaptable to multivariate approaches than factor-mimicking portfolios. Gowlland et al. (2009) provide examples of how quantile regression can be used to analyze the effectiveness of individual factors for quantitative investment purposes using a universe of US small caps, an asset class where insight into heteroskedastic factor effectiveness may have considerable significance for quantitatively based investment strategies.

Many additional topics in finance and numerous extensions of the analysis presented above can be conducted using quantile regressions. Many are currently under investigation.

Bibliography

T. Adrian and M. K. Brunnermeier. CoVaR. *American Economic Review*, 106:1705–1741, 2016.

A. Ang and G. Bekaert. International asset allocation with regime shifts. *Review of Financial Studies*, 15(4):1137–1187, 2002.

A. Ang and J. Chen. Asymmetric correlations of equity portfolios. *Journal of Financial Economics*, 63(3):443–494, 2002.

A. Ang, R. Hodrick, Y. Xing, and X. Zhang. The cross-section of volatility and expected returns. *Journal of Finance*, 61:259–299, 2006.

A. Ang, R. Hodrick, Y. Xing, and X. Zhang. High idiosyncratic volatility and low returns: International and further U.S. evidence. *Journal of Financial Economics*, 91:1–23, 2009.

P. Artzner, F. Delbaen, J.-M. Eber, and D. Heath. Coherent measures of risk. *Mathematical Finance*, 9:203–228, 1999.

N. Barberis and M. Huang. Stocks as lotteries: The implications of probability weighting for security prices. *American Economic Review*, 98:2066–2100, 2008.

M. L. Barnes and A. W. Hughes. A quantile regression analysis of the cross section of stock market returns. Working Paper, 2002.

K. Bartkiewicz, A. Jakubowski, T. Mikosch, and O. Wintenberger. Stable limits for sums of dependent infinite variance random variables. *Probability Theory and Related Fields*, 150:337–372, 2011.

S. Basak and A. Shapiro. Value-at-risk-based risk management: Optimal policies and asset prices. *Review of Financial Studies*, 14:371–405, 2001.

Basel Committee on Banking Supervision. *A New Capital Adequacy Framework*. Bank for International Settlements, Basel, 1999.

M. Basili and A. Chateauneuf. Extreme events and entropy: A multiple quantile utility model. *International Journal of Approximate Reasoning*, 52:1095–1102, 2011.

G. W. Bassett, R. Koenker, and G. Kordas. Pessimistic portfolio allocation and Choquet expected utility. *Journal of Financial Econometrics*, 2:477–492, 2004.

T. Bollerslev. Generalized autoregressive conditional heteroskedasticity. *Journal of Econometrics*, 31:307–327, 1986.

T. Bollerslev, R. Y. Chou, and K. F. Kroner. ARCH modeling in finance. *Journal of Econometrics*, 52:5–59, 1992.

J. Campbell, M. Lettau, B. Malkiel, and Y. Xu. Have individual stocks become more volatile? An empirical exploration of idiosyncratic risk. *Journal of Finance*, 56:1–43, 2001.

S. X. Chen. Nonparametric estimation of expected shortfall. *Journal of Financial Econometrics*, 1:87–107, 2008.

V. Chernozhukov. Extremal quantile regression. *Annals of Statistics*, 33:806–839, 2005.

V. Chernozhukov and I. Fernández-Val. Subsampling inference on quantile regression processes. *Sankhyā*, 67:253–276, 2005.

V. Chernozhukov and I. Fernández-Val. Inference for extremal conditional quantile models with an application to market and birthweight risks. *Review of Economic Studies*, 78: 559–589, 2011.

V. Chernozhukov and H. Hong. An mcmc approach to classical estimation. *Journal of Econometrics*, 115:293–346, 2003.

V. Chernozhukov and L. Umantsev. Conditional value-at-risk: Aspects of modeling and estimation. *Empirical Economics*, 26:271–292, 2001.

P. Christoffersen and F. X. Diebold. Financial asset returns, direction-of-change forecasting, and volatility dynamics. *Management Science*, 52:1273–1287, 2006.

C.-C. Chuang, C.-M. Kuan, and H.-Y. Lin. Causality in quantiles and dynamic stock return–volume relations. *Journal of Banking & Finance*, 33:1351–1360, 2009.

A. Cowles and H. E. Jones. Some a porteriori probabilities in stock market action. *Econometrica*, 5:280–294, 1937.

J. G. de Lara Resende and G. Wu. Competence effects for choices involving gains and losses. *Journal of Risk and Uncertainty*, 40:109–132, 2010.

G. De Rossi and A. Harvey. Quantiles, expectiles and splines. *Journal of Econometrics*, 152:179–185, 2009.

J. Dhaene, R. J. Laeven, S. Vanduffel, G. Darkiewicz, and M. J. Goovaerts. Can a coherent risk measure be too subadditive? *Journal of Risk and Insurance*, 75:365–386, 2008.

K. Dowd. Assessing VAR accuracy. *Derivatives Quarterly*, 34:61–63, 2000.

P. Embrechts, S. I. Resnick, and G. Samorodnitsky. Extreme value theory as a risk management tool. *North American Actuarial Journal*, 3:30–41, 1999.

R. Engle. Autoregressive conditional heteroscedasticity with estimates of variance of united kingdom inflation. *Econometrica*, 50:987–1008, 1982.

R. Engle and S. Manganelli. CAViaR: Conditional autoregressive value at risk by regression quantiles. *Journal of Business and Economic Statistics*, 22:367–381, 2004.

C. B. Erb, C. R. Harvey, and T. E. Viskanta. Forecasting international equity correlations. *Financial Analysts Journal*, 50:32–45, 1994.

C. Gowlland, Z. Xiao, and Q. Zeng. Beyond the central tendency: Quantile regression as a tool in quantitative investing. *Journal of Portfolio Management*, 3:106–119, 2009.

C. Granger. Investigating causal relations by econometric models and cross-spectral methods. *Econometrica*, 37:424–438, 1969.

C. Granger. Testing for causality: A personal viewpoint. *Journal of Economic Dynamics and control*, 2:329–352, 1980.

H. Han, O. Linton, T. Oka, and Y. Whang. The cross-quantilogram: Measuring quantile dependence and testing directional predictability between time series. Working Paper, September, 2015.

J. B. Hill. On tail index estimation for dependent, heterogeneous data. *Econometric Theory*, 26:1398–1436, 2010.

Y.-H. Hsu. *Applications of quantile regression to estimation and detection of some tail characteristics*. PhD thesis, University of Illinois, 2010.

L. Hu. Dependence patterns across financial markets: A mixed copula approach. *Applied Financial Economics*, 16:717–729, 2006.

R. Ibragimov and J. Walden. The limits of diversification when losses may be large. *Journal of Banking & Finance*, 31:2551–2569, 2007.

R. Ibragimov, D. Jaffee, and J. Walden. Nondiversification traps in catastrophe insurance markets. *Review of Financial Studies*, 22:959–993, 2009.

R. Koenker and J. A. F. Machado. Goodness of fit and related inference processes for quantile regression. *Journal of the American Statistical Association*, 94:1296–1310, 1999.

R. Koenker and Z. Xiao. Inference on the quantile regression process. *Econometrica*, 70: 1583–1612, 2002.

R. Koenker and Q. Zhao. Conditional quantile estimation and inference for ARCH models. *Econometric Theory*, 12:793–813, 1996.

S. Lee and J. Noh. Quantile regression estimator for GARCH models. *Scandinavian Journal of Statistics*, 40:2–20, 2013.

T.-H. Lee and W. Yang. Money-income Granger-causality in quantiles. University of California, Riverside Working Paper, 2006.

H. Levy. *Stochastic Dominance: Investment Decision Making under Uncertainty*. Springer, Berlin, 2016.

J. Lintner. Security prices, risk and maximal gains from diversification. *Journal of Finance*, 20:587–615, 1965.

O. Linton and Y. Whang. A quantilogram approach to evaluating directional predictability. *Journal of Econometrics*, 141:250–282, 2009.

O. Linton and Z. Xiao. Estimation of and inference about the expected shortfall for time series with infinite variance. *Econometric Theory*, 29:771–807, 2013.

O. Linton, E. Maasoumi, and Y.-J. Whang. Consistent testing for stochastic dominance under general sampling schemes. *Review of Economic Studies*, 72:735–765, 2005.

F. Longin and B. Solnik. Extreme correlations in international equity markets. *Journal of Finance*, 56:649–676, 2001.

H. Markowitz. Portfolio selection. *Journal of Finance*, 7(1):77–91, 1952.

R. Merton. A simple model of capital market equilibrium with incomplete information. *Journal of Finance*, 42:483–510, 1987.

J.P. Morgan and Reuters. RiskMetrics – Technical Document. http://bit.ly/2oo0xgv, 1996.

P. T. Ng, W.-K Wong, and Z. Xiao. Stochastic dominance via quantile regression. Technical Report, Northern Arizona University, 2011.

J. Quiggin. A theory of anticipated utility. *Journal of Economic Behavior and Organization*, 3:323–343, 1982.

R. T. Rockafellar and S. Uryasev. Optimization of conditional value-at-risk. *Journal of Risk*, 2:21–42, 2000.

R. T. Rockafellar and S. Uryasev. Conditional value-at-risk for general loss distributions. *Journal of Banking & Finance*, 26:1443–1471, 2002.

D. Roman, K. Darby-Dowman, and G. Mitra. Mean-risk models using two risk measures: A multi-objective approach. *Quantitative Finance*, 7:443–458, 2007.

M. J. Rostek. Quantile maximization in decision theory. *Review of Economic Studies*, 77: 339–371, 2010.

M. Sarma, S. Thomas, and A. Shah. Performance evaluation of alternative VaR models. Working paper, Indira Gandhi Institute of Development Research, 2000.

A. Saunders. *Financial Institutions Management: A Modern Perspective*. Irwin, Boston, 1999.

O. Scaillet. Nonparametric estimation and sensitivity analysis of expected shortfall. *Mathematical Finance*, 14:115–129, 2004.

D. Schmeidler. Subjective probability and expected utility without additivity. *Econometrica*, 57:571–587, 1989.

U. Segal. Nonlinear decision weights with the independence axiom (no. 353), 1984. UCLA Working Paper #353, Department of Economics, Los Angeles.

W. F. Sharpe. Capital asset prices: A theory of market equilibirum under conditions of risk. *Journal of Finance*, 19:425–442, 1964.

N. Sim and Z. Xiao. Modeling quantile dependence: Estimating the correlations of international stock returns. Working Paper, Boston College, 2009.

M. Statman. The diversification puzzle. *Financial Analysts Journal*, 60:44–53, 2004.

A. Tversky and D. Kahneman. Advances in prospect theory: Cumulative representation of uncertainty. *Journal of Risk and Uncertainty*, 5:297–323, 1992.

C. Wan and Z. Xiao. Portfolio optimization: A multi-quantile approach. Working Paper, Boston College, 2013.

C. Wan and Z. Xiao. Idiosyncratic volatility, expected windfall, and the cross-section of stock returns. In Y. Chang, T. B. Fomby, and J. Y. Park, editors, *Advances in Econometrics: Essays in Honor of Peter Phillips*, volume 33, pages 713–749. Emerald Group Publishing, Bingley, 2014.

Z. Xiao. Right tail information in financial markets. *Econometric Theory*, 30:94–126, 2014.

Z. Xiao and R. Koenker. Conditional quantile estimation and inference for GARCH models. *Journal of the American Statistical Association*, 104:1696–1712, 2009.

M. Yaari. The dual theory of choice under risk. *Econometrica*, 55:95–115, 1987.

K. Yu and R. A. Moyeed. Bayesian quantile regression. *Statistics & Probability Letters*, 54: 437–447, 2001.

Y. Zheng, Q. Zhu, G. Li, and Z. Xiao. Hybrid quantile regression estimation for time series models with conditional heteroscedasticity. arXiv:1610.07453, 2016.

21

Quantile Regression for Genetic and Genomic Applications

Laurent Briollais

Tanenbaum Lunenfeld Research Institute, Toronto, Canada

Gilles Durrieu

Laboratoire de Mathématiques de Bretagne Atlantique, Vannes, France

CONTENTS

21.1 Introduction

Recent advances in biological research have seen the emergence of high-throughput investigations with numerous applications in genetic, genomic and other "-omic" disciplines (transcriptomic, proteomic, metabolomic, etc.). The goal of these studies is to provide a richer and unbiased view of the cell at different biological levels. The cell is often represented as a complex system with various levels of biological information communicating with each other. The DNA found within each cell contains the genetic blueprint for the entire organism. Each gene contains the information necessary to instruct the cellular machinery how to make mRNA, and in turn the protein encoded by the order of DNA bases constituting the gene. Each one of these proteins is responsible for carrying out one or more specified molecular functions within the cell. Differing patterns of gene expression (i.e. different mRNA and protein levels) in different tissues can explain differences in both cellular function and appearance. The new high-throughput technologies allow the study of biological mechanisms at an unprecedented depth and scale. The challenge for statisticians is now to extract relevant biological insights from the vast volume of data generated and exploit this exhaustive information to guide the future development of new diagnostics and therapies for human diseases.

In this fast evolving field of -omic science, quantile regression (QR) offers a versatile and robust alternative approach to many competing statistical methods. Several aspects make QR an attractive statistical approach in genetic and genomic applications. Geneticists have long been interested in sampling from the extremes of a trait distribution as a way to leverage the detection of genetic signals. Population geneticists have, for example, considered the tail of a "liability normal distribution" as a way of identifying those individuals more susceptible to a certain disease condition (Falconer and McKay, 1996). Extreme sampling has also been an efficient approach in linkage analysis and association studies to discover new genetic loci (Risch and Zhang, 1995; Gu et al., 1997; Huang and Lin, 2007), and has recently been proposed to investigate rare genetic variants (Li et al., 2011). With the QR approach, instead of sampling from the extremes of the distribution, one can estimate specific parameters related to the extremes with the expectation that it will enhance genetic discoveries. QR is not restricted to the extremes of the distribution and one can estimate regression parameters for any given quantile of the distribution. QR also avoids the need to decide an arbitrary threshold to define the "extremes." The other feature that makes QR appealing for geneticists is its robustness to distribution assumptions and to outlying observations. Any development of new high-throughput technology always comes with measurement errors and systematic biases often difficult to control. The use of a robust approach such as QR makes the inference less biased and less subject to false positives.

Methods developed in the context of -omic data are numerous and entail many fields of statistics (e.g. hypothesis testing, linear and nonlinear regression, analysis of variance, multivariate analysis and data mining approaches). QR is a relatively new approach applied to the -omic fields but has become a widely used and accepted technique in many areas of theoretical and applied statistics (Koenker and Bassett, 1978). QR can be used at various stages of the analysis from the data pre-processing (e.g. normalization) to the data representation and data modeling. It has been developed in the context of independent and identically distributed -omic data, but recent extensions focus also on nonindependent and identically distributed data. It has been applied to many and various genetic and -omic areas, including genetic association studies, gene expression, CGH array and RNA-sequencing experiments, methylation, and proteomics (Briollais and Durrieu, 2014). In this review, our goal is to describe recent applications of QR to genetic and various -omic research areas.

21.2 Genetic applications

21.2.1 Background and definitions

The aim of genetic association studies is to identify the relationships between a response variable (phenotype), which is, in many cases, a binary disease or quantitative trait (QT) and one or more genetic markers. In the majority of applications, the genetic markers are single-nucleotide polymorphisms (SNPs), as they are common and well mapped throughout the genome. In genetic applications, QR has been used to find a genetic locus that influences the distribution of a continuous QT, such as body mass index (BMI), systolic blood pressure, or birthweight. Several types of association studies haven been conducted: candidate gene and genome-wide studies.

In a *candidate gene* approach, the aim is to test whether a limited number of pre-selected genes or genetic variants are associated with a QT of interest, the simplest case being to perform a test with a single gene or single genetic variant. These candidates are chosen because of biological function or previous study results, and some of them have strong genetic effects. This selection can involve genes or genetic variants belonging to the same genetic pathway or that are biologically likely to be causally related to the QT.

In contrast, the goal of a *genome-wide association study* (GWAS) is to perform an exhaustive search among all genetic variants present in the genome and test an association with the QT of interest without any prior selection. The entire human genome has a length of approximately 3.3×10^9 base-pairs, and approximately 500,000 well-chosen SNPs, termed tagSNPs, are required to capture most of the genomic variation in subjects of European ancestry. The rationale underlying GWAS is based on the premise that common genetic variants (i.e. present in more than 1–5% of the population) could explain most of the heritability of common human diseases and QTs, which is often referred to as the common disease, common variant hypothesis. The present paradigm for GWAS involves the collection of several thousand individuals and an exhaustive search among more than 500,000 SNPs of those associated with the QT of interest using simple univariate statistics. Therefore, the main difference between a candidate gene approach and a GWAS is that the former is considered as a *hypothesis based* approach while the latter is an *agnostic* approach.

Another important distinction to be made in association studies is that between single-marker and multiple-marker association tests. While most applications of QR concern single genetic marker analyses, recent studies based on multi-marker QR have also recently been proposed (Kong et al., 2016; Burgette and Reiter, 2012). Multi-marker approaches can be applied to gene based, pathway based or specific groups of markers defined according to specific biological relationships. The main motivations for using QR in genetic association studies entail the need to have a method robust to distribution assumptions (Olivieri et al., 2005; Scher et al., 2006; Hecker et al., 2009; Beyerlein et al., 2011; Haring et al., 2013) and/or estimate associations at various quantiles of the QT (Williams, 2012). To illustrate the interest of QR in the context of candidate gene association (Section 21.2.2) and GWAS (Section 21.2.3), we present below two recent applications that we performed on the Western Australian Pregnancy Cohort. Section 21.2.4 is related to genetic association involving a set of markers.

21.2.2 Candidate gene association study of child BMI

Obesity is a major global public health problem. The World Health Organization estimated in 2010 that there were at least 42 million overweight children under the age of 5 years and 1 billion overweight adults globally World Health Organization (2006). Rates of obesity and overweight are rising in nearly all countries around the globe (de Onis et al., 2010). Treating childhood obesity is a major public health challenge (Whitlock et al., 2005). Genetic factors have an important role in childhood obesity. Recent GWAS and candidate gene studies have identified hundreds of genetic variants contributing to the risk of obesity. The search for obesity-related genes is often conducted indirectly using phenotypes such as BMI, body fat percentage, waist circumference, waist–hip ratio, subcutaneous fat area, or visceral fat area. Because obesity is defined as the 95th percentile of BMI, QR can be a valuable approach to determining directly the genetic factors contributing to obesity. We illustrate this idea using the Western Australian Pregnancy Cohort (RAINE) from Australia and a candidate gene variant located in the *TCF4* gene.

The RAINE Cohort. Recruitment of the RAINE Cohort is described in detail by Newnham et al. (1993). In brief, between 1989 and 1991, 2,900 pregnant women were recruited prior to 18-weeks gestation into a randomized controlled trial to evaluate the effects of repeated ultrasound in pregnancy. Recruitment predominantly took place at King Edward Memorial Hospital (Perth, Western Australia). Women were randomized to repeat ultrasound measurements at 18, 24, 28, 34, and 38 weeks' gestation or to a regular ultrasound assessment at 18 weeks. Children were comprehensively phenotyped from birth to 18 years of age by trained members of the RAINE research team. Data up to 14 years only was available at the time of our study and included 1,048 children with genetic information available (549 boys and 499 girls).

TCF4 *gene.* This gene encodes transcription factor 4, a basic helix–loop–helix transcription factor. The encoded protein recognizes an Ephrussi-box ('E-box') binding site ('CAN-NTG'), a motif first identified in immunoglobulin enhancers. This gene is broadly expressed, and may play an important role in nervous system development. Defects in this gene are a cause of Pitt–Hopkins syndrome. Multiple alternatively spliced transcript variants that encode different proteins have been described. Gene variants in the *TCF4* gene have been associated with early onset obesity (Wheeler et al., 2013).[1] We studied one particular SNP within the gene, rs8086784, which has a minor allele frequency of 1.2% in the European population. An allelic dosage coding was used for this SNP, which counts the number of minor alleles and takes values between 0 and 2.

Analysis. We estimated BMI curves from birth to 14 years of age with respect to gender and the SNP rs8086784 in the *TCF4* gene, using the linear quantile mixed model (LQMM) approach (Geraci and Bottai, 2014) with cubic spline functions. Let $y_i = (y_{i1}, y_{i2}, \dots, y_{in_i})$ denote the longitudinal BMI measurements for the ith individual with n_i measurements. Child i's BMI at time j can be expressed in the LQMM as

$$y_{ij} = \beta_0^{(\tau)} + sp(t_{ij})\beta_{sp(t)}^{(\tau)} + X_i\beta_X^{(\tau)} + \left(sp(t_{ij}) \cdot X_i\right)\beta_{X \cdot sp(t)}^{(\tau)} + Z_i b_i + \varepsilon_{ij}^{(\tau)},$$

where $sp(t_{ij})$ is the cubic spline function of age that is used to catch the nonlinear BMI growth curve, X_i represents the variables of interest, including gender and the SNP rs8086784, and $sp(t_{ij}) \cdot X_i$ refers to the interaction terms of spline function of age and the variables of interest X_i that measure time-varying effects of X_i on BMI. The βs are fixed effects and b_i is a vector of random effects including random intercept. Finally, the parameter τ is set *a priori* and defines the quantile level to be estimated. We used τ equal to 95%, 85%, 50%, and 10%, which correspond to the cut-off values for obese, overweight, normal, and underweight children. The predicted BMI values as function of the SNP are depicted in Figure 21.1. The symbols "SNP +" and "SNP −" indicate children carrying and not carrying the minor allele of the SNP rs8086784, respectively. Figure 21.1 clearly illustrates that the effect of the SNP varies with the quantile level. At age 14 years, carriers of the SNP variant have their BMI decreased by 0.38 at $\tau = 0.1$ ($p = 0.21$), by 0.23 at $\tau = 0.5$ ($p = 0.61$), by 3.07 at $\tau = 0.85$ ($p < 10^{-6}$), and by 6.98 at $\tau = 0.95$ ($p < 10^{-6}$). Therefore, the results indicate a very substantial protective effect of the SNP minor allele in obese children and could guide further intervention targetting the *TCF4* gene. The results also stress that an SNP can have very low effect at the median of the response distribution and a much stronger effect at the more extreme quantiles. GWASs that usually focus on the mean or median of the response distribution could therefore miss important genetic factors contributing to the heritability of complex traits such as BMI. This is illustrated in the next subsection.

21.2.3   GWAS of birthweight

The emergence of high-throughput technologies for SNP genotyping and their application to large-scale GWASs have generated promises that the genetic basis of many common human

---

[1]See https://www.ncbi.nlm.nih.gov/gene/6925

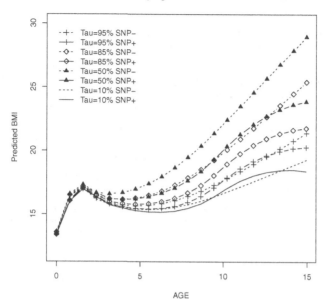

FIGURE 21.1
Predicted BMI by age and SNP rs8086784 status

diseases could be elucidated (Risch and Merikangas, 1996; Risch, 2000; Hirschhorn and Daly, 2005; Kingsmore et al., 2008; Kruglyak, 2008; McCarthy and Hirschhorn, 2008). These GWASs have identified hundreds of genetic variants implicated in various human diseases and complex traits, providing valuable insights into their genetic mechanisms (Hindorff et al., 2009a,b). The present paradigm for GWAS involves the collection of more than 1,000 cases and 1,000 controls and an exhaustive search among more than 500,000 SNPs of those associated with the disease outcome using simple univariate statistics. Despite its relative merits in identifying new genetic variants, GWASs have also given rise to many criticisms. For example, the SNPs selected through univariate statistics generally have a low predictive value, explain a fairly modest proportion of the genetic variability of the disease and, perhaps more importantly, do not usually provide much understanding of the biological process.

GWAS of birthweight in the RAINE study. Being in the extreme tail of the birthweight distribution is generally associated with high risks of perinatal morbidity and mortality. In addition, there are well-documented associations between lower birthweight and later-life chronic disease, including type 2 diabetes, cardiovascular disease and higher blood pressure. The mechanisms underlying these associations are still not well understood. Birthweight is a complex multifactorial trait that can involve genetic factors and gene–environment interactions. The importance of genetic factors acting independently of the intra-uterine environment is illustrated by correlations between paternal height or weight and offspring birthweight, and genetic variants that are associated both with low birthweight and increased risk of type 2 diabetes may account for some of the observed correlation between these phenotypes. However, the genetic loci that influence birthweight are largely unknown. Our goal in using QR is to find associations between the extremes of birthweight and genetic variants.

Analysis. We analyzed five chromosomes (18 to 22) from the RAINE study. A total of

1,043 babies had GWAS data available at birth. We fitted QRs with quantile levels corresponding to extreme high birthweight ($\tau = 0.9$ and $\tau = 0.8$) and extreme low birthweight ($\tau = 0.1$ and $\tau = 0.2$). We then performed a joint test of the slope parameters (i.e. for the SNP variable) at the two extreme high and the two extreme low quantiles with the Wald statistic (Koenker and Bassett, 1982) as implemented in the anovar.rq function of the quantreg R package. The results were compared with the Wald test for slope parameter from the standard least-squares (LS) linear regression. The genome-wide significance level is usually considered to be 5×10^{-8}. The results are represented by the Manhattan plot in Figure 21.2. The Manhattan plot is a standard representation of GWAS results where the $-\log_{10} p$-values are plotted against the chromosomal position for all SNPs analyzed. The results clearly indicate that at the extreme quantiles of the birthweight distribution, either low or high, a higher proportion of SNPs aresignificant at the genome-wide level, compared to LS regression. The most significant SNPs can differ between the low and high extreme quantiles, and the ranking of the best SNPs also differs across quantiles of the distribution and between QR and LS regression. The QR GWAS approach leads therefore to the detection of significant SNPs that could not be determined by fitting a model to the mean of the birthweight distribution.

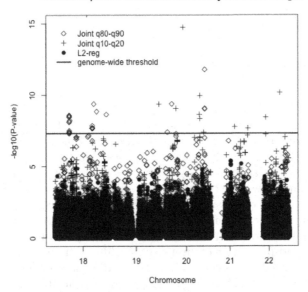

FIGURE 21.2

Manhattan plot of GWAS results. The diamond and plus signs represent the $-\log_{10} p$-values from QR based on a joint test of the slope parameters at $\tau = 0.9$ and $\tau = 0.8$ and at $\tau = 0.1$ and $\tau = 0.2$, respectively. The black bullets represent the $-\log_{10} p$-values obtained from the LS regression.

21.2.4   Genetic association with a set of markers

Candidate gene and genome-wide association studies typically perform single-SNP analysis on one genetic marker at a time. Recent approaches have been proposed to perform associations with a set of genetic markers, where the set could represent a gene, a gene network or a genetic pathway, for example. Kong et al. (2016) proposed a marker-set association test based on QR. They considered a linear model for a response y_i that depends on genetic variables $\mathbf{X}_i$ and nongenetic variables $\mathbf{Z}_i$ for an individual i,

$$y_i = f(\mathbf{X}_i) + \boldsymbol{\beta}^T \mathbf{Z}_i + \beta_0 + \epsilon_i,$$

where $f(\cdot)$ quantifies the genetic effects. The estimation of the parameters $\boldsymbol{\beta}$, β_0, and $f(\cdot)$ is obtained from the minimization problem

$$\min_{\boldsymbol{\beta}, \beta_0, f} \sum_{i=1}^n \rho_\tau(y_i - f(\mathbf{X}_i) - \boldsymbol{\beta}^T \mathbf{Z}_i - \beta_0) + P_\lambda(f),$$

where $P_\lambda(f)$ is a penalty function depending on the parameter λ that controls the smoothness of $f(\cdot)$. The function $f(\cdot)$ depends on a Kernel matrix $K(\cdot, \cdot)$ that quantities the distance between two individuals explained by the set of genetic markers. The form of $f(\cdot)$ is therefore

$$f(\mathbf{X}) = \frac{1}{\lambda} \sum_{i=1}^n \theta_i K(\mathbf{X}, X_i),$$

where $\boldsymbol{\theta}$ is a vector of parameters to be estimated. An example of gene set analysis is given below.

An alternative approach that considered sets of variables, including genetic variables, was introduced by Burgette and Reiter (2012). This approach is based on confirmatory factor analysis and uses QR to estimate quantile-specific parameters related to the latent factors. The inference is done within the Bayesian framework. The covariates are latent factors that summarize other individual covariates. In their application, Burgette and Reiter (2012) sought association between the latent factors summarizing psychological and social health factors, smoking behaviours, and genetic variables, with the quantiles of birthweight. Because this approach is not specific to a genetic problem, it is not described further.

Analysis. Kong et al. (2016) provided an application of their QR approach to the Vitamin Intervention for Stroke Prevention trial. In brief, the trial enrolled patients who were 35 years of age or older with a nondisabling cerebral infarction within 120 days of randomization and homocysteine (Hcy) levels in the top quartile for the US population. Subjects were randomly assigned to receive daily doses of either a high-dose or a low-dose formulation. Genetic information was collected from nine candidate genes related to Hcy metabolism where each gene could include from 3 to 20 SNPs. The aim of the analysis was to study the genetic influence on the Hcy level at baseline. In all, 1,587 subjects were available for analysis. The effect of each of the nine genes was assessed with QR (with $\tau = 0.1$, 0.5, 0.8), adjusting for age, sex, and population stratification variables (10 principal components) and an identity-by-state kernel was used to summarize the multi-locus information. Interestingly, the analysis found that the gene *CBS* was significant only at $\tau = 0.5$ and 0.8, and the gene *TCN1* was significant only at $\tau = 0.8$ after Bonferroni correction using the QR method. In comparison, none of the genes were significant with a standard LS approach.

21.3 Genomic and other -omic applications

21.3.1 Background

High-throughput experiments (HTEs) have radically changed the nature of genomics assays by allowing the simultaneous screening of thousands of genes in a single experiment.

Statistical analysis of HTE data has led to competing applications of numerous statistical approaches and computing algorithms for several problems: class comparison, class discovery, and class prediction (Simon et al., 2003). HTE methods in genomics include experimental data on gene expression, methylation, comparative genomic hybridization (CGH) and copy number variation (CNV), RNA-sequencing and proteomics, which we now briefly describe. In *gene expression* analysis, investigators are typically interested in whether gene expression varies with respect to various experimental conditions (e.g. disease subgroups or other classes of observations/individuals) either individually or in coordinated manner. Usually the response variable is the gene expression in a regression model and the explanatory variables describe the experimental conditions. In *methylation* studies, epigenetic changes that lead to the addition of a methyl group (CH_3) to the carbon 5 position of the cytosine at a CpG site are measured. Methylation plays an important role in gene expression regulation, transposon silencing, and transcription factor binding inhibition. Promoter hyper-methylation acts together with specific histone modifications to silence genes. *Array-based comparative hybridization* (array-CGH) is a technique offering genome-wide study of chromosomal aberrations such as deletions, amplifications, and structural rearrangements, which are the landmark of many cancers (Pinkel and Albertson, 2005). Amplification and deletion of chromosomal segments can lead to abnormal mRNA transcript levels and result in the disfunction of key cellular processes. The identification of genomic regions associated with systematic aberrations can provide insights into the initiation and progression of cancer, and ultimately improve diagnosis, prognosis, and therapy strategies. Chromosomal aberrations can be detected in tumor cells but can also be found in normal cells and measured by *copy number variations* (CNVs), which are related to phenotypic variation in the normal population. CNVs can be detected by different technologies including CGH and array-CGH, but also SNP arrays and new-generation sequencing technologies. *RNA-sequencing* allows the quantification of gene expression levels on a genome-wide scale. RNA transcripts are converted into cDNA fragments, which are sequenced to produce millions of sequences (i.e. reads). Gene expression is quantified by counting the number of reads that map back to each gene. This next-generation sequencing can provide information on differential expression of genes, including gene alleles and differently spliced transcripts; noncoding RNAs; post-transcriptional mutations and gene fusions. *Proteomics* uses mass spectrometry techniques to measure the relative abundance of thousands of peptides in a single experiment (Callister et al., 2006) by precise determination of the molecular mass of peptides as well as their sequences. This information can be used for protein identification, de novo sequencing, and identification of post-translational modifications, which in turn can reveal important biological processes.

The application of QR to HTEs entails several aspects of the data analysis, including data pre-processing (Section 21.3.2), sample size determination (Section 21.3.3), analysis of chromosomal region aberrations (Section 21.3.4), robust estimation and outlier determination (Section 21.3.5), and the genomic analysis of a set of genes (Section 21.3.6).

21.3.2   Genomic data pre-processing

Normalization procedures are often applied to HTEs to remove systematic biases and noise, which can result from many factors, including systematic errors in experimentation, sample preparation, and instrumentation. The biases and noise can lead to extraneous variability among replicate samples and can affect the accuracy and precision of biological research conclusions. QR was proposed to perform normalization of HTE data. The premise of this approach is that the distribution of particular HTE data (e.g. gene expression, peptide abundance) in different samples is expected to be similar and can be adjusted to be made similar when they are found to be different. The default algorithm implementing quantile normalization (Bolstad et al., 2003) maps the distribution of each sample, estimated by the empirical cumulative distribution function (cdf) of each sample, to a target distribution, estimated by the empirical distribution of the averaged order statistics. The normalized measure (e.g. gene expression, peptide abundance) m is obtained by the transformation $m^* = F^{-1}(G(m))$, where G represents the estimated empirical distribution of each array and F the empirical distribution of the averaged sample quantiles. This normalization technique has been applied to proteomic (Callister et al., 2006) and RNA-sequencing experiments (Bolstad et al., 2003; Hansen and Irizarry, 2012).

For RNA-sequencing, Hansen and Irizarry (2012) noticed that the percentage of C or G nucleotides (GC content) can vary significantly both across genes and within the same gene across samples. They proposed an algorithm to correct the distributional distortion in count reads and systematic bias due to GC content. The algorithm returns an offset value that can be used to adjust the observed read count for each gene and data sample. The normalized value can then be fitted using linear regression where design effects can be added as covariates such as GC content and gene length and a corrected value can be computed to remove the effect of GC content and gene length. If we let θ_g represent the gene expression of a gene g and Y_g the number of reads that map back to each gene g, the normalization model proposed by Hansen and Irizarry (2012) has the form

$$\log(Y_{g,i}) = \mathrm{Cst}_i + h(\theta_{g,i}) + \sum_{j=1}^{p} f(X_{g,j,i}),$$

where, for sample i, h is a smooth function and f is estimated by median QR as a function of p covariates (e.g. GC content). The gene expression value is then obtained by inverting the function h.

21.3.3 Sample size determination in gene expression studies

A critical aspect in the design of microarray studies is the determination of the sample size necessary to declare genes differentially expressed across different experimental conditions. We recently proposed a sequential approach where the decision to stop the experiment depends on the accumulated microarray data (Durrieu and Briollais, 2009). The study could stop whenever sufficient data have been accumulated to identify gene expression changes across several experimental conditions. The general form of the model is

$$\mathbf{Y}_n = \mathbf{X}_n \boldsymbol{\beta} + \boldsymbol{\varepsilon}_n$$

where, for all $n \geqslant 1$, $\mathbf{Y} \equiv \mathbf{Y}_n = (Y_1, \ldots, Y_n)'$ is the $\log_2$ ratio of gene expression from all slides, $\mathbf{X} \equiv \mathbf{X}_n \equiv \mathbf{X}_{n \times p}$ is a known $n \times p$ design matrix of experimental conditions, $\boldsymbol{\beta} = (\beta_1, \ldots, \beta_p)'$ is the vector of unknown parameters to be estimated, and $\boldsymbol{\varepsilon} \equiv \boldsymbol{\varepsilon}_n = (\varepsilon_1, \ldots, \varepsilon_n)'$ is the error term. The principle of this approach is to compare the means of the log-gene expression distribution under different conditions (treatments). We then construct a confidence interval I_n for the first component of the linear model (i.e. the gene expression at baseline) based on a robust estimator of β_1 which satisfies

$$L_n \leqslant 2d \quad \text{and} \quad P_F(I_n \ni \beta_1) \geqslant 1 - \alpha,$$

where L_n denotes the length of I_n. Starting with an initial sample size n_0, the experiment stops when the number of replicates N_d is the smallest $n > n_0$ such that the length $L(N_d) < 2d$. Then the stopping variable N_d satisfying the previous conditions on n can be defined by

$$N_d = \min\left\{ n \geqslant n_0 \,\middle|\, n \geqslant \frac{z_{1-\alpha/2}^2 \widehat{W}_n^2(1/2)}{d^2} \right\}, \quad d > 0,$$

where d is a precision parameter fixed for each gene and $\widehat{W}_n^2(1/2)$ is a kernel-type estimator of the variance for the median.

Application. The gene expression levels of a hormone-responsive breast cancer cell line (MCF-7) were measured after stimulation with various concentrations of estrogen, above (high-dose effect) and below (low-dose effect) normal physiologic levels and compared with the corresponding levels of expression in control samples (Lobenhofer et al., 2004). The cell lines were treated with estrogen at four concentrations (10^{-8} M, 10^{-10} M, high doses) and (10^{-13} M, 10^{-15} M, low doses) or with concentration-matched ethanol solvent (control samples). We had four technical replicates and two biological replicates at each dose, so eight measurements in total. In this analysis, we focused on seven genes that were validated by real-time polymerase chain reaction (PCR). Because the threshold dose can be observed below the level 10^{-8} M, we just included in our analysis the two low doses and only one high dose (10^{-10} M). Thus, there were a total of 24 measurements available for each gene (3 doses $\times$ 8 measurements). To analyze the data, we fitted the regression model above with the dose and dye effects as covariates using QR on the transformed response and covariates. The model can be written as

$$Y = \beta_1 + \beta_2 Dose + \beta_3 Dye + \varepsilon,$$

where β_1 is the intercept denoting the gene expression at an initial dose, β_2 is the slope parameter coding for the change in gene expression with dose effect, and β_3 is the parameter for the dye effect.

The goal of the analysis was to compute a robust confidence interval I_n for the model intercept β_1, the gene expression at the low dose. We start the experiment with an initial sample size $n_0 = 16$, which corresponds to the number of measurements at the two low doses. Then we determine sequentially the number of groups of technical replicates at the high dose, where each group had two observations (duplicated spots). The experimental process stops for N_d the smallest $n > n_0$ is such that $L_n < 2d$. The results are presented in Table 21.1 for $\alpha = 0.001$, the value chosen to adjust for multiple comparisons in this study (Lobenhofer et al., 2004).

TABLE 21.1
Results from group sequential discrete monitoring microarray experiment.

Label	Name	d	N_d	$\widehat{\beta}_1^{N_d}$	$SD\left(\widehat{\beta}_1^{N_d}\right)$	99.9% CI
Gene 1	*SDF1*	0.375	22	1.175	0.283	$(0.976, 1.374)$
Gene 2	*MYB*	0.410	20	0.971	0.457	$(0.578, 1.364)$
Gene 3	*CDC28*	0.243	18	0.668	0.111	$(0.582, 0.755)$
Gene 4	*LRP8*	0.480	22	0.479	0.547	$(0.094, 0.864)$
Gene 5	*CDC25A*	0.265	18	0.081	0.042	$(0.048, 0.113)$
Gene 6	*TMP3*	0.215	18	-0.369	0.257	$(-0.569, -0.169)$
Gene 7	*PRKCZ*	0.226	20	-0.445	0.260	$(-0.669, -0.221)$

Note: d, precision parameter; N_d, stopping variable estimator; $\widehat{\beta}_1^{N_d}$, least absolute deviation intercept estimator; $SD\left(\widehat{\beta}_1^{N_d}\right)$, kernel regression estimator of the standard deviation of $\widehat{\beta}_1^{N_d}$; CI, confidence interval for the intercept.

These results indicate the minimal number of technical replicates N_d required to estimate the gene expression level with a fixed precision. We observe that for a relatively small precision parameter (d values), the stopping value remains lower than 24 (the maximum number of technical replicates available for the threes doses). Genes 1–5 were significantly upregulated differentially expressed ($p < 0.001$) and genes 6 and 7 are downregulated differentially expressed. These seven genes were confirmed to be differentially expressed by quantitative PCR. In conclusion, our analysis showed that the experimenter could have stopped the experiment after collecting only 22 replicates and would have reached the same conclusion as in Lobenhofer et al. (2004).

21.3.4 Determination of chromosomal region aberrations

The determination of chromosomal aberrations (e.g. amplification or deletion of a whole chromosomal segment) is a common problem in genomics that can be assessed through CGH or CNV experiments (see above). QR has been proposed to analyze CGH data (Eilers and de Menezes, 2005). The idea is to minimize the sum of absolute values subject to a given penalty that provides better estimates of CGH profiles (in particular, sudden jumps and flat plateaus). QR was applied to obtain median smoothed curves. An alternative approach to Eilers and de Menezes (2005) was proposed in Li and Zhu (2007), the main difference being the form of the penalty function which is based on the fused penalty (Tibshirani et al., 2005) and incorporates information on the physical location of clones. A variant of Eilers' approach (Eilers and de Menezes, 2005) has also been introduced to detect CNVs (Gao and Huang, 2010).

The approaches of Eilers and de Menezes (2005) and Li and Zhu (2007) are both based on penalized QR. The function to minimize in Eilers and de Menezes (2005) is given by

$$Q_1 = \sum_{i=1}^{m} |y_i - z_i| + \lambda \sum_{i=2}^{m} |z_i - z_{i-1}|,$$

where i represents the clones ordered by physical distance, y_i the outcome value for clone i, z_i the estimated value and λ a penalty parameter. The second term in the expression is the penalty function, discouraging changes in adjacent clones by penalizing $|z_i - z_{i-1}|$. In Li and Zhu's (2007) method, the penalty function is adjusted by physical distance between two clones. The penalty parameter is estimated by cross-validation in Eilers and de Menezes (2005), while it is determined by the Schwartz information criterion in Li and Zhu (2007). The penalty proposed by Eilers and de Menezes (2005) yields piecewise constant fitted functions. These two approaches are very similar since it is more the signs of the residuals that matter when minimizing the sum of absolute values (Eilers and de Menezes, 2005). In Gao and Huang (2010), an additional term corresponding to the lasso penalty is added to the function to minimize Q_1 to control the sparsity of the model. It should be noted also that in a recent paper, Rippe et al. (2012) proposed minimizing the residual sum of squares instead of the sum of absolute differences with the same penalty as in expression Q_1 for a better trade-off between robustness and sensitivity to detect CNVs.

Application. We applied penalized QR to a collection of CNV arrays that contains data on chromosome 14 from three colorectal tumors. Copy number values of 207 probes from Illumina GoldenGate Linkage IV data were downloaded from the R package quantsmooth. We display one tumor in Figure 21.3 that shows one region of loss. CNV arrays allow high-resolution mapping of deletions and amplifications, and eventually identification of the underlying disease-causing genes (Oosting et al., 2007). If changes in copy numbers occur, we expect these to be visible in segments that cover multiple probes, because fragments of chromosomes are generally affected. The spatial information can be used to reduce noise and increase the reliability of detecting changes. Using spatial information means that some form of smoothing is being applied. Figure 21.3 shows the results obtained by a smoothing method inspired by penalized QR (Eilers and de Menezes, 2005). The median quantile estimate captures well the chromosomal segment that is lost.

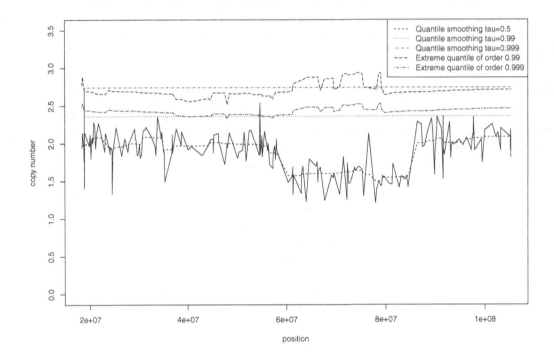

FIGURE 21.3
Copy number variation ($\log_2$ ratio intensity of tumor versus reference) as a function of
chromosomal position (Mb) for one colorectal tumor on chromosome 14. The solid line
represents the fitted median curves and the dashed lines the quantile smoothing curves for
$\theta \in \{0.5, 0.99, 0.999\}$ and the extreme quantile of order 0.99 and 0.999.

21.3.5 Robust estimation and outlier determination in genomics

One of the main motivations for using QR in genomics is its robustness against outliers and
flexibility to model nonlinear relationships. This is well illustrated by applications in gene
expression data. For example, a median QR was proposed by Yoon et al. (2007) to provide
robustness against distribution assumptions and violation of variance homogeneity of gene
expression. Inferential tools based on mixed model QR havw also been introduced by Wang
and He (2007, 2008). A random effect is added to the linear QR to account for correlations
of probes within arrays and a robust rank test statistic is used to detect differences in certain
quantiles of the gene expression distributions. QR has also been applied to model temporal
gene expression (Ho et al., 2009; Vinciotti and Yu, 2009). The originality of this latter work
was to combine quantile- and M-regression, which provides very good robustness against
outliers and distribution assumptions (the M-estimator) while allowing quantile-specific
estimates of regression coefficients.

The determination of outliers has generated fewer applications in genomics, and we
present here our recent work based on Durrieu et al. (2015) on extreme QR. We are interested in estimating the extreme quantile of gene expression or CNVs such as in Figure
21.3 where the expression varies with the chromosomal position t. Let $(X_t)_{t \in [0, T_{\max}]}$ denote
the gene expression process at t. The aim is to provide a pointwise estimate of the tail

probability $1 - F_t(x)$ where $F_t(x) = P(X_t \leqslant x)$ is the distribution of X_t, which is viewed as a random variable with values in the interval $[0, \infty)$. We also estimate the extreme τ-quantile processes on $[0, T_{\max}]$, where τ is a quantile order chosen to be close to 1 and the random variable X_t has a strictly positive density f_t. The distribution F_t is in the domain of attraction of the Fréchet distribution.

For a fixed threshold $\tau_0 \geqslant x_0$, the excess distribution function over this threshold τ_0 is

$$F_{t,\tau_0}(x) = \frac{F_t(x) - F_t(\tau_0)}{1 - F_t(\tau_0)},$$

where by the Fisher–Tippet–Gnedenko theorem (see Beirlant et al., 2004), $F_{t,\tau_0}(x)$ can be approximated by a Pareto distribution

$$G_{\tau_0,\theta_t}(x) = 1 - \left(\frac{x}{\tau_0}\right)^{-1/\theta_t}, \quad x \geqslant \tau_0 \text{ as } \tau_0 \to \infty,$$

where the unknown parameter θ_t is the conditional tail index.

Following Grama and Spokoiny (2008) and Durrieu et al. (2015), the distribution function F_t can be approximated as the threshold τ_0 converges to infinity by the semiparametric model

$$F_{t,\tau_0,\theta_t}(x) = \begin{cases} F_t(x), & \text{if } x \in [x_0, \tau_0], \\ 1 - (1 - F_t(\tau_0))(1 - G_{\tau_0,\theta_t}(x)), & \text{if } x > \tau_0. \end{cases} \tag{21.1}$$

Thus according to model (21.1), on the interval $[0, \tau_0]$ the distribution F_t is estimated by the empirical weighted distribution function, and on the interval $[\tau_0, \infty)$ the distribution F_t is fitted by the Pareto distribution with parameter θ_t.

We define by $W_{t,h}(t_i) = K\left(\frac{t_i - t}{h}\right)$ the weights where $h > 0$ is a bandwidth parameter and K is the kernel function. For any $t \in [0, T_{\max}]$, the maximum weighted log-likelihood estimator of θ_t is given by

$$\widehat{\theta}_{t,h,\tau_0} = \frac{1}{\widehat{n}_{t,h,\tau_0}} \sum_{i=1}^{n} W_{t,h}(t_i) 1_{\{X_{t_i} > \tau_0\}} \log\left(\frac{X_{t_i}}{\tau_0}\right),$$

where $\widehat{n}_{t,h,\tau_0} = \sum_{i=1}^{n} W_{t,h}(t_i) 1_{\{X_{t_i} > \tau_0\}}$ is the weighted number of observations beyond the threshold τ_0 which also depends on t. The distribution function $F_t(x)$ at time t is then estimated by

$$\widehat{F}_{t,h,\tau_0}(x) = \begin{cases} \widehat{F}_{t,h}(x), & \text{if } x \in [x_0, \tau_0], \\ 1 - \left(1 - \widehat{F}_{t,h}(\tau_0)\right)\left(\frac{x}{\tau_0}\right)^{-1/\widehat{\theta}_{t,h,\tau_0}}, & \text{if } x > \tau_0, \end{cases}$$

which combines the weighted empirical distribution function

$$\widehat{F}_{t,h}(x) = \frac{1}{\sum_{j=1}^{n} W_{t,h}(t_j)} \sum_{i=1}^{n} W_{t,h}(t_i) 1_{\{X_{t_i} \leqslant x\}}$$

and the fitted Pareto distribution. For any $\tau \in (0, 1)$ and for each $t \in [0, T_{\max}]$, the estimator of the τ-quantile of X_t is defined by

$$\widehat{q}_\tau(t, h, \tau_0) \equiv \begin{cases} \widehat{F}_{t,h}^{-1}(\tau), & \text{if } \tau < \widehat{p}_{\tau_0}, \\ \tau_0 \left(\frac{1 - \widehat{p}_{\tau_0}}{1 - \tau}\right)^{\widehat{\theta}_{t,h,\tau_0}}, & \text{otherwise}, \end{cases}$$

where $\widehat{p}_{\tau_0} = \widehat{F}_{t,h}(\tau_0)$, with τ_0 depending on t.

The estimators $\widehat{\theta}_{t,h,\tau_0}$, $\widehat{F}_{t,h,\tau_0}(x)$, and $\widehat{q}_\tau(t, h, \tau_0)$ depend on the unknown parameters τ_0 and h. Grama and Spokoiny (2008) and Durrieu et al. (2015) propose choosing the threshold parameter τ_0 by an adaptive procedure based on sequential multiple testing. A selection procedure for the bandwidth h is also given using a modified cross-validation method. The asymptotic properties (consistency and rates of convergence) of the estimators of θ_t, the extreme quantile of order τ, and the tail probability are studied in Durrieu et al. (2015). The properties of our approach have been studied through simulations that demonstrated the accuracy of the procedure for estimating the model parameters and its good performance for independent and dependent data.

Application. We applied extreme QR modeling to the CNV analysis from the previous subsection. We are interested here in the estimation of the extreme quantiles (e.g. $\tau = 0.99$ and $\tau = 0.999$). This analysis could be relevant when performing some quality control of the data and determining outliers as those data points above the extreme quantiles. It is clear from Figure 21.3 that the standard QR estimator does not perform well, showing no variation with t. On the other hand, the extreme QR approach is able to capture some of the variation with t and thus would provide a better determination of possible outliers.

21.3.6   Genomic analysis of set of genes

Similar to genetic analyses, in genomic applications one is interested in modeling a set of genes, for example, those belonging to the same genetic pathway or genetic network. A QR approach that has been developed specifically for this problem is support vector machine QR (SVM-QR; Sohn et al., 2008a,b; Villain et al., 2014)).

Let Y_i be the gene expression for observation i. The quantile function Y_i conditional on $X = x_i$ is given for $i = 1, \ldots, n$ by

$$Q(\tau|x_i) = \omega_\tau'\phi(x_i), \quad \text{for } \tau \in (0,1), \tag{21.2}$$

where ω_τ denotes the τ-quantile regression. SVM-QR can be defined by minimizing, for $\tau \in (0,1)$,

$$\frac{1}{2}\|\omega_\tau\|^2 + C \sum_{i=1}^{n} \rho_\tau(y_i - \omega_\tau'\phi(x_i)), \tag{21.3}$$

where C denotes the degree of penalization controlling the trade-off between the flatness of the quantile function estimate and the amount up to which deviations larger than zero are tolerated. A solution to the minimization problem (21.3) for $\tau \in (0,1)$ is obtained by optimizing its quadratic dual version. The τ-quantile regression for x^* can be written

$$\omega_\tau = \sum_{i=1}^{n} (\lambda_i^- - \lambda_i^+)\phi(x_i), \quad \text{for } Q(\tau|x^*) = \sum_{i=1}^{n} (\lambda_i^- - \lambda_i^+)K(x_i, x^*), \tag{21.4}$$

where λ_i^-, λ_i^+ are Lagrange multipliers and $K(x_i, x_j)$ denotes a kernel function. We consider the kernel Gaussian radial basis function given by

$$K(x_i, x_j) = \exp\left(-\frac{1}{2\sigma^2}\|x_i - x_j\|^2\right), \tag{21.5}$$

where σ corresponds to the bandwidth parameter which is estimated using the procedure developed in Caputo et al. (2002) and cross-validation. The parameter C determines the trade-off between the model complexity and the degree to which deviations larger than δ are tolerated in the optimization phase. The parameter δ controls in the SVM-QR the width of the δ-insensitive zone used to fit the data. Its values affect the number of support vectors. Larger values result in fewer support vectors and flatter regression estimates. To estimate C, we considered the approach of Cherkassky and Ma (2004). This choice of C is more robust than the approach of Mattera and Haykin (1999) when the data contain outliers.

Application. We applied the SVM-QR approach to a methylation study of prostate cancer (Kron et al., 2009). The aim is to compare the methylation profile of patients with advanced prostate cancer defined by a Gleason score of 8 to patients with less advanced cancer defined by score of 6. There were 10 patients in each group. To compare the methylation profiles of the two groups, we plotted the MA plot. The principle is to regress the relative abundance of methylation values (M) for the tissue of interest on the overall abundance of methylation for the different tissues considered (A). We therefore defined $M = \log_2 Y_{GS8}/Y_{GS6}$ and $A = \sqrt{Y_{GS8} \times Y_{GS6}}$. The experiment included about 20,000 genes. The corresponding regression coefficients, which measure tissue-specific methylations, were then evaluated at ten different quantiles of M. Figure 21.4 shows that the genes located below the 5% and above the 95% quantiles could be the most hypo- or hypermethylated (the dotted lines) The genes located within those lines corresponds to the normal profile. The solid and dashed are respectively associated with the median SVM-QR and to the classical SVM regression (Vapnik, 1998).

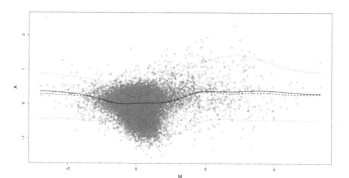

FIGURE 21.4
MA plot of prostate cancer methylation data, showing the difference in methylation between GS8 and GS6 tumors (M) with respect to the average methylation level (A). The plain line corresponds to the standard SVM regression estimator, the dashed line to the SVM-QR median estimator and the dotted lines to the SVM-QR estimator with $\tau = 0.05$ and $\tau = 0.95$.

21.4   Conclusion

In this chapter we have reviewed recent applications of QR to genetic, genomic and other - omic problems. The fields of application of QR are growing in response to the need for more robust and efficient statistical approaches to analyze complex biological data generated from high-throughput technologies. QR provides a compelling and versatile alternative to standard statistical approaches. In genetic research, QR can provide invaluable insights into associations with specific quantiles of a trait distribution and thus elucidate genetic factors that will not be discovered when looking at the mean of the distribution. This is clearly illustrated in our three examples related to a candidate gene association study, a genome-wide association study, and the analysis of a set of genetic markers. In genomic and other -omic problems, QR is a robust and sensitive approach that can generate new discoveries. As a robust approach, QR makes the inference less subject to biases and false positive discoveries. The results generated by QR are likely to be more reproducible, which is an important aspect given the cost of the new genomic technologies. As a sensitive approach, QR can help make new discoveries that would not be possible with alternative methods. For example, several studies have suggested that QR is superior to linear regression in terms of sensitivity and specificity to identify linear and nonlinear differential gene expressions (Ho et al., 2009). The application of QR to genetic and -omic problems should become easier with the development of software implementations, either as part of open-source R packages or within commercial software. New theoretical developments are also emerging and will provide better answers to complex -omic problems. Extensions of QR related to high-dimensional data and dependent data would be particularly relevant in genetic and -omic applications. Some of them have been briefly introduced in this chapter, such as the SVM-QR, and others are covered in other chapters in this book.

Bibliography

J. Beirlant, Y. Goegebeur, J. Segers, and J. Teugels. *Statistics of Extremes: Theory and Applications*. Wiley, Chichester, 2004.

A. Beyerlein, Kries V. R., A. R. Ness, and K. K. Ong. Genetic markers of obesity risk: Stronger associations with body composition in overweight compared to normal-weight children. *PLoS One*, 6(4), 2011.

B. M. Bolstad, R. A. Irizarry, M. Astrand, and T. P. Speed. A comparison of normalization methods for high density oligonucleotide array data based on variance and bias. *Bioinformatics*, 19(2):185–193, 2003.

L. Briollais and G. Durrieu. Application of quantile regression to recent genetic and -omic studies. *Human Genetics*, 133(8):951–966, 2014.

L. F. Burgette and J. P. Reiter. Modeling adverse birth outcomes via confirmatory factor quantile regression. *Biometrics*, 68(1):92–100, 2012.

S. J. Callister, R. C. Barry, J. N. Adkins, E. T. Johnson, W. Qian, B.-J. M. Webb-Robertson, R. D. Smith, and M. S. Lipton. Normalization approaches for removing systematic biases associated with mass spectrometry and label-free proteomics. *Journal of Proteome Research*, 5(2):277–286, 2006.

B. Caputo, K. Sim, F. Furesjo, and A. Smola. Appearance-based object recognition using SVMs: Which kernel should I use? In *Proceedings of NIPS Workshop on Statistical Methods for Computational Experiments in Visual Processing and Computer Vision*, 2002.

V. Cherkassky and Y. Ma. Practical selection of SVM parameters and noise estimation for SVM regression. *Neural Networks*, 17:113–126, 2004.

M. de Onis, M. Blossner, and E. Borghi. Global prevalence and trends of overweight and obesity among preschool children. *American Journal of Clinical Nutrition*, 92:1257–1264, 2010.

G. Durrieu and L. Briollais. Sequential design for microarray experiments. *Journal of the American Statistical Association*, 104(104):650–660, 2009.

G. Durrieu, I. Grama, Q. K. Pham, and J. M. Tricot. Nonparametric adaptive estimation of conditional probabilities of rare events and extreme quantiles. *Extremes*, 18:437–478, 2015.

P. H. C. Eilers and R. X. de Menezes. Quantile smoothing of array CGH data. *Bioinformatics*, 21(7):1146–1153, 2005.

D. S. Falconer and T. F. C. McKay. *Introduction to Quantitative Genetics*. Longmans Green, Harlow, 4th edition, 1996.

X. Gao and J. Huang. A robust penalized method for the analysis of noisy DNA copy number data. *BMC Genomics*, 11:517, 2010.

M. Geraci and M. Bottai. Linear quantile mixed models. *Statistics and Computing*, 24(3): 461–479, 2014.

I. Grama and V. Spokoiny. Statistics of extremes by oracle estimation. *Annals of Statistics*, 36(4):1619–1648, 2008.

C. Gu, A. A. Todorov, and D. C. Rao. Genome screening using extremely discordant and extremely concordant sib pairs. *Genetic Epidemiology*, 14:791–796, 1997.

K. D. Hansen and R. A. Irizarry. Removing technical variability in RNA-seq data using conditional quantile normalization. *Biostatistics*, 13(2):204–216, 2012.

R. Haring, H. Wallaschofski, A. Teumer, H. Kroemer, A. E. Taylor, C. H. L. Shackleton, M. Nauck, U. Volker, G. Homuth, and W. Arlt. A SULT2A1 genetic variant identified by GWAS as associated with low serum DHEAS does not impact on the actual DHEA/DHEAS ratio. *Journal of Molecular Endocrinology*, 50:73–77, 2013.

L. A. Hecker, A. O. Edwards, E. Ryu, N. Tosakulwong, K. H. Baratz, W. L. Brown, P. C. Issa, H. P. Scholl, B. Pollok-Kopp, K. E. Schmid-Kubista, K. R. Balley, and M. Oppermann. Genetic control of the alternative pathway of complement in humans and age-related macular degeneration. *Human Molecular Genetics*, 19:209–215, 2009.

L. A. Hindorff, H. A. Junkins, P. N. Hall, J. P. Mehta, and T. A. Manolio. A catalog of published genome-wide association studies. Available at www.genome.gov/gwastudies, 2009a.

L. A. Hindorff, P. Sethupathy, H. A. Junkins, E. M. Ramos, J. P. Mehta, F. S. Collins, and T. A. Manolio. Potential etiologic and functional implications of genome-wide association loci for human diseases and traits. *Proceedings of the National Academy of Sciences of the USA*, 106:9362–9367, 2009b.

J. N. Hirschhorn and M. J. Daly. Genome-wide association studies for common diseases and complex traits. *Nature Reviews Genetics*, 6:95–108, 2005.

J. W. K. Ho, M. Stefani, C. G. R. Remedios, and M. A. Charleston. A model selection approach to discover age-dependent gene expression patterns using quantile regression models. *BMC Genomics*, 10(3):1–18, 2009.

B. E. Huang and D. Y. Lin. Efficient association mapping of quantitative trait loci using selective genotyping. *American Journal of Human Genetics*, 80:567–576, 2007.

S. F. Kingsmore, I. E. Linquist, J. Mudge, D. D. Gessler, and W. D. Beavis. Genome-wide association studies: Progress and potential for drug discovery and development. *Nature Reviews*, 7:221–230, 2008.

R. Koenker and G. Bassett. Regression quantiles. *Econometrica*, 46:33–50, 1978.

R. Koenker and G. Bassett. Test of linear hypotheses and l_1 estimation. *Econometrica*, 50: 1577–1583, 1982.

D. Kong, A. Maity, F. C. Hsu, and J. Y. Tzeng. Testing and estimation in marker-set association study using semiparametric quantile regression kernel machine. *Biometrics*, 72:364–371, 2016.

K. Kron, V. Pethe, L. Briollais, B. Sadikovic, H. Ozcelik, A. Sunderji, V. Venkateswaran, J. Pinthus, N. Fleshner, T. van der Kwast, and B. Bapat. Discovery of novel hyper-methylated genes in prostate cancer using genomic CpG island microarrays. *PLoS One*, 4(3):e4830, 2009.

L. Kruglyak. The road to genome-wide association studies. *Nature Genetics*, 9:314–318, 2008.

D. Li, J. P. Lewinger, W. J. Gauderman, C. E. Murcray, and D. Conti. Using extreme phenotype sampling to identify the rare causal variants of quantitative traits in association studies. *Genetiic Epidemiology*, 35(8):790–799, 2011.

Y. Li and J. Zhu. Analysis of array CGH data for cancer studies using fused quantile regression. *Bioinformatics*, 23(18):2470–2476, 2007.

E. K. Lobenhofer, X. Cui, L. Bennett, P. L. Cable, B. A. Merrick, G. A. Churchill, and C. A. Afshari. Exploration of low-dose estrogen effects: Identification of no observed transcriptional effect level (NOTEL). *Toxicologic Pathology*, 2:482–92, 2004.

D. Mattera and S. Haykin. Support vector machines for dynamic reconstruction of a chaotic system. In C. J. C. Burges, B. Schölkopf, and A. J. Smola, editors, *Advances in Kernel Methods*, pages 211–241. MIT Press, Cambridge, MA, 1999.

M. I. McCarthy and J. N. Hirschhorn. Genome-wide association studies: Potential next steps on a genetic journey. *Human Molecular Genetics*, 17:R156–R165, 2008.

J. P. Newnham, S. F. Evans, C. A. Michael, F. J. Stanley, and L. I. Landau. Effects of frequent ultrasound during pregnancy: A randomised controlled trial. *Lancet*, 342:887–891, 1993.

O. Olivieri, N. Martinelli, M. Sandri, A. Bassi, P. Guarini, E. Trabetti, F. Pizzolo, D. Girelli, S. Friso, P. F. Pignatti, and R. Corrocher. Apolipoprotein C-III, n-3 polyunsaturated fatty acids, and "insulin-resistant" T–455C *APOC3* gene polymorphism in heart disease patients: Example of gene–diet interaction. *Clinical Chemistry*, 51(2):360–367, 2005.

J. Oosting, E. H. Lips, R. van Eijk, P. H. C. Eilers, K. Szuhai, C. Wijmenga, H. Morreau, and T. van Wezel. High-resolution copy number analysis of paraffin-embedded archival tissue using SNP BeadArrays. *Genome Research*, 17:368–76, 2007.

D. Pinkel and D. G. Albertson. Comparative genomic hybridization. *Annual Review of Genomics and Human Genetics*, 6:331–354, 2005.

R. C. Rippe, J. J. Meulman, and P. H. Eilers. Visualization of genomic changes by segmented smoothing using L_0 penalty. *PLoSone*, 7:e38230, 2012.

N. Risch. Searching for genetic determinants in the new millennium. *Nature*, 405:847–856, 2000.

N. Risch and K. Merikangas. The future of genetic studies of complex human diseases. *Science*, 273:1516–1517, 1996.

N. Risch and H. Zhang. Extreme discordant sib pairs for mapping quantitative trait loci in humans. *Science*, 268:1584–1589, 1995.

A. I. Scher, G. M. Terwindt, W. M. Verschuren, M. C. Kruit, H. J. Blom, H. Kowa, R. R. Frants, A. M. van den Maagdenberg, M. van Buchem, M. D. Ferrari, and L. J. Launer. Migraine and MTHFR C677T genotype in a population-based sample. *Annals of Neurology*, 59(2):372–375, 2006.

R. M. Simon, E. L. Korn, L. M. McShane, M. D. Radmacher, G. W. Wright, and Y. Zhao. *Design and Analysis of DNA Microarray Investigations*. Springer-Verlag, New York, 2003.

I. Sohn, S. Kim, C. Hwang, and J. W. Lee. New normalization methods using support vector machine quantile regression approach in microarray analysis. *Computational Statistics and Data Analysis*, 52:4104–4115, 2008a.

I. Sohn, S. Kim, C. Hwang, J. W. Lee, and J. Shim. Support vector machine quantile regression for detecting differentially expressed genes in microarray analysis. *Methods of Information in Medicine*, 5:459–467, 2008b.

R. Tibshirani, M. Saunders, S. Rosset, J. Zhu, and K. Knight. Sparsity and smoothness via the fused lasso. *Journal of the Royal Statistical Society, Series B*, 67(1):91–108, 2005.

V. N. Vapnik. *Statistical Learning Theory*. Wiley, New York, 1998.

J. Villain, S. Lozano, M.-P. Halm-Lemeille, G. Durrieu, and R. Bureau. Quantile regression model for a diverse set of chemicals: Application to acute toxicity for green algae. *Journal of Molecular Modeling*, 20(12):1–13, 2014.

V. Vinciotti and K. Yu. M-quantile regression analysis of temporal gene expression data. *Statistical Applications in Genetics and Molecular Biology.*, 8(1):1–20, 2009.

H. Wang and X. He. Detecting differential expressions in genechip microarray studies: A quantile approach. *Journal of the American Statistical Association*, 102:104–112, 2007.

H. Wang and X. He. An enhanced quantile approach for assessing differential gene expressions. *Biometrics*, 64:449–457, 2008.

E. Wheeler, N. Huang, E. G. Bochukova, J. M. Keogh, S. Lindsay, S. Garg, E. Henning, H. Blackburn, R. J. Loos, N. J. Wareham, S. O'Rahilly, M. E. Hurles, I. Barroso, and I. S. Farooqi. Genome-wide SNP and CNV analysis identifies common and low-frequency variants associated with severe early-onset obesity. *Nature Genetics*, 45(5):513–7, 2013.

E. P. Whitlock, S. B. Williams, R. Gold, P. R. Smith, and S. A. Shipman. Screening and interventions for childhood overweight: A summary of evidence for the US Preventive Services Task Force. *Pediatrics*, 116(1):125–144, 2005.

P.T. Williams. Quantile-specific penetrance of genes affecting lipoproteins, adiposity and height. *PLoS One*, 7(1):e28764, 2012.

World Health Organization. Obesity and overweight fact sheet. Available at `http://www.who.int/mediacentre/factsheets/fs311/en/`, 2006.

D. Yoon, E.-K. Lee, and T. Park. Robust imputation method for missing values in microarray data. *BMC Bioinformatics*, 8(Suppl. 2):S6, 2007.

22

Quantile Regression Applications in Ecology and the Environmental Sciences

Brian S. Cade

US Geological Survey, Fort Collins, Colorado, USA

CONTENTS

22.1 Introduction

Initial interest in quantile regression in ecology was primarily motivated by the desire to model boundary-type relationships between variables to provide an interpretation consistent with the concept of limiting factors based on Liebig's law of the minimum (Terrell et al., 1996; Scharf et al., 1998; Cade et al., 1999). In a sense this amounts to just focusing regression model estimation of heterogeneous responses (y) on the upper endpoint of a one-sided prediction interval where the lower endpoint is of no interest or conceded to be zero, for example, when organism abundance is the response variable (y) and the lowest value by definition is zero. It was rapidly recognized that modeling the entire regression quantile process (or a large subset) to estimate the heterogeneous conditional cumulative distribution functions was a useful analysis strategy for modeling ecological data regardless of the source of the heterogeneity (Cade and Noon, 2003; Cade et al., 2005). Heterogeneous statistical variation is a common feature of measurements of outcomes of biological processes underlying ecological relationships, and quantile regression provides a flexible regression methodology with minimal assumptions to model these relationships. Heterogeneity in measures of an outcome (y) from ecological or biological processes commonly occurs because there are multi-causal origins and pathways, often with complex interactions, thresholds, and nonlinearities operating (Quinn and Dunham, 1983). Rarely are all relevant processes known or measured so that they can be included in the statistical models (Cade et al., 2005; Lancaster and Belyea, 2006).

Statistical modeling relationships between organisms and their environment has especially benefited from quantile regression modeling because the heterogeneity observed in organism abundance conditional on environmental predictor variables is consistent with the knowledge that many other important biotic processes such as dispersal and competition were not included in these models focusing on resource constraints. Examples include Dunham et al. (2002), Eastwood et al. (2003), Sankaran et al. (2005), Schooley and Wiens (2005), Zoellick and Cade (2006), Thullen et al. (2008), Vaz et al. (2008), and Fornaroli et al. (2015). A quantile regression model for zero-inflated counts of organisms was used by Cade and Dong (2008) based on the transformation/back-transformation approach with randomly jittered discrete to continuous responses suggested by Machado and Santos Silva (2005). A similar focus on heterogeneity in organism responses, especially regarding upper limits on those relationships, has been found useful for developing and evaluating environmental monitoring and regulatory standards for contaminants (Pacheco et al., 2005; Linton et al., 2007; Brenden et al., 2008; Paul et al., 2009; Schmidt et al., 2012).

Analyses of other biological processes that have benefited from using quantile regression to model the heterogeneity in outcomes include body condition of fish based on allometric growth (Cade et al., 2008, 2011; Crane et al., 2015), nuclear DNA content of organisms related to environmental conditions (Knight and Ackerly, 2002; Knight et al., 2005), tree stem diameter growth related to age and other factors (Coomes and Allen, 2007; Mehtätalo et al., 2008), density-dependent survival of plants (Cade and Guo, 2000), changes in plant biomass (Visser et al., 2006; Daniels et al., 2010), and plant species diversity patterns (González-Espinosa et al., 2004; Rochinni et al., 2009; Grace et al., 2014; Simkin et al., 2016). Chavas (2015) developed an interesting threshold quantile autoregression model for animal population dynamics. In a meta-analysis of global data, Walters et al. (2016) used quantile regression to model the probability that persistent organic compounds (e.g., polychlorinated biphenyls - PCBs) introduced into aquatic systems will biomagnify in food webs based on the cumulative distribution of trophic magnification factors conditional on the metabolizability and water solubility of the chemicals. The effects of lead exposure on the reading ability of children were investigated with quantile regression by Magzamen et al. (2015). Ozone concentrations in the atmosphere (Baur et al., 2004), probabilities of precipitation in weather forecasting (Bremnes, 2004), and statistical downscaling of a global climate model for predicting extreme precipitation events (Friederichs and Hense, 2007) have all been modeled with quantile regression.

Intervals formed by quantile regression estimates often provide effective counterparts to conventional parametric distributional models for various statistical problems. Farmer et al. (2008) used quantile regression in a homogeneous model of deuterium isotopes in precipitation as a function of latitude to estimate prediction intervals for computing the minimum latitudinal separation of cumulative distributions of the isotopes that could be estimated well for assigning spatial origin of migratory animals. Intervals estimated from quantile regression models have been used in equivalence analyses where modeled outcomes are compared to intervals specified of substantive scientific importance or intervals required for regulatory compliance (Cade, 2011; Cade and Johnson, 2011). Confidence intervals for quantile regression estimates provide an easy approach for interpreting equivalence analyses in either a precautionary risk approach or a benefit-of-doubt risk approach, where the former is based on an inequivalence null hypothesis as in bioequivalence testing (minimizing consumer risk) and the latter is based on an equivalence null hypothesis (minimizing producer risk).

The short synopsis above was intended to give a sense of the breadth of applications of quantile regression in ecology and the environmental sciences. I now examine in greater detail the models, inferences, and interpretations for two example applications. The first example estimating trend in water quality (Bowen et al., 2015) makes use of nonlinear transformations for modeling multiplicative processes and censored quantile regression estimation for below-detection limit measurements. The second example estimating changes in herbaceous plant species richness related to atmospheric nitrogen deposition (Simkin et al., 2016) makes use of a parametric resampling approach on the linear quantile regression estimates to compute confidence intervals for a quantity that is a function of the estimates.

22.2   Water quality trends over time

Bowen et al. (2015) estimated changes in quantiles of two measures of water quality, specific conductance and chloride concentration, over the years 1970–2010 for surface water in watersheds in the United States where there had been different levels of unconventional oil and gas development. Contamination of surface and groundwater from activities associated with unconventional oil and gas development has been a growing environmental concern, especially with the increased use of hydraulic fracturing. Elevated levels of chloride concentrations and specific conductance often occur in produced waters associated with unconventional oil and gas development and can potentially degrade water quality when mixed with other surface water or groundwater. Here I focus on providing more details on the quantile regression (Koenker and Bassett, 1978; Koenker, 2005) estimates of chloride concentration trends over forty years for three of the watersheds analyzed in Bowen et al. (2015) to demonstrate features of the quantile regression modeling that make it a superior estimation procedure for water quality and flow monitoring.

Water-quality data were compiled from the US Geological Survey National Water Information System (NWIS) for streams within eight-digit hydrologic unit code (HUC-8) watersheds that intersected areas of unconventional oil and gas plays. Streams were then grouped by HUC-8. Once the quality assurance checks were completed, the number of samples from each site was reduced to no more than one sample per quarter per year to provide seasonal representation and to eliminate redundant data for those few sites that had greater sampling intensity. Additional details on the quality assurance and screening of data to be used in the analyses are provided in Bowen et al. (2015).

The estimated trends in quantiles of chloride (Cl^-) concentrations over time used a flow-adjusted model that allowed for seasonal fluctuations over time (Hirsch et al., 1991, 2010),

$$Q_Y(\tau|t,q) = \beta_0(\tau)q^{\beta_1(\tau)}10^{\beta_2(\tau)t+\beta_3(\tau)\sin(2\pi t)+\beta_4(\tau)\cos(2\pi t)}, \qquad (22.1)$$

where Y is Cl^- concentration in milligrams per liter, q is discharge in cubic feet per second, t is time in decimal year (centered so 0.0 is 1990), and the sine and cosine of $2 \times \pi \times$ time are used to adjust for seasonal fluctuations for quantiles $\tau \in \{0.10, 0.15, \ldots, 0.90\}$. Bowen et al. relied on a subset of these quantiles ($\tau \in \{0.10, 0.25, 0.50, 0.75, 0.90\}$) to provide a concise description of trends, but I used the larger set to provide a more comprehensive examination of the estimates. The model was estimated in its linear form by taking base-10 logarithms of both sides of the equation:

$$
\begin{aligned}
Q_{\log(Y)}(\tau|t,q) = {} & \log(\beta_0(\tau)) + \beta_1(\tau)\log(q) + \beta_2(\tau)t \\
& + \beta_3(\tau)\sin(2\pi t) + \beta_4(\tau)\cos(2\pi t).
\end{aligned} \tag{22.2}
$$

When two or more sampling sites were located within a watershed, model (22.2) was expanded to include orthogonal contrasts for the sites and their interactions with all other terms in the model so that flow adjustment, seasonal adjustment, and linear trends over time could differ among sites within the watershed and contribute to an average for each quantile for the watershed:

$$
\begin{aligned}
Q_{log(Y)}(\tau|t,q) = {} & \log(\beta_0(\tau)) + \beta_1(\tau)\log(q) + \beta_2(\tau)t + \beta_3(\tau)\sin(2\pi t) \\
& + \beta_4(\tau)\cos(2\pi t) + \log(\beta_{0j}(\tau))I_j + \beta_{1j}(\tau)\log(q)I_j \\
& + \beta_{2j}(\tau)tI_j + \beta_{3j}(\tau)\sin(2\pi t)I_j + \beta_{4j}(\tau)\cos(2\pi t)I_j,
\end{aligned} \tag{22.3}
$$

where the I_j are orthogonal contrasts coded as 1 for the $j = 1, \ldots, m - 1$ sites, as 0 if not, and as -1 for the mth site. Although Bowen et al. focused on reporting estimates in trend, $\beta_2(\tau)$, the value for a single site in model (22.2) or the average across multiple sites in model (22.3) in a watershed, I will provide additional estimates for multiple sites within a watershed to examine differences occurring within a watershed. Trends were expressed in the multiplicative form for 10 years (a decade) of change by back-transformation (exponentiating the estimate multiplied by 10) and in the text were reported as percentage change per decade ([proportionate changes] $- 1] \times 100\%$). This is done without bias or loss of information because quantile regression estimates are equivariant to nonlinear monotonic transformations (Koenker, 2005).

Below-detection limit (left-censored) measures occurred for some sites within some of the watersheds. Here I demonstrate the use of a censored quantile regression estimator that provides a generalization of the Kaplan–Meier estimator to linear models (Portnoy, 2003; Koenker, 2008). Quantile regression models were estimated with the quantreg package in R (R Development Core Team, 2012). Standard errors for all quantile regression estimates were obtained by using a weighted exponential bootstrap approach (Bose and Chatterjee, 2003) with $R = 1,000$ resamples.

22.2.1   A single site within a watershed

My first example of trends in chloride concentrations is for the most common situation analyzed in Bowen et al. (2015) where there was a single gauged stream within a watershed with measurements of flows and chemical concentrations. This example is for the Lower Missouri River watershed (HUC-8 = 10300200) which has flows ranging from 19,100 to 418,000 cubic feet per second (cfs) over the 1970–2010 period of interest (Figure 22.1). There was a strong negative relationship between increasing discharge and chloride concentrations, with estimates varying between -0.5 and -0.4 across the quantiles (Figure 22.2), indicating a 68% to 60% reduction in chloride concentrations with a 10× increase in discharge. The flow-adjusted trend estimates varied between 0.003 and 0.002 at $\tau \leqslant 0.75$, decreasing to near zero at $\tau > 0.75$, indicating a 7% to 5% increase in chloride concentrations per decade for the lower three fourths and approaching no increase in the upper fourth of the chloride distribution (Figure 22.2). This pattern is reflected in the flatter surface of the 0.90 quantile compared to lower quantiles estimates in the year dimension of Figure 22.1.

It is often easier to evaluate the flow-adjusted temporal trends in chloride concentrations by examining partial effects of the quantiles of trend estimates at selected values of discharge. In Figure 22.3 I have provided estimates across 1970–2010 for five selected quantiles at the

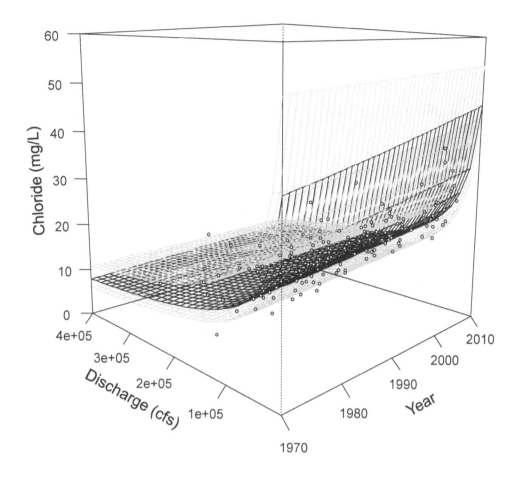

FIGURE 22.1
Chloride concentrations (milligrams per liter) by discharge (cubic feet per second) and decimal year of measurement for HUC-8 = 10300200, Lower Missouri River watershed (n = 158). The 0.90 (upper gray), 0.50 (black), and 0.10 (lower gray) quantile regression estimates of flow-adjusted chloride concentration trends 1970–2010 are graphed ignoring the estimated cyclical seasonal changes. (Color version available online.)

median (73,650 cfs), first (48,175 cfs) and third (119,000 cfs) quartiles of discharge. The seasonal fluctuations in chloride concentrations estimated with the sine/cosine function are graphed for the median discharge estimates. In this watershed, the seasonal fluctuations were rather small. Although all the flow-adjusted trend estimates included these cyclical seasonal fluctuations in concentration, I suppressed graphing these seasonal fluctuations to clarify trends over 1970–2010 for the first and third quartiles of discharge in Figure 22.3. The greater proportionate increases in chloride concentrations across 1970–2010 at lower discharges and for the lower three fourths of the quantiles ($\tau \leqslant 0.75$) at a given discharge are readily apparent in Figure 22.3.

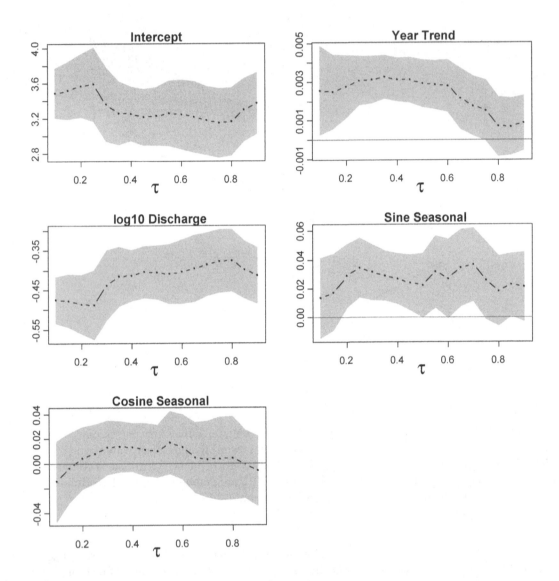

FIGURE 22.2
Parameter estimates and 95% confidence intervals for the flow- and seasonal adjusted annual trends in quantiles ($\tau \in \{0.10, 0.15, \ldots, 0.90\}$) of chloride concentrations (milligrams per liter) for HUC-8 = 10300200, Lower Missouri River watershed ($n = 158$).

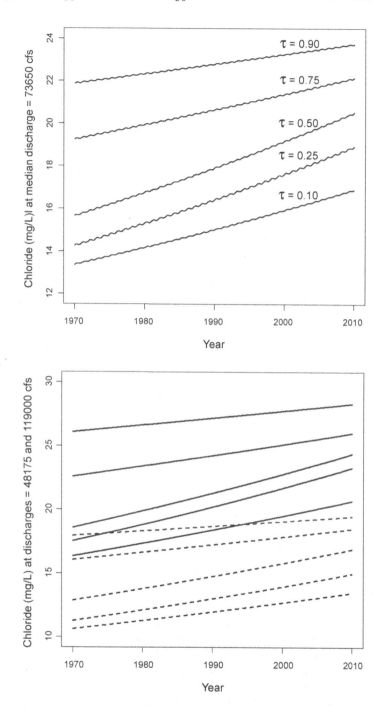

FIGURE 22.3
Estimated trends in $\tau \in \{0.10, 0.25, 0.50, 0.75, 0.90\}$ quantiles of chloride concentrations (milligrams per liter) 1970–2010 at the median (upper panel), and first (solid lines, lower panel) and third (dashed lines, lower panel) quartiles of flow for HUC-8 = 10300200, Lower Missouri River watershed ($n = 158$). Cyclical seasonal changes are shown in the upper panel and suppressed in the lower panel.

22.2.2   Multiple sites within a watershed

There were four stream sites measured in the Lower Souris River watershed in North Dakota (HUC-8 = 9010003) where discharges ranged from 0.03 to 5,700 cfs, with one site having considerably less discharge than the other three (Figure 22.4). Because Bowen et al. (2015) had estimates of unconventional oil and gas wells aggregated only down to the watershed level, they required a flow-adjusted estimated of chloride concentration trends at the level of the watershed. We obtained those estimates by model (22.3), which is just the usual linear model approach of including orthogonal contrasts for a categorical predictor (sites within the watershed) and its interactions with continuous predictors. This provides one set of estimates that are the averages across the four sites by quantile by predictor and three other sets of estimates that are differences from the average for three of the sites (Figure 22.5).

The quantile estimates plotted in Figure 22.4 are averages across the four sites. The 1970–2010 trends averaged across sites varied around 0.007 to 0.008 across the quantiles (Figure 22.5), indicating a 17% to 20% increase in chloride concentrations per decade. There appears to be considerable variation among sites that is not represented well by these averages across sites. Much of the among-site variation in chloride concentrations was associated with differences in the magnitude of chloride concentrations as evidenced by differences in intercepts in Figure 22.5. Sites 1 and 4 had higher overall concentrations of chloride and site 2 had much lower concentrations of chloride (Figures 22.5 and 22.6). However, there also were some differences among sites in chloride concentration trends. Sites 3 and 4 had decadal increases of chloride concentrations that varied from 17% to 15% and 20% to 17%, respectively, across lower to higher quantiles similar to the average across sites (Figures 22.5 and 22.6). Site 1 had increases of chloride concentrations of 26% per decade at lower quantiles to 5% increases at higher quantiles. Site 2 had 12% increases at lower quantiles to 35% increases per decade at higher quantiles (Figures 22.5 and 22.6). The different changes associated with lower to higher concentrations of chloride over time at the four sites suggest that different factors may be impacting the increase in chloride concentrations at different locations within the watershed. The heterogeneity in chloride concentration changes within a site estimated by quantile regression would have been ignored in a conventional mean regression analysis and the variation among sites would have been attenuated.

22.2.3   Estimation with below-detection limit values in a single site within a watershed

There were a small proportion of watersheds that had a small proportion of observations that were below-detection limit measurements (Bowen et al., 2015). Here I describe the use of censored quantile regression with left-censoring for below detection limit estimation using the Portnoy (2003) estimator for chloride concentration measurements in the Lake Sakakawea watershed in North Dakota (HUC-8 = 10110101). Three of the $n = 126$ observations had below-detection limit records (Figure 22.7). Here I used the option of estimating all possible quantile solutions (grid="pivot"), whereas the estimates in Bowen et al. (2015) were obtained by using the default solution associated with a grid of equally spaced quantiles.

The flow- and seasonal adjusted trends in chloride concentrations for this watershed varied between −0.009 and −0.006 across quantiles (Figure 22.8), indicating a decrease in chloride concentrations of 19% to 13% per decade. This is indicative of a minor improvement in water quality, at least as related to chloride concentrations, although chloride concentrations are low in this stream throughout the 40-year period analyzed. Given that there were few below-detection limit values actually observed in this watershed, I artificially in-

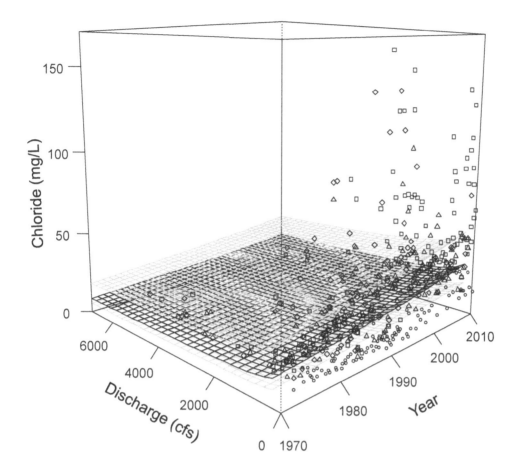

FIGURE 22.4
Chloride concentrations (milligrams per liter) by discharge (cubic feet per second) and decimal year of measurement for four stream sites in HUC-8 = 9010003, Lower Souris River watershed, North Dakota. The 0.90 (upper gray), 0.50 (black), and 0.10 (lower gray) quantile regression estimates of flow-adjusted chloride concentration trends 1970–2010 averaged across the four site estimates are graphed ignoring the estimated cyclical seasonal changes. Site 1 (squares, $n = 160$) is NWIS site 05120000, site 2 (circles, $n = 95$) is NWIS site 05120500, site 3 (triangles, $n = 103$) is NWIS site 05122000, and site 4 (diamonds, $n = 112$) is NWIS site 05124000. (Color version available online.)

creased the proportion of below-detection limit values to 11% by assigning all 14 of $n = 126$ chloride concentrations less than or equal to $2.5 \, \mathrm{mg \, L^{-1}}$ as below-detection limits. This was done to demonstrate impacts on estimates when there are greater proportions of below-detection limit measures. With the increased proportion of below-detection limit measures the estimated flow- and seasonal adjusted trends in chloride concentrations for $\tau \geqslant 0.20$ vary around -0.009 and -0.006, similar to the previous estimates with only three below-detection limit values, although the sampling variation quantified by the 95% confidence intervals has increased slightly (Figure 22.9). However, the lower quantiles ($\tau < 0.20$) have decreased to less than -0.010 with substantially larger 95% confidence intervals consistent with the lower information content of the below-detection limit values present in the lower quantiles.

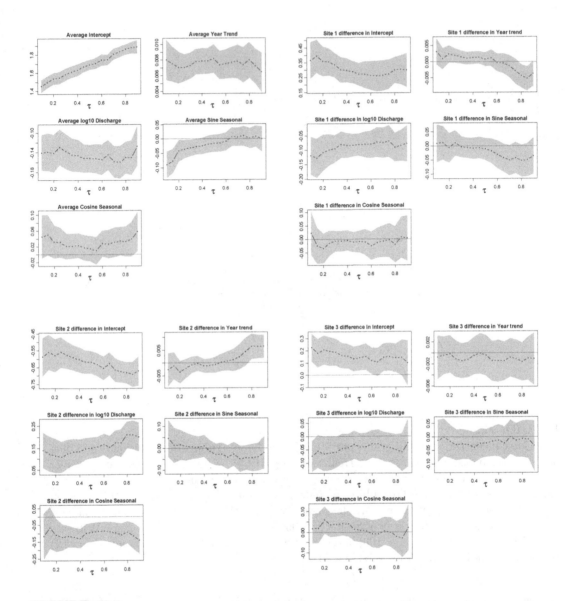

FIGURE 22.5

Parameter estimates and 95% confidence intervals for the flow- and seasonal adjusted annual trends in quantiles ($\tau \in \{0.10, 0.15, \ldots 0.90\}$) of chloride concentrations (milligrams per liter) for four stream sites in HUC-8 = 9010003, Lower Souris River watershed, North Dakota ($n = 470$). Estimates are for the average across the four sites and for differences from the average for each of three sites (the fourth can be obtained algebraically).

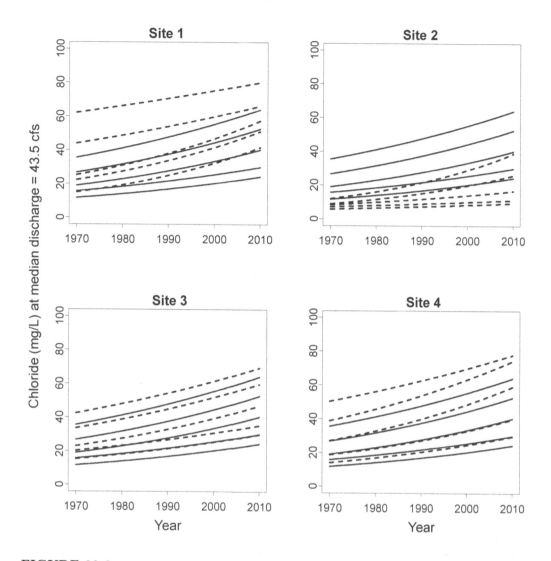

FIGURE 22.6
Estimated trends in $\tau \in \{0.10, 0.25, 0.50, 0.75, 0.90\}$ quantiles of chloride concentrations (milligrams per liter) 1970–2010 at the median flow for HUC-8 = 9010003, Lower Souris River watershed, North Dakota for each of four stream sites. Black lines are the average of the estimated quantiles across the sites and the dashed lines correspond to estimates for each site. Cyclical seasonal changes were suppressed.

22.2.4   Additional extensions possible for water quality and flow trend analyses

There have been recent modifications of the seasonal flow-adjusted models of water quality measures that recognize the possibility of shifting states over time by allowing the temporal trend relationships modeled similarly to equation (22.2) to be estimated differently for observations at different points in the temporal domain or by conditions modeled by other covariates. Hirsch et al. (2010) originally proposed this approach for mean regression and Beck and Hagy (2015) extended it to quantile regression. This approach is a "super-

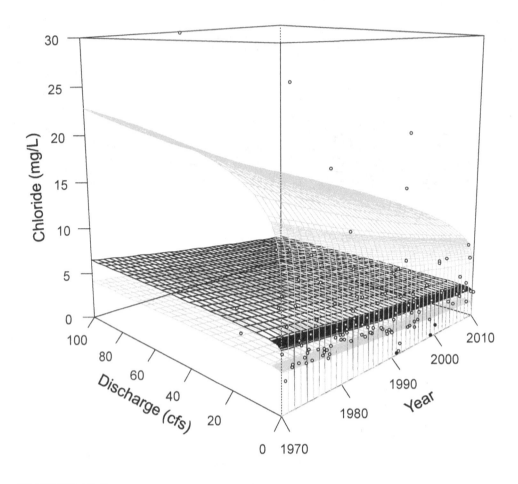

FIGURE 22.7
Chloride concentrations (milligrams per liter) by discharge (cubic feet per second) and decimal year of measurement for HUC-8 = 10110101, Lake Sakakawea watershed, North Dakota
($n = 126$). The 0.90 (upper gray), 0.50 (black), and 0.10 (lower gray) quantile regression
estimates of flow-adjusted chloride concentration trends 1970–2010 are graphed ignoring
the estimated cyclical seasonal changes. Three below-detection limit (left censored) chloride measurements between 1990 and 2000 are shown in solid black. Two extreme chloride
measurements of 64 and 180 mg L^{-1} at discharges of 0.13 and 0.21 cfs in 1992 and 1988,
respectively, are not graphed although they were used in estimation. (Color version available
online.)

smoothing" procedure where every observation potentially can have a different regression
relationship estimated depending on the local weighting of the observations. This highly
parameterized model obviously creates some issues with statistical inference procedures.
An approach that would not involve as extreme local smoothing and that would be more
amenable to statistical inferences (e.g. estimating confidence intervals) would be to allow
for different regression relationships in different regions of the temporal domain (or other
covariate conditions) by modeling those predictor variables with splines (e.g. piecewise linear B-splines on time with knots selected based on *a priori* knowledge or optimized across
some grid of values). This would allow the quantile regression relationships to differ in

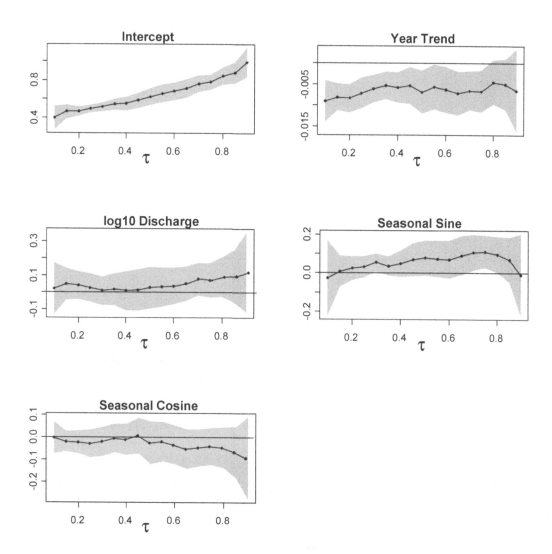

FIGURE 22.8
Parameter estimates and 95% confidence intervals for the flow- and seasonal adjusted annual trends in quantiles ($\tau \in \{0.10, 0.15, \ldots, 0.90\}$) of chloride concentrations (milligrams per liter) for HUC-8 = 10110101, Lake Sakakawea watershed, North Dakota ($n = 126$). Estimates were obtained from censored (left) quantile regression using the Portnoy (2003) estimator.

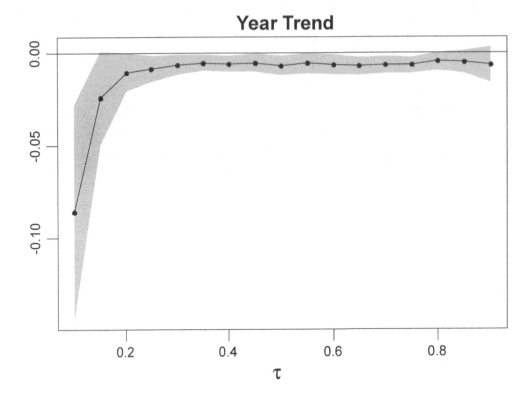

FIGURE 22.9
Parameter estimate and 95% confidence interval for the annual trends in quantiles ($\tau \in \{0.10, 0.15, \ldots, 0.90\}$) of chloride concentrations (milligrams per liter) for a hypothetical example corresponding to the data for HUC-8 = 10110101, Lake Sakakawea watershed, North Dakota ($n = 126$), except where 14 observations of $Cl^- \leq 2.5\,\mathrm{mg}\,\mathrm{L}^{-1}$ are considered below detection limits. Estimates were obtained from censored (left) quantile regression using the Portnoy (2003) estimator.

different regions of the predictor domain space but without forcing the models into the excessively parameterized form associated with the extreme local smoothing implemented by Hirsch et al. (2010) or Beck and Hagy (2015). The additive smoothing procedure of Koenker (2011) that selects the number and location of knots for piece-wise linear quantile regression functions also could be employed.

22.3 Herbaceous plant species diversity and atmospheric nitrogen deposition

Atmospheric deposition of nitrogen into terrestrial ecosystems has increased in the last century. This increased N deposition is a potential threat to plant diversity and ecosystem function. Nitrogen inputs can affect plant diversity through multiple processes and environmental contingencies, leading some studies to find decreases, some increases, and some no change in plant diversity with increasing N deposition (summarized in Simkin et al., 2016).

To evaluate large-scale patterns of plant diversity and N deposition, Simkin et al. (2016) compiled herbaceous (grasses and forbs) plant species diversity measures from studies across the continental United States to estimate relationships with site estimates of N deposition rates, annual precipitation, mean annual temperature, and soil pH. Simkin et al. used quantile regression to model herbaceous plant species diversity changes with N deposition and these other environmental covariates for open (grassland, shrubland, and woodland) and closed (deciduous, evergreen, and mixed forests) canopy vegetation types. Here I focus on the closed canopy vegetation analyses and expand on the quantile regression analyses to demonstrate their approach and additional interpretations possible.

Herbaceous plant species richness (number of species) and suitable environmental covariate data were available for $n = 11{,}819$ closed canopy vegetation sites. Nitrogen deposition rates (in kilograms per hectare per year) for 1985–2011), annual precipitation (millimeters), mean annual temperature (degrees Celsius), and soil pH were extracted from geospatial estimates for each site. Exploration of appropriate model forms for estimating quantiles of herbaceous plant species richness on the environmental covariates proceeded by examining nonparametric smoothed estimates based on using B-splines on combinations of all four environmental predictors, and then considering parametric functions of the predictors that included linear and quadratic terms for each predictor variable and two-way interactions between N deposition and the other three predictors. Akaike information criteria (AICs) across the quantiles were used to select the best supported models. For the closed canopy vegetation the following model with an interaction between N deposition and soil pH was selected as the best supported model:

$$Q_Y(\tau|N, pH, T, P) = \beta_0(\tau) + \beta_1(\tau)N + \beta_2(\tau)N^2 + \beta_3(\tau)pH$$
$$+ \beta_4(\tau)P + \beta_5(\tau)T + \beta_6(\tau)N \times pH, \qquad (22.4)$$

where the response Y is the number of herbaceous plant species, N is nitrogen deposition, pH is soil pH, P is annual precipitation, T is mean annual temperature, and $\tau \in \{0.05, 0.06, \ldots, 0.94, 0.95\}$. Simkin et al. (2016) only obtained and reported estimates for $\tau \in \{0.10, 0.50, 0.90\}$, but here I expand the analysis to smaller increments and a greater range of quantiles to provide a more comprehensive analysis. I also provide estimates and their 95% confidence intervals for all parameters across all quantiles to expand on the results provided in Simkin et al. (2016).

Although the response variable in model (22.4), herbaceous plant species richness, is a discrete count, the counts varied from 1 to 85 and, thus, were thought to cover a sufficient range of values to be reasonably modeled as a continuous rather than a discrete random variable. However, because of the large sample size, there was the potential that mass densities of observations at each count value might impact estimates and their standard errors. I performed additional analyses adding small, uniform random numbers on $[-0.5, 0.5]$ to the herbaceous plant species richness values to randomly jitter the herbaceous plant species counts into a more continuous distribution. Model (22.4) was then estimated with the randomly jittered responses to evaluate the sensitivity of the quantile regression estimates of the species counts to violations of the continuity assumption.

An important feature of the analyses of Simkin et al. (2016) was to determine whether there were values of N deposition where the modeled relationships indicated decreasing herbaceous plant species richness (i.e. critical loads for N deposition). Critical loads for N deposition were estimated from the peak of the quadratic nitrogen relationship determined from solving for the value where the partial derivative equals zero:

$$N \, critical \, load \, (\tau) = \frac{\hat{\beta}_1(\tau) + \hat{\beta}_6(\tau)pH}{-2\hat{\beta}_2(\tau)}, \qquad (22.5)$$

using estimates from the quantile regression (22.4). The critical loads for N deposition depend on soil pH because of the interaction with N deposition in this model. Confidence intervals (95%) for the critical loads were estimated by computing equation (22.5) on each of 10,000 randomly sampled parameter estimates obtained from a multivariate normal distribution with means and variance–covariance matrices from the regression quantile estimates for this model (22.4) and finding the 2.5th and 97.5th percentiles of this distribution. The variance–covariance matrices for the quantile regression estimates were based on assuming nonidentically distributed errors using the Hendricks–Koenker sandwich approach (Hendricks and Koenker, 1991; Koenker, 2005, pp. 79–80). Confidence intervals for quantile regression parameter estimates were also based on standard errors obtained from this variance–covariance approach. Simkin et al. focused on critical loads for N deposition based on the $\tau = 0.5$ quantile regression estimates for brevity and to be more consistent with previous studies that made similar estimates based on mean regression. I provide additional insight into critical loads for N deposition by examining estimates based on $\tau \in \{0.10, 0.50, 0.90\}$ quantile regression model estimates.

22.3.1   Quantile regression estimates

The quantile regression estimates of herbaceous species richness with respect to N deposition and soil pH interacting with N deposition indicated considerable heterogeneity between the lower and upper quantiles (Figure 22.10). This heterogeneity in response of herbaceous species richness is also evident in the heterogeneity of the estimates for these parameters across the quantiles (Figure 22.11). The strength of the interaction between soil pH and N deposition increased from lower to central ($\tau < 0.5$) quantiles, with similar strong effect across the upper ($\tau \geqslant 0.5$) portion of the herbaceous species distribution. It is also evident that the effect of annual precipitation had considerable heterogeneity, being more strongly negative at lower and higher quantiles and less negative at central quantiles (Figure 22.11). Average annual temperature had negative effects on herbaceous species richness at lower quantiles ($\tau < 0.6$) and positive effects at highest quantiles ($\tau > 0.85$), indicating an increase in variation of herbaceous species richness with increasing average annual temperature (Figure 22.11).

The randomly jittered herbaceous species counts provided estimated effects across quantiles very similar to those shown in Figure 22.11. Most estimates of standard errors for parameter estimates by quantile were within plus or minus 10% of those obtained with the species counts, with no consistent pattern of smaller or larger values by quantiles. Most estimates of the regression parameters also were within plus or minus 10%, with a few more deviant values at lowest or highest quantiles. There appeared to be little reason to think that the estimates based on treating the herbaceous plant species counts as continuous responses were unreasonable, even though they were technically discrete random variables.

22.3.2   Partial effects of nitrogen deposition and pH and critical loads

The partial quadratic effects of N deposition are not the same across soil pH values, as indicated by the strong interaction between these two predictors (Figure 22.11). Although this is apparent in Figure 22.10, I explore this more thoroughly by graphing the quadratic effects of N deposition at selected soil pH values, holding effects of annual precipitation and mean annual temperature fixed at their means (Figure 22.12). It is apparent in this figure that the increase and decrease in herbaceous plant species richness is much more homogeneous across lower to higher quantiles at lower values of soil pH and becomes more heterogeneous at higher soil pH values such that there is less evidence of a decline of higher quantiles of species richness values at higher soil pH values (Figure 22.12).

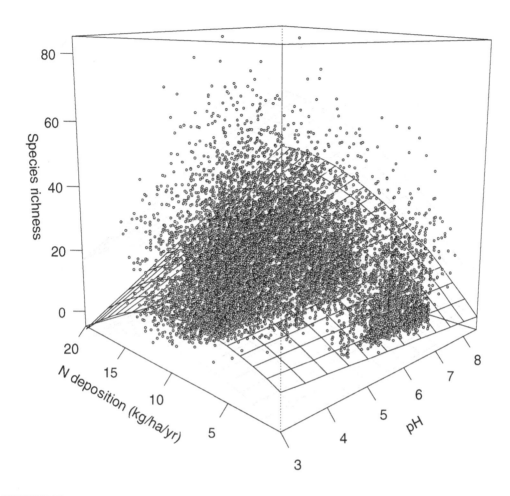

FIGURE 22.10

Herbaceous plant species richness in closed canopy sites (deciduous, evergreen, and mixed forests) as related to nitrogen (N) deposition (kilograms per hectare per year) and soil pH for $n = 11{,}819$ sites. The 0.90 (upper gray), 0.50 (black), and 0.10 (lower gray) quantile estimates were for the model that included quadratic effects of N, an interaction between N and pH, and additive effects of precipitation (millimeters) and temperature (degrees Celsius), with the last two terms fixed at their means for graphing purposes. (Color version available online.)

At soil pH $= 4$ most of the 95% confidence intervals for estimates of the critical load of N deposition are centered around $10.5\,\mathrm{kg\,ha^{-1}\,yr^{-1}}$ across all quantiles (Figures 22.12 and 22.13). The interval of values formed by the $\tau = 0.05$ and $\tau = 0.95$ quantile estimates can be used as a central 90% prediction interval for a single new observation. At a critical load of N deposition of $10.5\,\mathrm{kg\,ha^{-1}\,yr^{-1}}$ and soil pH $= 4$ (at means for annual precipitation and average temperature) the 90% prediction interval of herbaceous plant species richness is $[3, 40]$. When N deposition increases to $17\,\mathrm{kg\,ha^{-1}\,yr^{-1}}$ (the 99th percentile of N deposition values) the 90% prediction interval decreases to $[0, 35]$. At soil pH $= 6$ the 95% confidence intervals for estimates of critical loads of N deposition have intervals centered around values increasing from 12.1 to 14.8 to 17.3 kg ha^{-1} yr^{-1} as τ increases from 0.10 to 0.50 to 0.90, respectively (Figures 22.12 and 22.13). At a critical load of N deposition of 15 kg ha^{-1} yr^{-1} and soil pH = 6 the 90% prediction interval for herbaceous plant species richness is [5, 49], and when N deposition increases to 17 kg ha^{-1} yr^{-1} the 90% prediction interval barely decreases to [4, 48]. At soil pH = 8 only the $\tau = 0.10$ quantile estimates of critical loads for N deposition indicate a decrease in herbaceous plant species richness within the domain of observed N deposition values, with a 95% confidence interval centered around 14 kg ha^{-1} yr^{-1} (Figures 22.12 and 22.13). The 95% confidence intervals of critical loads of N deposition for both the $\tau = 0.50$ and $\tau = 0.90$ quantile estimates substantially

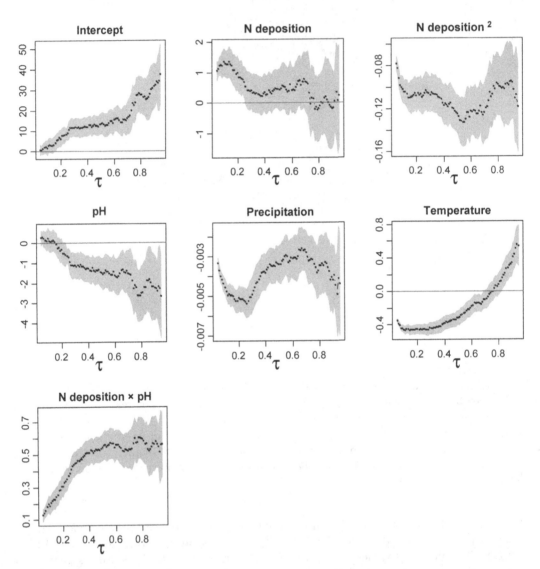

FIGURE 22.11
Parameter estimates and 95% confidence intervals for $\tau \in \{0.05, 0.06, 0.07, \ldots, 0.94, 0.95\}$ linear quantile regressions of herbaceous plant species richness in closed canopy sites as a quadratic function of nitrogen (N) deposition (kilograms per hectare per year) an interaction between N and soil pH, and an additive function of precipitation (millimeters) and temperature (degrees Celsius) for $n = 11,819$ closed canopy sites.

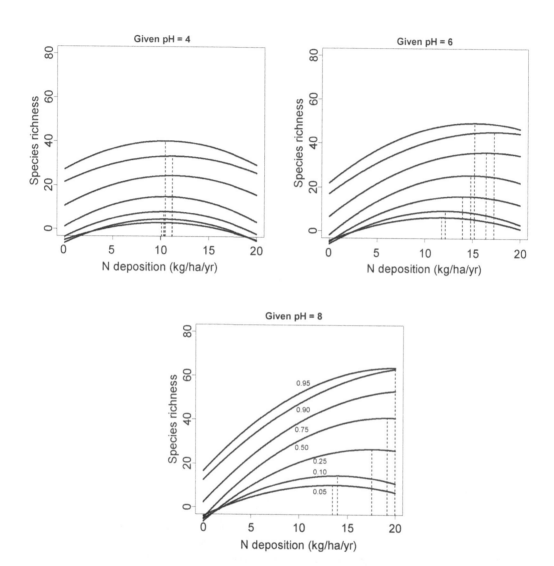

FIGURE 22.12

Partial effects of quadratic function of nitrogen (N) deposition (kilograms per hectare per year) at soil pH = 4, 6, and 8 for $\tau \in \{0.05, 0.10, 0.25, 0.50, 0.75, 0.90, 0.95\}$ linear quantile regressions (lower to higher curves) of herbaceous plant species richness as a quadratic function of N deposition, an interaction between N deposition and soil pH, and an additive function of precipitation (millimeters) and temperature (degrees Celsius) for $n = 11,819$ closed canopy sites. Estimates made with precipitation and temperature fixed at their means. Dashed vertical lines denote the peak of the quadratic function of N (the critical load of N deposition) for each quantile based on calculating partial derivatives.

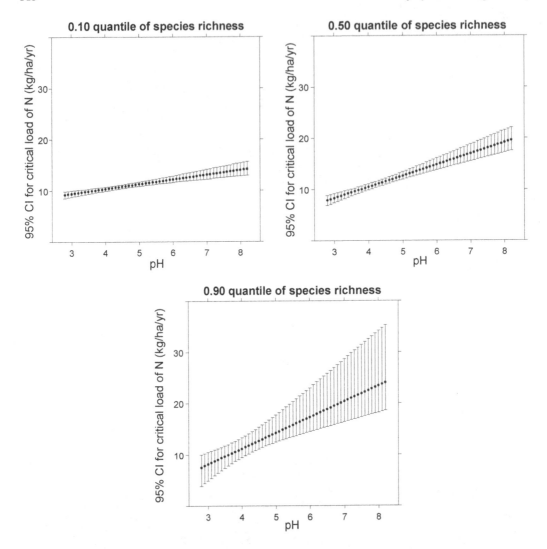

FIGURE 22.13
Estimates and 95% confidence intervals of critical loads of nitrogen (N) deposition (kilograms per hectare per year) as a function of soil pH (by increments of 0.1), where the critical loads were based on computing the partial derivatives for 10,000 random samples from a multivariate normal distribution with means and variances–covariances from the $\tau \in \{0.10, 0.50, 0.90\}$ linear quantile regressions estimates of herbaceous plant species richness as a quadratic function of N deposition, an interaction between N deposition and soil pH, and an additive function of precipitation (millimeters) and temperature (degrees Celsius) for $n = 11{,}819$ closed canopy sites.

exceed the maximum N deposition value observed, $18.95 \, \text{kg ha}^{-1} \, \text{yr}^{-1}$ (Figures 22.12 and 22.13), indicating no evidence of decreasing species richness at higher quantiles within the domain of observed N deposition values at soil pH $= 8$. At a critical load of N deposition of $14 \, \text{kg ha}^{-1} \, \text{yr}^{-1}$ and soil pH $= 8$ the 90% prediction interval for herbaceous plant species richness is $[10, 59]$, and when N deposition increases to $17 \, \text{kg ha}^{-1} \, \text{yr}^{-1}$ the 90% prediction interval is $[9, 62]$, indicating only a slight decrease in the lower endpoint with a slight increase in the upper endpoint of herbaceous plant species richness. These analyses reinforce the conclusion of Simkin et al. (2016) that herbaceous plant species richness in closed canopy vegetation types decreased with increasing N deposition only at sites with lower soil pH (more acidic), but in addition demonstrate that the magnitude of that reduction (5 species) is not great. Sites in open canopy vegetation types had greater reductions in species richness with increasing N deposition, but with additional interactions with the other environmental predictors.

22.3.3   Additional possible refinements to the model

Although my evaluation has suggested that there was minimal impact of treating the herbaceous species counts as continuous random variables in the quantile regression model, this model did not bound the estimates for herbaceous species counts as nonnegative for lower quantiles at more extreme parts of the domain of the predictor variables. Furthermore, the prediction intervals presented above based on the quantile estimates were actually for fractional values of species richness that I truncated to the integer values. This issue could be alleviated by considering an exponential model form ($Q_y(\tau|\mathbf{X}) = \exp(\mathbf{X}\boldsymbol{\beta})$) using the transformation and random jittering approach for quantile count models (Machado and Santos Silva, 2005; Cade and Dong, 2008). This would provide a refinement of estimates as made here and in Simkin et al. (2016) but not alter the overall pattern of changes in herbaceous plant species richness as related to N deposition.

22.4   Discussion

Quantile regression is one of the two major statistical approaches that ecologists and environmental scientists have come to use for dealing with heterogeneity in their linear statistical models. The other approach commonly used is linear mixed-effects models, which attempt to account for heterogeneity by modeling subgroupings of observations with random effects into more homogeneous units. There has been some advancement in developing mixed-effects models for quantile regression (e.g. Geraci and Bottai, 2014) that have been used in ecology (Grace et al., 2014; Fornaroli et al., 2015). Greater use of mixed-effects quantile regression that blends these two approaches may occur when the interpretation of these models is better established.

As the application on water quality trends demonstrated, the ability to use nonlinear (e.g. logarithmic) transformations with linear quantile regression to model inherently nonlinear, multiplicative relationships without any of the bias-correcting adjustments needed for modeling means in this same framework is a great advantage for ecological and environmental modeling. Furthermore, the quantiles (or percentiles) of chemical concentrations in water are commonly of much greater interest than the mean concentration for evaluating impacts on aquatic life or human health concerns. Stream flow analyses based on quantile regression are also potentially of great utility because of a similar focus on percentiles of discharge for evaluating trends or effects of lower and higher flow events. Other applications of quantile regression that benefit from the equivariance to monotonic transformation property include modeling continuous bounded outcomes such as proportions (Bottai et al., 2010) and modeling discrete outcomes such as counts of organisms (Cade and Dong, 2008). The use of quantile regression also immediately helps focus statistical modeling on prediction and tolerance intervals of outcomes, intervals that often are of much greater substantive scientific interest than confidence intervals on some parameter (Vardeman, 1992). Elements of this use of prediction and tolerance intervals from quantile regression estimates are in Zoellick and Cade (2006), Cade and Dong (2008), Farmer et al. (2008), Cade et al. (2011), and Walters et al. (2016). The potential for constructing prediction intervals for the changes in herbaceous plant species richness as modeled by Simkin et al. (2016) to aid interpretation of the effects of nitrogen deposition was demonstrated in the second example application. Greater use of quantile regression for constructing such intervals ought to be of substantial benefit in ecology and the environmental sciences, where estimating how some proportion of a statistical distribution of outcomes that is changing is often of greater relevance than estimating how the mean of that distribution is changing.

Acknowledgments

Both the water quality (Bowen et al., 2015) and the plant species richness as related to nitrogen deposition (Simkin et al., 2016) were studies conducted by working groups funded by the US Geological Survey, John Wesley Powell Center for Analysis and Synthesis `https://powellcenter.usgs.gov/`. G. Oelsner and S. Simkin reviewed drafts of this chapter. Simkin provided his R scripts that were modified for the quantile regression analyses included here. Data files and R scripts for performing the quantile regression analyses of water quality and plant species richness are available on ScienceBase (https://www.sciencebase.gov/catalog/). https://doi.org/10.5066/F75719S0. Any use of trade, firm, or product names is for descriptive purposes only and does not imply endorsement by the US Government.

Bibliography

D. Baur, M. Saisana, and N. Schulze. Modeling the effects of meterological variables on ozone concentration: A quantile regression approach. *Atmospheric Environment*, 38:4689–4699, 2004.

M. W. Beck and J. D. Hagy, III. Adaptation of a weighted regression approach to evaluate water quality trends in an estuary. *Environmental Modeling and Assessment*, 20:637–655, 2015.

A. Bose and S. Chatterjee. Generalized bootstrap for estimators of minimizers of convex functions. *Journal of Statistical Planning and Inference*, 117:225–239, 2003.

M. Bottai, B. Cai, and R. E. McKeown. Logistic quantile regression for bounded outcomes. *Statistics in Medicine*, 29:309–317, 2010.

Z. H. Bowen, G. P. Oelsner, B. S. Cade, T. J. Gallegos, A. M. Farag, D. N. Mott, C. J. Potter, P. J. Cinotto, M. L. Clark, W. M. Kappel, T. M. Kresse, C. P. Melcher, S. S. Paschke, D. D. Susong, and B. A. Varela. Assessment of surface water chloride and conductivity trends in areas of unconventional oil and gas development – Why existing national data sets cannot tell us what we would like to know. *Water Resources Research*, 51:704–715, 2015.

J. B. Bremnes. Probabilistic forecasts of precipitation in terms of quantiles using NWP model output. *Monthly Weather Review*, 132:338–347, 2004.

T. O. Brenden, L. Wang, and Z. Su. Quantitative identification of disturbance thresholds in support of aquatic resource management. *Environmental Management*, 42:821–832, 2008.

B. S. Cade. Estimating equivalence with quantile regression. *Ecological Applications*, 21: 281–289, 2011.

B. S. Cade and Q. Dong. A quantile count model of water depth constraints on Cape Sable seaside sparrows. *Journal of Animal Ecology*, 77:47–56, 2008.

B. S. Cade and Q. Guo. Estimating effects of constraints on plant performance with regression quantiles. *Oikos*, 91:245–254, 2000.

B. S. Cade and P. R. Johnson. Quantile equivalence to evaluate compliance with habitat management objectives. *Journal of Fish and Wildlife Management*, 2(2):169–182, 2011.

B. S. Cade and B. R. Noon. A gentle introduction to quantile regression for ecologists. *Frontiers in Ecology and the Environment*, 1:412–420, 2003.

B. S. Cade, J. W. Terrell, and R. L. Schroeder. Estimating effects of limiting factors with regression quantiles. *Ecology*, 80:311–323, 1999.

B. S. Cade, B. R. Noon, and C. H. Flather. Quantile regression reveals hidden bias and uncertainty in habitat models. *Ecology*, 86:786–800, 2005.

B. S. Cade, J. W. Terrell, and M. T. Porath. Estimating fish body condition with quantile regression. *North American Journal of Fisheries Management*, 28, 2008.

B. S. Cade, J. W. Terrell, and B. E. Neely. Estimating geographic variation in allometric growth and body condition of blue suckers with quantile regression. *Transactions of the American Fisheries Society*, 140:1657–1669, 2011.

J.-P. Chavas. Modeling population dynamics: A quantile approach. *Mathematical Biosciences*, 262:138–146, 2015.

D. A. Coomes and R. B. Allen. Effects of size, competition and altitude on tree growth. *Journal of Ecology*, 95:1084–1097, 2007.

D. P. Crane, J. M. Farrell, D. W. Einhouse, J. R. Lantry, and J. L. Markham. Trends in body condition of native piscivores following invasion of Lake Erie and Ontario by the round goby. *Freshwater Biology*, 60:111–124, 2015.

J. S. Daniels, B. S. Cade, and J. J. Sartoris. Measuring bulrush culm relationships to estimate plant biomass within a southern California treatment wetland. *Wetlands*, 30: 231–239, 2010.

J. B. Dunham, B. S. Cade, and J. W. Terrell. Influences of spatial and temporal variation on fish-habitat relationships defined by regression quantiles. *Transactions of the American Fisheries Society*, 131:86–98, 2002.

P. D. Eastwood, G. J. Meaden, A. Carpentier, and S. I. Rogers. Estimating limits to the spatial extent and suitability of sole (*Solea solea*) nursery grounds in the Dover Strait. *Journal of Sea Research*, 50:151–165, 2003.

A. H. Farmer, B. S. Cade, and J. Torres-Dowdall. Fundamental limits to the accuracy of deuterium isotopes for identifying the spatial origin of migratory animals. *Oecologia*, 158: 183–192, 2008.

R. Fornaroli, R. Cabrini, L. Sartori, F. Marazzi, D. Vracevic, V. Mezzanotte, M. Annala, and S. Canobbio. Predicting the constraint effect of environmental characteristics on macroinvertebrate density and diversity using quantile regression mixed model. *Hydrobiologia*, 742:153–167, 2015.

P. Friederichs and A. Hense. Statistical downscaling of extreme precipitation events using censored quantile regression. *Monthly Weather Review*, 135:2365–2378, 2007.

M. Geraci and M. Bottai. Linear quantile mixed models. *Statistics and Computing*, 24: 461–479, 2014.

M. González-Espinosa, J. M. Rey-Benayas, N. Ramírez-Marcial, M. A. Huston, and D. Golicher. Tree diversity in the northern Neotropics: Regional patterns in highly diverse Chiapas, Mexico. *Ecography*, 27:741–756, 2004.

J. B. Grace, P. B. Adler, W. S. Harpole, E. T. Borer, and E. W. Seabloom. Causal networks clarify productivity-richness interrelations, bivariate plots do not. *Functional Ecology*, 28: 787–798, 2014.

W. Hendricks and R. Koenker. Hierarchical spline models for conditional quantiles and demand for electricity. *Journal of the American Statistical Association*, 87:58–68, 1991.

R. M. Hirsch, R. B. Alexander, and R. A. Smith. Selection of methods for the detection and estimation of trends in water quality. *Water Resources Research*, 27(5):803–813, 1991.

R. M. Hirsch, D. L. Moyer, and S. A. Archfield. Weighted regressions on time, discharge, and season (WRTDS), with an application to Chesapeake Bay river inputs. *Journal of the American Water Resources Association*, 46(5):857–880, 2010.

C. A. Knight and D. D. Ackerly. Variation in nuclear DNA content across environmental gradients: A quantile regression analysis. *Ecology Letters*, 5:66–76, 2002.

C. A. Knight, N. O. Molinari, and D. A. Petrov. The large genome constraint hypothesis: Evolution, ecology, and phenotype. *Annals of Botany*, 95:177–190, 2005.

R. Koenker. *Quantile Regression*. Cambridge University Press, Cambridge, 2005.

R. Koenker. Censored quantile regression redux. *Journal of Statistical Software*, 27(6):1–25, 2008.

R. Koenker. Additive models for quantile regression: Model selection and confidence bandaids. *Brazilian Journal of Probability and Statistics*, 25:239–262, 2011.

R. Koenker and G. Bassett. Regression quantiles. *Econometrica*, 46(1):33–50, 1978.

J. Lancaster and L. R. Belyea. Defining the limits to local density: Alternative views of abundance-environment relationships. *Freshwater Biology*, 51:783–796, 2006.

T. K. Linton, M. A. W. Pacheco, D. O. McIntyre, W. H. Clement, and J. Goodrich-Mahoney. Development of bioassessment-based benchmarks for iron. *Environmental Toxicology and Chemistry*, 26:1291–1298, 2007.

J. A. F. Machado and J. M. C. Santos Silva. Quantiles for counts. *Journal of the American Statistical Association*, 100(472):1226–1237, 2005.

S. Magzamen, M. S. Amato, P. Imm, J. A. Havlena, M. J. Coons, H. A. Anderson, M. S. Kanarek, and C. F. Moore. Quantile regression in environmental health: Early life exposure and end-of-grade exams. *Environmental Research*, 137:108–119, 2015.

L. Mehtätalo, T. G. Gregoire, and H. E. Burkhart. Comparing strategies for modeling tree diameter percentiles from remeasured plots. *Environmetrics*, 19:529–548, 2008.

M. A. W. Pacheco, D. O. McIntyre, and T. K. Linton. Integrating chemical and biological criteria. *Environmental Toxicology and Chemistry*, 24:2983–2991, 2005.

M. J. Paul, D. W. Bressler, A. H. Purcell, M. T. Barbour, E. T. Rankin, and V. H. Resh. Assessment tools for urban catchments: Defining observable biological potential. *Journal of the American Water Resources Association*, 45:320–330, 2009.

S. Portnoy. Censored regression quantiles. *Journal of the American Statistical Association*, 98(464):1001–1012, 2003.

J. F. Quinn and A. E. Dunham. On hypothesis testing in ecology and evolution. *American Naturalist*, 122:602–617, 1983.

R Development Core Team. *R – A Language and Environment for Statistical Computing*. R Foundation for Statistical Computing, Vienna, 2012. URL http://www.R-Project.org.

D. Rochinni, H. Nagendra, R. Gate, and B. S. Cade. Spectral distance decay: Assessing species beta-diversity by quantile regression. *Photogrammetric Engineering & Remote Sensing*, 75:1225–1230, 2009.

M. Sankaran, N. P. Hanan, R. J. Scholes, J. Ratnam, D. J. Augustine, B. S. Cade, J. Gignoux, S. I. Higgins, X. Le Roux, F. Ludwig, J. Ardo, F. Banyikwa, A. Bronn, G. Bucini, K. K. Caylor, M. B. Coughenour, A. Diouf, W. Ekaya, C. J. Feral, E. C. February, P. G. H. Frost, P. Hiernaux, H. Hrabar, K. L. Metzger, H. H. T. Prins, S. Ringrose, W. Sea, J. Tews, J. Worden, and N. Zambatis. Determinants of woody cover in African savannas. *Nature*, 438(8):846–849, 2005.

F. S. Scharf, F. Juanes, and M. Sutherland. Inferring ecological relationships from the edges of scatter diagrams: Comparison of regression techniques. *Ecology*, 79:448–460, 1998.

T. S. Schmidt, W. H. Clements, and B. S. Cade. Estimating risks to aquatic life using quantile regression. *Freshwater Science*, 31:709–723, 2012.

R. I. Schooley and J. A. Wiens. Spatial ecology of cactus bugs: Area constraints and patch connectivity. *Ecology*, 86:1627–1639, 2005.

S. M. Simkin, E. B. Allen, W. D. Bowman, C. M. Clark, J. Belnap, M. L. Brooks, B. S. Cade, S. L. Collins, L. H. Geiser, F. S. Gilliam, S. E. Jovan, L. H. Pardo, B. K. Schulz, C. J. Stevens, K. N. Suding, H. L. Throop, and D. M. Waller. Conditional vulnerability of plant diversity to atmospheric nitrogen deposition across the United States. *Proceedings of the National Academy of Sciences of the USA*, 113:4086–4091, 2016.

J. W. Terrell, B. S. Cade, J. Carpenter, and J. M. Thompson. Modeling stream fish habitat limitations from wedged-shaped patterns of variation in standing stock. *Transactions of the American Fisheries Society*, 125:104–117, 1996.

J. S. Thullen, M. Nelson, B. S. Cade, and J. J. Sartoris. Macrophyte decomposition in a surface-flow ammonia dominated constructed wetland: Rates associated with environmental and biotic variables. *Ecological Engineering*, 32:281–290, 2008.

S. B. Vardeman. What about the other intervals? *American Statistician*, 46:193–197, 1992.

S. Vaz, C. S. Martin, P. D. Eastwood, B. Ernande, A. Charpentier, G. J. Meaden, and F. Coppin. Modeling species distributions using regression quantiles. *Journal of Applied Ecology*, 45:204–217, 2008.

J. M. Visser, C. E. Sasser, and B. S. Cade. The effect of multiple stressors on salt marsh end-of-season biomass. *Estuaries and Coasts*, 29:331–342, 2006.

D. M. Walters, T. D. Jardine, B. S. Cade, K. A. Kidd, D. C. G. Muir, and P. Leipzig-Scott. Trophic magnification of organic chemicals: A global synthesis. *Environmental Science and Technology*, 50:4650–4658, 2016.

B. W. Zoellick and B. S. Cade. Evaluating redband trout habitat in sagebrush desert basins in southwestern Idaho. *North American Journal of Fisheries Management*, 26:268–281, 2006.

Index

Printed in the United States
by Baker & Taylor Publisher Services